# Essentials of Polymer Science and Engineering

# Essentials of Polymer Science and Engineering

Editor: Jaden Baker

STATES
ACADEMIC PRESS
www.statesacademicpress.com

Published by States Academic Press
109 South 5th Street,
Brooklyn, NY 11249, USA
www.statesacademicpress.com

Essentials of Polymer Science and Engineering
Edited by Jaden Baker



International Standard Book Number: 978-1-63989-189-4 (Hardback)



**Cataloging-in-Publication Data**

Essentials of polymer science and engineering / edited by Jaden Baker.
p. cm.
Includes bibliographical references and index.
ISBN 978-1-63989-189-4
1. Polymers. 2. Polymerization. 3. Macromolecules. I. Baker, Jaden.
QD381 .E87 2022
547.7--dc23

# Contents

# Preface

This book has been a concerted effort by a group of academicians, researchers and scientists, who have contributed their research works for the realization of the book. This book has materialized in the wake of emerging advancements and innovations in this field. Therefore, the need of the hour was to compile all the required researches and disseminate the knowledge to a broad spectrum of people comprising of students, researchers and specialists of the field.

The study of polymers is known as polymer science. It comprises polymer physics, polymer chemistry, biophysics, and materials science and engineering. Polymer science and engineering is concerned with polymerization chemistry, polymerization catalysis, materials characterization, structure-property relationships, etc. It also deals with biomass, biorenewables, conducting polymers, biomimetic polymers, degradability and life cycle analysis, and controlled release formulations. Polymer science and engineering plays an important role in energy security, access to clean water, protection of the environment, and affordable healthcare. It focuses on every single process in the life cycle of a polymer ranging from monomer synthesis to product development. This book elucidates the concepts and innovative models around prospective developments with respect to polymer science and engineering. It unravels the recent studies in this field. This book will provide comprehensive knowledge to the readers.

At the end of the preface, I would like to thank the authors for their brilliant chapters and the publisher for guiding us all-through the making of the book till its final stage. Also, I would like to thank my family for providing the support and encouragement throughout my academic career and research projects.

**Editor**

1

# Research of Flow Field of Alkaline/Surfactant/ Polymer Solution in the Annular Depressurization Slot by PIV Experiment


Cheng Fu,[1,2] Tingting Zhu,[1] Di Wu,[3] Ying Wang,[4] and Bin Huang[1]

[1]*College of Petroleum Engineering, Northeast Petroleum University, China*
[2]*Post-Doctoral Scientific Research Station, Daqing Oilfield Company, China*
[3]*Hangzhou Urban & Rural Construction Design Institute Limited Company, China*
[4]*Aramco Asia, China*

Correspondence should be addressed to Bin Huang; huangbin502a@163.com



Academic Editor: Domenico Acierno



In order to study the flow field changes of ASP (alkaline/surfactant/polymer) solutions with different molecular weights and flow rates when flowing through the annular depressurization slot in the laboratory, the annulus pipeline model was designed based on the principle of similarity. The particle image velocimetry (PIV) system was used to continuously capture the transient images when the ASP solution flows in the model under different experimental conditions. Tecplot software was used to analyze and process the nephrograms of the velocity distribution of ASP solution when flowing through the partially pressured injection tool with different depressurization slots. The experimental results showed that the vortex occurred at the bottom of the depressurization slots; the greater the flow rate, the closer the vortex center to the outer wall; higher molecular weight of the polymer in the ASP solution caused larger velocity gradient towards the wall. The number of the slots has no significant effect on the position of the vortex. This experiment provides a new research method for the velocity distribution of the internal flow field in the partially pressured injection tool.


## 1. Introduction

In past years, the application of the partially pressured injection tool used in Indonesia Limau oil has been limited due to a lack of sufficient theoretical basis for the runner shape design. This also restricted the design and manufacture of pressure relief groove. Therefore, studying the internal flow field of the partially pressured injection tool is necessary [1, 2].

Compared with the point measurement by Laser Doppler Velocimetry (LDV), PIV can perform direct surface measurements for the flow field to ensure the accuracy of the measurement and certain spatial resolution. Therefore, PIV is suitable to analyze the change of the flow field. PIV has two distinct characteristics: it does not disturb the flow field and the entire picture of the flow field can be simultaneously obtained through instantaneous flow-velocity vector. It is a powerful tool to study complex flow structures such as eddy currents and turbulence [3–5].

Feng et al. [6] simulated the flow process and flow field of the polymer solution through the layered injector and found that the flow of the polymer in the layered injector was relatively stable. This could be a benefit for preventing the mechanical shearing and reducing the viscosity loss of the polymer solution during the polymer injection process. Meng et al. [7] used the software PHOENICS to calculate the flow field in the layered injector and found that the shape of the annular channel of the layered injector had great influence on the flow field characteristics. Cai et al. [8] analyzed the flow field of the ASP solution in the eccentric injector by the commercial software Fluent. The simulation results showed that the velocity and pressure of the flow were periodic, and each period was similar. Due to a lack of experimental methods, the optimization of the structure of the partially pressured injection tool and the flow-field distribution in the depressurization slot of the injection tool were mainly simulated and analyzed by using Fluent numerical simulation software. Our research is the first trial to study the flow field of the injection tool by experimental method.

In this paper, the interior flow field of the injection tool is studied by laboratory experiments. A set of experimental models of partially pressured injection tools for PIV system are designed by using the principle of similarity [9–11]. Three experiments with different flow rate, component, and number of slots were performed and compared with the experimental model [12]. A high-speed camera was used to capture the images of the flow field within the partially pressured injection tool. Different postprocessing had been used to handle the experimental data by Tecplot software. The flow field conditions within the partially pressured injection tool had been determined under different experimental conditions. The influence regulations for the inner flow field of the polymer with the weak base have been obtained for three element solutions with different molecular weight, flow rate, and number of slots. It will help provide guidance for the design of the depressurizing channel in the future.

## 2. The Principle of Similarity

### 2.1. Similar Proportions

*2.1.1. Geometric Similarity.* The shape of the model used in this experiment is exactly the same as the prototype of the partially pressured injection tool. In order for good quality observations, the model was enlarged to a certain proportion [13]. Assuming that the length of the depressurization slot in the model is $l_m$, the diameter is $D_m$, and the length of the actual depressurization slot of the partially pressured injection tool is $l_p$, the diameter is $D_p$.

The proportion of geometric similarity is expressed as

$$\lambda_l = \frac{l_p}{l_m} = \frac{D_p}{D_m}. \quad (1)$$

*2.1.2. Kinematic Similarity.* Under similar geometrical conditions, the scale of time can be described as

$$\lambda_l = \frac{l_p}{l_m} = \frac{D_p}{D_m}, \quad (2)$$

and the scale of velocity can be described as

$$\lambda_v = \frac{v_p}{v_m} = \frac{l_p/t_p}{l_m/t_m} = \frac{\lambda_l}{\lambda_t}, \quad (3)$$

where $v$ is the average velocity when the solution flows through the depressurization slot; $t$ is the time for the solution to pass through the length $l$ of the depressurization slot.

*2.1.3. Dynamic Similarity.* The scale of pressure can be described as

$$\lambda_F = \frac{F_p}{F_m} = \frac{m_p a_p}{m_m a_m} = \frac{\rho_p V_p a_p}{\rho_m V_m a_m} = \lambda_\rho \lambda_l^2 \lambda_v^2. \quad (4)$$

The solution used in the experiment is the same as that used in the field. It means that $\rho_p$ is equal to $\rho_m$. Therefore, the following formulation can be obtained:

$$\lambda_F = \lambda_l^2 \lambda_v^2. \quad (5)$$

### 2.2. Criterion of Similarity.

Since the viscous force plays an important role in this experiment, the criterion of similarity for viscous force is used to judge similar situations. The similitude criterion of the viscous force is the Reynolds number. Therefore, if the Reynolds number of the prototype is equal to the model, the dynamic similarity can be achieved.

The generalized Reynolds number of non-Newtonian fluid can be described as

$$\mathrm{Re}' = \frac{Dv\rho}{K'(8v/D)^{n'-1}} = \frac{D^{n'} v^{2-n'} \rho}{8^{n'-1} K'}. \quad (6)$$

The equation can be concluded by calculating as follows:

$$K' = K\left(\frac{3n+1}{4n}\right)^n, \quad (7)$$

where the value of $n$ is equal to the value of $n'$, namely,

$$\mathrm{Re} = \frac{D^n v^{2-n} \rho}{8^{n-1} K((3n+1)/4n)^n} = \frac{D^n v^{2-n} \rho}{K/8((6n+2)/n)^n}. \quad (8)$$

The experiment needs to

$$\frac{\mathrm{Re}_p}{\mathrm{Re}_m} = \frac{v_p d_p \rho_p}{v_m d_m \rho_m} = 1. \quad (9)$$

Because the cross section of the model is a concentric circle and the hydraulic diameter of the model is the characteristic length of the model. The characteristic length can be described as follows:

$$R_h = \frac{A}{\chi} = \frac{4\pi(R^2 - \bar{r}^2)}{\chi} = \frac{4\pi(R^2 - \bar{r}^2)}{2\pi(R + \bar{r})} = 2(R - \bar{r}). \quad (10)$$

Where $\chi$ is the wetted perimeter, $A$ is the cross section area of passage, $R$ is the outer radius of the flow section, and $\bar{r}$ is the inner radius of the flow section.

Since $G = v \cdot A$, $A = 4\pi(R^2 - \bar{r}^2)$ the following are obtained:

$$\frac{\mathrm{Re}_p}{\mathrm{Re}_m} = \frac{G_p d_p \rho_p / A_p}{G_m d_m \rho_m / A_m} = \frac{G_p/\chi_p}{G_m/\chi_m} = \frac{G_p/(R_p + \bar{r}_p)}{G_m/(R_m + \bar{r}_m)} = 1. \quad (11)$$

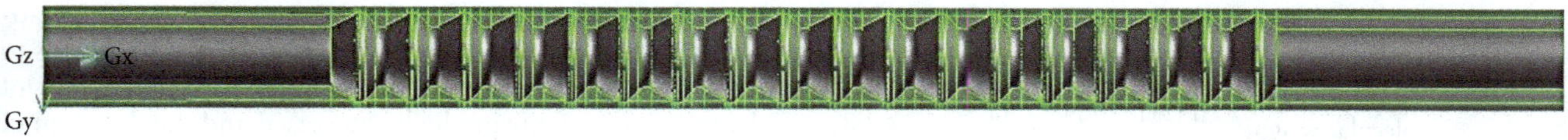


FIGURE 1: Geometric structure of streamlined partially pressured injection tool.

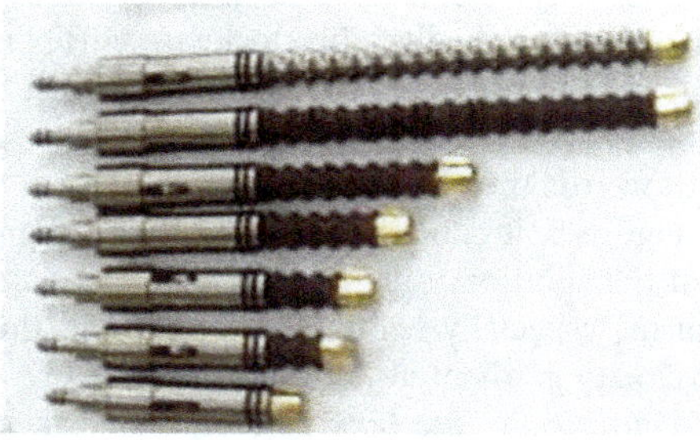

FIGURE 2: Modified shuttle rod partially pressured injection tool.

When the ratio of the flow rate of the experimental model vs. the field condition reaches the above conditions, the dynamic similarity can be considered.

*2.3. Streamlined Partially Pressured Injection Tool.* The main objective of the simulation research is to optimize the internal structure of the depressurizing channel in the streamlined partially pressured injection tool. The structure of streamlined-separating injection tool is shown in Figure 1, and the modified shuttle rod partially pressured injection tool is shown in Figure 2.

As shown in Figures 1 and 2, the depressurizing channel is a solid column in the middle with a waveform. By changing the cross-sectional area of the channel, a casing pipe allows the solution to flow through the gap to cause pressure loss so as to achieve the aim of reducing pressure. The various structure parameters such as slot length and slot distance of the injection tools with differential pressure type flow are suitable for different conditions.

*2.4. Experimental Models.* In order to ensure good quality particle image, the acrylic material was used to prepare the experiment model. The size of the depressurization slot is shown in Figure 3.

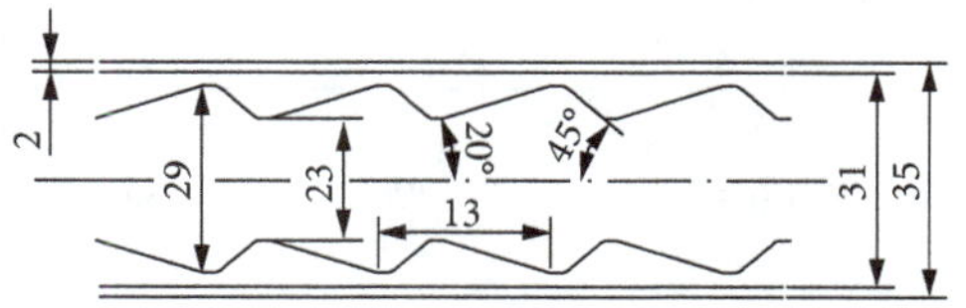


FIGURE 3: The size of the depressurizing channel in the experimental model.

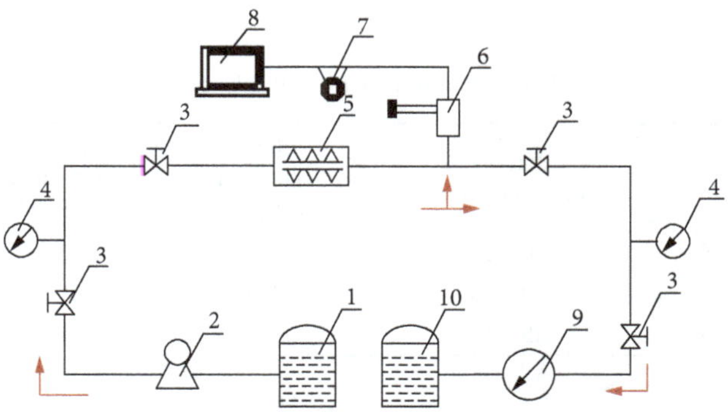


FIGURE 4: Schematic diagram of the experiment setup. 1 liquid storage tank, 2 screw pump, 3 valve, 4 pressure gauge, 5 injection tool model, 6 laser emitter, 7 camera, 8 data processing system, 9 flow meter, and 10 return liquid tank.

## 3. Experimental Scheme and Procedure

*3.1. Solution Compositions.* In this experiment, the water for the polymer solution was synthetic brine (the salinity was 6778 mg/L). The surfactant was 0.3% alkylbenzene sulfonate. The weak alkali was 1.2% $Na_2CO_3$. The polymer was polyacrylamide (HPAM) produced in Daqing Refinery and Chemical Company with molecular weights of $1.6 \times 10^6$, $1.9 \times 10^6$, and $2.5 \times 10^6$, respectively.

*3.2. Experimental Setup.* The PIV used in the experiment was MiniPIV. The exposure time of high-speed camera was 1/128000 sec. The tracer particle was hollow glass bead ($d \leq 10\ \mu m$, 0.1 g/L).

Figure 4 shows the schematic diagram of the experimental setup. All the experiments were carried out in a room with no interference from light source.

The camera in Figure 4 captures images of the solution in a reciprocal second slot of the depressurizing channel in Figure 1. It is a high-speed camera and part of PIV system. In the experiment, at least 100 images of the flow field should be continuous shot with the middle 50 images were selected to do cross-correlation calculation. Then, the vector files were poured into Tecplot 10.0 software, and finally, the velocity nephograms were generated with shooting accuracy of 0.01.

*3.3. Experiment Schemes.* A total of 3 groups of PIV test were carried out for the same molecular weight, flow rate, and the number of depressurization slots of the model. The experiment schemes are shown in Table 1.

*3.4. Experimental Procedures.* The experimental steps are as follows:

(1) Connect the experimental setup as shown in Figure 4

(2) Make the polymer solutions which were used in the experiment and add the tracer particles

Table 1: Experimental schemes of the PIV test.

| | Polymer molecular weight | Polymer concentration (mg/L) | Flow rate ($m^3/d$) | Number of slots |
|---|---|---|---|---|
| Scheme 1 | $1.6 \times 10^6$<br>$1.9 \times 10^6$<br>$2.5 \times 10^6$ | 2000 | 30 | 18 |
| Scheme 2 | $1.6 \times 10^6$ | 2000 | 20<br>30<br>50 | 18 |
| Scheme 3 | $1.6 \times 10^6$ | 2000 | 30 | 5<br>10<br>18 |

Table 2: Viscosity of ASP solution.

| | Polymer molecular weight | Polymer concentration (mg/L) | Solution viscosities (mPa·s) |
|---|---|---|---|
| Scheme 1 | $1.6 \times 10^6$ | | 42.3 |
| Scheme 2 | $1.9 \times 10^6$ | 2000 | 45.6 |
| Scheme 3 | $2.5 \times 10^6$ | | 51.4 |

(3) Start the experiment and photograph the images of the solution flowing in a depressurization slot

(4) Analyze the collected images using the Tecplot software

*3.5. Solution Viscosity.* The relationship between viscosity and concentration of the polymer in the ASP systems are listed in Table 2.

According to Table 2, the molecular weight of the polymer has a significant effect on the viscosity of ASP system. With the increase of polymer molecular weight, the viscosity of the ASP system increased.

## 4. Experimental Results and Analysis

*4.1. Velocity Nephograms.* The velocity nephograms of different molecular weight, flow rate, and number of slots (the length of the depressurizing channel increases with the number of slots) were obtained by the Tecplot software which are shown in Figures 5–7. The different colors in the velocity nephograms represent different speeds. For the same graph, red color indicates maximum speed and the blue color indicates zero velocity.

From Figure 5, the maximum velocity appears at the minimum clearance of the annular. The overall flow directions of the solution are the same. This is due to the minimum flow area of the decompression slot in the partially pressured injection tool, where the flow rate of the solution is constant with the greatest flow velocity at that point. As the solution continues to flow in the partially pressured injection tool, the flow area in the pressure-reducing groove increases gradually, resulting in the velocity decreasing gradually. The flow direction of the solution gradually spreads to the center of the pressure reducing groove. At about 3/4 slot length, the flow rate gradually increased and the direction of flow gather up again. The reason for that is as the solution in the decompression slot continues to move forward, the flow cross-sectional area becomes smaller, thus the liquid flow velocity increased gradually. The clockwise vortex is formed at the center of the pressure reducing groove with the minimum velocity at the center of the vortex and gradually increasing towards the channel wall.

From Figure 5, it can be concluded that the greater the viscosity of the solution, the greater the velocity around the vortex will be, especially below the vortex, and the smaller the axial velocity gradient at the top of the vortex.

From Figure 6, as the flow rate of the weak base ASP solution flows through the partially pressured injection tool decreased, the center of the vortex approached more to the left; the higher the flow rate caused the center in the groove of the vortex offset more away from the right.

As can be observed from Figure 7, with the same flow rate and composition, by changing the number of slot, the vortex position and velocity distribution change is not obvious.

*4.2. The Distribution Images of the Axial Velocity.* The velocity distributions of different polymer molecular weight are shown in Figure 8. The cutline represents the distance between the bushing and the bottom of the depressor trough of 10 mm, 23 mm, 32 mm, 37 mm, 40 mm, and 42 mm, respectively.

Figure 8 shows that, for weak base ASP solution with larger molecular weight when flowing through the partially pressured injection tool, the flow rate becomes smaller when approaching to the outer wall of the pressure injection tool but bigger in the bottom. With the decrease of the molecular weight of the polymer in solution, the smaller the velocity gradient is, the smaller the difference of the velocity of the solution with adjacent height will be. At the same time, the maximum velocity at the minimum gap of the annulus decreases with the increase of molecular weight of the polymer. The velocity gradient at the minimum clearance of the annulus increases with the increase of molecular weight of the polymer in weak base ASP system.

The velocity distribution of different flow rate is shown in Figure 9.

From Figure 9, the velocity is less, and the flow rate variation of the solution near the outer wall is greater than that at the center of the annulus gap of the pressure relief groove. With the decrease of the flow rate of the pressure injection tool, the location where the velocity changes from a decrease to an elevation shifts gradually to the right. The flow rate in the buck groove bottom is greater than that at the center of the vortex. The closer to the center of the pressure relief groove, the lower the velocity will be. With the buck groove bottom near the inner wall, the velocity changes from increase to decrease, which is due to the viscous force in the partially pressured injection tool device. In addition, when the flow rate is low, the change of flow velocity at the groove is more obvious.

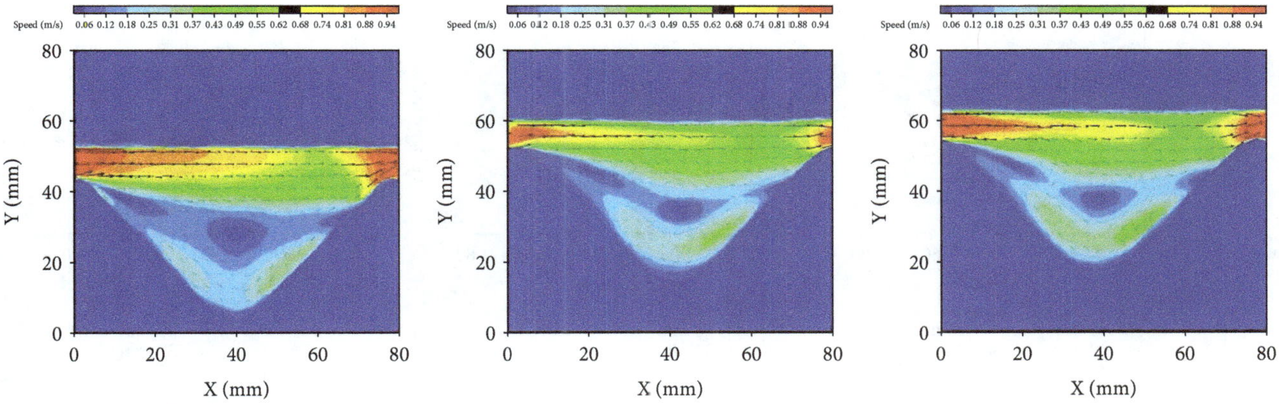


(a) $1.6 \times 10^6$ dalton (b) $1.9 \times 10^6$ dalton (c) $2.5 \times 10^6$ dalton

FIGURE 5: Velocity nephograms of different polymer molecular weight when flowing through the partially pressured injection tool.

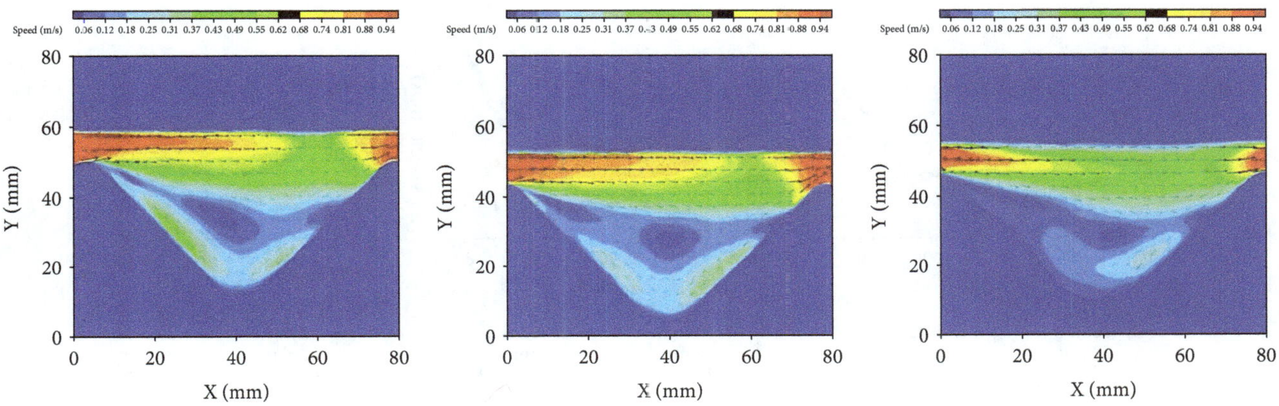


(a) $20\ m^3/d$ (b) $30\ m^3/d$ (c) $50\ m^3/d$

FIGURE 6: Velocity nephograms of different flow rate of polymer solution flowing through the partially pressured injection tool.

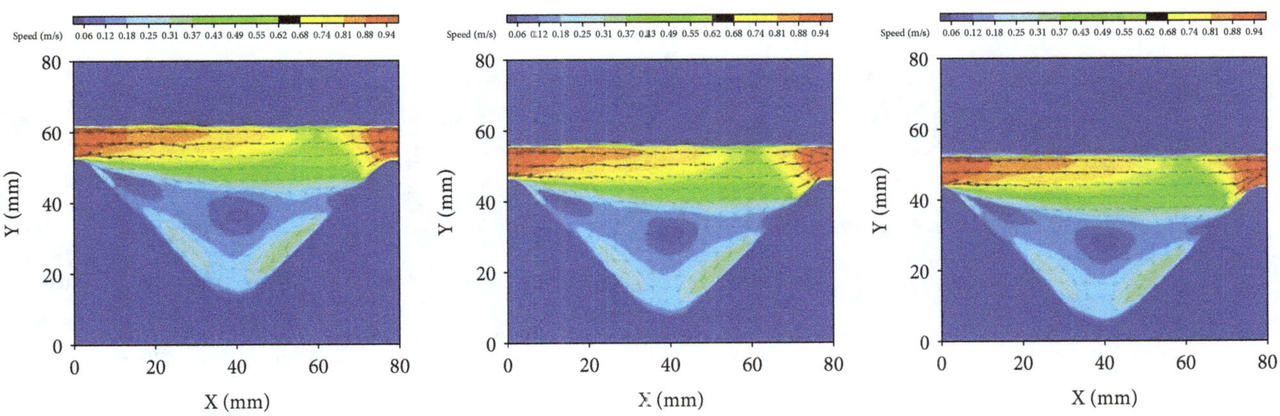


(a) 5 pressure reduction trough (b) 10 pressure reduction trough (c) 18 pressure reduction trough

FIGURE 7: Velocity nephograms of the polymer solution that flowed through the partially pressured injection tool with different number of slots.

The velocity distribution of different number of slots are shown in Figure 10.

The same can be seen from Figure 11. When the weak base ASP solution with the same composition and velocity flows through the partially pressured injection tool with different slots, the change of the end groove of the flow field is not obvious. Comparing to Figure 10, the velocity difference at the relative position is very small. The reason is that the partially pressured injection tool device has a shearing action to the weak base ASP solution. Therefore, the viscosity changes slightly leading to the internal flow field changes. However, the change was not obvious.

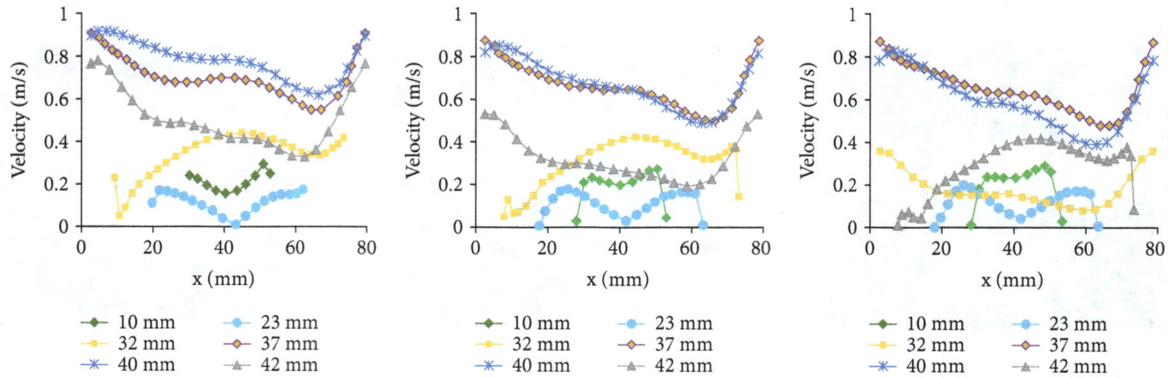


(a) $1.6 \times 10^6$ dalton (b) $1.9 \times 10^6$ dalton (c) $2.5 \times 10^6$ dalton

FIGURE 8: Velocity distribution of the polymer solution at $30\,m^3/d$ through the partially pressured injection tool with 18 slots.

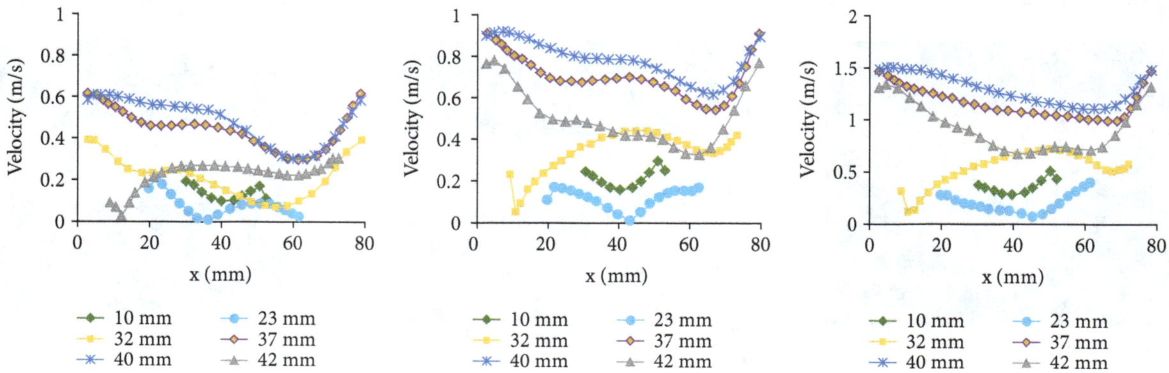


(a) (b) (c)

FIGURE 9: Velocity distribution of the polymer solution with a molecular weight of $1.6 \times 10^6$ dalton at $50\,m^3/d$ when flowing through the partially pressured injection tool with 18 slots.

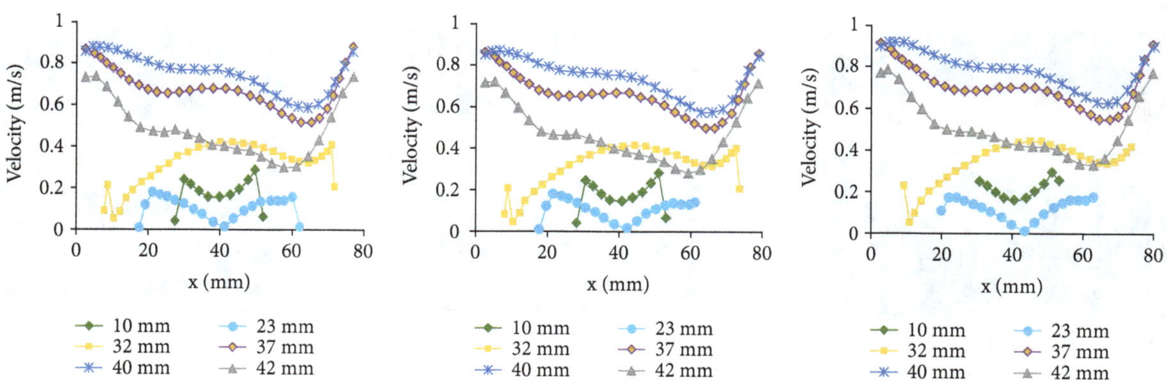


(a) (b) (c)

FIGURE 10: Velocity distribution of the polymer solution with a molecular weight of $1.6 \times 10^6$ dalton at $50\,m^3/d$ when flowing through the partially pressured injection tool with 5 slots.

## 5. Conclusions

(1) The similarity between the experimental model and the prototype was analyzed. The experimental model was designed based on the theory of similarity.

(2) Using PIV system and Tecplot software, the flow-field distribution of the polymer solution was obtained for different flow rates, molecular weight polymer, and number of slots. It can provide the basis for optimization of the flow-channel structure in the future.

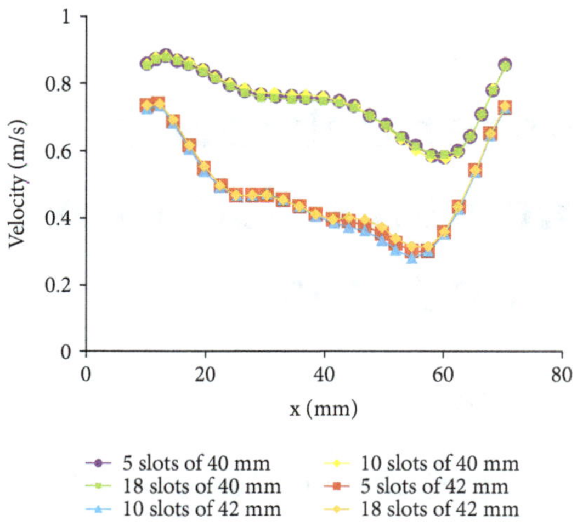


FIGURE 11: Contrast chart of flow velocity under different number of slots.

(3) The viscosity and flow rate of the solution have a significant influence on the velocity distribution in the depressurization slot. The effect of the slot number on the velocity distribution is not obvious.

(4) In this paper, the flow field inside the depressurizing channel of the partially pressured injection tool was studied. The variation rules under different working conditions were analyzed so as to design and optimize the structure of the depressurizing channel in the partially pressured injection tool in the future.

## Conflicts of Interest

The authors declare that they have no conflicts of interest.

## Acknowledgments

This work is financially supported by the Science and Technology Research Project of Heilongjiang Education Department (No. 12541092).

## References

[1] H. C. Li, "Overview of separate injection technique for polymerflooding in Daqing oilfield," *Oil and Gasgeology*, vol. 33, no. 2, pp. 296–301, 2012.

[2] C. Ruan, "The characteristics of the tracer particles used in water flow field for PIV system," *Journal of Experiments in Fluid Mechanics*, vol. 20, no. 2, pp. 72–77, 2006.

[3] D. Wu, "Structural optimization and internal flow field analysis of the mass separate injection tool for weak base ASP solution," Northeast Petroleum University, Daqing, China, 2017.

[4] L. Wang, "Polymer separate injection technology with separate mass and pressure," *Oil Drilling & Production Technology*, vol. 29, no. 2, pp. 73–75, 2007.

[5] X. H. Chen, "Principle and application of polymer single tube multilayer separate injection technique," *Petroleum Geology and development in Daqing*, vol. 23, no. 4, pp. 53–55, 2004.

[6] D. Feng, X. M. Li, and C. B. Huang, "Research on the flow of the annular polymer in separate-layer polymer injection allocator," *Petroleum Machinery*, vol. 37, no. 10, pp. 7–10, 2009.

[7] L. Z. Meng, H. Q. Cui, and C. B. Wang, "Numerical simulation for corrugation and annular flow field of a stratification injection allocator," *Petroleum Machinery*, vol. 34, no. 2, pp. 15–17, 2006.

[8] M. Cai, Y. Sun, and L. D. Liu, "The flow flied numerical analysis of ternary composite solution in eccentric injection allocation device flows through the nozzle," *Oil Production Engineering*, vol. 3, no. 1, pp. 12–16, 2013.

[9] T. Schmitt, J. N. Koster, and H. Hamacher, "Particle design for displacement tracking velocimetry," *Measurement Science and Technology*, vol. 6, no. 6, pp. 682–689, 1995.

[10] A. Melling, "Tracer particles and seeding for particle image velocimetry," *Measurement Science and Technology*, vol. 8, no. 12, pp. 1406–1416, 1997.

[11] R. Y. Zhang, "Technologies of separated layer water flooding: an overview," *Oil Drilling and Production Technology*, vol. 33, no. 2, pp. 102–107, 2011.

[12] F. S. Wang, "Based on the technology of PIV flow field characteristics of the research in layered injection tools and experiment comparison," *Oil Field Equipment*, vol. 45, no. 6, pp. 16–19, 2016.

[13] S. R. Yang, *Engineering Fluid Mechanics*, Petroleum Industry Press, Beijing, 2006.

# 2

# Phosphorus Flame Retardants for Polymeric Materials from Gallic Acid and other Naturally Occurring Multihydroxybenzoic Acids

**Bob A. Howell, Kendahl L. Oberdorfer, and Eric A. Ostrander**

*Center for Applications in Polymer Science, Department of Chemistry and Biochemistry, Central Michigan University, Mt. Pleasant, MI 48859-0001, USA*

Correspondence should be addressed to Bob A. Howell; bob.a.howell@cmich.edu

Academic Editor: Jui-Yang Lai

The development of polymer and polymer additives from renewable biosources is becoming increasingly prominent. This reflects increasing concerns about sustainability, environmental quality, and human health. Bioproducts produced in nature are generally inexpensive and benign in the environment. Moreover, degradation of derivatives does not yield toxic products. Gallic acid (3,4,5-trihydroxybenzoic acid) is found widely in nature and has long been touted for its medicinal qualities. 3,5-Dihydroxybenzoic acid is also produced by several plants, most notably buckwheat. Both compounds, as the anilide and methyl ester, respectively, have been converted to a series of phosphorus esters, both phosphonate and phosphate. Esters have been fully characterized using spectroscopic and thermal methods. These compounds display good flame retardancy at low loadings in DGEBA epoxy resin.

## 1. Introduction

There is an increasing reliance on natural biosources as precursors to polymers and polymer additives [1–11]. This is particularly the case for plasticizers [12–14] and flame retardants [15–17]. These developments are driven by an increasing concern for sustainability, environmental quality, and human health. The release of potentially toxic agents into the environment is increasingly unacceptable. Traditionally, organohalogen compounds, particularly brominated aromatics, have been widely used as effective flame retardants [18]. However, these compounds often migrate from the polymer matrix into which they have been incorporated. Brominated aromatics are stable in the natural environment, tend to bioaccumulate, are increasingly finding their way into the human diet, and may pose health risks [18–22]. To reduce the propensity for migration from the polymer matrix, oligomeric flame retardants containing brominated units have been developed [23, 24]. These materials can function as effective flame retardants both in the presence and in the absence of an antimony oxide promoter. However, they suffer from the perception that halogens are undesirable established by the behavior of lower molecular weight counterparts. These smaller compounds continue to face increasing regulatory pressure and legislative restriction [25–27]. Replacements for these compounds are actively being sought. Phosphorus compounds offer great potential as acceptable flame retardants [28]. In general, phosphorous compounds are less toxic than the materials that they are replacing [29, 30]. Phosphorus flame retardants derived from renewable biomolecules are particularly attractive [17]. Starting materials obtained from natural sources are usually nontoxic and environmentally benign. Costs are independent of fluctuations in petrochemical markets.

Gallic acid (3,4,5-trihydroxybenzoic acid) is widely produced in nature and may be extracted from gallnuts, oak bark, several legumes, grapes, berries, hickory nuts, and walnuts [31, 32]. Gallic acid has long been touted for its medicinal properties. When consumed, it acts as an antioxidant to help prevent potential cardiac dysfunction and the formation of radical oxygen species which disrupt cell membranes [33]. Many pharmaceuticals utilize the anti-inflammatory, anticarcinogenic, and radio-protective nature of gallic acid for the treatment of a variety of diseases [34–36].

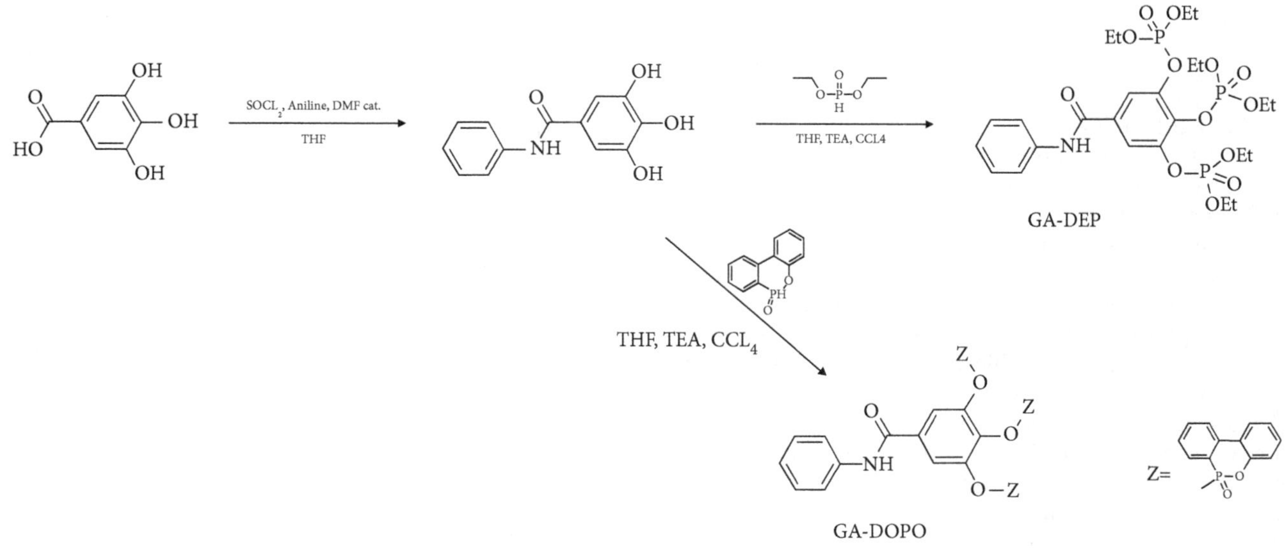

SCHEME 1: Synthesis of phosphorus esters of gallic acid.

TABLE 1: Thermal decomposition of phosphorous esters of gallic acid anilide.

| Compound | Temperature of decomposition onset (°C)[a] | | Temperature of maximum decomposition rate (°C) | | Decomposition residue (%)[b] | |
|---|---|---|---|---|---|---|
| | Nitrogen | Air | Nitrogen | Air | Nitrogen | Air |
| GA-DEP | 239 | 175 | 253 | 205 | 37 | 13 |
| GA-DOPO | 316 | 313 | 373 | 366 | 18 | 7 |

[a]Extrapolated onset temperature from the derivative plot of mass loss versus temperature. [b]Percentage of the initial sample mass.

As a precursor to flame retardants, it offers three hydroxyl groups which may be subjected to esterification. Because the three hydroxyls occupy adjacent positions on the aromatic nucleus, it is not always possible to generate the corresponding triester, depending on the steric requirements of the groups being introduced. This may be overcome by first converting a gallic ester derivative to triacrylate followed by the Michael addition of phosphite [37]. Alternatively, a polyhydroxy aromatic acid with nonadjacent hydroxyl groups can be used. 3,5-Dihydroxybenzoic acid is a naturally occurring acid found in numerous plants and grains, most notably buckwheat [38–41]. Both gallic acid and 3,5-dihydroxybenzoic acid have been converted to derivatives, analide and methyl ester, respectively, suitable for use as phosphorus ester precursors. The phosphorus esters generated display good flame-retarding properties in polymeric matrices.

## 2. Experiment

*2.1. Materials.* Common solvents and reagents were obtained from Thermo Fisher Scientific or the Aldrich Chemical Company. Tetrahydrofuran (THF) was distilled from lithium aluminum hydride prior to use, and methylene chloride from calcium hydride. Gallic acid, 3,5-dihydroxybenzoic acid, carbon tetrachloride, triethylamine, diethylphosphite, and aniline were obtained from the Aldrich Chemical Company. 9,10-Dihydro-9-oxa-10-phosphaphenanthrene-10-oxide (DOPO) was from TCI. Diphenyl chlorophosphate was provided by ICL-IP America Inc. The diglycidyl ether of *bis*-phenol A (DGEBA) was supplied by the Dow Chemical Company.

*2.2. Methods and Instrumentation.* General methods and instrumentation have been previously described [17, 42, 43]. Nuclear magnetic resonance (NMR) spectra were obtained using a Varian Mercury 300 MHz or INOVA 500 MHz spectrometer. Tetramethylsilane was used as an internal reference ($\delta = 0.00$) for proton and carbon spectra, and triphenylphosphate ($\delta = -18.00$) for phosphorus spectra. Infrared spectra were recorded using a Thermo Scientific Nicolet 380 FTIR spectrometer. Thermal transitions were determined by differential scanning calorimetry (DSC) using a TA Instruments Q2000 instrument. Thermal stability was assessed by thermogravimetry using a TA Instruments Q500 instrument. Peak heat release rates were determined using a microscale combustion calorimeter (ASTM D7309a) (Fire Testing Technology Ltd. (FTT)). Values reported are the averages of five determinations with a deviation of less than 5%. Limiting oxygen index values were determined using an FTT Oxygen Index unit (ASTM D2863-13). Vertical burn tests were conducted in an FTT test chamber (ASTM D2863-06).

*2.3. Test Specimen.* Samples for flammability testing were prepared as previously described [42].

### 2.4. *Synthesis*

*2.4.1. 3,4,5-Trihydroxybenzanilide.* Gallic acid chloride was synthesized by the dropwise addition, over a period of 0.25 hr, of 3.16 ml (44.3 mmol) of thionyl chloride to a solution of 5.03 g (29.6 mmol) of gallic acid and three drops of dimethylformamide in 20 ml of anhydrous tetrahydrofuran (THF) maintained near 0°C (external ice bath). Upon completion of the addition, the mixture was stirred near 0°C for an additional 0.25 hr. The progress of the reaction was monitored by periodic removal of aliquots for analysis using infrared spectroscopy. Upon observation of the complete conversion of acid to acid chloride, 10.8 ml (118.0 mmol) of aniline was added dropwise over a period of 0.25 hr to the stirred solution. Upon completion of the addition, the solution was allowed to slowly warm to room temperature. The progress of the reaction was monitored by periodic removal of an aliquot of the mixture for analysis using infrared spectroscopy. The reaction was completed within 12 hr. Anilinium chloride was removed by filtration at reduced pressure. The filtrate was diluted with 200 ml of ethyl acetate and washed, successively, with 50 ml of water, 50 ml of 10% aqueous hydrochloric acid solution, 50 ml of 10% aqueous sodium hydroxide solution, and 50 ml of saturated aqueous sodium chloride solution. The ethyl acetate solution was dried over anhydrous sodium sulfate, and the solvent was removed by rotary evaporation at reduced pressure. The residual solid was recrystallized from water to afford 6.37 g (88.1% yield) of 3,4,5-trihydroxybenzanilide as a pale yellow solid: mp 210°C (DSC); ESI-MS, 246 g/mol; IR ($cm^{-1}$) 3490 (s) N-H, 3533 (s, broad) O-H, 1636 (s) amide C=O, 1593 (s) aromatic nucleus; $^1$H-NMR ($\delta$ DMSO-$d_6$) 7.26 (m, 7H), 8.78 (s, 1H), 9.14 (s, 2H), 9.87 (s, 1H); $^{13}$C NMR ($\delta$, DMSO-$d_6$) 127.2 (m, aromatic carbon atoms), 166.0 (s, carbonyl carbon atom).

*2.4.2. Methyl 3,5-Dihydroxybenzoate.* A solution of 10.0 g (64.9 mmol) of 3,5-dihydroxybenzoic acid in 200 ml of methanol containing (0.35 ml, 0.64 g, 6.49 mmol) concentrated aqueous sulfuric acid solution was stirred at solvent reflux. The progress of the reaction was monitored by periodic removal of an aliquot of the reaction mixture for analysis using infrared spectroscopy (changes in the carbonyl region of the spectrum). The reaction was completed within 10 hr. Excess methanol was removed by rotary evaporation at reduced pressure to provide oil which was dissolved in 450 ml of ethyl acetate. The resulting solution was washed, successively, with two 40 ml portions of saturated aqueous sodium bicarbonate solution and 40 ml of saturated aqueous sodium chloride solution. The solution was dried over anhydrous sodium sulfate, and the solvent was removed by rotary evaporation at reduced pressure to afford a fine white solid. This material was dried overnight at 15 torr and 50°C to provide 9.08 g (90.8% yield) of methyl 3,5-dihydroxybenzoate: IR ($cm^{-1}$) 3229 (s, br) phenolic hydroxyl, 3089 (m) $C_{sp2}$-H, 2952 (m) $C_{sp3}$-H, 1687 (s) ester C=O, 1600 (s) aromatic nucleus; $^1$H NMR ($\delta$, DMSO-$d_6$) 3.76 (s, 3H, methoxyl protons), 6.41 (t, 1H) and 6.79 (d, 2H) aromatic protons, 9.10 (br, s, 2H, hydroxyl protons); $^{13}$C ($\delta$, DMSO-$d_6$) 52.4

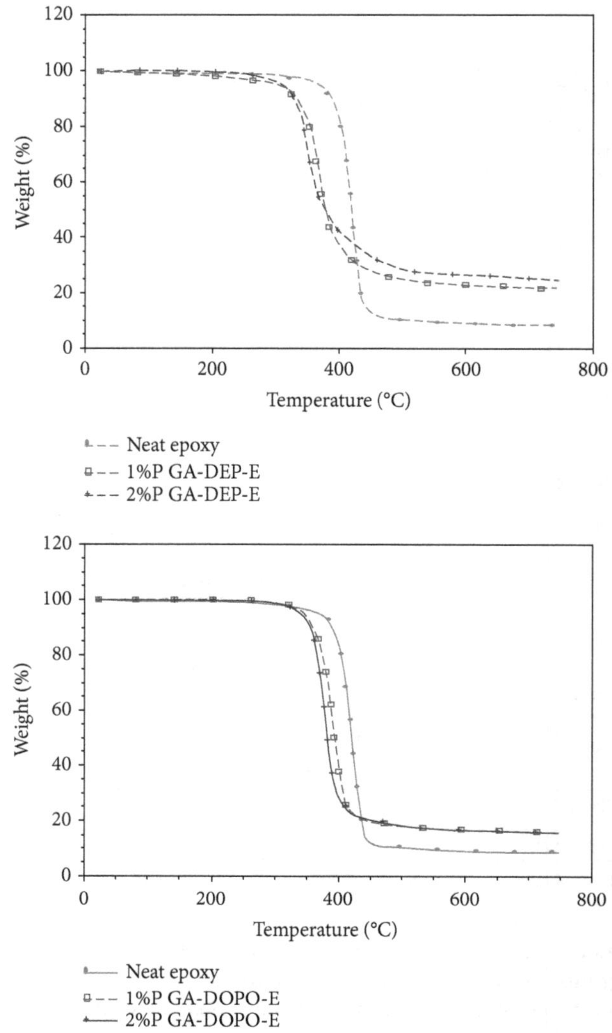


FIGURE 1: Thermal stability of blends of phosphorus esters of gallic acid anilide with DGEBA epoxy.

(methoxy carbon atom), 107.5, 131.7, 158.9 (aromatic carbon atoms), 166.7 (ester carbonyl carbon atom).

Both the anilide and benzoate ester were converted to the corresponding phosphorus esters using methods previously described [42]. In general, the appropriate phenolic compound was treated with a phosphite in the presence of carbon tetrachloride (Atherton-Todd procedure). For the preparation of the diphenyl phosphate ester, diphenyl chlorophosphate was used as a reagent.

*2.4.3. 3,4,5-Tri(diethylphosphato)benzanilide.* Into a 500 ml three-necked, round-bottomed flask fitted with a magnetic stirring bar, a dropping funnel, and a Liebig condenser bearing a gas-inlet tube were placed 5.53 g (49.0 mmol) of diethylphosphite, 2.71 g (11.0 mmol) of 3,4,5-trihydroxybenzanilide, 5.98 ml (42.9 mmol) of triethylamine, and 300 ml of ethyl acetate. The solution was stirred near 0°C (external ice bath) as 5.37 ml (55.9 mmol) of carbon tetrachloride was added, dropwise, over a period of five minutes. The progress of the reaction was monitored by periodic removal of an

Table 2: Thermal degradation of blends of phosphorous esters of gallic acid anilide with DGEBA epoxy.

| Blend | Level of additive in blend (%) | Level of phosphorous in blend (%) | Decomposition onset temperature (°C)[a] | | Temperature of maximum decomposition rate (°C) | | Decomposition residue (%)[b] | |
|---|---|---|---|---|---|---|---|---|
| | | | Nitrogen | Air | Nitrogen | Air | Nitrogen | Air |
| DGEBA epoxy | 0 | 0 | 390 | 374 | 423 | 408 | 8 | 8 |
| GA-DEP-E | 7.5 | 1 | 342 | 349 | 370 | 371 | 22 | 4 |
| GA-DEP-E | 15.0 | 2 | 327 | 334 | 352 | 354 | 25 | 6 |
| GA-DOPO-E | 9.6 | 1 | 360 | 372 | 394 | 392 | 16 | 3 |
| GA-DOPO-E | 19.1 | 2 | 348 | 353 | 381 | 375 | 16 | 3 |

[a]Extrapolated onset temperature from the derivative plot of mass/loss versus temperature. [b]Percentage of the initial sample mass.

TABLE 3: Glass transition temperatures for blends of phosphorus esters of gallic acid anilide with DGEBA epoxy resin.

| Blend | Phosphorous level (%) | Additive by weight (%) | $T_g$ (°C) |
|---|---|---|---|
| DGEBA epoxy | 0 | 0 | 110 |
| GA-DEP-E | 1 | 7.5 | 129 |
| | 2 | 14.0 | 148 |
| GA-DOPO-E | 1 | 9.6 | 145 |
| | 2 | 19.1 | 142 |

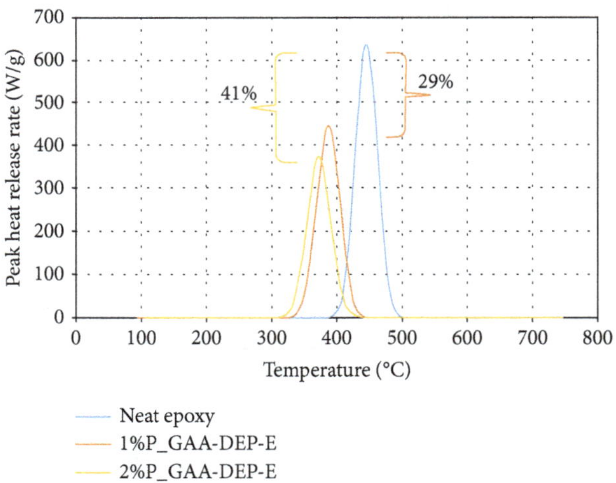


FIGURE 2: Peak heat release rates for blends of 3,4,5-tri(diethylphosphato)benzanilide with DGEBA epoxy resin.

aliquot of the reaction mixture for analysis using infrared spectroscopy. When the spectrum of the mixture no longer contained any phenolic absorption, the mixture was filtered to remove anilinium chloride which had formed. The filtrate was washed, successively, with 50 ml of water, 50 ml of 10% aqueous hydrochloric acid solution, 50 ml of 10% aqueous sodium hydroxide solution, and 50 ml of saturated aqueous sodium chloride solution. The solution was dried over anhydrous magnesium sulfate, and the solvent was removed by rotary evaporation at reduced pressure to afford 5.86 g (93.1% yield) of the triphosphate as a light tan solid: mp 79°C (DSC); ESI-MS 654 g/mol; IR ($cm^{-1}$) 3302 (s) N-H, 2988 (s) $C_{sp2}$-H, 2912 (s) $C_{sp3}$-H, 1677 (s) amide carbonyl, 1590 (m) aromatic nucleus, 1271 (s) P=O, 1013 (s) P-O-C; $^1$H NMR ($\delta$, $CDCl_3$) 1.35 (m, 18H, methyl protons), 4.06 (m, 4H, methylene protons), 4.28 (m, 8H, methylene protons), 7.35 (m, 7H, aromatic protons), 8.63 (s, 1H, amide proton); $^{13}$C NMR ($\delta$, $CDCl_3$) 15.1, 64.5, 128.4 (aromatic carbon atoms), 163.9 (carbonyl carbon atoms); $^{31}$P NMR ($\delta$, $CDCl_3$) 6.88, 7.20.

*2.4.4. 3,4,5-Tri(dopyloxy)benzanilide.* 3,4,5-Tri(dopyloxy)-benzanilide was prepared in a manner similar to that described above from 5.83 g (27.1 mmol) of 9,10-dihydro-9-oxa-10-phosphaphenathrene-10-oxide (DOPO) and 1.47 g (6.0 mmol) of 3,4,5-trihydroxybenzanilide. The triester, 3,4,5-tri(dopyloxy)benzanilide (4.94 g, 93.2% yield), was obtained as a white solid: mp 187°C (DSC); ESI-MS 890 g/mol; IR ($cm^{-1}$) 3301 (w) N-H, 3066 (s) $C_{sp2}$-H, 1670 (s) amide carbonyl, 1596 (s) aromatic nucleus, 1293 (s) P=O, 910 (s) P-O-C; $^1$H NMR ($\delta$, DMSO-$d_6$) 7.36 (m, 31H), 10.47 (s, 1H); $^{13}$C NMR ($\delta$, DMSO-$d_6$) 129.8 (aromatic carbon atoms), 163.0 (carbonyl carbon atom); $^{31}$P NMR ($\delta$, DMSO-$d_6$) 5.76, 6.63.

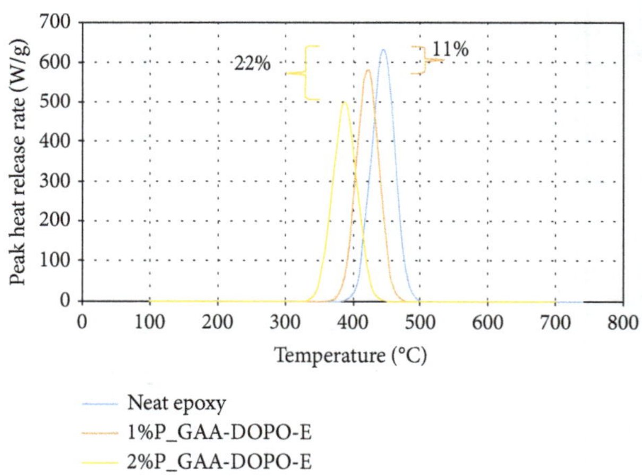


FIGURE 3: Peak heat release rates for blends of 3,4,5-tri(dopyloxy)benzanilide with DGEBA epoxy resin.

*2.4.5. Methyl 3,5-Di(diethylphosphato)benzoate.* IR ($cm^{-1}$) 3094 (w) $C_{sp2}$-H, 2986 (m) $C_{sp3}$-H, 1728 (s) ester C=O,

Table 4: Flammability characteristics of blends of phosphorous esters of gallic acid anilide with DGEBA epoxy resin.

| Additive | Additive level (%) | Phosphorous level (%) | Peak heat release rate (W/g) | Limiting oxygen index (%) | UL 94 rating |
|---|---|---|---|---|---|
| DGEBA epoxy | 0 | 0 | 692 | 19.0 | NR |
| GA-DEP-E | 7.5 | 1 | 491 | 22.5 | NR |
| GA-DEP-E | 14.0 | 2 | 411 | 23.0 | NR |
| GA-DOPO-E | 9.6 | 1 | 617 | 23.8 | V2 |
| GA-DOPO-E | 19.1 | 2 | 541 | 27.0 | V0 |

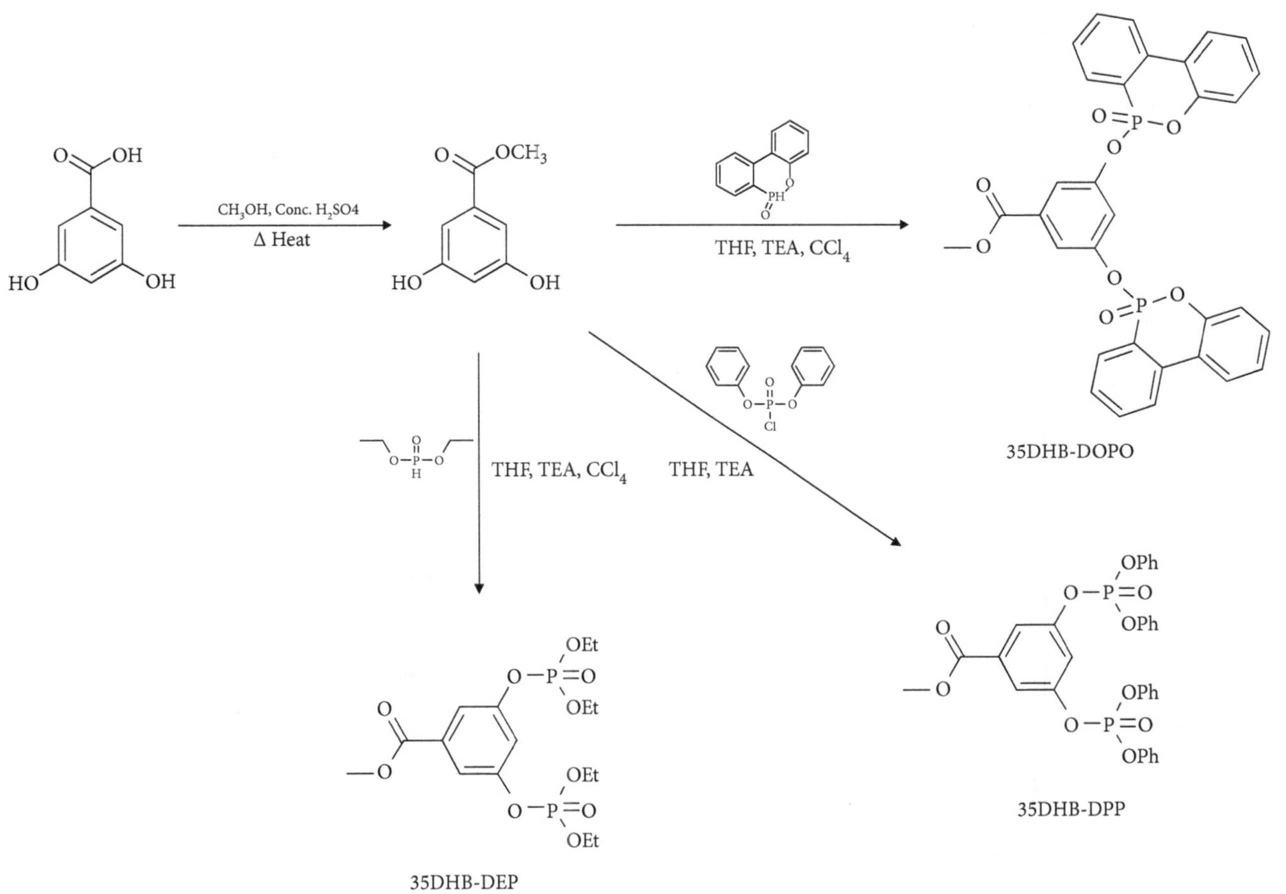


Scheme 2: Synthesis of phosphorus esters of methyl 3,4-dihydroxybenzoate.

1596 (s) aromatic nucleus, 1278 (s) P=O, 1006 (s) P-O-C; $^1$H NMR ($\delta$, DMSO-$d_6$) 1.25 (t, $J = 7.1$ Hz, 12H), 3.86 (s, 3H), 4.17 (q, $J = 7.1$ Hz, 8H), 7.37 (m, 1H) and 7.61 (m, 2H) aromatic protons; $^{13}$C NMR ($\delta$, DMSO-d$_6$) 16.3, 53.2 (ethyl carbon atoms), 117.2, 117.5, 132.8, 151.5 (aromatic carbon atoms), 164.9 (ester carbonyl carbon atom), $^{31}$P NMR ($\delta$, DMSO-d$_6$) −7.47.

*2.4.6. Methyl 3,5-Di(diphenylphosphato)benzoate.* mp 68°C (DSC); IR (cm$^{-1}$) 3083 (w) $C_{sp2}$-H, 2948 (w) $C_{sp3}$-H, 1724 (s) ester C=O, 1588 (s) aromatic nucleus, 1300 (s) P=O, 1183 (s), 903 (s) P-O-C; $^1$H NMR ($\delta$, DMSO-d$_6$) 3.85 (s, 3H) 7.24-7.47 (m, 20H, protons of diphenylphosphato groups), 7.54 (m, 1H) and 7.69 (m, 2H) (aromatic protons of trihydroxybenzoate nucleus); $^{13}$C NMR ($\delta$, DMSO-d$_6$) 53.3 (methoxy carbon atom), 117.4, 118.4, 133.4, 150.5 (aromatic carbon atoms of trihydroxybenzoate nucleus), 116-152 (additional peaks for carbon atoms of the diphenylphosphato groups), 164.5 (ester carbonyl carbon atom); $^{31}$P NMR ($\delta$, DMSO-d$_6$) −17.6.

*2.4.7. Methyl 3,5-Di(dopyloxy)benzoate.* mp 145°C (DSC); IR (cm$^{-1}$) 3061 (w) $C_{sp2}$-H, 2953 (w) $C_{sp3}$-H, 1731 ($\delta$) ester C=O, 1594 (s) aromatic nucleus, 1308 (s) P=O, 1195 (s) and 929 (s) P-O-C; $^1$H NMR ($\delta$, DMSO-d$_6$) 3.81 (s, 3H), 6.90-8.41(m, 20H); $^{13}$C NMR ($\delta$, DMSO-d$_6$) 53.2 (methoxy

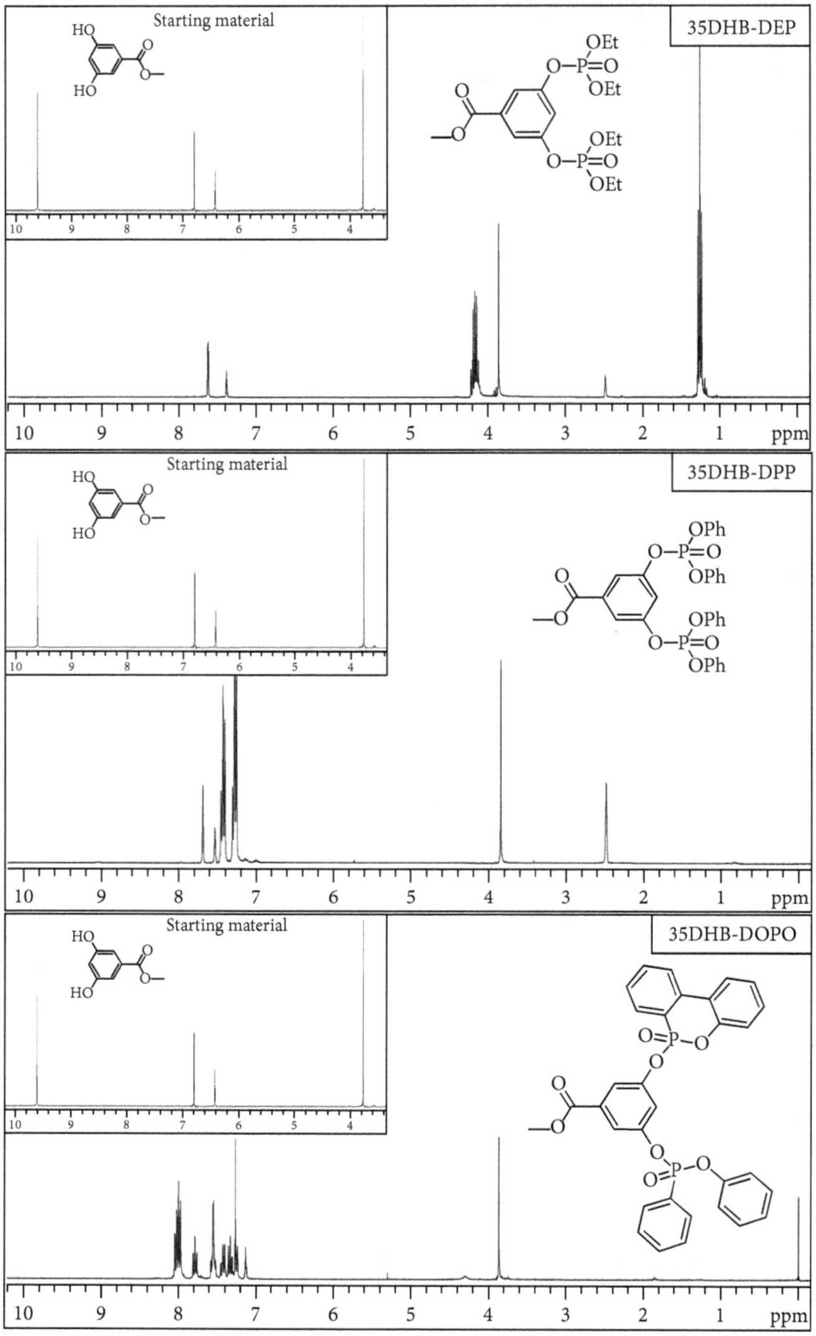


FIGURE 4: $^1$H NMR spectra of phosphorus esters of methyl 3,5-dihydroxybenzoate. All the esters display good thermal stability (Table 5). Two esters, the diphenyl phosphate and the DOPO phosphonate, display decomposition onset temperatures above 300°C.

carbon atom), 116.2-151.1 (aromatic carbon atoms), 164.5 (ester carbonyl carbon atom); $^{31}$P NMR ($\delta$, DMSO-$d_6$) 6.50.

## 3. Results and Discussion

Gallic acid and other multihydroxybenzoic acids are found widely in nature and represent annually renewable platforms for the generation of phosphorus flame retardants. The multiple hydroxyl groups provide ready sites for the incorporation of relatively high levels of phosphorus. The hydroxyls may be converted to a range of phosphorus esters using the well-known Atherton-Todd procedure [43]. This is illustrated in Scheme 1 for gallic acid. The acid is first converted to the corresponding anilide, and then the hydroxyls are utilized to generate both the *tris*-DOPO phosphonate and the *tris*-(diethylphosphate). Both are solids melting at 187°C and 79°C, respectively. The $^{31}$P NMR spectra of both compounds contain two resonances reflecting two different

TABLE 5: Thermal decomposition of phosphorus esters of methyl 3,5-dihydroxybenzoate.

| Compound | Temperature of decomposition onset (°C)[a] | | Temperature of maximum decomposition rate (°C) | Decomposition residue (%)[b] | |
|---|---|---|---|---|---|
| | Nitrogen | Air | Nitrogen | Nitrogen | Air |
| 35DHB-DEP | 224 | 224 | 272 | 5 | <1 |
| 35DHB-DPP | 328 | 303 | 366 | <1 | 5 |
| 35DHB-DOPO | 332 | 319 | 393 | 22 | <1 |

[a]Extrapolated onset temperature from the derivative plot of mass loss versus temperature. [b]Percentage of the initial sample mass.

environments for the phosphorus moieties. The spectrum for the *tris*-(diethylphosphate) contains peaks at $\delta = -6.88$ and $-7.20$ for the esters at C-4 and at C-3 and C-5, respectively. The corresponding peaks for the DOPO phosphonate appear at $\delta = 6.63$ and 7.06. Both compounds are thermally stable to relatively high temperatures. The degradation onset temperature for the *tris*-(diethylphosphate) (GA-DEP) is 239°C, and that for the DOPO phosphonate (GA-DOPO) is 316° C. The complete decomposition characteristics are listed in Table 1.

For the assessment of flame-retarding impact, the esters were incorporated into DGEBA resin at levels sufficient to provide one and two percent phosphorus (7–19% additive). The thermal stability of the blends is reflected in Figure 1. It can be seen that the presence of the phosphorus esters does not significantly lower the thermal stability of the epoxy polymer. Numerical data for the degradation of the phosphorus ester/epoxy blends are collected in Table 2.

Glass transition temperatures for the phosphorus ester blends with epoxy are collected in Table 3. As may be seen, the glass transition temperature for the resin is altered only slightly by the presence of the esters.

The flammability of the blends was assessed using limiting oxygen index (LOI) measurements, UL 94 vertical burn test, and microscale combustion calorimetry (MCC). Plots for the peak heat release rates for blends of 3,4,5-tri(diethylphosphato)benzanilide in epoxy are displayed in Figure 2, and those for the analogous DOPO ester blends in Figure 3. In both cases, a significant peak heat release rate reduction for the combustion of epoxy is achieved by incorporation of either of the phosphorus esters at a level sufficient to provide one or two percent phosphorus. This is particularly true for the incorporation of 3,4,5-tri(diethylphosphato)benzanilide.

All the flammability data are collected in Table 4. Most notably, incorporation of two percent 3,4,5-tri(dopyloxy)-benzanilide into DGEBA epoxy afforded a material with a 22% peak heat release rate reduction, an LOI value of 27, and a UL 94 rating of V0.

The preparation of phosphorus esters of methyl 3,5-dihydorxybenzoate is illustrated in Scheme 2. The progress of the reaction may conveniently be monitored using infrared spectroscopy. The band for hydroxyl absorption (3229 $cm^{-1}$) in the spectrum of the starting material gradually disappears. The spectra for the phosphorus esters contain prominent absorptions for ester carbonyl (1728, 1724, and 1731 $cm^{-1}$), aromatic nuclei (1596, 1588, and 1594 $cm^{-1}$), P=O (1278, 1300, and 1308 $cm^{-1}$), and P-O-C (1006, 1183, 903, 1185, and 929 $cm^{-1}$). The proton NMR spectra for the phosphorus esters may be found in Figure 4. All contain distinct absorptions for the aromatic nucleus of the starting material as well as those expected for the ester moieties.

TABLE 6: Glass transition temperatures for blends of phosphorus esters of methyl 3,5-dihydroxybenzoate with DGEBA epoxy resin.

| Additive | Additive level (wt%) | Phosphorus level (wt%) | $T_g$ (°C) |
|---|---|---|---|
| None | 0 | 0 | 109 |
| 35DHB-DEP | 7.1 | 1 | 126 |
| 35DHB-DEP | 14.2 | 2 | 137 |
| 35DHB-DPP | 10.2 | 1 | 119 |
| 35DHB-DPP | 20.4 | 2 | 116 |
| 35DHB-DOPO | 9.6 | 1 | 128 |
| 35DHB-DOPO | 19.2 | 2 | 142 |

For the assessment of the impact of these esters on flammability, blends of these esters at a level to provide one or two percent phosphorus with DGEBA epoxy were prepared. As may be seen from the data presented in Table 6, the incorporation of the esters at these levels does not appreciably alter the glass transition temperature for the polymer. The thermal stability for the blends is reflected in Table 7. The thermal stability of the resin is not much impacted by low loadings of the esters.

The peak heat release rates for blends of the esters with DGEBA epoxy are shown in Figures 5-7. The reduction in peak heat release is most notable for the *bis*-diethylphosphato ester. For this compound, a loading sufficient to provide 2% phosphorus induces a 50% reduction in the peak heat release rate.

Flammability data for all the blends are collected in Table 8. The most effective compound is the di(dopyloxy) ester. At a loading sufficient to provide 2% phosphorus, combustion of the blend reflects a reduction in the peak heat release rate (678 W/g to 519 W/g), an LOI of 33, and a UL 94 rating of V2.

For both series of phosphorus esters, their incorporation into DGEBA epoxy resin at levels sufficient to provide one or two percent phosphorus does not significantly impact the properties or thermal stability of the polymer but does impart significant flame retardancy. The DOPO phosphonate is the most effective. This is clear from both LOI and UL 94

TABLE 7: Thermal degradation characteristics for blends of phosphorus esters of methyl 3,5-dihydroxybenzoate in DGEBA epoxy resin.

| Additive | Additive level (wt%) | Phosphorus level (wt%) | $T_{onset}$ (°C)[a] | $T_{max}$ (°C)[b] | Char[c] (wt%) |
|---|---|---|---|---|---|
| None | 0 | 0 | 390 | 423 | 10 |
| 35DHB-DEP | 7.1 | 1 | 345 | 373 | 23 |
| 35DHB-DEP | 14.2 | 2 | 329 | 355 | 26 |
| 35DHB-DPP | 10.2 | 1 | 349 | 375 | 24 |
| 35DHB-DPP | 20.4 | 2 | 328 | 356 | 27 |
| 35DHB-DOPO | 9.6 | 1 | 369 | 394 | 15 |
| 35DHB-DOPO | 19.2 | 2 | 354 | 379 | 18 |

[a]Extrapolated onset temperature from the derivative plot of mass loss versus temperature. [b]Percentage of the initial sample mass. [c]Nitrogen atmosphere.

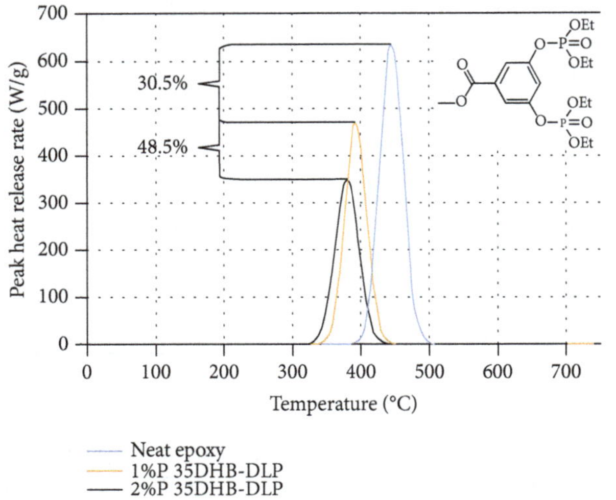


FIGURE 5: Peak heat release rates for blends of methyl 3,5-di(diethylphosphato)benzoate in DGEBA epoxy.

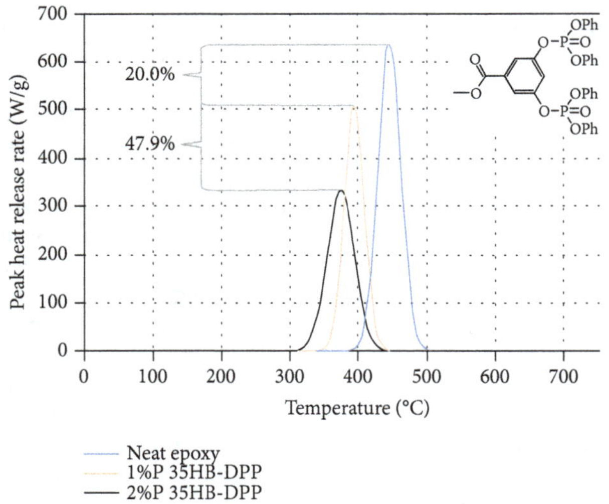


FIGURE 6: Peak heat release rates for blends of methyl 3,5-di(diphenylphosphato)benzoate in DGEBA epoxy.

measurements. The phosphates promote char formation and strongly lower the peak release rate for the combustion of the polymer. The peak heat release rate reduction for the

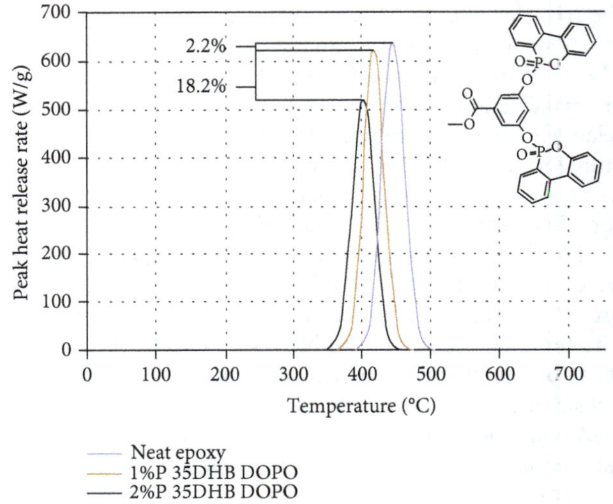


FIGURE 7: Peak heat release rates for blends of methyl 3,5-di(dopyloxy)benzoate in DGEBA epoxy.

polymer containing the DOPO phosphonate is much smaller than that for the polymer containing the corresponding phosphates. This is probably reflective of the different modes of action. DOPO derivatives are generally thought to act predominately by liberating species to the gas phase which inhibits flame-propagating reactions [44, 45]. The results from MCC may underestimate the effectiveness of flame retardants that function in the gas phase [46–48]. The action of a phosphorus ester additive is dependent on the level of oxygenation at phosphorus [49–52]. In general, increasing gas phase activity is observed for a decreasing level of oxygenation at phosphorus [48–51]. Compounds with a high level of oxygenation at phosphorus tend to promote char formation in the solid phase. These observations are consistent with those presented here. It would appear that the phosphates function largely in the solid phase while the phosphonates display predominately gas phase activity.

## 4. Conclusions

Gallic acid and other multihydroxybenzoic acids are found widely in nature and represent a renewable biosource for the generation of effective phosphorus flame retardants. Gallic acid and 3,5-dihydroxybenzoic acid have been

TABLE 8: Flammability characteristics for blends of phosphorus esters of methyl 3,5-dihydroxybenzoate with DGEBA epoxy.

| Compound | % phosphorus | PHRR (W/g) | THR (kJ/g) | LOI (% $O_2$) | UL 94 rating | % additive |
|---|---|---|---|---|---|---|
| Epoxy | 0% | 677.7 | 29.8 | 19 | NR | 0 |
| 35DHB-DEP | 1% | 470.9 | 25.7 | 23.6 | NR | 7.1 |
| | 2% | 348.7 | 18.9 | 23.4 | NR | 14.2 |
| 35DHB-DPP | 1% | 505.4 | 23.3 | 24.6 | NR | 10.2 |
| | 2% | 330.1 | 20.1 | 24.6 | NR | 20.4 |
| 35DHB-DOPO | 1% | 620.3 | 22.5 | 30.8 | NR | 9.6 |
| | 2% | 518.8 | 22.6 | 33.3 | V2 | 19.2 |

converted to a series of phosphorus esters that display good flame retardancy in DGEBA epoxy resin. Phosphonates have a greater inhibitory effect on the flammability of the resin than do the corresponding phosphates and probably function predominately in the gas phase.

## Conflicts of Interest

The authors declare that there is no conflict of interest regarding the publication of this paper.

## Acknowledgments

This research was conducted at Central Michigan University. Diphenyl chlorophosphate was provided by ICL-IP America. Epoxy (DGEBA) was supplied by the Dow Chemical Company.

## References

[1] M. A. Hillmyer, "The promise of plastics from plants," *Science*, vol. 358, no. 6365, pp. 868–870, 2017.

[2] D. K. Schneiderman and M. A. Hillmyer, "50th anniversary perspective: there is a great future in sustainable polymers," *Macromolecules*, vol. 50, no. 10, pp. 3733–3749, 2017.

[3] R. Mulhaupt, "Green polymer chemistry and biobased plastics: dreams and reality," *Macromolecular Chemistry and Physics*, vol. 214, no. 2, pp. 159–174, 2013.

[4] M. M. Reddy, S. Vivekanandhan, M. Misra, S. K. Bhatia, and A. K. Mohanty, "Biobased plastics and bionanocomposites: current status and future opportunities," *Progress in Polymer Science*, vol. 38, no. 10-11, pp. 1653–1689, 2013.

[5] M. Malinconico, P. Cerruti, G. Santagata, and B. Immirzi, "Natural polymers and additives in commodity and specialty applications: a challenge for the chemistry of future," *Macromolecular Symposia*, vol. 337, no. 1, pp. 124–133, 2014.

[6] J. Kuczynski and D. J. Boday, "Biobased materials for high-end electronics applications," *International Journal of Sustainable Development and World Ecology*, vol. 19, no. 6, pp. 557–563, 2012.

[7] T. Zhang, B. A. Howell, A. Dumitrascu, S. J. Martin, and P. B. Smith, "Synthesis and characterization of glycerol-adipic acid hyperbranched polyesters," *Polymer*, vol. 55, no. 20, pp. 5065–5072, 2014.

[8] T. Zhang, B. A. Howell, and P. B. Smith, "Rational synthesis of hyperbranched polyesters," *Industrial and Engineering Chemistry Research*, vol. 56, no. 6, pp. 1661–1670, 2017.

[9] P. F. H. Harmsen, M. M. Hackmann, and H. L. Bos, "Green building blocks for biobased plastics," *Biofuels, Bioproducts and Biorefining*, vol. 8, no. 3, pp. 306–324, 2014.

[10] A. Pellis, E. Herrero Acero, L. Gardossi, V. Ferrario, and G. M. Guebitz, "Renewable building blocks for sustainable polyesters: new biotechnological routes for greener plastics," *Polymer International*, vol. 65, no. 8, pp. 861–871, 2016.

[11] T. J. Farmer, J. W. Comerford, A. Pellis, and T. Robert, "Post-polymerization modification of biobased polymers: maximizing the high functionality of polymers derived from biomass," *Polymer International*, vol. 67, no. 7, pp. 775–789, 2018.

[12] B. A. Howell, *Developing Alternative Plasticizers to Overcome Issues of Migration, Toxicity and Flammability*, AMI PVC Formulation, Pittsburgh, PA, USA, 2018.

[13] Z. Yu, J. Zhou, J. Zhang, K. Huang, F. Cao, and P. Wei, "Evaluating effects of biobased 2,5-furandicarboxylate esters as plasticizers on the thermal and mechanical properties of poly(vinyl chloride)," *Journal of Applied Polymer Science*, vol. 131, no. 20, article 40938, 2014.

[14] X. Li, X. Nie, J. Chen, and Y. Wang, "Preparation of epoxidized cardanol butyl ether as a novel renewable plasticizer and its application for poly(vinyl chloride)," *Polymer International*, vol. 66, no. 3, pp. 443–449, 2017.

[15] B. A. Howell, K. E. Carter, and H. Dangalle, "Flame retardants based on tartaric acid: a renewable by-product of the wine industry," in *Renewable and Sustainable Polymers, (ACS Symposium Series 1063), Chapter 9*, G. F. Payne and P. B. Smith, Eds., pp. 133–152, American Chemical Society, Washington, D.C., USA, 2011.

[16] B. A. Howell and Y. G. Daniel, "Phosphorus flame retardants from esters of isosorbide and 10-undecanoic acid," in *Green Polymer Chemistry: Biobased Materials and Biocatalysis, Chapter 21*, ACS Symposium Series 1192, H. N. Cheng, R. A. Gross, and P. B. Smith, Eds., pp. 339–367, American Chemical Society, Washington, D.C, USA, 2015.

[17] Y. G. Daniel and B. A. Howell, "Flame retardant properties of isosorbide *bis*-phosphorus esters," *Polymer Degradation and Stability*, vol. 140, pp. 25–31, 2017.

[18] A. Pettigrew, "Halogenated flame retardants," in *Encyclopedia of Polymer Science and Technology, Volume 10*, J. I. Kroschwitz and M. Howe-Grant, Eds., John Wiley and Sons, New York, NY, 4th edition, 1993.

[19] M. Alaee and R. J. Wenning, "The significance of brominated flame retardants in the environment: current understanding, issues and challenges," *Chemosphere*, vol. 46, no. 5, pp. 579–582, 2002.

[20] P. O. Darnerud, "Brominated flame retardants as possible endocrine disrupters," *International Journal of Andrology*, vol. 31, no. 2, pp. 152–160, 2008.

[21] J. B. Herbstman, A. Sjödin, M. Kurzon et al., "Prenatal exposure to PBDEs and neurodevelopment," *Environmental Health Perspectives*, vol. 118, no. 5, pp. 712–719, 2010.

[22] S. Shaw, "Halogenated flame retardants: do the fire safety benefits justify the risks?," *Reviews on Environmental Health*, vol. 25, no. 4, pp. 261–305, 2010.

[23] S. Narayan and M. Moore, "Flame retardants - a new, versatile flame retardant for olefinic and styrenic polymers," *Popular Plastics and Packaging*, vol. 75, no. 3, pp. 58–60, 2012.

[24] S. Levchik, E. Eden, Y. Hirschsohn, M. Leifer, and A. Ben-Zvi, *Antimony Trioxide Free Solutions for Engineering Thermoplastics*, AMI Fire Retardants in Plastics, Pittsburgh, PA, USA, 2018.

[25] G. Hess, "Industry drops flame retardant," *Chemical and Engineering News*, vol. 88, no. 1, p. 10, 2010.

[26] C. Hogue, "U.S. chemical regulation shifts," *Chemical and Engineering News*, vol. 94, no. 27, pp. 18–20, 2016.

[27] C. Hogue, "Two U.S. states ban flame retardants in furniture," *Chemical and Engineering News*, vol. 95, no. 41, p. 19, 2017.

[28] M. M. Velencoso, A. Battig, J. C. Markwart, B. Schartel, and F. R. Wurm, "Molecular firefighting-how modern phosphorus chemistry can help solve the challenge of flame retardancy," *Angewandte Chemie International Edition*, vol. 57, no. 33, pp. 10450–10467, 2018.

[29] I. van der Veen and J. de Boer, "Phosphorus flame retardants: properties, production, environmental occurrence, toxicity and analysis," *Chemosphere*, vol. 88, no. 10, pp. 1119–1153, 2012.

[30] C. Hirsch, B. Striegl, S. Mathes et al., "Multiparameter toxicity assessment of novel DOPO-derived organophosphorus flame retardants," *Archives of Toxicology*, vol. 91, no. 1, pp. 407–425, 2017.

[31] M. Vazirian, M. Khanavi, Y. Amarzadeh, and H. Hajimehdipoor, "Quantification of gallic acid in fruits of three medicinal plants," *Iranian Journal of Pharmaceutical Research*, vol. 2, pp. 233–236, 2011.

[32] D. C. Vu, P. H. Vo, M. V. Coggeshall, and C.-H. Lin, "Identification and characterization of phenolic compounds in black walnut kernels," *Journal of Agricultural and Food Chemistry*, vol. 66, no. 17, pp. 4503–4511, 2018.

[33] T.-R. Su, J.-J. Lin, C.-C. Tsai et al., "Inhibition of melanogenesis by gallic acid: possible involvement of the PI3K/Akt, MEK/ERK and Wnt/$\beta$-catenin signaling pathways in B16F10 cells," *International Journal of Molecular Sciences*, vol. 14, no. 10, pp. 20443–20458, 2013.

[34] S.-H. Kim, C.-D. Jun, K. Suk et al., "Gallic acid inhibits histamine release and pro-inflammatory cytokine production in mast cells," *Toxicological Sciences*, vol. 91, no. 1, pp. 123–131, 2006.

[35] J. Ma, X.-D. Luo, P. Protiva et al., "Bioactive novel polyphenols from the fruit of *Manilkara zapota* (sapodilla)," *Journal of Natural Products*, vol. 66, no. 7, pp. 983–986, 2003.

[36] C. Yang, X. Xie, H. Tang, X. Dong, X. Zhang, and F. Huang, "Transcriptome analysis reveals GA induced apoptosis in HCT 116 human colon cancer cells through calcium and p53 signal pathways," *RSC Advances*, vol. 8, no. 22, pp. 12449–12458, 2018.

[37] J. A. Bahry and B. A. Howell, "Gallic acrylates as a base for the development of nontoxic flame retardants," in *47th Central Regional Meeting of the American Chemical Society*, p. 208, Covington, KY, USA, May, 2016.

[38] K. Dziedzic, D. Górecka, A. Szwengiel et al., "The content of dietary fibre and polyphenols in morphological parts of buckwheat (Fagopyrum tataricum)," *Plant Foods for Human Nutrition*, vol. 73, no. 1, pp. 82–88, 2018.

[39] M. Holasova, V. Fiedlerova, H. Smrcinova, M. Orsak, J. Lachman, and S. Vavreinova, "Buckwheat-the source of antioxidant activity in functional foods," *Food Research International*, vol. 35, no. 2-3, pp. 207–211, 2002.

[40] P. V. Hung and N. Morita, "Distribution of phenolic compounds in the graded flours milled from whole buckwheat grains and their antioxidant capacities," *Food Chemistry*, vol. 109, no. 2, pp. 325–331, 2008.

[41] A. Ahmed, N. Khalid, A. Ahmad, N. A. Abbasi, M. S. Z. Latif, and M. A. Randhawa, "Phytochemicals and biofunctional properties of buckwheat: a review," *The Journal of Agricultural Science*, vol. 152, no. 03, pp. 349–369, 2014.

[42] Y. G. Daniel and B. A. Howell, "Phosphorus flame retardants from isosorbide bis-acrylate," *Polymer Degradation and Stability*, vol. 156, pp. 14–21, 2018.

[43] Y. G. Daniel and B. A. Howell, "Synthesis and characterization of isosorbide bis-phosphorus esters," *Heteroatom Chemistry*, vol. 28, no. 3, article e21369, 2017.

[44] A. Schafer, S. Seibold, W. Lohstroh, O. Walter, and M. Doring, "Synthesis and properties of flame-retardant epoxy resins based on DOPO and one of its analog DPPO," *Journal of Applied Polymer Science*, vol. 105, no. 2, pp. 685–696, 2007.

[45] S. Liang, P. Hemberger, N. M. Neisius et al., "Elucidating the thermal decomposition of dimethyl methylphosphonate by vacuum ultraviolet (VUV) photoionization: pathways to the PO radical, a key species in flame-retardant mechanisms," *Chemistry - A European Journal*, vol. 21, no. 3, pp. 1073–1080, 2015.

[46] R. N. Walters, N. Safronava, and R. E. Lyon, "A microscale combustion calorimeter study of gas phase combustion of polymers," *Combustion and Flame*, vol. 162, no. 3, pp. 855–863, 2015.

[47] H. Guo, R. E. Lyon, N. Safronava, R. N. Walters, and S. Crowley, "A simplified model on carbon monoxide yield in burning of polymeric solids containing flame retardants," *Fuel*, vol. 222, pp. 175–179, 2018.

[48] F. Raffan-Montoya, S. I. Stoliarov, S. Levchik, and E. Eden, "Screening flame retardants using milligram-scale flame calorimetry," *Polymer Degradation and Stability*, vol. 151, pp. 12–24, 2018.

[49] B. Schartel, B. Perret, B. Dittrich et al., "Flame retardancy of polymers: the role of specific reactions in the condensed phase," *Macromolecular Materials and Engineering*, vol. 301, no. 1, pp. 9–35, 2016.

[50] U. Braun, A. I. Balabanovich, B. Schartel et al., "Influence of the oxidation state of phosphorus on the decomposition and fire behaviour of flame-retarded epoxy resin composites," *Polymer*, vol. 47, no. 26, pp. 8495–8508, 2006.

[51] R. Sonnier, C. Negrell-Guirao, H. Vahabi, B. Otazaghine, G. David, and J.-M. Lopez-Cuesta, "Relationships between the molecular structure and the flammability of polymers: study of phosphonate functions using microscale combustion calorimeter," *Polymer*, vol. 53, no. 6, pp. 1258–1266, 2012.

[52] A. Lorenzetti, M. Modesti, S. Besco, D. Hrelja, and S. Donadi, "Influence of phosphorus valency on thermal behaviour of flame retarded polyurethane foams," *Polymer Degradation and Stability*, vol. 96, no. 8, pp. 1455–1461, 2011.

# 3

# Design of an Emulgel-Type Cosmetic with Antioxidant Activity using Active Essential Oil Microcapsules of Thyme (*Thymus vulgaris* L.), Cinnamon (*Cinnamomum verum* J.), and Clove (*Eugenia caryophyllata* T.)

**Glicerio Leon-Méndez,[1,2,3] Maria Osorio-Fortich,[1] Rodrigo Ortega-Toro,[4] Nerlys Pajaro-Castro,[5] Miladys Torrenegra-Alarcón,[2,3] and Adriana Herrera-Barros[6]**

[1]*Faculty of Pharmaceutical Sciences, University of Cartagena, Research in Technology Pharmaceutical, Cosmetic and of Food Group (GITFCA), Cartagena de Indias D.T. y C, Cartagena, Colombia*
[2]*Center of Commerce and Services, Regional Bolivar, SENA, Research in Innovation and Biotechnology Group (GIBEI), Cartagena de Indias D.T. y C, Cartagena, Colombia*
[3]*Doctorate Student in Engineering, University of Cartagena, Cartagena, Bolívar 130015, Colombia*
[4]*Faculty of Engineering, University of Cartagena, Grupo de Investigación de Ingeniería de Fluidos Complejos y Reología de Alimentos (IFCRA), Cartagena de Indias D.T. y C, Cartagena, Colombia*
[5]*Department of Medicine, School of Health Sciences, Research in Medical and Pharmaceutical Sciences Group, University of Sucre, Sincelejo, Sucre 700003, Colombia*
[6]*Faculty of Engineering, University of Cartagena, Research in Multi-functional Nanomaterials Group, Cartagena de Indias D.T. y C., Cartagena 130015, Colombia*

Correspondence should be addressed to Rodrigo Ortega-Toro; rortegap1@unicartagena.edu.co

Academic Editor: Mehdi Salami-Kalajahi

Essential oils (EO) are complex mixtures of biosynthesized chemicals basically by plants, which provide them their characteristic aroma. Many have biologically recognized activities as antioxidants and anti-inflammatory among others, and many of them are employed as cosmetic actives. Very often, these properties are not fully exploited because of their high volatility and tendency to oxidize, so it is necessary to attach them to a conveyor to provide them adequate stabilization and lifetime. One of the best alternatives to carry out this is microencapsulation, for which natural biopolymers can be used, such as the starches. Therefore, it used EO of thyme, cinnamon, and clove, which were obtained by conventional and assisted hydrodistillation by microwave radiation from the plant material. The chemical composition was evaluated by gas chromatography/mass spectrometry (GC/MS). The radical scavenging ability was determined by antiradical activity techniques including DPPH (2,2-diphenyl-1-picrylhydrazyl) and ABTS (2,2′-azino-bis-(3-ethylthiazoline-benzenesulfonic acid-6)), and the ORAC method was used for determining the antioxidant capacity. Also, starches of yam (*D. rotundata*), sweet potato (*I. batatas*), corn (*Z. mays*), and cassava (*M. esculenta*) were used, which were subjected to hydrolysis and lipophilization processes using dodecenyl succinic anhydride (DDSA); this chemical process achieves a significant increase in emulsifier capacity (surfactant) compared to its native state; that of cassava was the most promising starch which was used as an EO microencapsulating agent. Finally, from the EO microcapsules, an emulgel-type cosmetic was designed which maintained its antioxidant activity. The results of this work contribute to the development of stable and functional cosmetic formulations of essential oils, emphasizing that the extraction of EO by assisted microwave radiation hydrodistillation is considered a fast, efficient, green, and relatively economical method compared to conventional hydrodistillation.

# 1. Introduction

The World Health Organization estimates that 80% of the population in developing countries use plants to solve their health needs. The modern pharmacopoeia contains 25% of plant-derived medicines; these are used for multiple health problems and also in the treatment of snake bites. India is one of the countries where its tribal communities possess valuable and unique knowledge about the use of wild plants for the treatment of diseases [1, 2].

In traditional medicine, we find plants that contain antioxidant compounds, which protect cells from the damaging effects of reactive oxygen species (ROS). The reduction of oxygen occurs through the electrons that escape from the respiratory chain, giving rise to the super oxide ($O^{2-}$), which can easily dismantle and form hydrogen peroxide ($H_2O_2$), which with transition metals as iron ($Fe^{2+}$) and copper ($Cu^+$) produce radical hydroxyl ($OH^-$) by Fenton reaction, which is considered one of the most damaging oxidant species in biological systems and most responsible for the oxidative damage together with the radical peroxyl [3].

Oxidative damage can be prevented by antioxidant molecules which are capable of donating electrons to stabilize free radicals and neutralize its harmful effects; they can be of endogenous (synthesized by the body) or exogenous (from external sources) origins [3]. There are some synthetic antioxidant compounds like butylhydroxytoluene, butylhydroxyanisole, and tertiary butylhydroquinone. However, it has been suggested that these compounds have demonstrated toxic effects such as liver damage and mutagenesis [2]. Flavonoids and other phenolic plant origin compounds have been reported as free radical scavengers [1]. The plants as *Thymus vulgaris, Cinnamomum verum*, and *Eugenia caryophyllata* belonging to the families Lamiaceae, Lauraceae, and Myrtaceae, respectively, known as thyme, cinnamon, and clove in Colombia [4], containing $\beta$-caryophyllene, quercetin, triterpenic acids, $\alpha$-pinene, $\beta$-pinene, thymol, eugenol, carvacrol, 8-cineol, $\beta$-felandren, and p-cymene, with which many pharmacological properties have been reported, including antiepileptic, antitumor and antimutagenic, radioprotective, antioxidant, antibacterial, antifungal, diuretic, antipyretic, and analgesic [5]. This work is aimed at designing an emulgel-type cosmetic with antioxidant activity which was designed as active essential oil microcapsules of thyme (*Thymus vulgaris* L.), canela (*Cinnamomum verum* J.), and clove of odor (*Eugenia caryophyllata* T).

# 2. Experimental

*2.1. Extraction of Essential Oils.* The EO was obtained by two methods: hydrodistillation steam with conventional water and microwave-assisted hydrodistillation. For that, it was used a hydrodistillation equipment with 4 liters of capacity. 500 g of plant material was taken; then they were introduced into the extraction flask, which contained 500 mL of distilled water. In both the steam processes where a heating mantle was used in the microwave-assisted hydrodistillation, the extraction time was 3-4 hours. As a source of microwave radiation, a conventionally modified oven of the brand Samsung was used, with a 1-cycle irradiation of 60 minutes and a power of 70% [6]. In both cases, the EO was collected in a Dean-Stark-type vessel.

*2.2. Determination of the Major Components of EO by Gas Chromatography-Mass Spectrometry (GC/MS).* An Agilent 7890A/5975C chromatograph was used. Each EO sample (50 $\mu$L) was dissolved in 450 $\mu$L dichloromethane, the injector temperature was 250°C, and a HP-5MS 5% Phenyl-Methyl-Silox capillary column was used; helium was used as carrier gas at a constant flow rate of 1 mL/min, pressure of 7.6354 psi, and linear velocity of 36 cm/sec. The initial temperature was 45°C and the transfer line temperature was 280°C. Mass spectra were obtained by electron ionization (70 eV), with automatic scanning at a range of 30–400 u.m. (*m/z*), at 3.85 scans/s. The component identities were assigned by comparison of each spectrum with the database standards reported in the literature [6].

*2.3. Methods to Determine the Radical Scavenging Ability and Measurement of the Antioxidant Activity of EO.* Two methodologies were used to determine the *radical scavenging ability*—DPPH (2,2-diphenyl-1-picrylhydrazole) and ABTS (2,2′-azino-bis-(3-ethylthiazolin-benzenesulfonic acid-6)). The antioxidant capacity was measured by ORAC method.

*2.3.1. Radical Method DPPH•.* Free radical scavenging activity DPPH was determined using the method described by Silva et al. [7] (with some modifications of 75 $\mu$L of the sample were added to 150 $\mu$L of a methanol solution of DPPH (100 ppm), and they were incubated at room temperature for 30 min, after the disappearance of the DPPH radical was determined spectrophotometrically at 405 Nm in microplate reader Multiskan Ex (Thermo Scientific). Ascorbic acid (25 $\mu$g/mL) was used as a positive control.

$$\% \text{ inhibition} = \frac{(A_0 - A_f)}{A_0} * 100, \qquad (1)$$

where $A_0$ and $A_f$ are the absorbance values of blank (DPPH solution in alcohol) and the sample (DPPH solution plus alcohol dissolved), respectively.

*2.3.2. Radical Method ABTS•+.* The ABTS free radical scavenging activity was determined using the method described by Re et al. [8] with some modifications. The ABTS radical was formed following the reaction of 3.5 mM ABTS with 1.25 mM of potassium persulfate (final concentration). The samples were incubated at 5°C and in darkness for 16 h. Once the ABTS radical was formed, it was diluted with ethanol until having an absorbance of 0.7 ± 0.05 at 734 nm. The ABTS dilution (190 $\mu$L) was mixed with the EO sample (10 $\mu$L), and after, it was incubated at room temperature for 5 minutes. After this time, the disappearance of the ABTS radical at 734 nm was determined spectrophotometrically in the microplate reader Multiskan Ex (Thermo Scientific). Ascorbic acid (4 $\mu$g/mL) was used as a positive control for the uptake of ABTS radicals.

*2.3.3. ORAC Method (Hydrophilic and Lipophilic).* Hydrophilic ORAC, trolox (TX), was used as standard, and controlled conditions of temperature (37°C) and pH (7.4) were used in this evaluation. The readings were performed at an excitation of 493 nm and emission of 515 nm. For the technique development, $1 \times 10^{-2}$ M fluorescein solutions in PBS (75 mM) and 0.6 M AAPH (2,2′-azobis(2-amidinopropane) dihydrochloride) in PBS (75 mM) were used. The sample was prepared by mixing 21 $\mu$L fluorescein, 2899 $\mu$L PBS, 30 $\mu$L extract (test sample), and 50 $\mu$L AAPH. Trolox was used as reference. The protective effect of the antioxidant was calculated using the area differences under the decrease of the curve of fluorescein with blank and the sample, and the result was compared with curve obtained with trolox. The results were expressed in micromoles of trolox equivalents per 100 grams of the sample ($\mu$mol Tx/100 g sample), according to the following equation.

$$\text{ORAC} = \frac{(\text{AUC} - \text{AUC}^{\circ})}{\text{AUC}_{\text{trolox}-\text{AUC}^{\circ}}} f\,[\text{trolox}], \tag{2}$$

where AUC is the area under the curve of the sample, AUC° is the area under the curve of the control, $\text{AUC}_{\text{trolox}}$ is the area under the curve of trolox, and $f$ is the dilution factor of the extracts.

For lipophilic ORAC, the previous described procedure was followed except that the sample and trolox were prepared with methylated cyclodextrin (7%) and in a 50% acetone-water mixture. The solutions were shaken for 1 hour, and the analysis was performed as described in the hydrophilic ORAC method [9].

*2.4. Extraction of Native Starch.* The process of extracting the starch from each tuber (yam, cassava, and sweet potato) began with washing to remove dirt in general, followed by debarking and washing again. Then the tuber was chopped and liquefied with distilled water to obtain a slurry, which was filtered through a cloth. The filtrate could stand and be decanted, and the supernatant was removed. The sediment was washed with distilled water, then filtered under vacuum, and dried at 60°C for 12 hours; then it was ground and finally packed [10].

*2.5. Chemical Modification of Starch*

*2.5.1. Hydrolysis.* 3.4 g of hydrochloric acid was added to a suspension of native starch (40 g of starch in 100 mL of deionized water), with constant stirring for 6 hours, keeping the temperature at 50°C in a thermostated bath. Over time, the system was neutralized with 10% NaOH to pH 6.5 and vacuum filtered and the residue was washed 4 times with distilled water. It was dried at 50°C for 24 hours, and finally, the obtained hydrolyzed starch was ground and stored in hermetic plastic bags until use [10].

*2.5.2. Lipophilization (Esterification).* A suspension of natively hydrolyzed starch at 25°C (40 g of starch in 100 mL of deionized water) was adjusted to pH 8.5 with 3% NaOH with constant stirring for 30 minutes. In separate assays, 10 mL of dodecenyl succinic anhydride (DDSA) diluted in 50 mL of 96% ethanol was added to each starch to achieve a single level of substitution. Because, during the reaction, the pH of the solution tends to drop, it remained between 8.5 and 9.0 with small additions of 3% NaOH. Once the limiting reagent is finished, which happens when the pH did not drop further, the solution was adjusted to pH 6.5 with 10% HCl. The material was washed three times with deionized water and dried at 50°C for 24 hours [10].

*2.5.3. Determination of the Substitution Percentage (SP).* The degree of substitution (DS) was the average number of hydroxyl groups substituted per anhydroglucose unit. The DS of DDSA starch was determined by titration. 5.00 g starch derivative was accurately weighted into a 50 mL beaker. Then 10 mL of ethanol was added to the beaker and was allowed to stir. 25 mL of 0.1 N HCl ethanol solution was added, and the entire solution was allowed to stir for 30 min. The slurry was filtered and the wet cake was washed with deionized water until no $Cl^{-1}$ could be detected any longer (using 0.1 N $AgNO_3$ solution). The wet cake was quantitatively transferred to a 900 mL beaker, 100 mL deionized water was added, and then 200 mL of boiling deionized water was added. The solution was placed into a boiling water bath and cooked for 30 min. Immediately, after the solution was cooked, 6–10 drops of 1% thymolphthalein indicator were added and the solution was immediately titrated with 0.1 N sodium hydroxide to the thymolphthalein endpoint [10]. The DS was calculated by the following equation:

$$\text{DS} = \frac{162 \times C \times V \times 10^{-3}}{W - 266 \times C \times V \times 10^{-3}} \times 100, \tag{3}$$

where $C$ is the molarity of the NaOH solution used during the titration; $V$ is the consuming volume of NaOH solution; $W$ is the weight of the sample analyzed. 162 is the molecular weight of the anhydroglucose unit; 266 is the molecular weight of dodecenyl succinic anhydride.

*2.6. Infrared Spectroscopy with Fourier Transform.* For the analysis, tablets of 1 cm in diameter were prepared by homogeneously mixing dry potassium bromide (KBr) powder (with dry starch in the ratio of 1 : 50 (0.15 g and 0.003 g) and molding the mixture in a hydraulic press. FTIR spectra were obtained by collecting an average of 64 scans between 400 and 4000 $cm^{-1}$ [11].

*2.7. Properties of Modified Starches*

*2.7.1. Oil Absorption.* To determine the oil absorption capacity, 1 g of starch was taken and mixed with 10 mL of corn oil in vortex for 30 seconds. Samples could stand for 30 minutes at room temperature (25 ± 2°C). Then they were centrifuged at 3000 rpm for 15 minutes, and the volume of the supernatant obtained was measured in a 10 mL test tube. The results were obtained by subtracting the initial 10 mL of oil with the volume after centrifugation, reporting the oil absorption capacity in mL/g of the sample [10].

*2.7.2. Water Absorption.* The test was run in an analogous way to the oil absorption test, but using water instead. The

Table 1: Proposed formulations for the design of the emulgel (%). Formulations F1–F32.

| | F 1 | F 2 | F 3 | F 4 | F 5 | F 6 | F 7 | F 8 | F 9 | F 10 | F 11 | F 12 | F 13 | F 14 | F 15 | F 16 | F 17 | F 18 | F 19 | F 20 | F 21 | F 22 | F 23 | F 24 | F 25 | F 26 | F 27 | F 28 | F 29 | F 30 | F 31 | F 32 |
|---|---|---|---|---|---|---|---|---|---|---|---|---|---|---|---|---|---|---|---|---|---|---|---|---|---|---|---|---|---|---|---|---|
| Acrylate copolymer | 1 | 4 | 1 | 4 | 1 | 4 | 1 | 4 | 1 | 4 | 1 | 4 | 1 | 4 | 1 | 4 | — | — | — | — | — | — | — | — | — | — | — | — | — | — | — | — |
| Carbomer | — | — | — | — | — | — | — | — | — | — | — | — | — | — | — | — | 0.1 | 0.5 | 0.1 | 0.5 | 0.1 | 0.5 | 0.1 | 0.5 | 0.1 | 0.5 | 0.1 | 0.5 | 0.1 | 0.5 | 0.1 | 0.5 |
| Glycerin | 5 | 5 | 5 | 5 | 5 | 5 | 5 | 5 | 5 | 5 | 5 | 5 | 5 | 5 | 5 | 5 | 5 | 5 | 5 | 5 | 5 | 5 | 5 | 5 | 5 | 5 | 5 | 5 | 5 | 5 | 5 | 5 |
| Triethanolamine | 1 | 4 | 1 | 4 | 1 | 4 | 1 | 4 | 1 | 4 | 1 | 4 | 1 | 4 | 1 | 4 | 0.1 | 0.5 | 0.1 | 0.5 | 0.1 | 0.5 | 0.1 | 0.5 | 0.1 | 0.5 | 0.1 | 0.5 | 0.1 | 0.5 | 0.1 | 0.5 |
| Preserver | 0.1 | 0.1 | 0.1 | 0.1 | 0.1 | 0.1 | 0.1 | 0.1 | 0.1 | 0.1 | 0.1 | 0.1 | 0.1 | 0.1 | 0.1 | 0.1 | 0.1 | 0.1 | 0.1 | 0.1 | 0.1 | 0.1 | 0.1 | 0.1 | 0.1 | 0.1 | 0.1 | 0.1 | 0.1 | 0.1 | 0.1 | 0.1 |
| Glyceryl stearate, PEG-100 stearate | 5 | 5 | 10 | 10 | 5 | 5 | 10 | 10 | — | — | — | — | — | — | — | — | 5 | 5 | 10 | 10 | 5 | 5 | 10 | 10 | — | — | — | — | — | — | — | — |
| Self-emulsifying glyceryl stearate | — | — | — | — | — | — | — | — | 2 | 2 | 5 | 5 | 2 | 2 | 5 | 5 | — | — | — | — | — | — | — | — | 2 | 2 | 5 | 5 | 2 | 2 | 5 | 5 |
| Isopropyl myristate | 5 | 5 | 5 | 5 | — | — | — | — | 5 | 5 | 5 | 5 | — | — | — | | 5 | 5 | 5 | 5 | — | — | — | — | 5 | 5 | 5 | 5 | — | — | — | — |
| Oleyl erucate | — | — | — | — | 5 | 5 | 5 | 5 | — | — | — | — | 5 | 5 | 5 | 5 | — | — | — | — | 5 | 5 | 5 | 5 | — | — | — | — | 5 | 5 | 5 | 5 |
| Active (microcapsule) | 5 | 5 | 5 | 5 | 5 | 5 | 5 | 5 | 5 | 5 | 5 | 5 | 5 | 5 | 5 | 5 | 5 | 5 | 5 | 5 | 5 | 5 | 5 | 5 | 5 | 5 | 5 | 5 | 5 | 5 | 5 | 5 |
| Water | c.s.p. 100% | | | | | | | | | | | | | | | | | | | | | | | | | | | | | | | |

results were obtained by subtracting the initial 10 mL of water with the volume after centrifugation. Water absorption capacity is reported in mL/g of the sample [10].

*2.7.3. Emulsifying Capacity.* To determine the ability of the starches to hold a stable emulsion, 1 g of each starch was mixed with 25 mL of distilled water, stirred for 15 minutes with a magnetic system. This solution was added with 25 mL of an oily substance used in the cosmetic industry (corn oil, isopropyl myristate, octyldodecanol, and mineral oil) and stirred/homogenized with ULTRA-TURRAX® IKA T10 for 3 min. Afterwards, the product was centrifuged at 1300 rpm for 5 min [10]. The emulsifying capacity (EC) was expressed in terms of percentage, considering the volume of the layer which is still remaining emulsified (VLE) to the total liquid volume (TLV).

$$EC = \frac{VLE}{TLV} \times 100. \quad (4)$$

*2.8. Microencapsulation of Essential Oils.* All essential oils were microencapsulated using polymeric matrix-modified starch with higher emulsifying capacity using the drying method by spray dryer. Emulsions were prepared with 30% solids (*w/w*) of the encapsulating material (for this, the succinated cassava starch was screened at mesh 80 and a 20% (*w/w*) ratio of the essential oil to the modified starch solids); then it was homogenized with an ULTRA-TURRAX IKA T-10 Basic at 14000 rpm and for 5 min. The mixture was microencapsulated in a BUCHI Mini Spray Dryer Model B-290 (BÜCHI Labortechnik, Germany). The input temperature and output temperature were maintained between180°C ± 5°C and 116°C ± 5°C, respectively. Microcapsules obtained were collected in a self-sealing polyethylene package and stored in a room with humidity and temperature controlled at 45% and 20 ± 5°C [12].

*2.9. Encapsulation Efficiency.* The total oil content in the spray-dried microencapsulated products was determined in duplicate by distilling 20 g of encapsulated powder for 3 h in a Clevenger-type apparatus. The volume of peppermint EO collected in the trap was multiplied by a density factor (1.00 g/mL OE *C. verum* HD, 1.030 g/mL OE *C. verum* MWHD, 0.919 g/mL OE *T. vulgaris* HD, 0.921 g/mL OE *T. vulgaris* MWHD, 1.027 g/mL OE *E. caryophyllata* MWHD, and 1.039 g/mL OE *E. caryophyllata* MWHD) to calculate the weight of oil recovered from the sample. The ability of the microcapsules to retain the essential oil was evaluated; 10 bodies of Petri plates were taken, where 20 g of encapsulated powder was poured, which was left exposed to the environment (temperature 25 ± 2°C, HR 60%) for 15 days, evaluating in triplicate serially. The compositions of total (isolated from processed products) oils were analyzed by gas chromatography-mass spectrometry (GC-MS).

*2.10. Design of Phytocosmetics with Antioxidant Activity.* A preformulation study was carried out to determine that there were no incompatibilities between the active ingredient and the formulation auxiliaries, which affected the stability of the final product. This was done by reviewing the technical

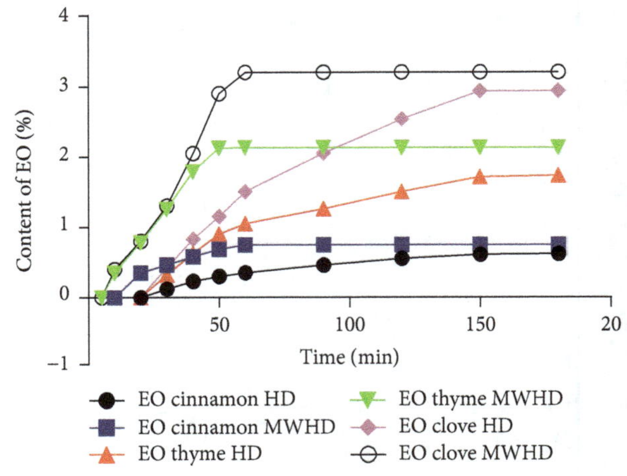


FIGURE 1: Kinetic extraction of essential oils (EO) obtained by the method of steam distillation (HD) and microwave-assisted hydrodistillation (MWHD).

data sheets of each raw material to verify possible interactions between the components and take the necessary measures Table 1 [13].

To ensure that the formulation maintains its organoleptic, rheological, and chemical characteristics, they were controlled when the formulation was being fabricated; the *in vitro* antioxidant activity of the finished product was determined by the antiradical activity techniques DPPH (2,2-diphenyl-1-picrylhydrazole), ABTS (2,2′-azino-bis-(3-ethylthiazolin-benzenesulfonic acid-6)), and ORAC.

*2.11. Sensory Analysis.* The samples kept were studied for its sensory characteristics. Hedonic scale ratings were used to evaluate the sensory attributes of the formulation phytocosmetics by expert panel members. The measure of the degree of acceptance of the product was obtained by the use of the hedonic scale. Panelists were asked to their degree of likes or dislikes in terms of which best describes their perception about the product. The term may be given numerical values to enable the results to be scored.

*2.12. Statistical Analysis.* All trials were performed by sextupled. The results were expressed as the mean ± SD (standard deviation). Significant differences were determined by ANOVA analysis followed by Dunnett's or Tukey's test or as deemed appropriate.

## 3. Results and Discussion

Figure 1 shows the extraction kinetics of the obtained EO, after 40 minutes of extraction of EO by MWHD which resulted in a recovery similar to that obtained for 3 hours of HD. Reaching final yields by MWHD and HD methods of 0.747% and 0.617%, respectively, for cinnamon, 2.133% and 1.733%, respectively, for thyme, and 3.2% and 2.933%, respectively, for clove. These results indicate that the MWHD technique uses less extraction time; this is because of the breaking of the structures of the main components of greater abundance in the EO by electromagnetic radiations applied

TABLE 2: Major components detected in EO obtained by HD and MWHD.

| Component | % relative abundance ($t_R$, min) | | | | | |
|---|---|---|---|---|---|---|
| | *C. verum* | | *T. vulgaris* | | *E. caryophyllata* | |
| | HD | MWHD | HD | MWHD | HD | MWHD |
| γ-Terpinene | 3.20 (5.336) | | 15.36 (5.358) | 15.48 (5.343) | | |
| Linalool | 2.12 (5.856) | 2.13 (5.856) | 4.47 (5.870) | 4.59 (5.855) | | |
| Cinnamaldehyde | 67.97 (8.259) | 89.25 (8.568) | 1.14 (8.289) | 1.20 (8.290) | | |
| Thymol | 3.39 (8.855) | 3.40 (8.859) | 44.68 (8.952) | 49.38 (8.952) | | |
| Carvacrol | | | 8.55 (9.050) | 9.58 (9.043) | | |
| Eugenol | 2.73 (9.842) | | | | 60.32 (9.902) | 65.24 (9.955) |
| Caryophyllene | 1.46 (11.281) | | 6.47 (11.303) | 6.34 (11.296) | 11.57 (11.289) | 12.02 (11.296) |

TABLE 3: Radical scavenging ability and antioxidant capacity of extracted EO by the HD and MWHD methods analyzed by DPPH, ABTS, and ORAC.

| Essential oil | Extraction method | Methods to determine the radical scavenging ability and antioxidant activity | | | |
|---|---|---|---|---|---|
| | | $DPPH^{\bullet}$ $IC_{50}$ (μg/mL) | $ABTS^{\bullet+}$ $IC_{50}$ (μg/mL) | ORAC mmol trolox/100 g sample | |
| | | | | Hydrophilic | Lipophilic |
| *T. vulgaris* | HD | 165.5 ± 1.05 | 29.07 ± 0.07 | 255.0 ± 0.5 | 586.3 ± 0.2 |
| | MWHD | 150.3 ± 0.60 | 22.08 ± 0.05 | 403.3 ± 0.2 | 693.95 ± 0.05 |
| *C. verum* | HD | 290.5 ± 0.35 | 59.00 ± 0.07 | 213.3 ± 0.7 | 235.2 ± 0.9 |
| | MWHD | 240.3 ± 0.60 | 53.13 ± 0.09 | 299.39 ± 0.15 | 303.1 ± 0.8 |
| *E. caryophyllata* | HD | 170.5 ± 0.35 | 30.97 ± 0.05 | 234.4 ± 0.6 | 565.7 ± 0.6 |
| | MWHD | 167.3 ± 0.60 | 24.19 ± 0.10 | 384.4 ± 0.2 | 662.63 ± 0.05 |

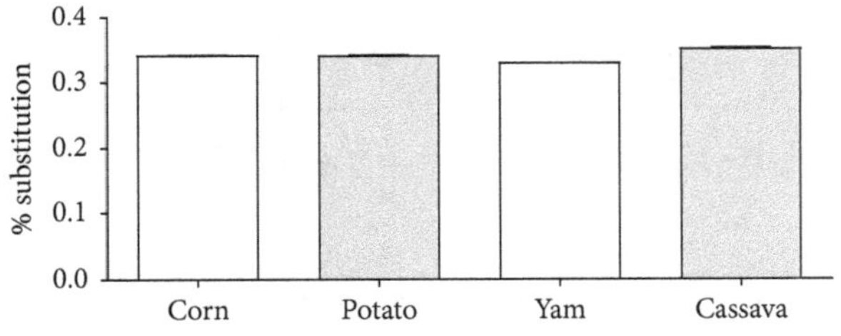


FIGURE 2: Graphical representation of the substitution percentages of the lipophilized starches.

in the extraction; the microwaves involve a more efficient heat flow, and they can heat the entire sample almost simultaneously at a high rate, generating higher performance and lower power consumption in the MWHD method compared to HD. Microwave-assisted hydrodistillation uses three forms of heat transfer within the sample: irradiation, conduction, and convection. As a result, it produces heat more quickly inside and outside the glands. With HD, this heat transfer can only occur through conduction and convection, which makes it less effective [6].

Table 2 presents the majority of components present in the EO obtained by the steam distillation hydrodistillation and microwave-assisted hydrodistillation methods.

Retention time ($t_R$) and relative abundance (%) of the essential oils are identified by comparison with the reference mass spectrum of the NIST-2008 database.

Ranasinghe et al. [5] demonstrated that eugenol, carvacrol, thymol, and cinnamaldehyde are the main components of EO, which provide antifungal and antioxidant activities. It is important to note that thyme EO obtained by the two extraction methods have thymol as the major component and it also presents, to a lesser extent, γ-terpinene, which is a cyclic monoterpene biosynthetic precursor of thymol and carvacrol. Castillo et al. [14] found that EO thyme antioxidant activity depends on the content of thymol and carvacrol. It should be noted that the MWHD method is an excellent alternative because it is green and eco-friendly compared to other solvent extraction techniques such as Soxhlet, since it avoids the use of organic solvents [6].

The antioxidant capacity evaluated *in vitro* can be used as an indirect indicator of *in vivo* activity. Most methods for determining antioxidant capacity consist of accelerating oxidation in a biological system. The antioxidant capacity of an EO is determined by interactions between different compounds with several mechanisms of action [15].

The radical scavenging ability and antioxidant activity of EO of *C. verum* (cinnamon), *T. vulgaris* (thyme), and *E. caryophyllata* (clove) were evaluated by DPPH, ABTS, and ORAC. Antioxidants can act by multiple mechanisms; it depends on the reaction system or the radical or oxidizing source used [16]. The results are expressed as antiradical activity or $IC_{50}$ (effective concentration 50), which is defined as the concentration of the antioxidant which decreases the radical absorption to 50% of the initial amount. There is an inversely proportional relationship indicating that the higher is the $IC_{50}$ value, the lower is the antiradical activity [17].

Table 3 presents the results obtained in the determination of the total antioxidant capacity of the samples from

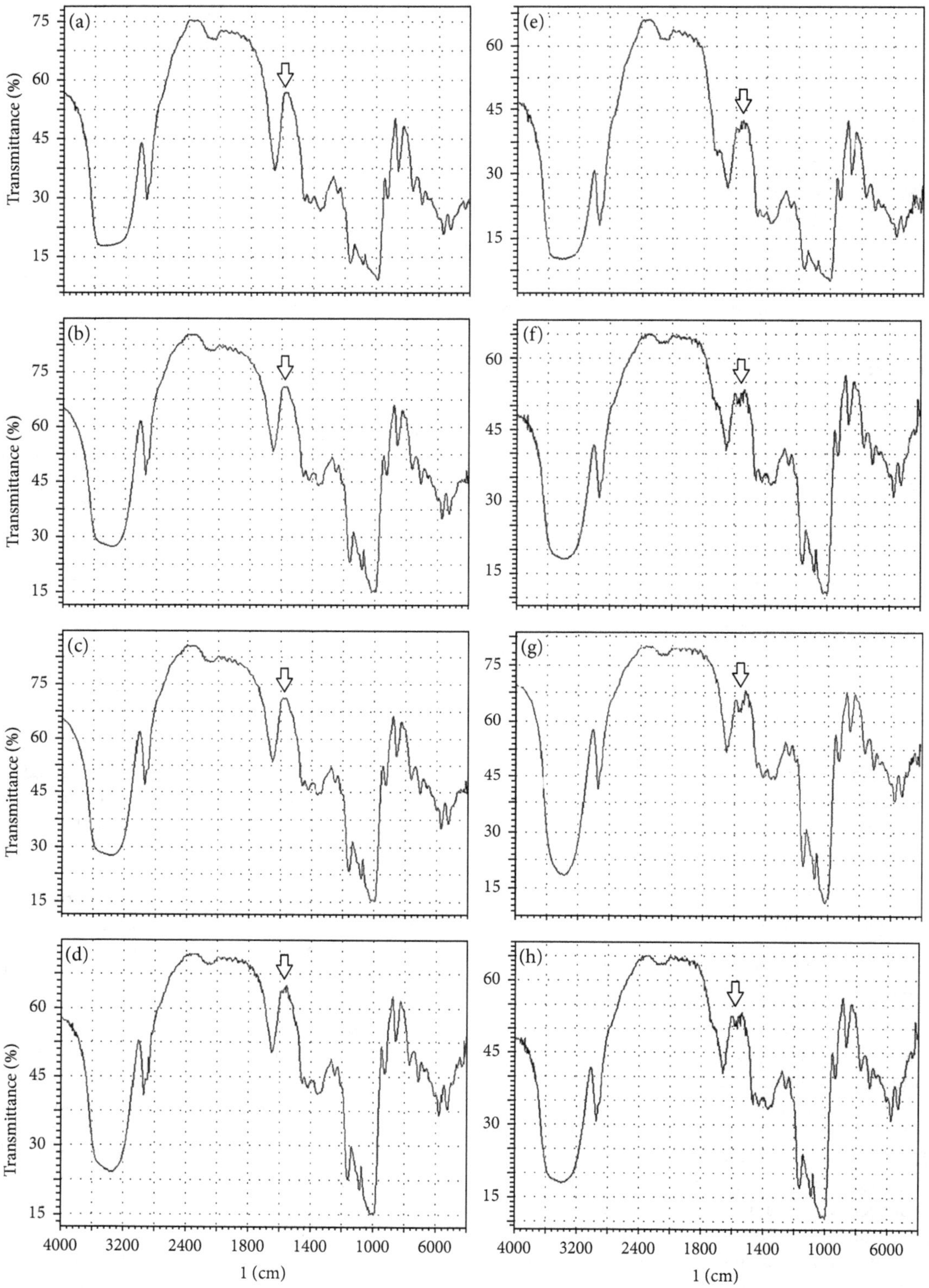


FIGURE 3: FTIR diagram for native and modified starches: (a) native yam; (b) native corn; (c) native potato; (d) native cassava; (e) modified yam (esterification); (f) modified corn (esterification); (g) modified potato (esterification); (h) modified cassava (esterification).

the EO extracted by the HD and MWHD methods analyzed by the spectrophotometric methods DPPH, ABTS, and ORAC.

Yanishlieva et al. [18] report on some aspects of the mechanism and kinetics of the oxidation of different lipid systems inhibited by the presence of thymol, carvacrol, and eugenol, finding out the greater effectiveness of thymol due to the greater stability of the thymol radical; it is originated by the inductive effect of the isopropyl group located at the ortho position relative to the hydroxyl group.

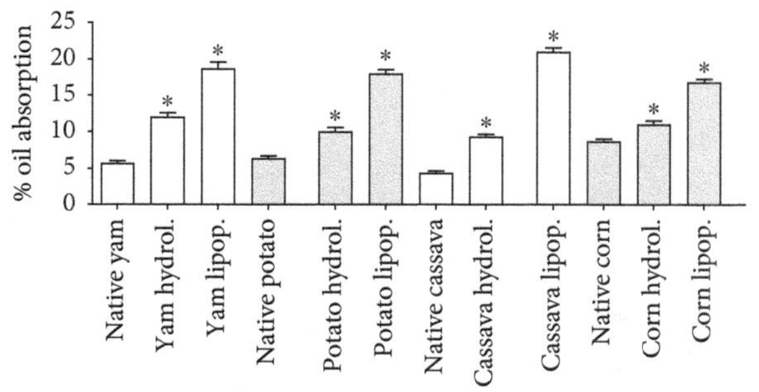


FIGURE 4: Graphical representation of the percentages of oil capture of all starches. *Significant differences compared to native corn ($p < 0.05$).

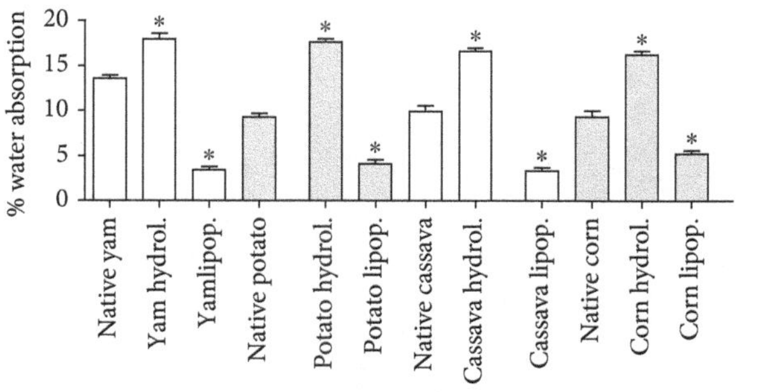


FIGURE 5: Graphical representation of the percentages of water capture of all starches. *Significant differences compared to native corn ($p < 0.05$).

The EO that were studied as the most promising, because they showed a high activity, were the EO of thyme and clove obtained by MWHD. It is possible to infer that each technique of extraction by its conditions of reaction and solubility carries the existence of diverse compounds with antioxidant potential in the mixture.

According to the chemical modification of the starches, the results show that the substitution was very similar, reaching substitution values between 0.33 and 0.35%. The substitution percentage corresponds to the effective number of the OH groups that were replaced in the starch polymer chain; these values were below the standard that is allowed by the US FDA which has a maximum of 3%, because it is a material to be used in the food industry [10]. This can be seen more clearly in Figure 2.

The qualitative characterization of the native starches and modified starches by infrared spectroscopy is shown in Figure 3.

One of the methods used to structurally characterize a chemical is FTIR analysis that is able to provide both qualitative and quantitative information. Figure 3 shows the infrared spectra of the native starches. Among the most important signals observed within these spectra are the bands between 570 and 860 $cm^{-1}$ attributed to the CC stretches and the deformations of the CH and C-$H_2$ bonds which form the glycosidic ring; between 900 and 1500 $cm^{-1}$ is the fingerprint region $Alm_n$ where the peaks between 900 and 1160 $cm^{-1}$ corresponding to the stretching and deformation of the anhydroglucose-COC bonds are highlighted; then a peak at 1640 $cm^{-1}$ is found which is due to a small amount of water present in the amorphous region of the $Alm_n$ at 2920 $cm^{-1}$, and the characteristic stretching of the CH bonds is observed and at 3400–3600 $cm^{-1}$ which is a band referred to the OH groups.

In the spectra related to modified starches, the appearance of two new peaks can be clearly observed: 1580 $cm^{-1}$ corresponds to the asymmetric deformation of the bonding vibrations of the carboxylate group $RCOO^-$ which is formed from the opening of the succinic ring and ionically bound to sodium; in 1720 $cm^{-1}$, another peak appears generated by the deformation of the bond of the carbonyl ester group that joins the functional group with the starch molecule.

Commonly, the introduction of the aliphatic hydrophobic group of the DDSA with a high degree of substitution within the starch structure modifies its surface properties, which is observed in Figures 4 and 5. The capacity of oil capture is presented in Figure 4. According to expectations, the oil capture capacity of each native starch increased considerably by hydrolyzation and lipophilization.

Starches in their native state tend to trap and retain similar amounts of water and oil. With lipophilization, the balance changes and it is able to trap more oil, but without losing its hydrophilic capacity. This constitutes some starches in excellent surfactants for water/oil- (W/O-) type emulsions, that is a low hydrophilic/lipophilic balance (HLB).

The emulsifying capacity of the DDSA-modified starches was more or less the same for all four starches: corn, sweet potato, yam, and cassava; however, the latter showed the highest percentage against the oily substances evaluated and native starches had a zero emulsification capacity (Figure 6).

The emulsifying agents are colloidal stabilizers, which contain the hydrophilic and hydrophobic groups. The hydrophilic part may be ionic (e.g., sodium lauryl sulphate (SLS) having a clearly electrostatic stabilizing mechanism), nonionic (ethoxylated nonyl phenol), or amphoteric (proteins), emphasizing that the mechanism by which they perform the stabilization can be electrostatic or steric. DDSA-modified starches may function as electrostatic stabilizers, but because of the size of the molecules in relation to the number of charged groups, their mechanism is primarily steric. It should be noted that esterification (succination) of starches has been used during the last years as a method to obtain surfactants that serve to stabilize water-oil biphasic systems. When the hydrophobic group of the aliphatic chain is incorporated into the hydrophilic starch molecule, it allows it to establish a strong interaction at the water/oil interface giving the emulsion resistance to the rebonding. It has also been shown that the incorporation of twelve carbon chains into starch improves the thermoplastic characteristics and thermal stability of starch esters [10, 19].

For the encapsulation of the essential oils of thyme, cinnamon, and clove extracted by both methods, hydrolyzed/lipophilized cassava starch was selected, which showed the highest lipophilic and emulsifying capacity, obtaining efficiency percentages of the encapsulation process between 97.72 and 99.22% for all evaluated samples (Table 4).

The spray-drying technique is based on the entrapment of the essential oil compounds on a solid matrix of starches

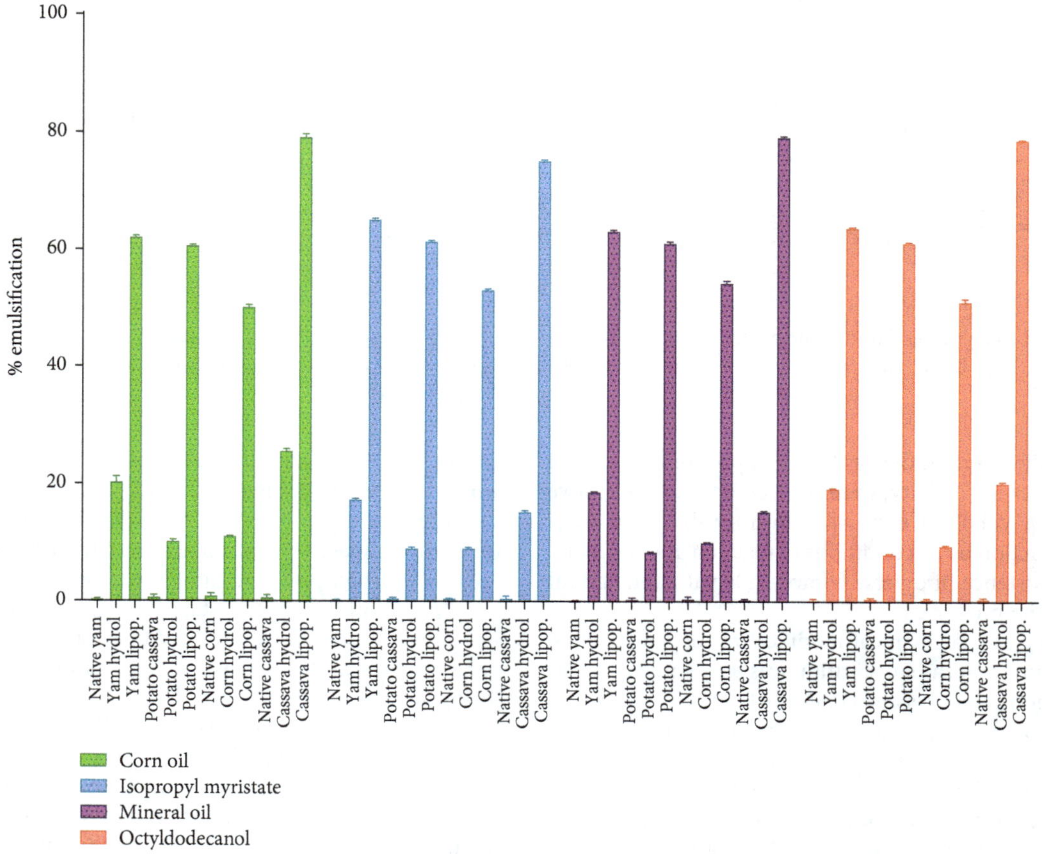


Figure 6: Graphical representation of emulsification percentages of all starches.

Table 4: Efficiency of the encapsulation process of different essential oils.

| Essential oil | Extraction method | Average |
|---|---|---|
| Cinnamon | HD | 97.72 ± 0.02 |
| | MWHD | 98.1 ± 0.07 |
| Thyme | HD | 98.89 ± 0.05 |
| | MWHD | 98.99 ± 0.033 |
| Clove | HD | 98.92 ± 0.006 |
| | MWHD | 99.22 ± 0.011 |

to reduce their mobility [11]. The incorporation of aromas to different matrices in the food, cosmetic, or pharmaceutical industry has taken on great importance since the use of microencapsulation allows to improve the retention of volatile compounds as well as their controlled release [11]. The encapsulation of aromatic components entails an obligatory period of storage before its consumption in food or other sectors. Therefore, when choosing the encapsulation method to be employed, it is important to consider not only the initial retention of the active material but also the amount retained during the storage of the microcapsules.

Retention of flavors depends on the physicochemical properties and the molecular weight of the encapsulating agent. The molecular weight of such materials (amylose < 0.5 million Daltons; amylopectin 50–500 million Daltons) reduces the diffusivity of solute during spray drying and increases the formation of a dry crust on the surface of the microcapsule, increasing the encapsulation efficiency [11]. The hydrophobic nature of groups in DDSA-modified starches provides emulsifying properties to such starches. Fine emulsions are usually more stable during the atomization and spray drying, and for the properly selected matrix, the size of emulsion droplets is a significant factor for oil essential retention. Researchers have showed that the concentration of solids and their ratio with aroma compounds are very important factors for volatile retention during spray drying. Evaporation of volatiles during spray drying is associated both with the interactions of droplets being dried with hot air and with the process of droplet formation (atomization).

The effectiveness of microencapsulation is calculated either by subtracting the nonentrapped into capsule surface oil from the total oil retained after drying or by hydrodistillation of EO from the matrix after washing out the surface oil by the solvent. Figures 7, 8, and 9 show the results. Day zero corresponds to the initial composition of the spray-dried microencapsulated products, where it is shown that the efficiency of the encapsulation is greater in all cases to 97%. At day 3, about 2.8% of essential oils have been lost. By day 15, there is an excess of 91% of the essential oils, that is, a loss of approximately 6%. These results demonstrate the

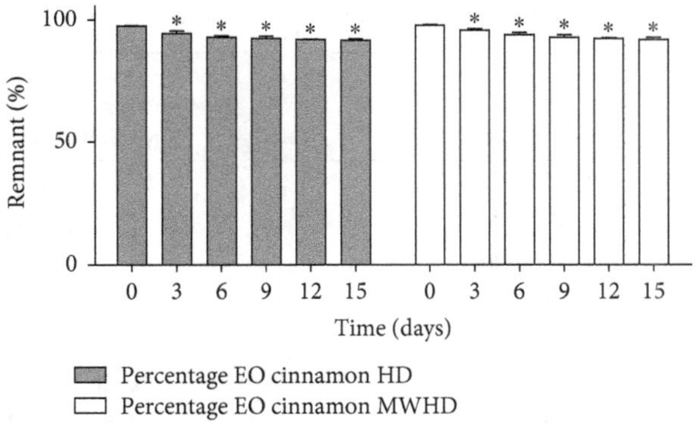


FIGURE 7: Remnant percentage of cinnamon essential oil in a 15-day aging, compared to the original composition. *Significant differences compared to day 0 ($p < 0.05$).

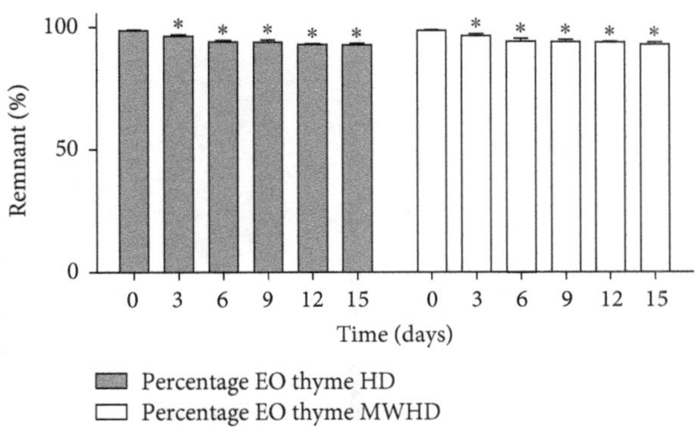


FIGURE 8: Remnant percentage of thyme essential oil in a 15-day aging, compared to the original composition. *Significant differences compared to day 0 ($p < 0.05$).

efficiency of the system to retaining the volatile component under extreme conditions of free evaporation.

From the main components of EO extracted from *C. verum* by both methods, it is noted that cinnamaldehyde is reduced by 2.89% and 0.39% for HD and MWHD, respectively, whereas for EO of *T. vulgaris* obtained by HD and MWHD, it does at 1.522% and 0.769%, respectively, for thymol, whereas eugenol does at 0.5305% and 0.3678% for EO of *E. caryophyllata* achieved by HD and MWHD, respectively, and the appearance of new chemical species is not observed, which means that there was a small loss of material but there was no evidence of deterioration of the constituent components (Table 5).

Retention time ($t_R$) and relative abundance (%) of the essential oils are identified by comparison with the reference mass spectrum of the NIST-2008 database.

The major challenge is to achieve an effective emulgel-type phytocosmetic design that achieves percutaneous permeation without presenting harmful effects [20]. It is essential to select the formulation that presents extensibility, texture, and rheological characteristics suitable for topical administration; it must be aesthetically acceptable and easy to use. It is important to keep in mind that, in skin application preparations, the pH should be between 4.5 and 8.5 [21], in order not to cause irritation and damage to the skin.

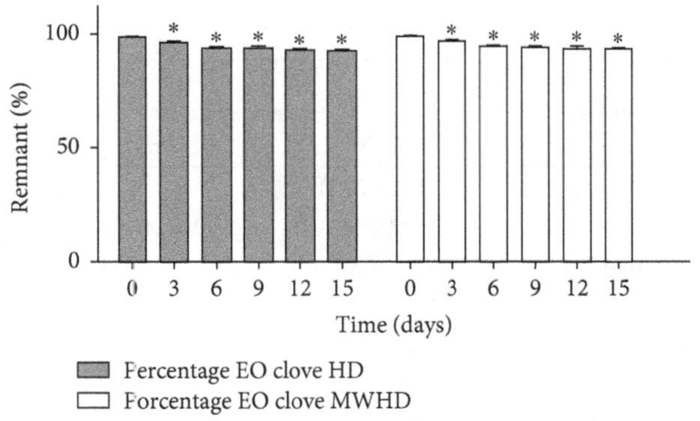


FIGURE 9: Remnant percentage of clove essential oil in a 15-day aging, relative to the original composition. *Significant differences compared to day 0 ($p < 0.05$).

The pH values of the different formulations were maintained in a range of 5.307–8.472, which are within the recommended values and ensures that the emulgel does not cause irritation at the time of application.

In the dynamic viscosity analysis, a characteristic behavior of the emulgels was observed, with values between 175.66 and 15905 Cp. The centrifugation test, primordial assay, during its development stage, since, if there is any instability of the product, it must be reformulated. It was observed that after the test time, the characteristics of 24 of the 32 formulations were unchanged compared to the ones presented at the beginning of the test. Therefore, formulations F1, F5, F9, F13, F17, F21, F25, and F29 were rejected because the emulgels exhibited phase separation after performing this assay.

It should be noted that sensory assessments represent an essential method for measuring pleasing, perception of efficacy, stability parameters, and the most important application characteristics of the product. This allows the optimization of the cost of a formula, to find out the consumer's complacency; they are also an instrument of nonverbal dialogue between the formulator and the consumer, which can be a guide for the correct application of the product [22], being an instrument for the perception of total quality. Figures 10, 11, 12, 13, and 14 show the values observed in the acceptability tests using the facial hedonic scale method by panelists for each of the base formulations named above.

Using daily cosmetic actives that in one way or another allow to keep the skin properly hydrated is most pertinent, since the skin maintains its moisture thanks to water from the deeper layers (transepidermal water) and normal sweat secretion. Due to various factors, for example, lack of substances that retain water, excessive drying of the air, or a damaged barrier function, the loss of water to the outside can be increased. Below 10%, the skin dries and it becomes more fragile, rough, dull, and more exposed to skin diseases, making wrinkles more visible, which are directly related to free radicals; making a comparison with the values obtained from the wetting degree of different formulations, it is indicated that F4 was the one that exhibited the best results. It is also observed that the panelists were able to differentiate the formulations in terms of ease of application and texture and again, F4 presented greater acceptability. The results

Table 5: Major components detected in EO microcapsules obtained by HD and MWHD after 15 days of exposure to air.

| Component | % relative abundance ($t_R$, min) | | | | | |
|---|---|---|---|---|---|---|
| | *C. verum* | | *T. vulgaris* | | *E. caryophyllata* | |
| | HD | MWHD | HD | MWHD | HD | MWHD |
| γ-Terpinene | 3.15 (5.222) | | 15.00 (5.087) | 15.1 (5.234) | | |
| Linalool | 2.10 (5.336) | 2.11 (5.443) | 4.00 (5.670) | 4.50 (5.665) | | |
| Cinnamaldehyde | 66.00 (8.443) | 88.90 (8.555) | 1.0 (8.373) | 1.20 (8.566) | | |
| Thymol | 3.2 (8.666) | 3.33 (8.789) | 44.00 (8.977) | 49.00 (8.789) | | |
| Carvacrol | | | 8.44 (9.010) | 9.50 (9.000) | | |
| Eugenol | 2.70 (9.766) | | | | 60.00 (9.900) | 65.00 (9.902) |
| Caryophyllene | 1.40 (11.002) | | 6.44 (11.222) | 6.30 (11.555) | 11.57 (11.119) | 12.00 (11.278) |

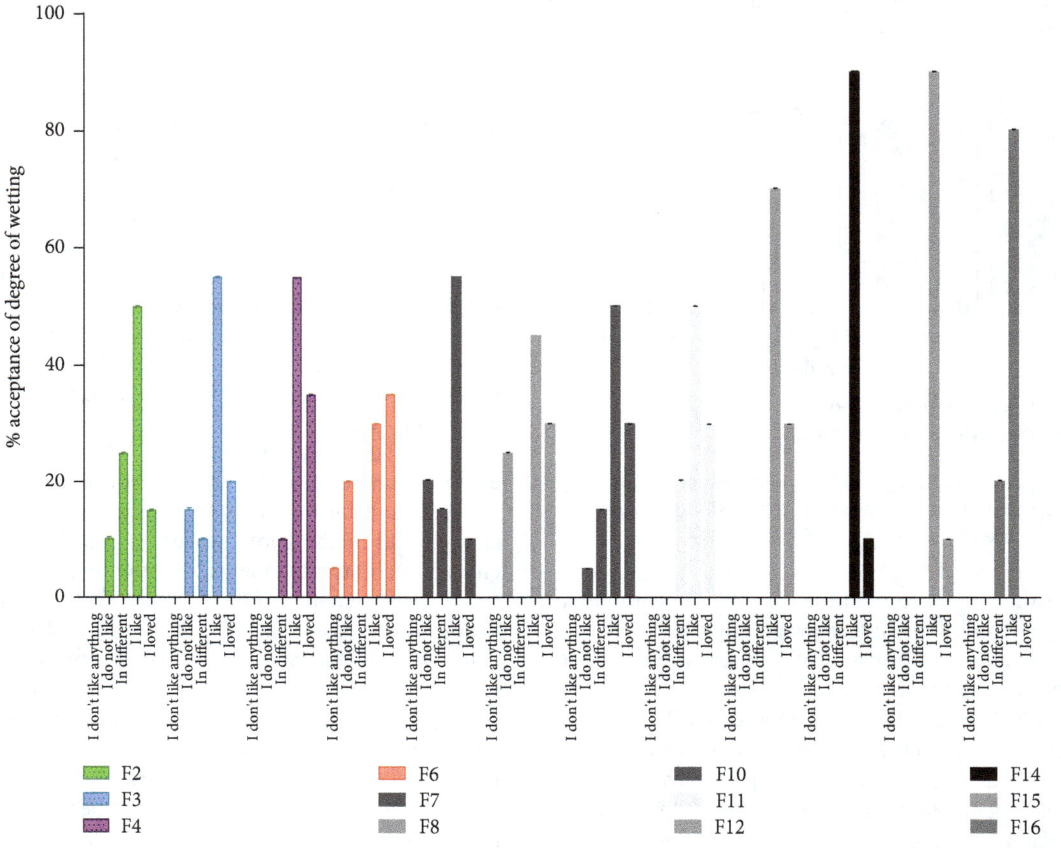


Figure 10: Acceptance percentage of the degree of wetting by panelists for different formulations.

indicate that, from the point of view of color perception, the F4 formulation could be used to make a commercially available emulgel, achieving a 35% acceptance rate in the charm option.

The analysis of the intention to buy, which evaluated the willingness of the people to buy the designed cosmetic product (emulgel), presents significant differences, being the F4 the best formulation with 90% acceptance.

When performing a detailed analysis of the sensory evaluation, we can indicate that there is a direct relationship between the color and the intention to buy. Studies say that 80% of the final purchase of a product is motivated by irrational stimuli of people. Among the latter, we should highlight the emotional component that we associate with the product. One of the added values that contributes to forging that emotional sphere is color. Apparently, its functionality is associated with factors more aesthetic than practical. But nothing further from reality, color psychology acts as a doctrine for brand marketing professionals. Each tone hides its own meaning. And that is why the brands take advantage of this powerful visual tool, to reinforce the meaning they intend to convey. In contrast, our results reflect that the F4 formulation presented the highest acceptability in the different variables evaluated by panelists (degree of wetting, texture, ease of application, color, and intention to buy). Also, the behavior of the area of extensibility ($cm^2$) of the emulgel (F4) was

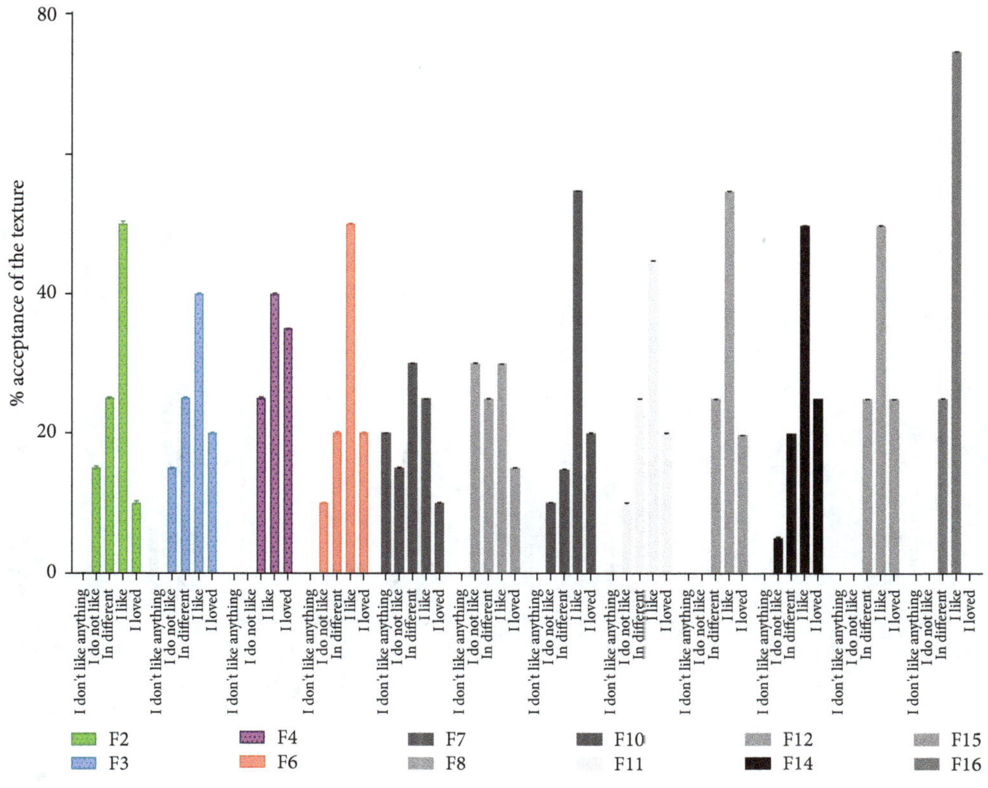


FIGURE 11: Acceptance percentage of the texture by the panelists for the different formulations.

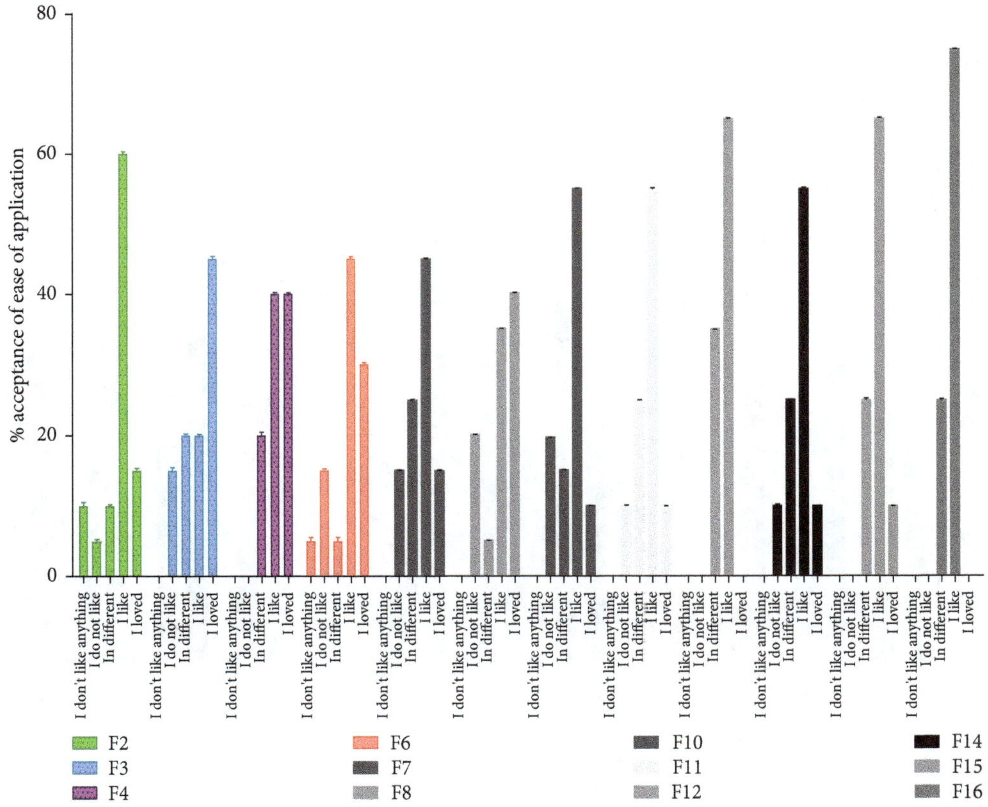


FIGURE 12: Acceptance percentage of ease of application by panelists for different formulations.

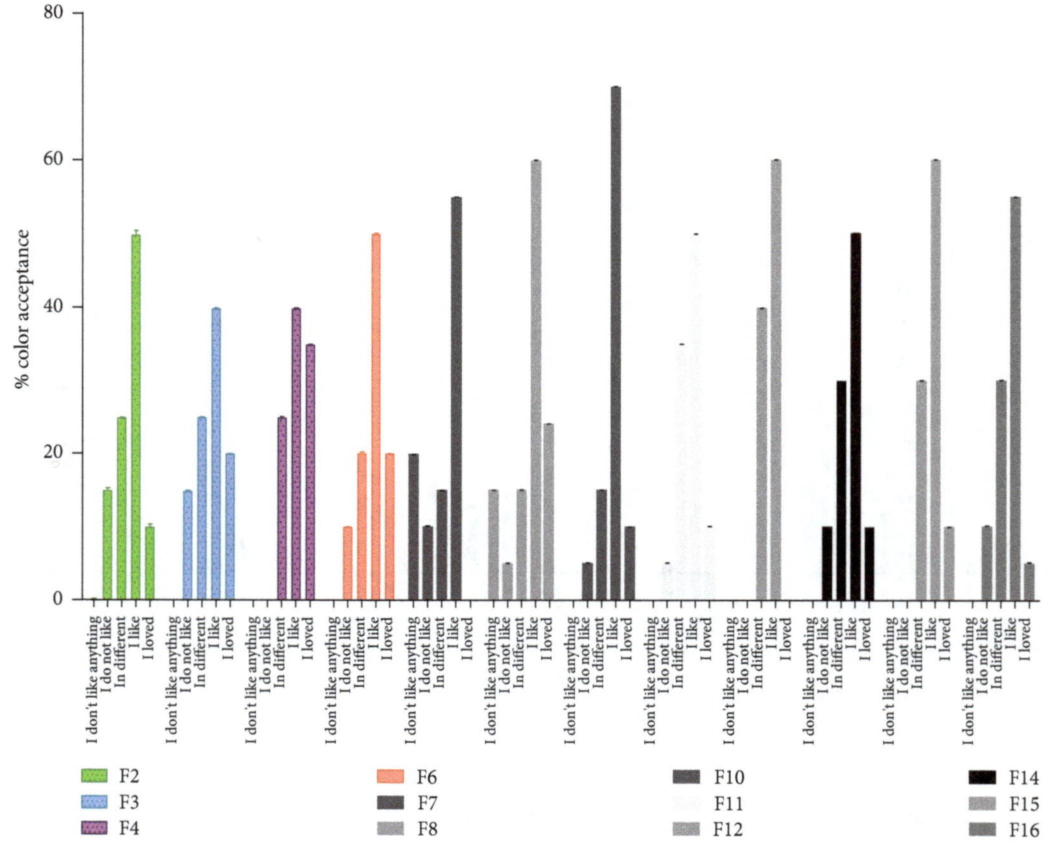


Figure 13: Percentage of color acceptance by panelists for different formulations.

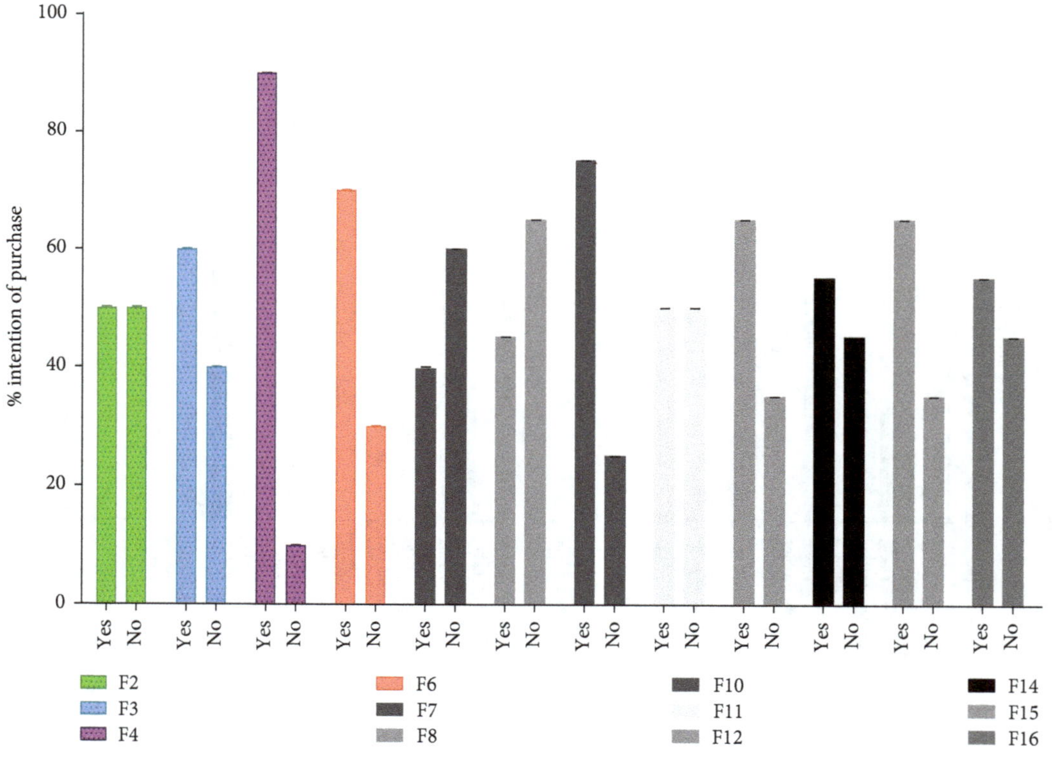


Figure 14: Percentage of intention of purchase of the product by the panelists for the different formulations.

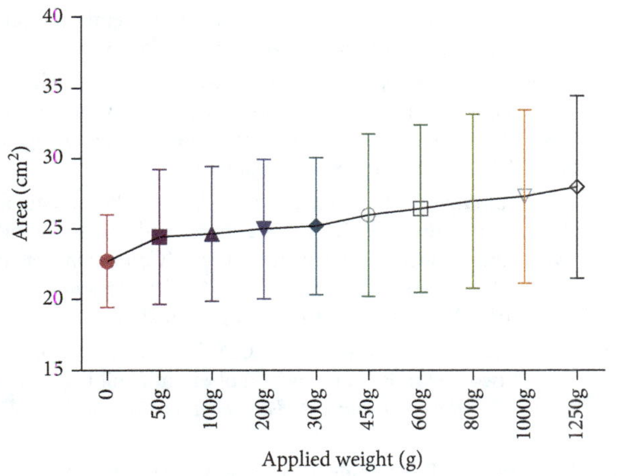


FIGURE 15: Behavior of the extensibility area ($cm^2$) of the emulgel with greater acceptability (F4).

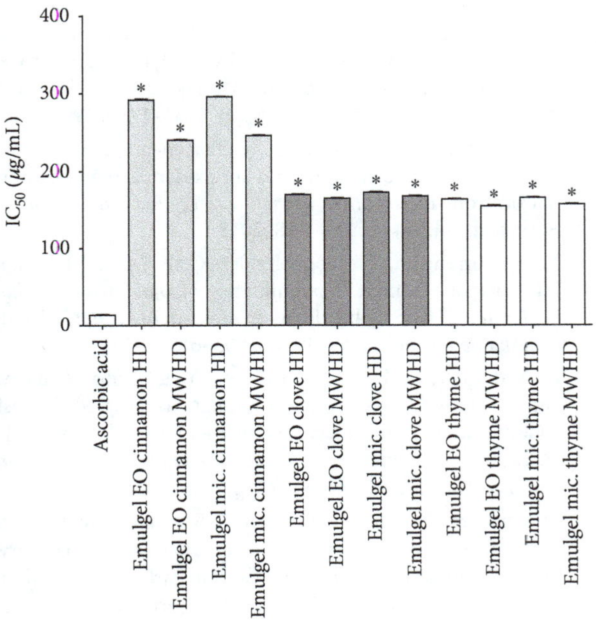


FIGURE 16: $IC_{50}$ of the EO without and with microcapsule emulgels evaluated by the DPPH method. *Significant differences compared to positive control ($p > 0.05$). $IC_{50}$ corresponds to the value of the essential oil concentration that stabilizes 50% of the free radical. Lower values indicate greater antioxidant activity.

evaluated as shown in Figure 15. In the determination of the area of extensibility, no significant differences were shown ($p = 0.05$) between the weights evaluated.

The antioxidant activity of a cosmetic is the expression of the different components, which behave through different mechanisms of reducing action in their interactions with reactive oxygen species (ROS) or other radicals. The measurement of the antioxidant capacity of cosmetics has had a lot of relevance in the last years, due to the quality of the information that can be obtained. The results of the radical scavenging ability and antioxidant activity by DPPH$^{\bullet}$, ABTS$^{\bullet +}$, and ORAC for the emulgel designed with EO microcapsules are shown graphically in Figures 16, 17, and 18 using a polymeric matrix-modified cassava starch.

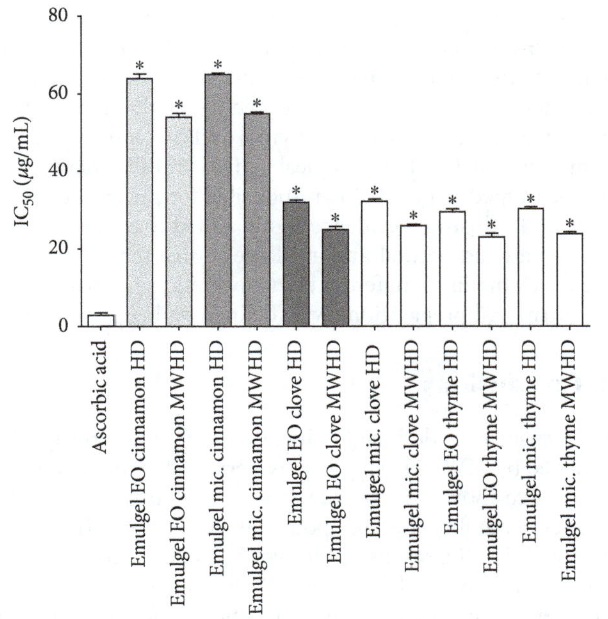


FIGURE 17: $IC_{50}$ of the EO without and with microcapsule emulgels evaluated by the ABTS method. *Significant differences compared to positive control ($p > 0.05$).

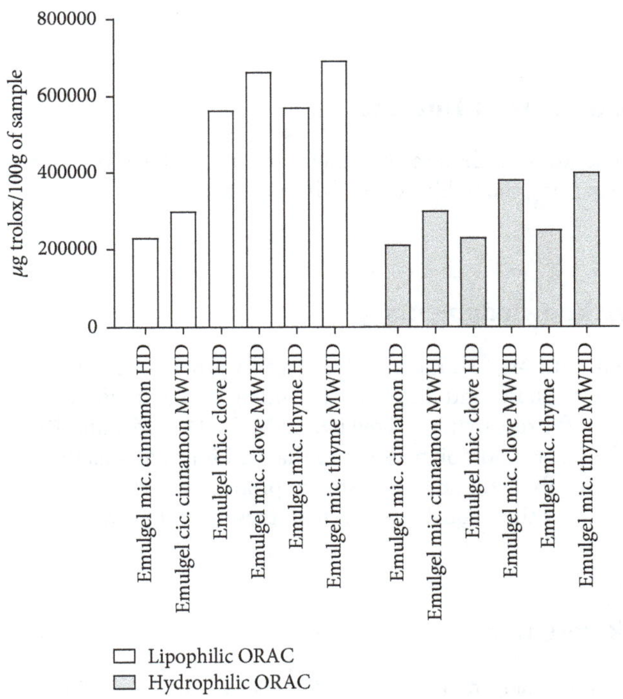


FIGURE 18: Antioxidant activity of EO microcapsule emulgels determined by the hydrophilic and lipophilic ORAC method.

In cosmetics, antioxidants are ingredients for antiwrinkle treatment; they can be of natural or synthetic origin. Its mechanism of action is to counteract the free radicals that can mutate the skin cells. Evaluated emulgels, the one that presented the best activity, was the one designed using active EO microcapsules of thyme obtained by MWHD; it presents a high content of monoterpenes: carvacrol, thymol, and linalool $\gamma$-terpinene; the latter is a cyclic biosynthetic precursor molecule of the two isomers (thymol and carvacrol); all these compounds define the biological activity of EO, where it can not be framed to one, being responsible for pharmacological actions. In this way, evidence is still added that the essential oils are a good natural and available source that will enable the development of different phytocosmetic, pharmaceutical, or nutritional preparations with defined biological activity.

## 4. Conclusions

Phytocosmetics designed with thyme EO microcapsules obtained by MWHD using as polymeric matrix-modified cassava starch showed antioxidant activity, stability, extensibility, and acceptability by the potential consumers. It should be noted that the results of this work contribute to the development of stable and functional phytocosmetic formulations of essential oils, emphasizing that the extraction of EO by microwave radiation-assisted hydrodistillation is considered a novel alternative which contributes to the reduction of environmental impact in the cosmetic industry and is relatively economic compared to conventional hydrodistillation.

## Conflicts of Interest

The authors declare that there are no conflicts of interest regarding the publication of this paper.

## Acknowledgments

The authors thank the University of Cartagena, the University of Sucre, and SENA for providing space, resources, and time for researchers. Likewise, SUMIQUIM Ltda. and BASF Química Colombiana S.A. are thanked for giving us the raw materials to design the cosmetic product, as well as ACCYTEC for their collaboration in all these research works.

## References

[1] A. Manjamalai, T. Alexander, and V. M. Grace, "Bioactive evaluation of the essential oil of *Plectranthus amboinicus* by GC-MS analysis and its role as a drug for microbial infections and inflammation," *International Journal of Pharmacy and Pharmaceutical Sciences*, vol. 4, pp. 205–211, 2012.

[2] S. S. El-Hawary, R. H. El-sofany, A. R. Abdel-Monem, R. S. Ashour, and A. A. Sleem, "Polyphenolics content and biological activity of *Plectranthus amboinicus* (Lour.) spreng growing in Egypt (Lamiaceae)," *Pharmacognosy Journal*, vol. 4, no. 32, pp. 45–54, 2012.

[3] B. Uttara, A. V. Singh, P. Zamboni, and R. T. Mahajan, "Oxidative stress and neurodegenerative diseases: a review of upstream and downstream antioxidant therapeutic options," *Current Neuropharmacology*, vol. 7, no. 1, pp. 65–74, 2009.

[4] A. Muñoz-Acevedo, V. V. Kouznetsov, and E. E. Stashenko, "Composition and *in-vitro* antioxidant capacity of essential oils rich in thymol, carvacrol, *trans*-anethole or estragole," *Salud UIS*, vol. 41, pp. 287–294, 2009.

[5] P. Ranasinghe, S. Pigera, G. A. S. Premakumara, P. Galappaththy, G. R. Constantine, and P. Katulanda, "Medicinal properties of 'true' cinnamon (*Cinnamomum zeylanicum*): a systematic review," *BMC Complementary and Alternative Medicine*, vol. 13, no. 1, p. 275, 2013.

[6] M. G. León, F. M. R. Osorio, and U. S. R. Martínez, "Comparison of two methods for extraction of essential oil from *Citrus sinensis* L," *Revista cubana de Farmacia*, vol. 49, pp. 742–750, 2015.

[7] B. M. Silva, P. B. Andrade, P. Valentão, F. Ferreres, R. M. Seabra, and M. A. Ferreira, "Quince (*Cydonia oblonga* miller) fruit (pulp, peel, and seed) and jam: antioxidant activity," *Journal of Agricultural and Food Chemistry*, vol. 52, no. 15, pp. 4705–4712, 2004.

[8] R. Re, N. Pellegrini, A. Proteggente, A. Pannala, M. Yang, and C. Rice-Evans, "Antioxidant activity applying an improved ABTS radical cation decolorization assay," *Free Radical Biology & Medicine*, vol. 26, no. 9-10, pp. 1231–1237, 1999.

[9] M. G. León, F. M. R. Osorio, M. E. Torrenegra, and G. J. Gil, "Extraction, characterization and antioxidant activity of essential oil from *Plectranthus amboinicus* L," *Revista Cubana de Farmacia*, vol. 49, pp. 708–718, 2015.

[10] M. E. Torrenegra, G. León-Méndez, G. E. Matiz-Melo, and J. D. Sastoque Gómez, "Lipophilitation of *Dioscorea rotundata* P. starch and its possible use as an emulsifying agent," *Revista Cubana de Farmacia*, vol. 49, pp. 605–617, 2015.

[11] D. Rodríguez, M. Espitia, Y. Caicedo, Y. Córdoba, Y. Baena, and C. Mora, "Caracterización de algunas propiedades fisicoquímicas y farmacotecnicas del almidón de Arracacha (Arracacia xanthorriza)," *Revista Colombiana de Ciencias Químico-Farmacéuticas*, vol. 34, p. 141, 2005.

[12] M. Gil-Garzón, L. Alzate-Tamayo, A. Sánchez-Camargo, and L. Millán-Cardona, "Spray drying: an alternative to conserve bioactive and aromatic compounds from garlic extract (*Allium sativum L.*)," *Revista Lasallista de Investigación*, vol. 8, pp. 40–52, 2011.

[13] G. E. M. Melo, K. F. López, and G. L. Méndez, "Microencapsulación de aceite esencial de tomillo (*Thymus vulgaris*) en matrices poliméricas de almidón de ñame (*Dioscorea rotundata*) modificado," *Revista Colombiana de Ciencias Químico-Farmacéuticas*, vol. 44, no. 2, pp. 189–207, 2015.

[14] G. Castillo-Herrera, J. García-Fajardo, and M. Estarrón-Espinosa, "Extraction method that enriches phenolic content in oregano (Lippia graveolens H.B.K.) essential oil," *Journal of Food Process Engineering*, vol. 30, no. 6, pp. 661–669, 2007.

[15] E. M. Kuskoski, A. G. Asuero, A. M. Troncoso, J. Mancini-Filho, and R. Fett, "Aplicación de diversos métodos químicos para determinar actividad antioxidante en pulpa de frutos," *Ciência e Tecnologia de Alimentos*, vol. 25, no. 4, pp. 726–732, 2005.

[16] R. L. Prior, X. Wu, and K. Schaich, "Standardized methods for the determination of antioxidant capacity and phenolics in foods and dietary supplements," *Journal of Agricultural and Food Chemistry*, vol. 53, no. 10, pp. 4290–4302, 2005.

[17] A. Floegel, D. O. Kim, S. J. Chung, S. I. Koo, and O. K. Chun, "Comparison of ABTS/DPPH assays to measure antioxidant

capacity in popular antioxidant-rich US foods," *Journal of Food Composition and Analysis*, vol. 24, no. 7, pp. 1043–1048, 2011.

[18] N. V. Yanishlieva, E. M. Marinova, M. H. Gordon, and V. G. Raneva, "Antioxidant activity and mechanism of action of thymol and carvacrol in two lipid systems," *Food Chemistry*, vol. 64, no. 1, pp. 59–66, 1999.

[19] A. Miladys Torrenegra, M. Glicerio León, M. German Matiz, C. Nerlis Pájaro, and G. Judith Sastoque, "Evaluación de un biorecubrimiento comestible a base de almidon de ñame modificado," *Revista Chilena de Nutrición*, vol. 43, no. 3, pp. 284–289, 2016.

[20] R. Alejandro Palma, L. Alejandra Hernández, B. Imelda Mejía, R. Laura Castrillón, D. Carmen Padilla, and U. Blanca Lilia Cejudo, "Control de calidad de un lote de emulgel con kanamicina, utilizado como un auxiliar en el tratamiento de micetomas por *Actinomadura madurae*," *Revista Mexicana de Ciencias Farmacéuticas*, vol. 36, pp. 16–25, 2005.

[21] T. Pérez, D. M. Soler, Y. Rodríguez et al., "Estabilidad en anaquel (segundo año) de un gel de *Rhizophora mangle* L," *Revista de Salud Animal*, vol. 34, pp. 178–183, 2012.

[22] L. Rigano and A. Bonfigli, *Cosmética eficaz y sensorial*, ISPE, Milano, 2015.

4

# Experimental Study on the Transport of Light Gas Molecules through Low-Density Polyethylene Films

**Riccardo Checchetto**

*Dipartimento di Fisica, Università di Trento, via Sommarive 14, Povo, 38123 Trento, Italy*

Correspondence should be addressed to Riccardo Checchetto; riccardo.checchetto@unitn.it

Academic Editor: Miriam H. Rafailovich

An original experimental procedure for the study of gas permeation process through thin polymer films is presented. Employing mass spectroscopy techniques, this procedure allows the detection of the permeation flux with a signal-to-noise ratio large enough to obtain accurate measurements of the gas diffusivity also in processes with transient transport conditions lasting for short-interval times (~few seconds). The procedure is validated using as test material a thin low-density polyethylene (LDPE) film: the transport of four test gases with different molecular sizes and condensation properties ($CO_2$, $N_2$, $D_2$, and He) is studied in the 295 to 350 K temperature interval. The $CO_2$ diffusivity values well compare with values previously obtained studying the same LDPE film samples by integral permeation technique measuring the time lag value. Original data on the diffusivity of the He and $D_2$ penetrant molecules are reported: in the examined temperature range, the diffusivity values of these small-size penetrants are in the $10^{-6}$ cm$^2$/s range and follow an Arrhenius behavior with temperature. The activation energy values for diffusion are $18.8 \pm 0.4$ and $10.0 \pm 0.4$ kJ/mol for $D_2$ and He, respectively.

## 1. Introduction

The study of mass transport process through polymer films is an important research area given its applicative interest for the development of innovative membranes for gas separation [1] and of gas barrier films to be used in the packaging technology [2]. This topic also presents fundamental interest: gas diffusion occurs by successive jumps of the penetrant molecules between cavities formed in the polymer matrix by the random thermal fluctuations of the macromolecular chains [3, 4]. These cavities constitute the free volume of the polymer which controls structural properties such as the glass transition temperature [5]. Penetrant molecules thus act as sensitive probe to explore the polymer-free volume and its variation, for example, with temperature and cross-linking density [6, 7] or with the addition of filler nanoparticles [8], and to analyze the structure of rigid cavities in biopolymer membranes [9].

The gas transport through polymer membranes obeys to the solution-diffusion mechanism [1] which assumes as transport parameters the molecular diffusivity $D$ (units: cm$^2$/s) and solubility $\Pi$ (units: mol·cm$^{-3}$·Pa$^{-1}$) in the polymer layers (the gas permeability $\Phi$ is given by the product of gas solubility and diffusivity, $\Phi = DS$). Measurements are carried out by gas-phase permeation technique often using planar membrane samples shaped in the form of a thin disc with a diameter much larger than the disc thickness $L$. Keeping the sample temperature $T$ constant, at time $t = 0$ (which is the trigger time of the experiment), one side of the planar membrane is exposed to the test gas kept at fixed pressure $P_{HPS}$. This side of the membrane is called the high-pressure side (HPS). The other side is called the low-pressure side (LPS) and faces an evacuated analysis chamber kept in vacuum conditions. The gas permeation process consists in the transfer of gas molecules from the HPS of the membrane to the analysis chamber through the membrane layers and occurs in three microscopic steps:

(i) Absorption: penetrant molecules in the gas phase are absorbed by the HPS surface layers of the membrane. Here, their concentration immediately reaches its equilibrium value $c_{HPS} = c(x = 0, t) = \Pi P_{HPS}$. Here, $x = 0$ represents the coordinate of the HPS

(ii) Diffusion step: molecules absorbed in the HPS of the membrane diffuse through the membrane layers to the LPS down to their concentration gradient according to the first Fick law, $j(x,t) = -D(\partial c(x,t)/\partial x)$; $c(x,t)$ is the gas concentration at time $t$ in the membrane layers at depth $x$ below the HPS

(iii) Desorption step: molecules reaching the LPS surface layers of the membrane desorb in the analysis chamber. When $P_{\text{LPS}}^{\text{gas}} \ll P_{\text{HPS}}$, the approximation $c_{\text{LPS}} = c\,(x = L, t) = \Pi\, P_{\text{LPS}}^{\text{gas}} \cong 0$ holds ($x = L$ is the coordinate of the LPS, that is, the membrane thickness)

The permeation flux $j(t) = j(x = L, t) = -D\,(\partial c(x = L, t)/\partial x)$ can be obtained solving Fick's second equation $\partial c(x,t)/\partial t = D\,(\partial^2 c(x,t)/\partial x^2)$.

The previous diffusion problem admits for $j(t)$ the following analytical solution:

$$j(t) = \frac{D\Pi}{L} P_{\text{HPS}} \left[ 1 + 2 \sum_{n=1}^{\infty} (-1)^n \exp \frac{-D n^2 \pi^2 t}{L^2} \right]. \quad (1)$$

Steady-state transport conditions are set when $j(t)$ assumes the constant, time-stationary value $J = (D\Pi/L)P_{\text{HPS}}$.

Gas molecules permeate through the membrane of surface area $A$ at a rate $Q(t) = A\,j(t)$. Molecules permeated in the analysis chamber form a rarefied gas at thermal equilibrium with the chamber walls at temperature $T_{\text{cham}}$.

The partial pressure $P_{\text{LPS}}^{\text{gas}}(t)$ of the test gas varies, as a function of time, according to the following relation:

$$Q(t) = \frac{1}{R\,T_{\text{cham}}} \left[ V \frac{dP_{\text{LPS}}^{\text{gas}}(t)}{dt} + S_p P_{\text{LPS}}^{\text{gas}}(t) \right], \quad (2)$$

where $R$ is the universal gas constant, $V$ is the volume of the analysis chamber, and $S_p$ is the pumping speed of the vacuum system.

During the permeation experiments, the pressure $P_{\text{exp}}(t)$ in the analysis chamber is measured as a function of time: the analysis of the $P_{\text{exp}}(t)$ curve by (2) permits to measure the permeation flux $j(t)$ and to obtain the values of the $D$ and $\Phi$ transport parameters.

Gas-phase permeation experiments are carried out by integral methods [10, 11]: at $t = 0$, the pumping system is excluded (i.e., $S_p = 0$) and permeated molecules accumulate in the analysis chamber. According to the solution-diffusion model, at time $t$, the total number of permeated moles $m(t)$ is given by [12]

$$m(t) = A \int_{t'=0}^{t'=t} j(t')\,dt' = A \left\{ \frac{D\Pi P_{\text{HPS}}}{L} \left[ t - \frac{L^2}{6D} \right] - 2 \frac{L\Pi P_{\text{HPS}}}{\pi^2} \sum_{n=1}^{\infty} \frac{(-1)^n}{n^2} \exp \frac{-D n^2 \pi^2 t}{L^2} \right\}, \quad (3)$$

and the partial pressure of the test gas in the analysis chamber $P_{\text{LPS}}^{\text{gas}}(t)$ increases as a function of time as $P_{\text{LPS}}^{\text{gas}}(t) = (R\,T_{\text{cham}}/V)\,m(t)$.

In the integral permeation experiments, the pressure in the analysis chamber can be recorded by using low-pressure gauges (in fact, $S_p = 0$ and low-vacuum conditions are rapidly set). Gas transport parameters are obtained by the analysis of the experimental permeation curves $P_{\text{exp}}(t)$ when steady-state transport conditions are set, that is, when the $P_{\text{exp}}(t)$ value linearly increases with time $t$.

According to the model, in steady state conditions $j(t) = J = (D\Pi/L)P_{\text{HPS}}$, thus, $m(t) = A(D\Pi\, P_{\text{HPS}}/L)[t - (L^2/6D)]$ and

$$P_{\text{LPS}}^{\text{gas}}(t) = \frac{R\,T_{\text{cham}}}{V} \frac{A}{L} D\,S\,P_{\text{HPS}} \left[ t - \frac{L^2}{6D} \right]. \quad (4)$$

The measured rate of pressure increase $dP_{\text{exp}}(t)/dt$ provides the gas permeability $\Phi = \Pi D$ by the relation $dP_{\text{exp}}(t)/dt = dP_{\text{LPS}}^{\text{gas}}(t)/dt = (R\,T_{\text{cham}}/V)(A/L)DSP_{\text{HPS}}$. The intercept of the straight line fitting the measured pressure data with the time axis, according to (4), defines a parameter called time lag given by $\tau_{\text{lag}} = L^2/6\,D$: this characteristic time is a measure of the lasting of transient transport conditions and permits the evaluation of the gas diffusivity $D$.

The accuracy of this procedure in the evaluation of the gas transport parameter is limited by the background flow $Q_{\text{back}}$ due to degassing from the chamber walls ($H_2O$ and CO adsorbed during the venting procedure), to small vacuum leaks ($H_2O$ vapor, $N_2$, and $O_2$), and to air permeation through Viton O-rings ($H_2O$ vapor, $N_2$, and $O_2$). During the permeation test, $Q_{\text{back}}$ increases the pressure $P_{\text{LPS}}^{\text{back}}(t)$ of the gases forming the residual vacuum: the pressure $P_{\text{LPS}}^{\text{tot}}(t)$ measured in the analysis chamber is thus given by two contributions $P_{\text{exp}}(t) = P_{\text{LPS}}^{\text{back}}(t) + P_{\text{LPS}}^{\text{gas}}(t)$ which cannot be resolved by the total pressure gauge.

For a given sample, the accuracy in the evaluation of the gas permeability $\Phi$ rises by improving the signal-to-noise ratio $S/N = (dP_{\text{LPS}}^{\text{gas}}/dt)/(dP_{\text{LPS}}^{\text{back}}(t)/dt)$, that is decreasing the sample thickness $L$. The $P_{\text{LPS}}^{\text{back}}(t)$ term affects the accuracy in the measure of $P_{\text{LPS}}^{\text{gas}}(t)$ and introduces large relative indetermination in the evaluation of the $\tau_{\text{lag}}$ parameter which impedes to obtain accurate measures of the gas diffusivity studying the fast transport process as it occurs with thin film samples or analyzing the transport of penetrants with high $D$ values or membrane samples made of low-permeability materials [13–16]. Note also that in permeation processes with penetrant diffusivities $D$ in the $10^{-6}$ cm$^2$/s order, using film samples with thickness $L$ in the 10 $\mu$m range, time lag values are in the few-second order in which the result is thus comparable to that of the response time of the low-pressure gauges.

In this paper, we present an original experimental procedure for the study of the gas permeation process which permits accurate evaluation of the gas diffusivity (and permeability) of thin polymer films and processes characterized by

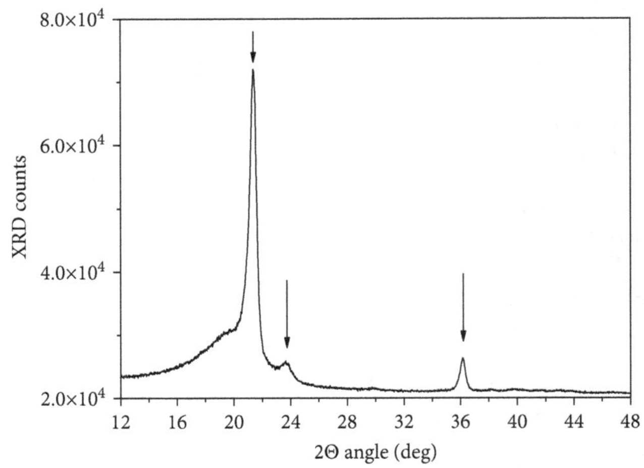


FIGURE 1: XRD spectrum of the LDPE film ($Cu\ K_\alpha$ radiation, $\lambda = 1.5414$ Å). The arrow indicates the LDPE reflection peaks.

short transient times: this often occurs in the permeation of small-size penetrants and in high-temperature processes. The validation of the proposed experimental approach will be carried out using thin films made of low-density polyethylene (LDPE) which is a polymer used as gas barrier in packaging applications [17].

Measurements were carried out using test gases with different molecular sizes, molecular masses, and condensation properties and analyzing the transport process up to temperatures close to the LDPE melting temperature. Our measurements permitted us to obtain original data on the diffusivity of the light and small-size $D_2$ and He penetrant molecules, and analysis of the diffusivities permitted the evaluation of the $D_2$ and He activation energy for diffusion which are original data here presented.

## 2. Experimental Part

The study was carried out using membrane samples made of commercial semicrystalline low-density polyethylene (LDPE) kindly provided by ICBM-CNR (Naples, Italy). The X-ray diffraction pattern (XRD) of the LDPE films is presented in Figure 1. The spectrum shows sharp, narrow diffraction peaks due to the crystalline fraction of the material at $2\theta = 21.1°$, $2\theta = 23.8°$, and $2\theta = 36°$ (see vertical arrows) while the amorphous component gives a very broad peak. The Bragg-Brentano analysis of the XRD peak at $2\theta = 21.1°$ indicated that the average size of PE crystallites is about 12 nm. The crystalline fraction of the LDPE samples, as evaluated by DSC analysis, was 29 ± 1%. The melting temperature $T_m$ of the present LDPE samples was ~380 K [18].

The study of the gas transport properties of the present samples was previously carried out by the procedure described in the introduction, and measurements were carried out in the 290 to 350 K temperature range using $H_2$, $CO_2$, and $N_2$ [18]. Reliable measurements of the permeability and diffusivity values were possible only with the $CO_2$ test gas. Diffusivity values ranged from $\sim 1 \times 10^{-7}$ cm$^2$/s at 290 K to $\sim 1 \times 10^{-6}$ cm$^2$/s at 350 K while the activation energy values for diffusion and permeation were 29.7 ± 0.4 kJ/mol and 36.4 ± 0.4 kJ/mol, respectively. Using the $H_2$ and $N_2$ test gases, it was possible to evaluate only the gas permeability because of the large relative indetermination in the short time lag values [18].

In the procedure, we are here describing that the permeation experiment is carried out under continuous pumping of the analysis chamber which is done by turbomolecular or sputter-ion pumps: high vacuum conditions are thus maintained during the permeation test.

When the condition

$$\frac{S_p}{V} \gg \frac{1}{P_{\mathrm{LPS}}^{\mathrm{gas}}(t)} \frac{dP_{\mathrm{LPS}}^{\mathrm{gas}}(t)}{dt}, \tag{5}$$

is satisfied, then according to (2), the permeation flux $j(t)$ is directly proportional to the instantaneous value, $P_{\mathrm{LPS}}^{\mathrm{gas}}(t)$, of the test gas partial pressure

$$Q(t) = A\, j(t) = S_p P_{\mathrm{LPS}}^{\mathrm{gas}}(t). \tag{6}$$

In the present vacuum system, $(S_p/V) > 5\,s^{-1}$ and the condition $(S_p/V) \gg (1/P_{\mathrm{LPS}})(dP_{\mathrm{LPS}}(t)/dt)$ are satisfied for each test gas in the examined temperature range.

The $P_{\mathrm{LPS}}^{\mathrm{gas}}(t)$ curves can be thus converted to experimental permeation flux curves $j_{\exp}(t)$ through the relation

$$j_{\exp}(t) = \frac{1}{A} \frac{1}{R\,T_{\mathrm{cham}}} S_p P_{\mathrm{LPS}}^{\mathrm{gas}}(t). \tag{7}$$

Following this procedure,

(i) the measured value of the permeation flux $J_{\exp}$ in stationary transport conditions permits the evaluation of the penetrant permeability $\Phi = D\Pi$ by the relation $J_{\exp} = J = (D\Pi/L)P_{\mathrm{HPS}}$

(ii) the penetrant diffusivity $D$ can be obtained fitting the normalized $j_{exp}(t)/J_{exp}$ permeation curves with the curve $j_{norm}(t) = j(t)/J = 1 + 2\sum_{n=1}^{\infty}(-1)^n \exp(-D n^2 \pi^2 t/L^2)$; see (1)

Let us discuss now the advantages of this "*constant-pumping*" procedure. The continuous pumping of the analysis chamber during the experimental run maintains ultra high- (UHV) or high-vacuum (HV) conditions and, consequently, permits the use of a quadrupole mass spectrometer to detect the permeation signal $I_{LPS}^{gas}(t)$, that is the ionic current pertinent to the test gas. Knowing the QMS sensitivity for the test gas, $\alpha_{gas}$, the $I_{LPS}^{gas}(t)$ signal provides a measure of the partial pressure of the permeating gas, $P_{LPS}^{gas}(t) = I_{LPS}^{gas}/\alpha_{gas}$. The $P_{LPS}^{gas}(t)$ value is thus obtained, as a function of time, without noise contributions due to background flows of any other residual gas component.

This experimental procedure consequently allows

(i) very large the signal-to-noise ratio $S/N$. In fact, $S/N = P_{LPS}^{gas}(t)/\delta P_{back}^{gas} = I_{LPS}^{gas}(t)/\delta I_{back}^{gas}$ = where the noise $\delta P_{back}^{gas} = \alpha_{gas}\delta I_{back}^{gas}$ is due to the fluctuations of the background partial pressure $P_{back}^{gas}$ of the test gas, only. The $S/N$ is very large for all gases, in particular for those not forming the residual vacuum of the analysis chamber

(ii) very low detection limit of the permeation flux. For each test gas, in fact, the $S_p\delta P_{back}^{gas} = \alpha_{gas}S_p\delta I_{back}^{gas}$ term defines the detection limit (A/Pa). Consequently, the permeation signal can be efficiently detected also studying using low-permeability materials and the permeation fluxes $j(t)$ can be detected also when $A\,j(t) < Q_{back}$

(iii) that the high sampling rate of the QMS, adjustable from 0.5 ms/amu to 60 s/amu (amu: atomic mass unit), permits to record $P_{LPS}^{gas}(t)$ data with high time resolution

To test the effectiveness of this "*constant-pumping*" (CP) approach and for its validation, we used disc-shaped, planar samples with 13.5 ± 0.2 mm diameter and thickness $L = 53 \pm 1\ \mu m$. Measurements were carried out in the 295 to 350 K temperature range with $P_{HPS}$ values of 40 kPa using test gas with different molecular sizes and condensation properties: carbon dioxide ($CO_2$), nitrogen ($N_2$), deuterium ($D_2$), and helium (He); see Table 1.

Experiments were carried out in a stainless steel analysis chamber pumped by turbomolecular pumps; the detection of the permeation signal is carried out by a QMS equipped with secondary electron multiplier (SEM) and grid-type ion source: this setup permits to detect partial pressures down to $10^{-12}$ Pa.

A schematic diagram of the system is shown in Figure 2(a); for the sake of simplicity, in this figure, the tubular oven used for sample heating is not reported. In Figure 2(b), we present the results of the QMS calibration procedure to evaluate the sensitivity $\alpha_{gas}$ for the examined test gases. Starting from background pressure levels $P_{back}$ lower than $10^{-6}$ Pa, through a calibrated leak valve connected to the gas reservoir, the pressure of the test gas in the pumped analysis chamber is increased by a factor~10 and the corresponding QMS and IG signals were recorded. Repeating this procedure, couples of $[P_{LPS}(t) \text{vs} L_{LPS}^{gas}(t)]$ data were obtained and the slope of the line fitting them provides the value of the $\alpha_{gas}$ parameter. We obtained the following values $\alpha_{He}$ = 125 ± 1 Pa/A, $\alpha_{D2}$ = 210 ± 1 Pa/A, $\alpha_{N2}$ = 342 ± 3 Pa/A, and $\alpha_{CO2}$ = 457 ± 1 Pa/A.

Table 1: Chemical-physical properties of the examined test gases [19].

| Gas | Critical temperature (K) | Kinetic diameter (Å) | Molecular mass (amu) |
|---|---|---|---|
| $D_2$ | 38.2 | 2.89 | 4 |
| He | 5.19 | 2.6 | 4 |
| $CO_2$ | 304.2 | 3.3 | 44 |
| $N_2$ | 126.2 | 3.64 | 28 |

In Figure 3(a), we present the mass spectrum of the background vacuum at $P_{back} = 1.2 \times 10^{-6}$ Pa, after one night of turbomolecular pumping: the spectrum reports the QMS mass signal $I_{back}^{gas}$ pertinent to the gases forming the residual vacuum. It can be seen that the background vacuum is mostly composed of water vapor ($H_2O^+$, m/e = 18, fragmentation products $OH^+$ at m/e = 17, and $O^+$ at m/e = 16), carbon monoxide ($CO^+$, m/e = 28, fragmentation product $C^+$ at m/e = 12), and molecular hydrogen ($H_2^+$, m/e = 2, fragmentation product $H^+$ at m/e = 1) [19]. In Figure 3(b), we report the QMS mass signal for $N_2$ (upper panel) and $D_2$ (lower panel) recorded in the permeation test carried out at 295 K with the described "constant-pumping" procedure: vertical arrows indicate the trigger time $t = 0$.

Looking at the curves for $t < 0$, we observe that the QMS background signals $I_{back}^{gas}$ are in the $10^{-12}$ and $10^{-10}$ A order for $D_2$ and $N_2$ while the $\delta I_{back}^{gas}$ fluctuations are ~$2 \times 10^{-12}$ A for both gases: the $S/N$ is thus larger than 10 already in the initial instants of the permeation transient ($t > 0$) and reaches values in the $10^3$ order in stationary transport conditions.

The $\delta I_{back}^{gas}$ fluctuations in the $10^{-12}$ A order permit to evaluate the detection limit of the present experimental approach by the relation $\delta j \sim (1/A)(1/R\,T_{cham})S_p\alpha_{gas}\delta I_{back}^{gas}$. Given the effective $S_p$ values in the 50 L/s order, QMS sensitivity $\alpha_{gas}$~200 Pa/A, see Figure 1(b), and membrane surface area $A$~1 cm$^2$, then $\delta j$ is in the $10^{-14}$ mol·cm$^{-2}$·s$^{-1}$ order ($T_{cham} = 300K$; $R = 8.3$ L·kPa·mol$^{-1}$K$^{-1}$): $\delta j$ well compare with the upper value of the oxygen transmission rate (OTR) for gas barrier films to be used in the organic light-emitting device (OLED) technology, ~$3 \times 10^{-14}$ mol·cm$^{-2}$·s$^{-1}$ [20].

The permeation flux is obtained from the curves in Figure 3(a) subtracting the background QMS signal

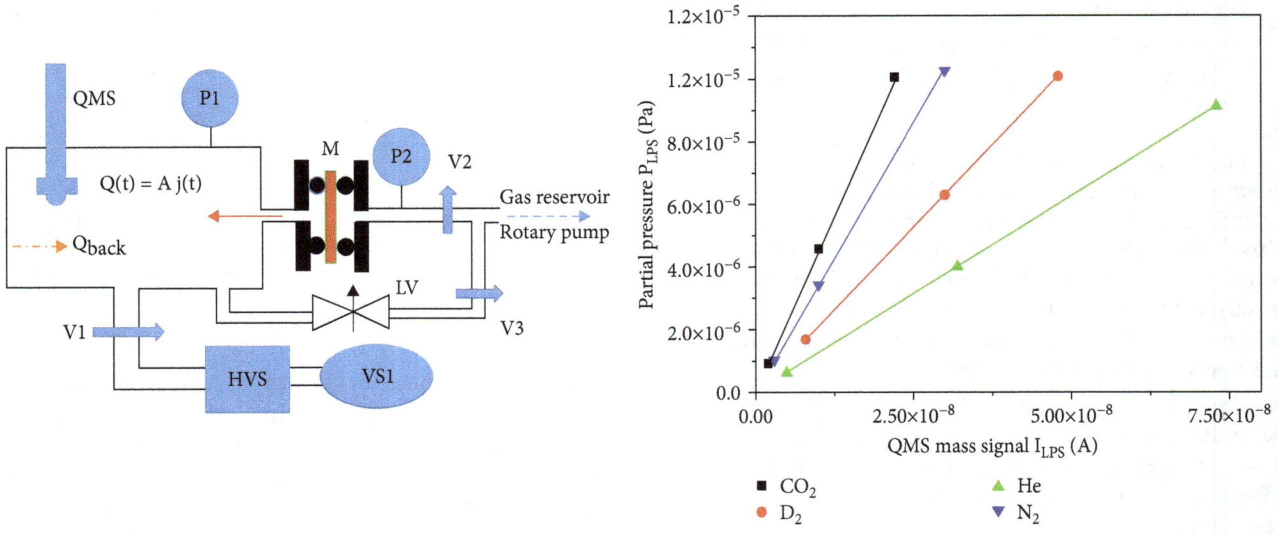


(a) (b)

FIGURE 2: (a) Schematic of the experimental apparatus. QMS: quadrupole mass spectrometer; HVS: ultra high-vacuum pumping system; VS1: medium vacuum pumping system (diaphragm pump and molecular drag pump); P1: ionization gauge; P2: low-pressure gauge; V1, V2, and V3: vacuum valves; LV: calibrated leak valve. The membrane sample (M) is vacuum sealed using two Viton O-rings. In this schematic diagram, the tubular oven for membrane sample heating is not reported. (b) QMS calibration curves. The QMS mass signal $I_{LPS}$ is reported as a function of the $P_{LPS}$ pressure value for the different test gases. Indetermination is inside the size of the symbol.

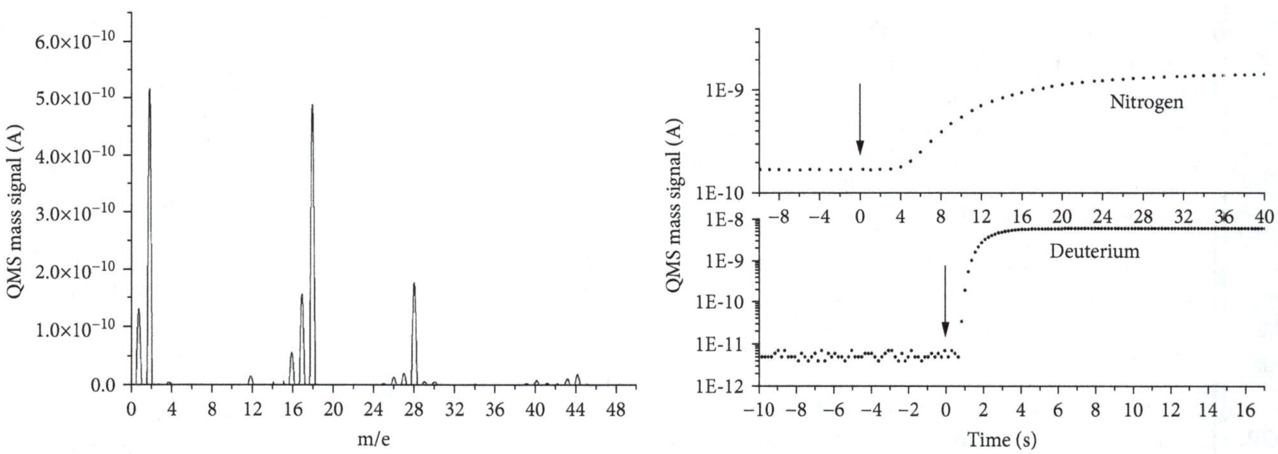


(a) (b)

FIGURE 3: (a) Mass spectrum of the background vacuum at $P_{back} = 1.2 \times 10^{-6}$ Pa after one night of turbomolecular pumping. The dominant contributions are related to the $I^{gas}_{back}$ signals pertinent to $H_2O$, CO, and $H_2$ (m/e = 18, 28, and 2, respectively). Their fragmentation products are also reported; see text. (b) QMS mass signals $I^{D2}_{LPS}$ and $I^{N2}_{LPS}$ (m/e = 4 and 28, respectively) recorded in the permeation test. Arrows indicate the trigger time of the experiment.

$I^{gas}_{back}$ from the detected QMS signal $I^{gas}_{LPS}(t)$ and using the relation

$$j_{exp}(t) = \frac{1}{A}\,\frac{1}{R\,T_{cham}}\,S_p\,\alpha_{gas}\left[I^{gas}_{LPS}(t) - I^{gas}_{back}\right]. \quad (8)$$

## 3. Results and Discussion

In Figure 4(a), we present the $j_{exp}(t)$ curves as obtained at 295 K exposing the HPS at 40 kPa pressure. In the figure, symbols are the experimental data (relative indeterminations are inside the size of the symbol) obtained by (8) while

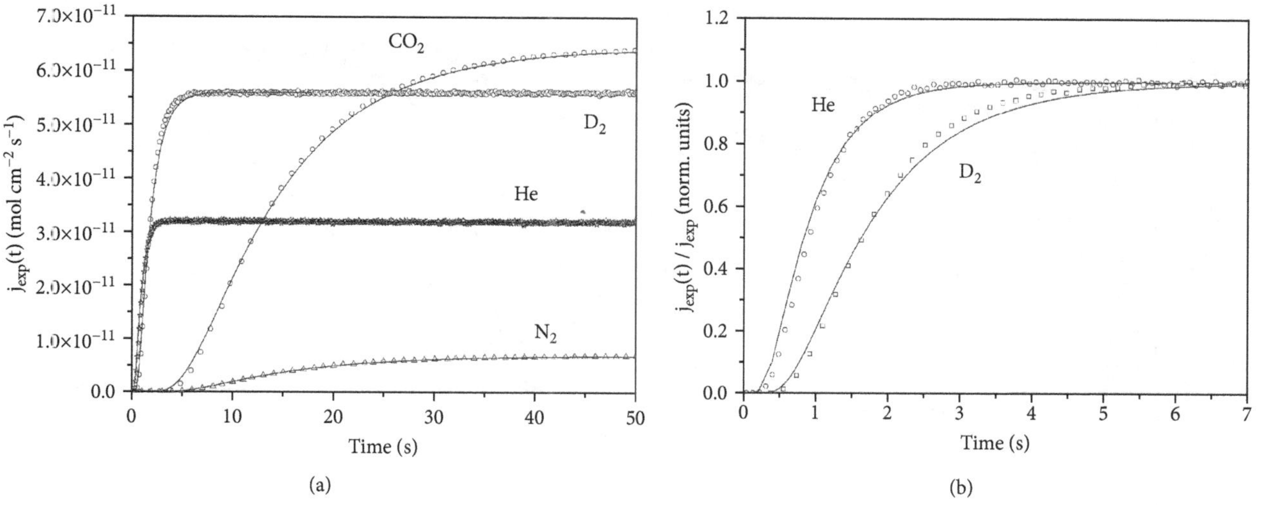

FIGURE 4: (a) Permeation flux curves $j_{exp}(t)$ (open symbols) pertinent to $CO_2$, $N_2$, $D_2$, and He obtained at $T = 295$ K and $P_{HPS} = 40$ kPa. Lines are fitting of the experimental curves; see text. (b) Normalized $j_{exp}(t)/J_{exp}$ curves (open symbols) pertinent to $D_2$ and He obtained at $T = 295$ K and $P_{HPS} = 40$ kPa. Lines are the $j_{exp}(t)/J_{exp}$ curve fittings by the $j_{norm}(t)$ function; see text.

continuous lines are their numerical fit. All permeation curves are characterized by an initial interval time where penetrant occurs in transient conditions; the permeation flux then reaches a constant value indicating that stationary transport conditions are set. Fits were obtained first measuring the value of the permeation flux in stationary transport condition relation $J_{exp} = (D\Pi/L)P_{HPS}$ to evaluate the gas permeability $\Phi = D\Pi$ value and then fitting the normalized $j_{exp}(t)/J_{exp}$ curve with the $j_{norm}(t)$ function to evaluate the gas diffusivity $D$.

Comparing the curves in Figure 4(a), we observe that transient transport conditions last for longer interval times with the $CO_2$ and $N_2$ penetrants than with $D_2$ and He which have smaller molecular size. The difference between the permeation curves in transient transport conditions pertinent to the small-size test molecules, namely, $D_2$ and He, can be appreciated by their plots in Figure 4(b) where the normalized $j_{exp}(t)/J_{exp}$ permeation flux curves are presented.

In Figures 5(a)–5(d), we present the experimental $j_{exp}(t)/J_{exp}$ permeation curves obtained at temperatures between 295 and 350 K for $CO_2$, $N_2$, $D_2$, and He with $P_{HPS} =$ 40 kPa (symbols): the continuous lines are their fitting by the $j_{norm}(t)$ function. Note that the high signal-to-noise ratio of the present procedure permits to clearly distinguish the permeation curves obtained in the 295 to 350 K temperature range also using the small-size penetrants.

The Arrhenius plots of the permeability and diffusivity values are reported in Figures 6(a) and 6(b), respectively. Permeability values are reported in practical *Barrer* units

$$1 \text{ Barrer} = 10^{-10} \frac{\text{cm}^3(\text{STP})\,\text{cm}}{\text{cm}^2\,\text{s}\,\text{cmHg}} = 3.34 \times 10^{-18} \frac{\text{mol cm}}{\text{cm}^2\,\text{s Pa}}. \tag{9}$$

The activation energy value for permeation and the apparent value of the activation energy for diffusion were obtained fitting the permeability and diffusivity data by the following relations for the gas diffusivity $D$ and permeability $\Phi$:

$$D(T) = D_0 e^{-(E_D/k_B T)}, \quad \Phi(T) = \Phi_0 e^{-(E_\Phi/k_B T)}. \tag{10}$$

$E_D$ and $E_\Phi$ values of the examined test gases are reported in Table 2.

It is worthy to observe that the diffusivity values have been obtained by the analysis of the whole permeation curve which is composed, given the high sampling rate of the QMS, by a large ensemble of experimental data. This procedure thus allows more accurate evaluation of the gas diffusivity than using the value of a single parameter, as in the time lag method.

The obtained diffusivity and permeability values for $CO_2$ as well as the pertinent activation energies are in well agreement with values previously obtained by the "*constant-volume*" approach [18] and with other literature data; see references therein. The same consideration holds for the permeability values and activation energy values for $H_2$ and $N_2$.

To conclude this discussion, it is worthy to comment the obtained results considering the physical-chemical properties of the employed test gases. Considering the diffusion data in Figure 6(b) and the penetrant properties in Table 1, we conclude that the penetrant diffusivity through the present LDPE samples decreases by increasing the molecular size. Moreover, comparing the permeability data in Figure 6(a) and the molecular physical-chemical properties in Table 1, we conclude that the gas permeability of the present LDPE samples increases with the critical temperature of the test gas evidencing the important role of the gas condensation properties. In fact, $CO_2$ and $N_2$ have similar molecular size (see Table 1) and diffusivity (see Figure 6(b)) but the $CO_2$ permeability is larger than that of $N_2$. Note that

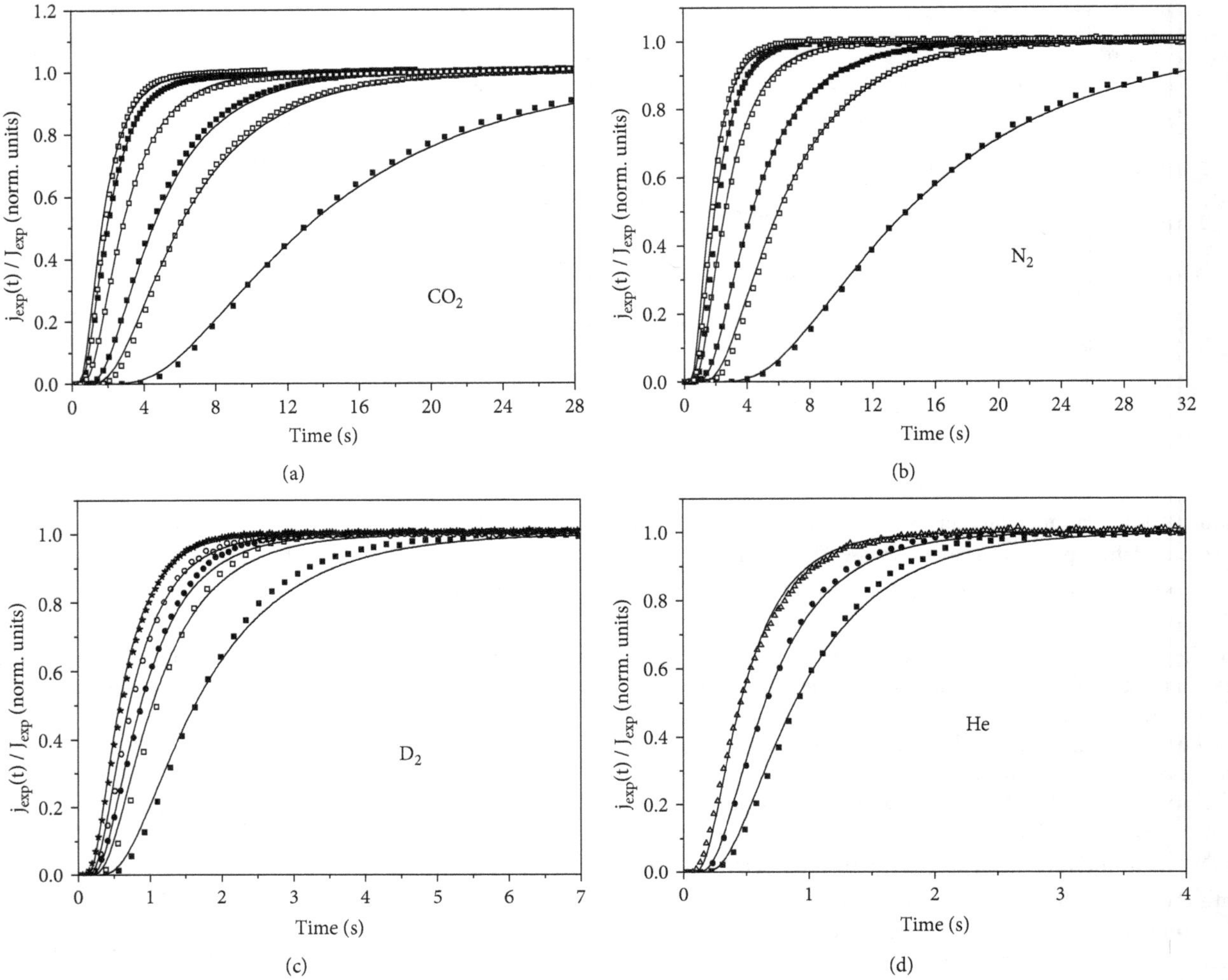


FIGURE 5: (a) Experimental $j_{exp}(t)/J_{exp}$ curves (symbols) pertinent to $CO_2$ obtained at temperatures between $T = 295$ K and $T = 351$ K with $P_{HPS} = 40$ kPa. Lines are fitting of the experimental curves; see text. (b) Experimental $j_{exp}(t)/J_{exp}$ curves (symbols) pertinent to $N_2$ obtained at temperatures between $T = 295$ K and $T = 351$ K with $P_{HPS} = 40$ kPa. Lines are fitting of the experimental curves; see text. (c) Experimental $j_{exp}(t)/J_{exp}$ curves (symbols) pertinent to $D_2$ obtained at temperatures between $T = 295$ K and $T = 345$ K with $P_{HPS} = 40$ kPa. Lines are fitting of the experimental curves; see text. (d) Experimental $j_{exp}(t)/J_{exp}$ curves (symbols) pertinent to He. To avoid confusion, in the figure, we only report curves obtained at 295, 330, and 351 K with $P_{HPS} = 40$ kPa. Lines are fitting of the experimental curves; see text.

the same considerations hold also comparing the $H_2$ and He transport parameters.

This scaling can be discussed considering the theoretical results presented in a recent paper of Dutta and Bhatia on the free volume structure and gas transport properties of polyethylene carried out by equilibrium molecular dynamic simulations [21]. The $CO_2$ and $N_2$ self-diffusivity values indicated by the authors are in the $10^{-6}$ cm$^2$/s range, in accord with the present experimental evaluation. They also suggested that the temperature dependence of the penetrant self-diffusivity follows the Arrhenius behavior, as also observed in the present study. Simulations also revealed that the free volume of PE is composed by voids with size ranging from 1.5 to 3 Å. The larger $D_2$ and He diffusivity compared with $CO_2$ and $N_2$ can be explained as consequence of the availability of diffusion sites with a diameter close to the $D_2$ and He molecular sizes and thus is more efficient in accommodating these small-size penetrants.

## 4. Conclusions

We have described a procedure for the experimental study of gas permeation through polymer films. The procedure is based on the measure of the permeation signal in a continuously pumped analysis chamber by a calibrated quadrupole mass spectrometer. This procedure allows (i) very low detection limit of the permeation flux, ~$10^{-14}$ mol·cm$^{-2}$·s$^{-1}$, and (ii) high signal-to-noise ratio in transient transport conditions also when studying polymer films having gas barrier properties. The improved performances permit to resolve the kinetics of the He and $D_2$ transport processes, to obtain their diffusivity values up to temperatures close to those of

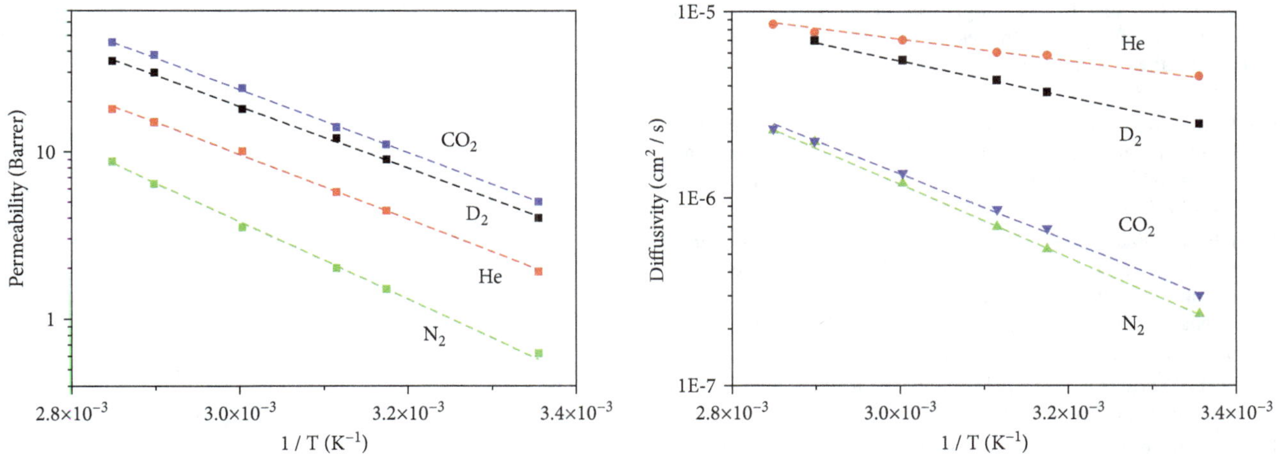


(a) (b)

FIGURE 6: (a) Arrhenius plot of the gas permeability as obtained by analysis of the permeation flux value in stationary transport conditions. Permeability values are presented in practical units (Barrer). (b) Arrhenius plot of the gas diffusivity as obtained by the fitting of $j_{exp}(t)/J_{exp}$ curves; see Figures 5(a)–5(d).

TABLE 2: Activation energy for permeation $E_\Phi$ and activation energy for diffusion $E_D$ as obtained by the Arrhenius plot of the permeability and diffusivity data in Figures 6(a) and 6(b).

| Test gas | $E_\Phi$ (kJ/mol) | $E_D$ (kJ/mol) |
|---|---|---|
| $CO_2$ | $36.5 \pm 0.5$ | $34.0 \pm 0.3$ |
| $N_2$ | $43.2 \pm 0.8$ | $38.2 \pm 0.5$ |
| $D_2$ | $33.2 \pm 0.8$ | $18.8 \pm 0.4$ |
| He | $37.2 \pm 0.7$ | $10.0 \pm 0.4$ |

the LDPE melting and the pertinent activation energies of diffusion. This procedure can be applied for the study of the gas transport processes with the recently developed polymer materials for packaging applications to be applied in fields such as OLED technology. It is also of interest for the community of polymer scientists interested to the study of the correlation between the polymer-free volume structure and transport properties.

## Conflicts of Interest

The author declares that there is no conflict of interests regarding the publication of this paper.

## Acknowledgments

The help of M. Zanolli in the gas transport experiments is gratefully acknowledged.

## References

[1] S. Matteucci, Y. Yampolskii, B. D. Freeman, and I. Pinneau, *Materials Science of Membranes for Gas and Vapor Separation*, John Wiley and Sons, Chichester, 2006.

[2] A. Lopez-Rubio, E. Almenar, P. Hernandez-Munoz, R. Catala, and R. Gavara, "Overview of active polymer-based packaging technologies for food applications," *Food Reviews International*, vol. 20, no. 4, pp. 357–387, 2004.

[3] J. S. Vrentas and J. M. Vrentas, "Evaluation of the free volume theory of diffusion," *Journal of Polymer Science Part B: Polymer Physics*, vol. 41, no. 5, pp. 501–507, 2003.

[4] H. Fujiita, "Notes on free volume theories," *Polymer Journal*, vol. 23, no. 12, pp. 1499–1506, 1991.

[5] R. P. White and J. E. G. Lipson, "Polymer free volume and its connection to the glass transition," *Macromolecules*, vol. 49, no. 11, pp. 3987–4007, 2016.

[6] P. N. Patil, D. Roilo, R. S. Brusa, A. Miotello, and R. Checchetto, "Influence of nano-level molecular packing on the gas transport properties in amine-modified epoxy resins," *Polymer*, vol. 58, pp. 130–138, 2015.

[7] P. N. Patil, D. Roilo, R. S. Brusa et al., "Free volumes and gas transport in polymers: amine-modified epoxy resins as a case study," *Physical Chemistry Chemical Physics*, vol. 18, no. 5, pp. 3817–3824, 2016.

[8] D. Roilo, P. N. Patil, R. S. Brusa et al., "Polymer rigidification in graphene based nanocomposites: gas barrier effects and free volume reduction," *Polymer*, vol. 121, pp. 17–25, 2017.

[9] D. Roilo, C. A. Maestri, M. Scarpa et al., "Cellulose nanofibrils films: molecular diffusion through elongated nanocavities," *Journal of Physical Chemistry C*, vol. 121, no. 28, pp. 15437–15447, 2017.

[10] B. Flaconnèche, J. Martin, and M. H. Klopffer, "Transport properties of gases in polymers: experimental methods," *Oil & Gas Science and Technology*, vol. 56, no. 3, pp. 245–259, 2001.

[11] R. M. Felder, "Estimation of gas transport coefficients from differential permeation, integral permeation, and sorption rate data," *Journal of Membrane Science*, vol. 3, no. 1, pp. 15–27, 1978.

[12] J. Crank, *The Mathematics of Diffusion*, Claredon, Oxford, 2nd edition, 1999.

[13] W. H. Lin and T. S. Chung, "Gas permeability, diffusivity, solubility, and aging characteristic of 6FDA-durene polyimide membranes," *Journal of Membrane Science*, vol. 186, no. 2, pp. 183–193, 2011.

[14] J. Ahn, W. J. Chung, I. Pinneau, and M. D. Guiver, "Polysulfone/silica nanoparticle mixed-matrix membranes for gas separation," *Journal of Membrane Science*, vol. 314, no. 1-2, pp. 123–133, 2008.

[15] J. P. G. Villaluenga and B. Seoane, "Experimental estimation of gas-transport properties of linear low-density polyethylene membranes by an integral permeation method," *Journal of Applied Polymer Science*, vol. 82, no. 12, pp. 3013–3021, 2001.

[16] W. J. Koros, A. H. Chan, and D. R. Paul, "Sorption and transport of various gases in polycarbonate," *Journal of Membrane Science*, vol. 2, pp. 165–190, 1977.

[17] K. Marsch and B. Bugusu, "Food packaging roles, materials and environmental issues," *Journal of Food Science*, vol. 72, no. 3, pp. R39–R55, 2007.

[18] R. Checchetto, A. Miotello, L. Nicolais, and G. Carotenuto, "Gas transport through nanocomposite membrane composed by polyethylene with dispersed graphite nanoplatelets," *Journal of Membrane Science*, vol. 463, pp. 196–204, 2014.

[19] P. A. Redhead, J. P. Hobson, and E. V. Kornelsen, *The Physical Basis of Ultrahigh Vacuum*, American Institute of Physics, New York, NY, USA, 1993.

[20] J. S. Lewis and M. S. Weaver, "Thin-film permeation-barrier technology for flexible organic light-emitting devices," *IEEE Journal of Selected Topics in Quantum Electronics*, vol. 10, no. 1, pp. 45–57, 2004.

[21] R. C. Dutta and S. K. Bhatia, "Transport diffusion of light gases in polyethylene using atomistic simulations," *Langmuir*, vol. 33, no. 4, pp. 936–946, 2017.

# 5

# Microscale Patterning of Electrochromic Polymer Films via Soft Lithography

**Jae-Yeong Jung,[1] Seokwon Joo,[2] Da-Seul Kim,[1] Kyoung-Hwan Kim,[1] Tae Soup Shim,[1,3] Soonmin Seo,[2] and Ju-Hyung Kim[1,3]**

[1]*Department of Energy Systems Research, Ajou University, Suwon 16499, Republic of Korea*
[2]*College of BioNano Technology, Gachon University, Gyeonggi 13120, Republic of Korea*
[3]*Department of Chemical Engineering, Ajou University, Suwon 16499, Republic of Korea*

Correspondence should be addressed to Tae Soup Shim; tsshim@ajou.ac.kr, Soonmin Seo; soonmseo@gachon.ac.kr, and Ju-Hyung Kim; juhyungkim@ajou.ac.kr

Academic Editor: Miriam H. Rafailovich

We present a direct fabrication technique of patterned polymeric electrochromic (EC) devices via soft lithography, enabling both negative patterning and positive patterning of the polymer. For this work, elastomeric polydimethylsiloxane (PDMS) molds were employed as not only stamps for direct contact printing of polymer inks but also templates for dewetting of polymer solutions under mild experimental conditions. We performed both negative patterning and positive patterning of a prototypical EC polymer and investigated the EC device characteristics according to solvents, solution concentrations, and pattern types. Eventually, the complex patterns, which cannot be realized by conventional shadow masking processes, and large-area structures were successfully demonstrated. We anticipate that these results will be applied to the development of various patterned devices and circuits, which may lead to further applications.

## 1. Introduction

In the past few decades, electrochromic (EC) materials and devices have attracted a lot of interest for the prospect of potential applications such as optical displays, optical switching devices, antiglare mirrors, and smart windows [1–5]. The EC materials reversibly change their optical properties (i.e., light transmittance, absorbance, and reflectance) in response to external bias voltages, resulting from electrochemical oxidation and reduction. Various classes of materials including transition metal oxides and conducting polymers have been widely studied as promising EC materials thus far [6–9]. In particular, considerable research efforts have been devoted to the EC behaviors of conducting polymers owing to their outstanding advantages of low cost, mechanical flexibility, and solution processability in comparison with inorganic materials [10–15].

For practical applications of EC devices, patterning processes of the EC layers in thin-film form are necessarily required in accordance with their intended uses [10–16]. However, conventional patterning processes such as photolithography and shadow mask deposition are mostly inappropriate for conducting polymers because the polymers are significantly damaged or even broken down under harsh processing conditions involved in etching or physical vapor deposition [17]. Solution processability of conducting polymers rather facilitates various soft lithographic patterning methods using elastomeric polydimethylsiloxane (PDMS) molds in a wide range of scales [17–20]. The PDMS molds can be employed as not only stamps for direct contact printing of polymer inks but also templates for dewetting of polymer solutions under mild experimental conditions, owing to the low modulus and surface energy of PDMS leading to conformal contact and easy release. Even though those methods enable both negative patterning and positive patterning of the polymers without any harmful damages, only a few attempts have been made to apply them to fabricating the patterned EC devices thus far.

Here, we demonstrate the fabrication of patterned polymeric EC devices via soft lithographic methods. For this work, poly(3-hexylthiophene-2,5-diyl) (P3HT) was used as a prototypical EC material in consideration of excellent EC properties (i.e., high coloration efficiency, high optical contrast, and rapid switching time), compatibility with solution processes, and functional versatility [13, 21–25]. We performed both negative patterning and positive patterning of the polymer using the PDMS molds consisting of pillar arrays and investigated the EC device characteristics according to solvents, solution concentrations, and pattern types. In addition, more complex patterns (such as island and ring arrays), which cannot be realized by conventional shadow masking processes, and large-area structures were successfully achieved. These results suggest great potential for the further development of various patterned EC devices and circuits, including display pixels, optical switching devices, and smart window applications.

## 2. Materials and Methods

*2.1. Preparation of PDMS Molds.* The PDMS molds used in this work were replicated from initially patterned photoresist layers. Spin-coated photoresist (SU-8 2025, MicroChem) layers (of 12 $\mu$m thickness) were engraved with the hole patterns in accordance with the previous reports [26, 27]. The diameters of the hole motifs were designed to be 50 and 100 $\mu$m. The patterned photoresist layers were then negatively replicated using a PDMS elastomer (Sylgard 184, Dow Corning). A mixture of PDMS prepolymer and curing agent (in a 10 : 1 weight ratio) was poured onto each patterned photoresist layer and subsequently degassed and cured in a vacuum oven at 70°C for 3 hr. After thermal curing, the PDMS replicas were easily peeled off from the photoresist layers due to the low surface energy of PDMS, and consequently, the PDMS molds with the pillar arrays (of 50 and 100 $\mu$m diameters) were prepared [28].

*2.2. Fabrication of Patterned Polymeric EC Devices.* Patterned polymeric EC devices using P3HT (Sigma-Aldrich) as an active component were fabricated on glass substrates coated with indium-tin-oxide (ITO) via both negative and positive patterning processes of the polymer. The ITO-coated glass substrates were cleaned before use by successive sonication in deionized water, ethyl alcohol, acetone, and isopropyl alcohol for 10 min each. P3HT was individually dissolved in chlorobenzene and chloroform with varied concentrations (i.e., 0.10, 0.25, 0.50, 1.00, 2.00, and 3.00 wt%) for solution processes, and all solutions were filtered through polyvinylidene fluoride (PVDF) membranes with 0.45 $\mu$m pore size.

To prepare the negatively patterned P3HT layers (i.e., negative hole arrays of the PDMS pillars) on the ITO surfaces, the P3HT solutions were drop-dispensed onto the ITO-coated glass substrates, and then the PDMS molds were brought into contact with the samples as schematically illustrated in Figure 1. Since structural deformation of the PDMS molds can be induced by solvent swelling [29], an external pressure of 0.1 bar was uniformly applied during the process for ensuring conformal contact between PDMS and ITO. The PDMS structure with low surface energy leads to dewetting of the polymer solutions, accompanying solvent evaporation under ambient conditions, and thus, the negatively patterned P3HT layers are simultaneously formed on the ITO surfaces. In the case of the positive patterns of P3HT (i.e., same patterns as the PDMS pillar arrays), the PDMS molds were used as stamps for direct contact printing of the polymer, as shown in Figure 1. The P3HT solutions (100 $\mu$L each) were preferentially spin-coated at 1000 rpm for 30 s onto the PDMS molds to form the patterned layers, and the solvents were subsequently removed in the vacuum oven at 80°C for 20 min to avoid solvent swelling effects [29]. For enhancing the adhesion properties between P3HT and ITO, both P3HT and ITO surfaces were treated using oxygen plasma for 1 min with a power of 10 W at a constant oxygen flow rate of 30 standard cubic centimeters per minute (sccm) [17], and then the patterned P3HT layers were directly transferred onto the ITO-coated glass substrates by applying an external pressure of 0.02 bar for 10 s. Note that the patterned P3HT layers were also observed using scanning electron microscopy (SEM) (COXEM, EM-30AX) at an accelerating voltage of 20 kV with a working distance of ~8 mm. For the SEM measurements, the samples were coated with gold (of ~10 nm thickness) using a sputter coater.

To complete the EC device structure, each prepared sample was covered with another ITO-coated glass, and a gap distance between the sample and the cover was fixed using adhesive tapes (of 270 $\mu$m thickness) as shown in Figure 1(b). The gap space was then filled with an electrolyte solution (0.1 M of $LiClO_4$ in propylene carbonate) by capillary action. The device characteristics were measured using a Keithley 2636 source/measure unit.

## 3. Results and Discussion

Figure 2(a) shows the negatively patterned P3HT layer using the PDMS mold with the pillar array of 100 $\mu$m diameter, which was successfully achieved on the ITO-coated glass substrate with chloroform as a solvent. Since chloroform is rapidly evaporated and easily permeates into PDMS, the complete negative pattern (i.e., open-hole array) was distinctly formed at a high concentration of 3 wt%. On the other hand, since the surface energy of P3HT (~36.0 mJ/m$^2$) is significantly higher than that of the PDMS mold [17, 30–32], the P3HT-enriched solutions induced by solvent evaporation enhance the dewetting behaviors. Such dewetting behaviors of the P3HT-enriched solutions are also confirmed by the contact angle measurements. The contact angle of the P3HT solution on the PDMS surface gradually increases with increasing the solution concentration (see Supplementary Figure 1). Thus, accompanying with solvent evaporation, the less-viscous solutions with longer evaporation times (i.e., diluted solutions) are expelled from the molds before completing the formation of the patterns. These phenomena are intensified with diluting the solution, and accordingly,

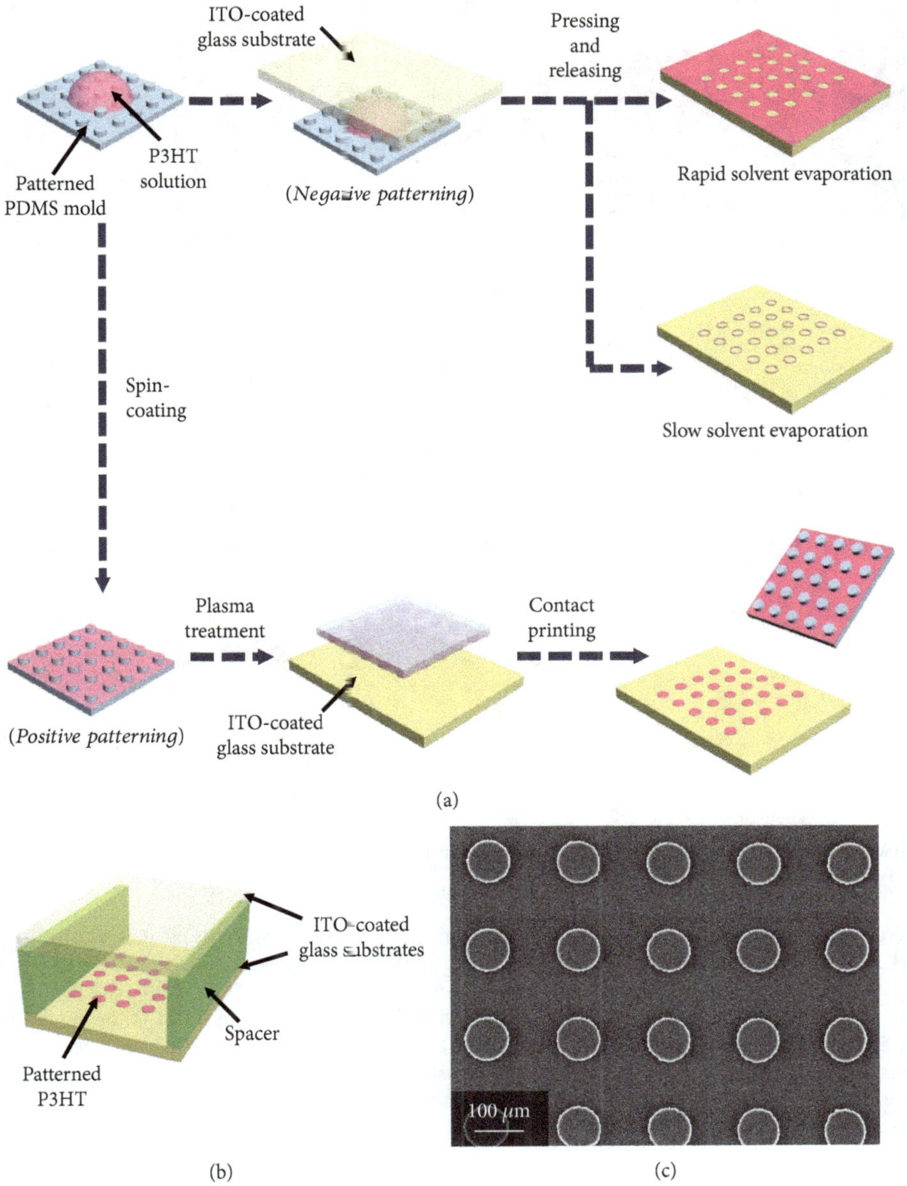


FIGURE 1: (a) Schematic illustration of the fabrication of patterned P3HT layers via both negative and positive patterning processes. (b) Configuration of EC display device. The gap space is filled with an electrolyte solution by capillary action. (c) SEM image of the PDMS mold with the pillar array of 100 $\mu$m diameter.

smaller amounts of P3HT are aggregated along the corners of the pillars due to the surface tension at the edges (see Figures 2 and 3) [33]. Note that the solution with a high concentration of 3 wt% using chloroform as a solvent was not seriously dewetted by solvent evaporation, which may originate from high viscosity and rapid evaporation of the solvent. These results imply that well-defined patterning of the polymer corresponding to the sharp outlines of the patterns can be performed if the residual layers of P3HT are sufficiently removed. In this context, the formation of the residual layers is progressively suppressed with increasing the solvent evaporation time. For instance, in the case of chloroform, the solutions at low concentrations (less than 0.25 wt%) rarely formed the residual layers as shown in Figure 2(e). This tendency is more clearly observed using a less-volatile solvent such as chlorobenzene. Chlorobenzene exhibits lower volatility and less permeates into PDMS in comparison with chloroform [34], which can provide enough time to expel the solutions without remaining residues even at high concentrations. Figure 2 also shows the negatively patterned P3HT layers which were prepared with chlorobenzene as a solvent following the same procedure as for chloroform. Almost no residual layers were observed as the concentration increased up to 1 wt%, and the well-defined patterns were distinctly formed along the outlines of the pillars (i.e., ring array). The fabrication of the complete negative pattern (i.e., open-hole array) was not achievable using chlorobenzene, and only the residual layers slightly

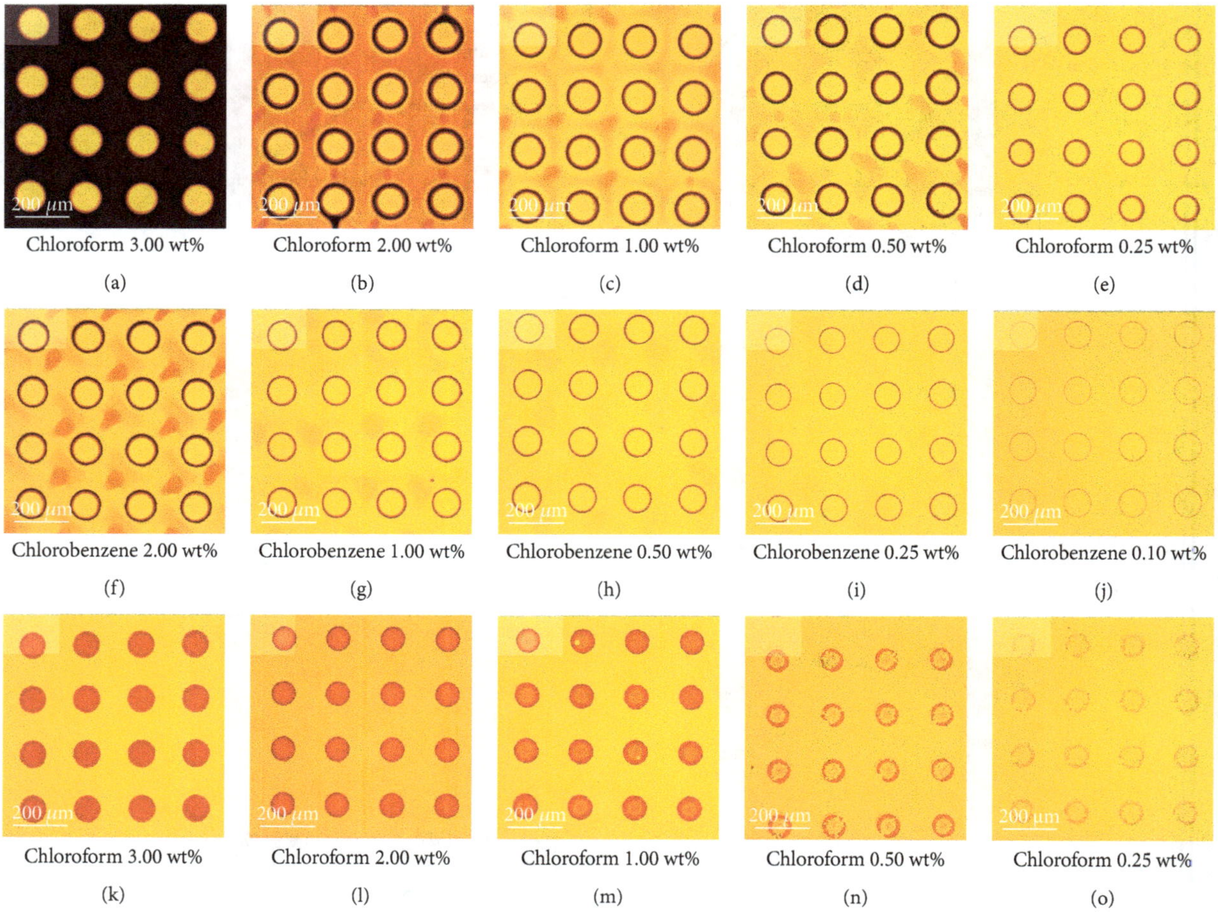


FIGURE 2: Optical microscopy images of the patterned P3HT layers using the PDMS mold with the pillar array of 100 $\mu$m diameter. (a–e) The negatively patterned P3HT layers using chloroform as a solvent. The PDMS molds were used as templates for dewetting of the polymer solutions, accompanying with solvent evaporation. The solution concentrations were (a) 3.00, (b) 2.00, (c) 1.00, (d) 0.50, and (e) 0.25 wt%, respectively. (f–j) The negatively patterned P3HT layers, prepared with chlorobenzene as a solvent following the same procedure as for chloroform. The solution concentrations were (f) 2.00, (g) 1.00, (h) 0.50, (i) 0.25, and (j) 0.10 wt%, respectively. (k–o) The positively patterned P3HT layers via direct contact printing. The chloroform solvent was solely used for spin-coating deposition of the polymer on the PDMS surface due to serious dewetting issues. The solution concentrations were (k) 3.00, (l) 2.00, (m) 1.00, (n) 0.50, and (o) 0.25 wt%, respectively.

remained even at higher concentrations above 2 wt% as shown in Figure 2(f). It is worth noting that the overall tendency for the polymer patterning (i.e., surface tension and dewetting effects) was not significantly changed according to the size of the pillars in a micrometer scale (see Supplementary Figure 2). In particular, the contact angles of water droplets on both flat and patterned PDMS surfaces were measured to be within the range of 100° to 110° without any distinct difference, of which corresponding surface energies (from 16.8 to 22.7 mJ/m$^2$) are sufficiently lower than that of P3HT [17]. Figure 3 also shows an exemplary P3HT pattern fabricated using the PDMS mold with a smaller pillar array (of ~1 $\mu$m thickness). In addition, such negative and closed-outline patterns (i.e., hole and ring arrays) cannot be achieved by typical shadow masking methods, and thus, our demonstration shows promise as an easy process for realizing various complex patterns and their related applications.

The positively patterned P3HT layers were successfully fabricated on the ITO-coated glass substrates via direct contact printing using the same PDMS molds as for the negative patterning (see Figure 2 and Supplementary Figure 2). For this work, the P3HT solutions with varied concentrations were spin-coated on the PDMS molds to form the patterned layers and subsequently dried in a vacuum oven for avoiding solvent swelling effects [29]. Note that chlorobenzene was found to be not an appropriate solvent for spin-coating deposition of the polymer on the PDMS surface due to dewetting issues, and the complete deposition of P3HT could be achieved only using chloroform at an appropriate spin speed (~1000 rpm). After oxygen plasma treatments for enhancing the adhesion properties [17], the deposited P3HT layers were brought into contact with the ITO-coated glass substrates and pressed (at 0.02 bar) to ensure the conformal contact. The patterned P3HT layers in the contact regions (i.e., filled circle array) were sharply transferred onto the substrates. Note that the edges of the transferred patterns were thicker than the other parts, because larger amounts of the polymer are piled up near

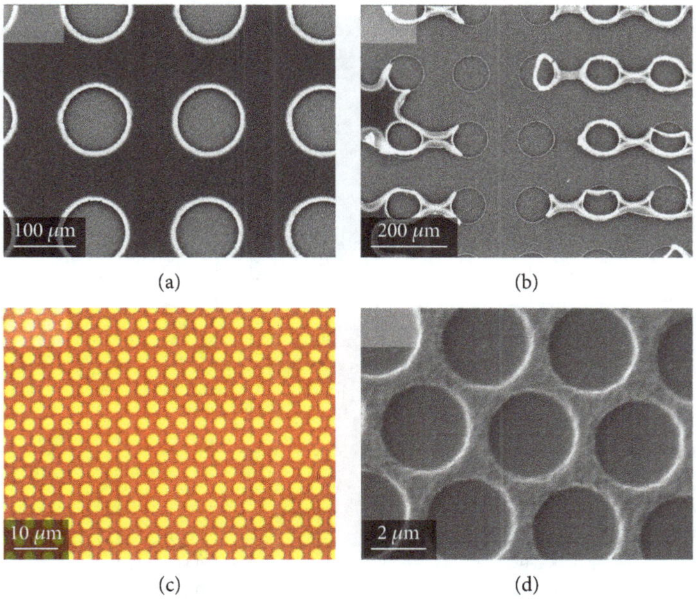


FIGURE 3: (a) SEM image of the complete negative pattern (i.e., open-hole array) of P3HT, distinctly formed at a high concentration of 3.00 wt% using chloroform as a solvent. (b) SEM image of the negatively patterned P3HT layer formed at a concentration of 2.00 wt% using chloroform as a solvent. The solution was expelled from the mold before completing the formation of the pattern, and accordingly, a small amount of P3HT was aggregated along the corners of the pillars in comparison with (a). (c) Optical microscopy image of an exemplary P3HT pattern fabricated using the PDMS mold with a smaller pillar array (of ~1 μm thickness). A significant change in the overall tendency for the polymer patterning was not observed according to the size of the pillars. (d) SEM image of (c).

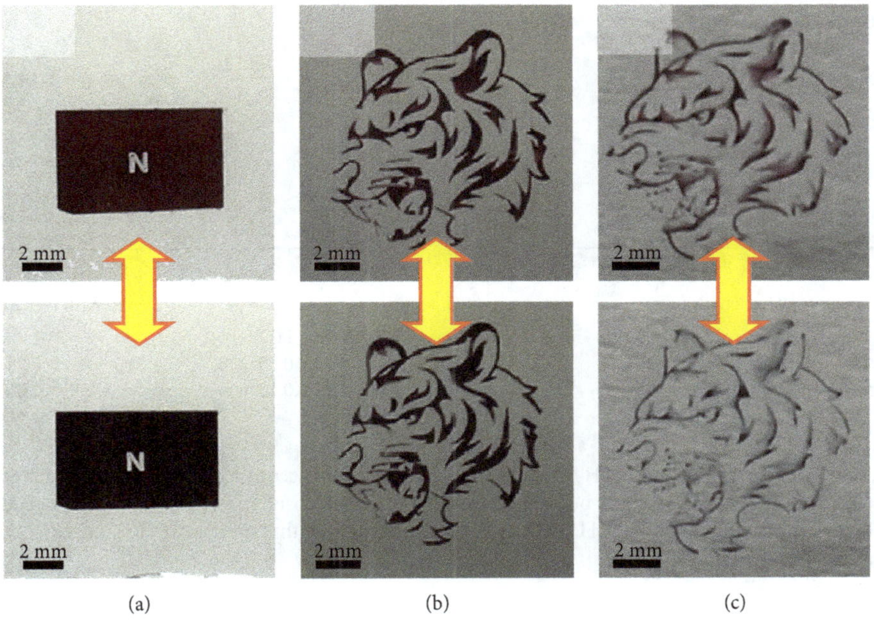


FIGURE 4: Photographs of the EC display devices with (a) negatively patterned P3HT layer at a concentration of 3.00 wt%, (b) positively patterned P3HT layer at a concentration of 3.00 wt%, and (c) positively patterned P3HT layers at a concentration of 1.00 wt%. All devices were fabricated using chloroform as a solvent. Changes in colors of the P3HT layers were clearly observed according to the reversible oxidation (upper: red) and reduction (lower: blue) reactions in the electrolyte solution.

the edges during the spin-coating processes owing to the surface tension [35, 36]. Such thickness deviation is mitigated as the overall thickness increases with increasing the solution concentration (see Figure 2(k)).

Figure 4 shows demonstrations of the polymeric EC display devices with large-area patterns (~1 $cm^2$), which were prepared by both negative and positive patterning processes. Changes in the color of the polymer were clearly

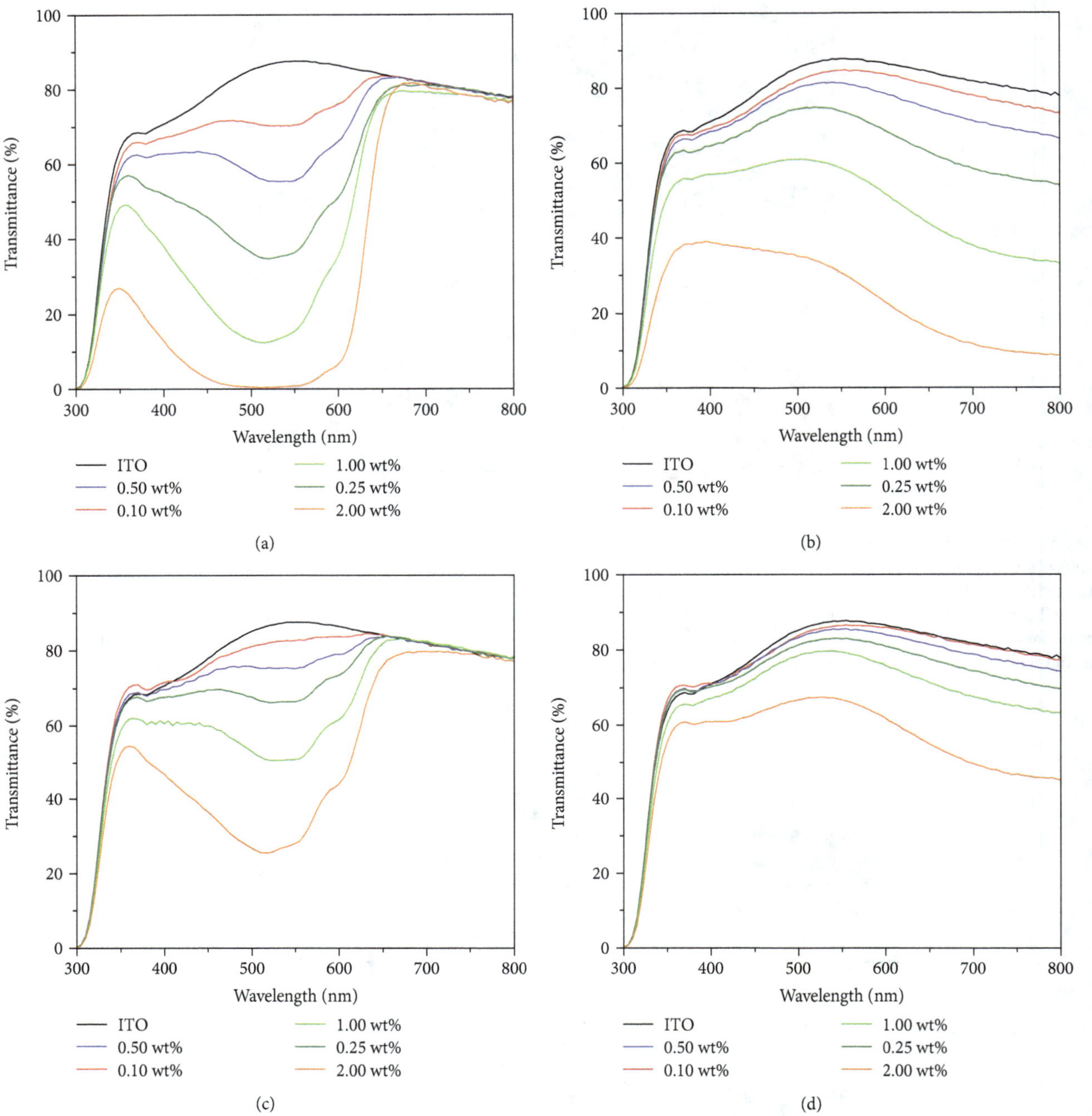


FIGURE 5: Light transmittances of the spin-coated P3HT films on the ITO substrates according to the solution concentrations, in the range of 300 nm to 800 nm. The spin-coating processes were performed using (a, b) chloroform and (c, d) chlorobenzene as solvents, respectively. The spectra in (a) and (c) correspond to the P3HT oxidation (red) states, and the spectra in (b) and (d) correspond to the P3HT reduction (blue) states.

observed according to the reversible oxidation (red) and reduction (blue) reactions in the electrolyte solution (see Supplementary Movie 1). For the EC operations, the sample bias voltage between the two electrodes was kept at −3.0 V for reduction and +3.0 V for oxidation. As indicated in Figure 5, light transmittance of the P3HT film, leading to color contrast, significantly varies with the film thickness rather than the type of solvents [37]. The film thickness can be easily modified by the solution concentration, and the rapid solvent evaporation of chloroform enables the formation of the thicker films in comparison with chlorobenzene. For the 0.10, 0.25, 0.50, 1.00, and 2.00 wt% solutions using the chloroform (chlorobenzene) solvent, the film thicknesses were measured to be <10 nm (<10 nm), 13 nm (<10 nm), 30 nm (14 nm), 100 nm (25 nm), and 320 nm (50 nm), after spin-coating deposition onto the substrates. The EC response time of P3HT was measured to be less than a second.

## 4. Conclusions

In conclusion, we demonstrate the fabrication of patterned polymeric EC devices via soft lithographic methods, enabling both negative patterning and positive patterning of the polymer. The PDMS molds with pillar arrays were employed for the patterning of P3HT as the prototypical EC polymer with the excellent EC properties (i.e., high coloration efficiency, high optical contrast, and rapid switching time). For the negative patterning, the PDMS molds were used as templates for dewetting of the polymer solutions, accompanying with solvent evaporation. The complete negative (i.e., hole arrays) and the distinct closed-outline (i.e., ring arrays) patterns were selectively achieved without significant residual defects according to the volatility and permeability of the solvent. For the positive patterning, the direct contact printing method using the PDMS molds as the stamps was successfully performed via suppressing the solvent swelling effects and enhancing the adhesion properties. Eventually, the large-area structures were successfully fabricated, and the EC device characteristics were also investigated according to solvents, solution concentrations, and pattern types. In addition, the complex patterns (such as island and ring arrays) presented in this work cannot be realized by typical shadow masking methods. Thus, it is anticipated that these results will significantly contribute not only to the fabrication of various polymer patterns but also to the development of more diversified EC devices and circuits leading to a wide range of applications.

## Conflicts of Interest

There are no conflicts to declare.

## Acknowledgments

This work was supported by the Ajou University research fund.

## Supplementary Materials

*Supplementary 1.* Supplementary Figure 1: contact angles of the P3HT solutions with varied concentrations (i.e., 0.00, 1.00, 2.00, and 3.00 wt%) on the PDMS surface using chloroform and chlorobenzene as solvents. Supplementary Figure 2: optical microscopy images of the patterned P3HT layers using the PDMS mold with the pillar array of the 50 $\mu$m diameter.

*Supplementary 2.* Supplementary Movie 1: demonstration of the polymeric EC display device with a large-area pattern (~1 cm$^2$).

## References

[1] C. M. Lampert, "Electrochromic materials and devices for energy efficient windows," *Solar Energy Materials*, vol. 11, no. 1-2, pp. 1–27, 1984.

[2] R. D. Rauh, "Electrochromic windows: an overview," *Electrochimica Acta*, vol. 44, no. 18, pp. 3165–3176, 1999.

[3] E. S. Lee and D. L. DiBartolomeo, "Application issues for large-area electrochromic windows in commercial buildings," *Solar Energy Materials & Solar Cells*, vol. 71, no. 4, pp. 465–491, 2002.

[4] R. J. Mortimer, A. L. Dyer, and J. R. Reynolds, "Electrochromic organic and polymeric materials for display applications," *Displays*, vol. 27, no. 1, pp. 2–18, 2006.

[5] P. Andersson, R. Forchheimer, P. Tehrani, and M. Berggren, "Printable all-organic electrochromic active-matrix displays," *Advanced Functional Materials*, vol. 17, no. 16, pp. 3074–3082, 2007.

[6] R. J. Mortimer, "Electrochromic materials," *Chemical Society Reviews*, vol. 26, no. 3, pp. 147–156, 1997.

[7] P. R. Somani and S. Radhakrishnan, "Electrochromic materials and devices: present and future," *Materials Chemistry and Physics*, vol. 77, no. 1, pp. 117–133, 2003.

[8] S. K. Deb, "Opportunities and challenges in science and technology of WO3 for electrochromic and related applications," *Solar Energy Materials & Solar Cells*, vol. 92, no. 2, pp. 245–258, 2008.

[9] P. M. Beaujuge and J. R. Reynolds, "Color control in $\pi$-conjugated organic polymers for use in electrochromic devices," *Chemical Reviews*, vol. 110, no. 1, pp. 268–320, 2010.

[10] D. Hohnholz, H. Okuzaki, and A. G. MacDiarmid, "Plastic electronic devices through line patterning of conducting polymers," *Advanced Functional Materials*, vol. 15, no. 1, pp. 51–56, 2005.

[11] G. H. Shim, M. G. Han, J. C. Sharp-Norton, S. E. Creager, and S. H. Foulger, "Inkjet-printed electrochromic devices utilizing polyaniline-silica and poly (3,4-ethylenedioxythiophene)-silica colloidal composite particles," *Journal of Materials Chemistry*, vol. 18, no. 5, pp. 594–601, 2008.

[12] J. Kim, Y. Kim, and E. Kim, "Electrochromic pattern formation by photo cross-linking reaction of PEDOT side chains," *Macromolecular Research*, vol. 17, no. 10, pp. 791–796, 2009.

[13] Y. Kim, Y. Kim, S. Kim, and E. Kim, "Electrochromic diffraction from nanopatterned poly(3-hexylthiophene)," *ACS Nano*, vol. 4, no. 9, pp. 5277–5284, 2010.

[14] J. Kim, J. You, B. Kim, T. Park, and E. Kim, "Solution processable and patternable poly(3,4-alkylenedioxythiophene)s for large-area electrochromic films," *Advanced Materials*, vol. 23, no. 36, pp. 4168–4173, 2011.

[15] H. Shin, S. Seo, C. Park, J. Na, M. Han, and E. Kim, "Energy saving electrochromic windows from bistable low-HOMO level conjugated polymers," *Energy & Environmental Science*, vol. 9, no. 1, pp. 117–122, 2016.

[16] S. Joo, J.-H. Kim, and S. Seo, "Direct fabrication of electrochromic devices with complex patterns on three-dimensional substrates using polymeric stencil films," *RSC Advances*, vol. 7, no. 68, pp. 43283–43288, 2017.

[17] H. Kim, B. Yoon, J. Sung, D.-G. Choi, and C. Park, "Micropatterning of thin P3HT films via plasma enhanced polymer transfer printing," *Journal of Materials Chemistry*, vol. 18, no. 29, pp. 3489–3495, 2008.

[18] H. Schmid and B. Michel, "Siloxane polymers for high-resolution, high-accuracy soft lithography," *Macromolecules*, vol. 33, no. 8, pp. 3042–3049, 2000.

[19] J.-H. Kim, H.-S. Hwang, S.-W. Hahm, and D.-Y. Khang, "Hydrophobically recovered and contact printed siloxane oligomers for general-purpose surface patterning," *Langmuir*, vol. 26, no. 15, pp. 13015–13019, 2010.

[20] L. Xue and Y. Han, "Pattern formation by dewetting of polymer thin film," *Progress in Polymer Science*, vol. 36, no. 2, pp. 269–293, 2011.

[21] A. Kumar, D. M. Welsh, M. C. Morvant, F. Piroux, K. A. Abboud, and J. R. Reynolds, "Conducting poly (3,4-alkylenedioxythiophene) derivatives as fast electrochromics with high-contrast ratios," *Chemistry of Materials*, vol. 10, no. 3, pp. 896–902, 1998.

[22] S. A. Sapp, G. A. Sotzing, and J. R. Reynolds, "High contrast ratio and fast-switching dual polymer electrochromic devices," *Chemistry of Materials*, vol. 10, no. 8, pp. 2101–2108, 1998.

[23] B. D. Reeves, C. R. G. Grenier, A. A. Argun, A. Cirpan, T. D. McCarley, and J. R. Reynolds, "Spray coatable electrochromic dioxythiophene polymers with high coloration efficiencies," *Macromolecules*, vol. 37, no. 20, pp. 7559–7569, 2004.

[24] S. W. Huang and K. C. Ho, "An all-thiophene electrochromic device fabricated with poly (3-methylthiophene) and poly (3,4-ethylenedioxythiophene)," *Solar Energy Materials & Solar Cells*, vol. 90, no. 4, pp. 491–505, 2006.

[25] Y. Pang, X. Li, H. Ding, G. Shi, and L. Jin, "Electropolymerization of high quality electrochromic poly(3-alkyl-thiophene)s via a room temperature ionic liquid," *Electrochimica Acta*, vol. 52, no. 20, pp. 6172–6177, 2007.

[26] J.-H. Kim, M. J. Han, and S. Seo, "Flexible, stretchable, and patchable organic devices integrated on freestanding polymeric substrates," *Journal of Polymer Science, Part B: Polymer Physics*, vol. 53, no. 6, pp. 453–460, 2015.

[27] K. G. Sharp, G. S. Blackman, N. J. Glassmaker, A. Jagota, and C.-Y. Hui, "Effect of stamp deformation on the quality of microcontact printing: theory and experiment," *Langmuir*, vol. 20, no. 15, pp. 6430–6438, 2004.

[28] S.-J. Choi, H. N. Kim, W. G. Bae, and K.-Y. Suh, "Modulus- and surface energy-tunable ultraviolet-curable polyurethane acrylate: properties and applications," *Journal of Materials Chemistry*, vol. 21, no. 38, pp. 14325–14335, 2011.

[29] J. N. Lee, C. Park, and G. M. Whitesides, "Solvent compatibility of poly (dimethylsiloxane)-based microfluidic devices," *Analytical Chemistry*, vol. 75, no. 23, pp. 6544–6554, 2003.

[30] Z. Yoshimitsu, A. Nakajima, T. Watanabe, and K. Hashimoto, "Effects of surface structure on the hydrophobicity and sliding behavior of water droplets," *Langmuir*, vol. 18, no. 15, pp. 5818–5822, 2002.

[31] C. Lv, C. Yang, P. Hao, F. He, and Q. Zheng, "Sliding of water droplets on microstructured hydrophobic surfaces," *Langmuir*, vol. 26, no. 11, pp. 8704–8708, 2010.

[32] Y. Yoon, D.-W. Lee, and J.-B. Lee, "Surface modified nano-patterned SU-8 pillar array optically transparent super-hydrophobic thin film," *Journal of Micromechanics and Microengineering*, vol. 22, no. 3, article 035012, 2012.

[33] C. Luo, F. Meng, and A. Francis, "Fabrication and application of silicon-reinforced PDMS masters," *Microelectronics Journal*, vol. 37, no. 10, pp. 1036–1046, 2006.

[34] S. Banerjee, P. H. Howard, and S. S. Lande, "General structure-vapor pressure relationships for organics," *Chemosphere*, vol. 21, no. 10-11, pp. 1173–1180, 1990.

[35] G. Carcano, M. Ceriani, and F. Soglio, "Spin coating with high viscosity photoresist on square substrates - applications in the thin film hybrid microwave integrated circuit field," *Micro-Electronics International*, vol. 10, no. 3, pp. 12–20, 1993.

[36] H. Lee, K. Lee, B. Ahn, J. Xu, L. Xu, and K. W. Oh, "A new fabrication process for uniform SU-8 thick photoresist structures by simultaneously removing edge bead and air bubbles," *Journal of Micromechanics and Microengineering*, vol. 21, no. 12, article 125006, 2011.

[37] D. E. Motaung, G. F. Malgas, and C. J. Arendse, "Comparative study: the effects of solvent on the morphology, optical and structural features of regioregular poly(3-hexylthiophene):fullerene thin films," *Synthetic Metals*, vol. 160, no. 9-10, pp. 876–882, 2010.

# 6

# Exploiting the Interplay between Liquid-Liquid Demixing and Crystallization of the PVDF Membrane for Membrane Distillation

**N. I. M. Nawi,[1] M. R. Bilad,[1,2] N. A. H. M. Nordin,[1] M. O. Mavukkandy,[3] Z. A. Putra,[1] M. D. H. Wirzal,[1] J. Jaafar,[4] and Asim L. Khan[5]**

[1]*Chemical Engineering Department, Universiti Teknologi PETRONAS, Bandar Seri Iskandar, 32610 Perak, Malaysia*
[2]*Jurusan Pendidikan Kimia, Institut Keguruan Ilmu Pendidikan, Jalan Pemuda No 59A, Mataram, Indonesia*
[3]*Institute Center for Water and Environment (iWater), Department of Chemical and Environmental Engineering, Masdar Institute, Khalifa University of Science and Technology, PO Box 54224, Abu Dhabi, UAE*
[4]*Advanced Membrane Technology Research Centre (AMTEC), Faculty of Chemical and Natural Resources Engineering, Universiti Teknologi Malaysia, 81310 Skudai, Johor, Malaysia*
[5]*Department of Chemical Engineering, COMSATS Institute of Information Technology, Lahore, Pakistan*

Correspondence should be addressed to M. R. Bilad; mroil.bilad@utp.edu.my

Academic Editor: Hossein Roghani-Mamaqani

Membrane distillation (MD) purifies water by transporting its vapor through a hydrophobic membrane. An ideal MD membrane poses high water flux and high fouling, scaling, and wetting resistances. In this study, we develop polyvinylidene fluoride (PVDF) membranes for MD by focusing on reduction of PVDF degree of crystallinity. We explore the roles of dope solution temperature in dictating the phase separation mechanisms as well as the structure and the performance of semicrystalline PVDF membranes. DSC spectra show that higher dope solution temperature depresses crystallinity via formation of imperfect crystal. Such findings were also supported by FTIR and XRD results. The SEM images reveal formation of spherulite-like morphology in the membrane matrices for membranes prepared from high temperature dope solutions. A good balance between solid-liquid and liquid-liquid phase separations that offers low degree of crystallinity was found at a dope solution temperature of 60°C (PVDF-60), which showed the MD flux of 18 $l/m^2$ h (vs. 6 $l/m^2$ h for temperature of 25°C, as a benchmark) and nearly complete salt rejection when run at hot and cold temperatures of 65°C and 25°C, respectively. The PVDF-60 shows a high wetting resistance and stable MD flux of 10.5 $l/m^2$ h over a 50 h test for treating brine solution as the feed (70 g NaCl/l).

## 1. Introduction

Membrane distillation (MD) is an emerging technology that, under correct circumstances, can play important roles in solving the water scarcity issue. MD exploits the vapor pressure difference between the hot feed and the cold permeate to drive water vapor transport across a hydrophobic membrane. The membrane is selective to water vapors, but thanks to its hydrophobic property, it repels liquid water and other non-volatile solutes including salts. The passing water vapor condenses at the cold side as permeate.

The ideal MD membrane should facilitate high flux, high salt rejection, high wetting resistance, low scaling and fouling, and low thermal conductivity [1]. High flux can be achieved by having the membrane with large enough pore size (not too large to maintain high liquid entry pressure, LEP: a minimum required pressure for water to penetrate through the membrane), high bulk porosity, abundant surface pore, and low tortuosity factor and a thin membrane. Wetting resistance can be improved by having the membrane with high hydrophobicity and small bubble point (BP, maximum pore size). The pore size distribution (PSD) must be narrow and the median pore must be close to the BP to obtain both high wetting resistance and high flux.

Membrane surface hydrophobicity is achieved by using materials of low surface energy (i.e., fluorinated). A low surface energy material facilitates a high water LEP. The indicators for determining membrane surface hydrophobicity include water contact angle (CA) and BP, as ascribed by the Laplace equation [2]. Meanwhile, a highly hydrophobic membrane also helps to increase scaling and fouling resistances. Membrane fouling refers to deposition of a particle or mineral, all referred as foulant, on the surface of the membrane surface blocking and plugging the pore mouth. Plugging then declines and diminishes the water vapor flux and lowers permeate quality by depressing the salt rejection or even promotes membrane wetting when applied for desalination [3]. Heat loss via conduction, despite often seen as the least concern, can be reduced by opting for materials with low thermal conductivity or generating thick and porous membranes [4, 5]. In MD, heat transfers through both pores and membrane matrix from feed to the permeate side; thus, employing a thin and conductive membrane may cause great conductive heat losses. Severe heat transfer from the hot to the cold side can decrease flux because of loss in driving force.

Substantial research focuses on development of a polyvinylidene fluoride- (PVDF-) based membrane for MD. The PVDF polymer has a good chemical resistance, a high thermal stability, and excellent structural matrices [6]. It also poses outstanding morphologies which facilitate formation of permeable and highly porous films [1]. Currently, PVDF polymers have extensively been used as material for the MD membrane and other porous-based membrane applications [6].

Recent researches on the PVDF membrane for MD focus on development of highly hydrophobic materials. Although PVDF poses a relatively low surface energy, a phase-inverted one from a strong nonsolvent (i.e., water) usually has low hydrophobicity [7]. The low CA originates from the absence of the *lotus effect*. As reported by Peng et al. [8], hydrophobicity is promoted by combination of both micro- and nanoscales onto the membrane surface. Application of either strong solvent or strong nonsolvent in the phase inversion flattens the membrane surface and reduces the surface CA.

Controlling water vapor supply from humid air in vapor-induced phase separation (VIPS) has been proven to reduce the extent of flattening effect [7]. In this approach, the immediate contact of the film surface with water can be evaded. Moreover, employing very low PVDF concentrations using the VIPS technique may produce a super hydrophobic surface (CA > 140°) which has a discontinuous hierarchical structure due to the occurrence of spinodal decomposition during phase separation [9]. Regardless of the discontinuous structure, crystallization also contributes to nanoscale roughness in the morphology of the surface [10]. This will eventually increase surface CA and hydrophobicity of the membrane if the crystals reside on the membrane surface, as reported elsewhere [7].

As a semicrystalline polymer, crystallization of the PVDF polymer also occurs during membrane formation via phase inversion. It occurs at a relatively slower rate relative than the liquid-liquid demixing. The degree of PVDF crystallization can be altered by manipulating phase inversion parameters, in particular by selection of a nonsolvent type and adjusting its temperature [11]. The type and the shape of PVDF crystals can be manipulated by changing bath temperature [12] or by selecting a weak nonsolvent that hinders liquid-liquid demixing and thus allows higher extent of crystallization.

The membrane with a low crystallinity is desirable to improve overall porosity and its hydraulic performance. To our best knowledge, limited study explores the role of PVDF crystallinity in its MD performance. Our previous study only focuses on the effect of dope solution temperature on the structure and short-term performance of MD membranes [13]. This study exploits semicrystalline behavior of the PVDF polymer to develop PVDF membranes and evaluate their MD performances in short- and long-term tests. We manipulate the dope solution temperature to lower its degree of crystallinity. It is hypothesized that high dope solution temperature increases the rate of solvent and nonsolvent exchanges to expedite liquid-liquid demixing and restrict crystallization. Five PVDF membranes were prepared, 25, 40, 60, 80, and 100°C of dope solutions, further referred to as PVDF-25, PVDF-40, PVDF-60, PVDF-80, and PVDF-100, respectively, via the phase inversion. The resulting membranes were thoroughly characterized, and their performances were assessed by treating simulated seawater and brine in short- and long-term direct contact membrane distillation (DCMD) tests.

## 2. Methods

*2.1. Membrane Preparation.* To prepare the phase-inverted PVDF membranes, PVDF (534 kDa, Sigma-Aldrich), dimethylacetamide (DMAC, Sigma-Aldrich), and deionized water were used as polymer, solvent, and nonsolvent, respectively. The summary of the prepared membranes and their respective codes is given in Table 1. The dope solution was prepared by stepwise dissolution of 15% *w/w* of PVDF into DMAC. The polymer concentration was selected as an optimum value obtained from [14]. The dopes were simultaneously stirred and heated at 25, 40, 60, 80, and 100°C until they were homogeneous. To ensure thorough dissolution, the stirrings were done at least for 24 h at 150 rpm, followed by degassing to achieve bubble-free solutions. Each of them was then casted atop a nonwoven support (NWS, Novatexx 24413, Freudenberg-filter, Germany) using a casting blade at a wet thickness of 0.240 mm. After casting, the cast films were immediately immersed in a DI bath at 25°C. The solidified film was then maintained in the coagulation bath overnight to ensure complete removal of solvent followed by drying at room temperature.

*2.2. Membrane Characterization.* The membrane thickness of each sample was measured (at least at 5 locations) using a micrometer (Mitutoyo, Japan). The membrane pore size including mean pore size (MPS), BP, and PSD was determined using capillary flow porometer (CFP, Porous Materials Inc. Ithaca, USA). Porosity of each samples was measured gravimetrically using the dry-wet method. Fourier transform infrared spectrometer (FTIR) in the mode of attenuated total

TABLE 1: Summary of the developed PVDF membrane properties.

| Membrane code | Dope solution temperature (°C) | Thickness ($\mu$m) | Porosity (%) | Contact angle (°) Top | Contact angle (°) Bottom |
|---|---|---|---|---|---|
| PVDF-25 | 25 | 333.0 ± 5.3 | 83.2 ± 0.4 | 77.4 ± 1.8 | 104.6 ± 3.5 |
| PVDF-40 | 40 | 280.4 ± 8.6 | 78.0 ± 1.1 | 77.2 ± 3.3 | 113.8 ± 8.4 |
| PVDF-60 | 60 | 250.2 ± 6.1 | 73.3 ± 1.1 | 80.4 ± 3.0 | 121.8 ± 4.1 |
| PVDF-80 | 80 | 265.8 ± 6.9 | 76.0 ± 1.0 | 82.8 ± 3.7 | 120.2 ± 7.3 |
| PVDF-100 | 100 | 276.0 ± 10.0 | 77.5 ± 1.3 | 75.6 ± 6.1 | 114.2 ± 4.1 |

TABLE 2: Heat of fusion and degree of crystallinity of the developed PVDF membranes.

| Membrane code | Heat of fusion, $\Delta H_f$ (J/g) | Degree of crystallinity, $X_c$ (%) |
|---|---|---|
| PVDF-25 | 54.37 | 51.93 |
| PVDF-40 | 48.97 | 46.77 |
| PVDF-60 | 46.87 | 44.77 |
| PVDF-80 | 48.01 | 45.85 |
| PVDF-100 | 49.20 | 46.77 |

reflection (ATR) was used to study the chemical bonds of the membrane samples. The measurements were done at wavenumbers ranged from 650 to 4000 cm$^{-1}$. The morphology of the membranes was obtained by a scanning electron microscopy (SEM, Quanta-250, FEI) on the cross section and top surface. To obtain a representative cross section structure, the samples were cut under liquid nitrogen before they were coated with gold and palladium at 50 Å to avoid overcharging. The CA was measured using goniometer and the membrane surface using sessile drop of 7 $\mu$l (5 replicates). An X-ray diffraction (XRD) was used to investigate the type of PVDF crystals in the membrane matrix. Differential scanning calorimeter (DSC) was used to evaluate crystallinity degree of the PVDF. The heat of fusion and total crystallinity of the developed membranes were summarized in Table 2. Heat of fusion was obtained by integrating the area under the DSC curves. Total crystallinity, $X_c$ of PVDF, was calculated by using the following formula [15]:

$$X_c = \frac{\Delta H_f}{\Delta H_f^*} \times 100\%, \quad (1)$$

where $\Delta H_f^*$ is the melting enthalpy for 100% crystalline PVDF (104.7 J/g) and $\Delta H_f$ is the melting enthalpy of samples obtained from DSC.

*2.3. DCMD Set-Up and Operational Parameter.* A lab-scale DCMD set-up (Figure 1) was used to evaluate membrane performance. It consisted of a filtration cell in which both feed and permeate were flown at a fixed crossflow velocity of 2.22 cm s$^{-1}$. The system was fed using 65°C of synthetic seawater (comprising of 3.5 wt% NaCl in DI water) and brine (comprising of 7.0 wt% NaCl in DI water) coupled with 25°C distilled water as the cold stream. The flux was collected as an overflow, and the weight was measured using a high precision weighing balance.

To study the performance stability and wetting resistibility of the samples, a long-term DCMD test (50 h) was conducted for a selected membrane (the PVDF-60 that shows the best performance in the short-term test). The simulated brine solution was used as the feed, which was fed at temperature of 65°C. DI water fed at temperature of 25°C was used as the cold stream in the permeate side. The conductivity of permeate was periodically measured every 1 hour during the operation to calculate the salt rejection involving salt balance calculation.

## 3. Results and Discussion

### *3.1. Effect of Dope Solution Temperature on Membrane Properties*

*3.1.1. Contact Angle.* The selected properties of the PVDF membranes, including top layer CA, bottom layer CA, membrane thickness, MPS, porosity, and BP, are shown in Table 1. Results show that temperature of dope solution does not affect the CA of the resulting membranes. The CA values of the top (75.6–82.8°) are much lower than the bottom surfaces (104.6–121.8°). The top surface CAs are also relatively low compared to [7] but still fall within reported values using the same systems (employing strong solvent and nonsolvent) reported elsewhere [16].

The low CA of the top membrane surface is largely a result of surface *flattening effect* as shown by the surface SEM images (Figure 2). Flattening effect refers to the polymer chain mobility atop the cast film as a result of a large interfacial tension gap between the cast film and water (nonsolvent). Such a large gap forces the polymer chain to move downwards into the bulk. This leads to formation of a flat and smooth surface contributing in lowering the CA [17]. A flat surface reduces the presence of structural features and thus depresses hydrophobicity [7, 18]. Despite being categorized as hydrophilic (CA < 90°), such membranes can still be employed in MD because the sublayers under the top skin are much more hydrophobic (see CA value of the bottom surface), as also proven in Sections 3.2 and 3.3. As demonstrated by the cross section image, a morphology like the bottom surface is expected across the PVDF matrix, except for the one on the top surface.

*3.1.2. Thickness, Porosity, and Pore Size.* The resulting membrane morphology can explain the effect of dope solution

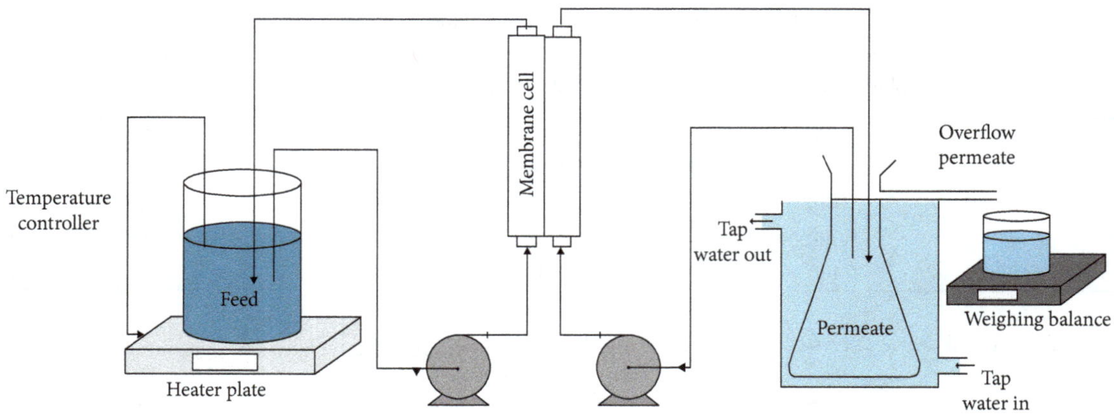


Figure 1: Experimental set-up for direct contact membrane distillation.

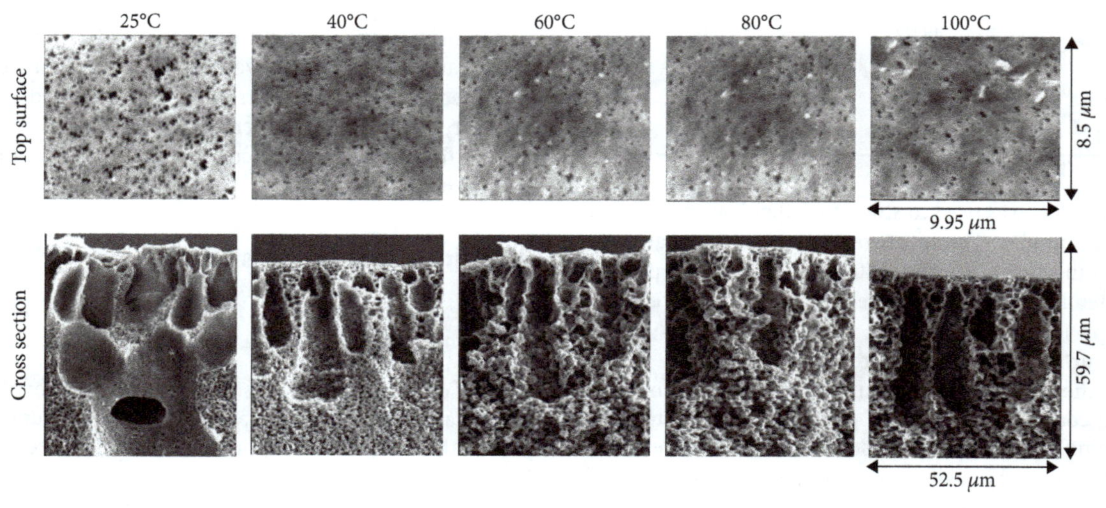


Figure 2: SEM images of the PVDF membranes (magnification: 30,000x).

temperature on thickness, porosity, and pore size. The dope solution temperature dictates the type of the crystalline phase that constructs the membrane matrix [11, 19]. Membrane thickness is greatly reduced as the dope solution temperature is increased (from 333 $\mu$m for the PVDF-25 to 250.2–280.4 $\mu$m for the rest) but has less prominent effect on the porosity (Table 1). Higher temperatures reduce the size of macrovoids underneath the top skin layer. This can be explained by reduction of solution viscosity that, at high temperature, facilitates exchange of solvent and nonsolvent during phase separation. Increasing the dope solution temperature accelerates the dissolution of solvent, which also explains why the membrane is thinner. The faster the speed of the double diffusion, the faster is the membrane formation rate [20], which makes the macrovoid structure not have enough time to develop. Therefore, the sizes of macrovoids are relatively small.

The trend on overall porosity can be explained as below. The overall structure of the membrane matrix can be distinguished as macrovoid and sponge or particulate-like morphology. For the membrane with large macrovoid (PVDF-25), the porosity of the sponge-like region is low and thus compensate the large space contributed by the macrovoids. On the other hand, there is only a lower contribution of smaller macrovoid to overall porosity, due to their smaller sizes (Figure 3). For higher dope solution temperature (>25°), they are compensated by large contribution of spherulite-like zone in the matrix.

Figure 3 shows the PSD of the PVDF-100, PVDF-60, and PVDF-25°C with their BP of 0.153, 0.139, and 0.989 $\mu$m and MPS of 0.099, 0.138, and 0.238 $\mu$m, respectively. PVDF-25 also shows larger size of pore mouths from the SEM images of the membrane surface (Figure 2). For PVDF-40, PVDF-60, and PVDF-100, the pore sizes have insignificant differences. However, the surface pore of PVDF-100 has larger size, but the size is still smaller than PVDF-25.

Increasing the temperature of dope solution decreases the MPS and BP because higher temperature accelerates exchange of solvent and nonsolvent. Rapid exchange of solvent/nonsolvent results in rapid formation of the solidified top skin layer. The solidification process is so fast, as such, the expansion of pore nucleus that manifests the polymer lean-phase is restricted thus depressing the surface pore. Instantaneous liquid-liquid demixing also suppresses macrovoids, which then restrict expansion of the cast firm to form a thinner membrane (Table 1), as discussed earlier. It is worth

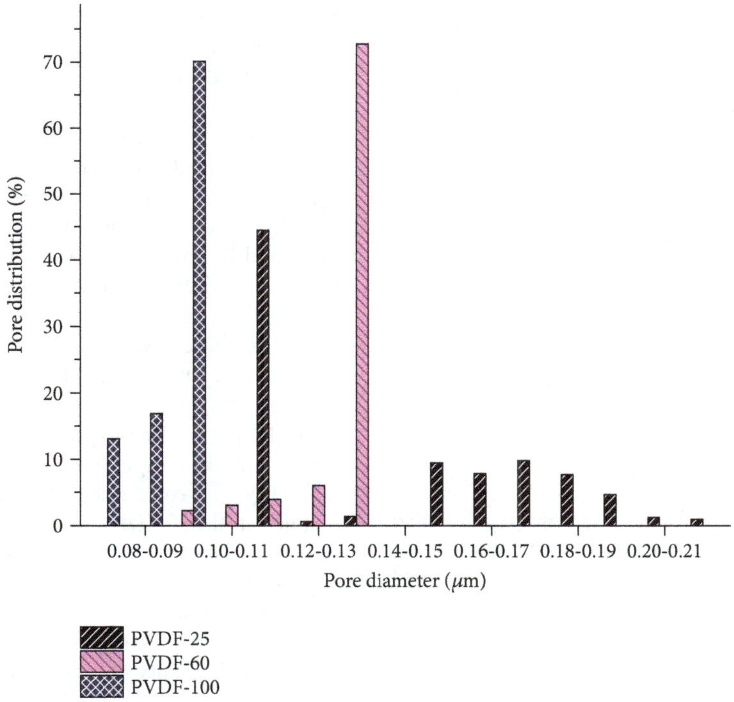


FIGURE 3: Pore size distribution of PVDF-25, PVDF-60, and PVDF-100.

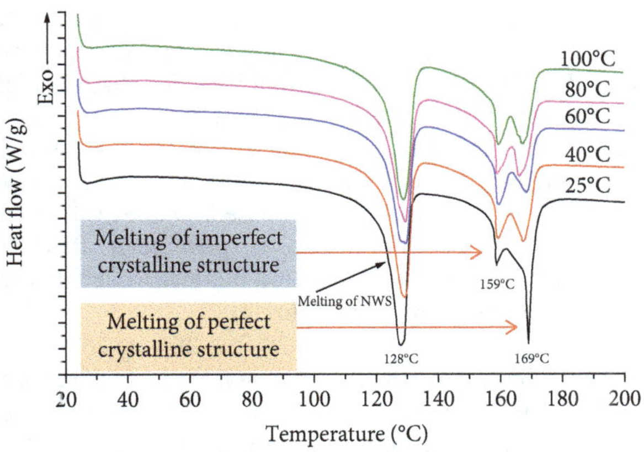


FIGURE 4: Differential scanning calorimeter (stacked) curves of PVDF membranes cast from different dope solution temperatures.

noting that majority of pore sizes (about 80%) have values close to BP, as showed by PVDF-60. Such profile promotes water vapor flux, and since the BP-value is small, it enhances wetting resistance, a desired MD membrane property.

*3.1.3. Morphology.* Figure 2 shows the effect of dope solution temperatures on the morphologies of the membrane surface (first row) and cross section (second row). Surface SEM images show quite smooth and flat topology, a result of the flattening effect as discussed earlier.

The cross section morphology is clearly affected by dope solution temperature (Figure 2). There is no spherulite-like structures over the cross section of the PVDF-25. This membrane poses large macrovoids underneath the top layer, sandwiched by sponge-like morphology below them. The spherulite-like structures are only formed in the matrices of PVDF-40, PVDF-60, and PVDF-100. The size of the spherulite-like structure increases toward PVDF-60 and remains almost the same at higher dope solution temperature (PVDF-80 and PVDF-100). The spherulite-like structure is believed to be the imperfect crystals from the semicrystalline PVDF, as also suggested from DSC results (Section 3.1.4). The cross-sectional images also show that they only exist underneath the top skin layer, leaving the top surface in flat topology.

*3.1.4. Crystallinity.* The PVDF membranes prepared from high temperature dope solution have lower degree of crystallinity with different crystal morphologies as shown from DSC curves in Figure 4. This occurs because of the more rapid demixing process. High cast film temperature reduces viscosity and facilitates faster exchange of solvent and nonsolvent during phase inversion. Such situation leads to overwhelming effect of liquid-liquid demixing and hence curbs crystallization.

The first peak at 128°C indicates the melting of NWS used as baking material for membrane fabrication. Two peaks at temperatures of 159 and 169°C indicate the melting point of the PVDF polymer. The coexistence of two peaks at these temperatures occurs because of the different crystalline phases that coexist in the membrane matrix [15, 19]. The peak at 159°C represents the melting of the imperfect crystalline region, while the melting of the perfect region occurs at 169°C. Imperfect crystallization is facilitated by a rapid liquid-liquid demixing and solidification process. Fast demixing, in combination with rapid solidification of the top film, restricts expansion of the cast film to upper direction. Under such circumstances, rearrangement of PVDF chain during crystallization is thus limited.

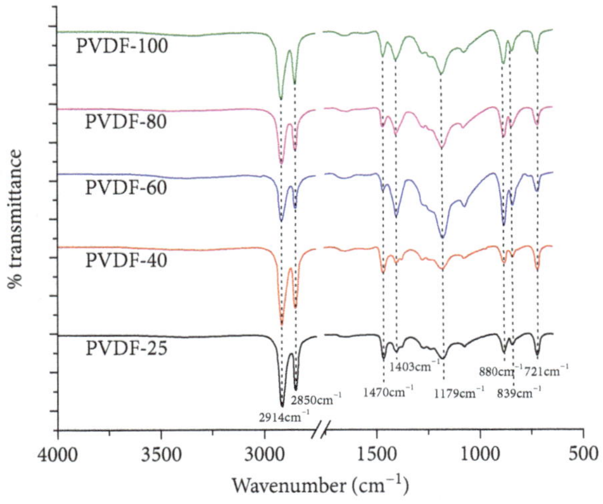


FIGURE 5: Fourier transform infrared spectroscopy-attenuated total reflectance (stacked) spectra of PVDF membranes casted from different dope solution temperatures of 25, 40, 60, 80, and 100°C for PVDF-20, PVDF-40, PVDF-60, PVDF-80, and PVDF-100, respectively.

PVDF-25 shows that majority of the crystal consists of a perfect crystalline structure. Lower dope solution temperature (i.e., 25°C), despite being categorized as instantaneous demixing, provides ample time for formation of perfect crystal during phase inversion. Remarkably, when comparing all samples prepared from heated dope solutions (PVDF-40 to PVDF-100), all membranes have similar peak intensities at temperatures of 159 and 169°C, except the PVDF-60 that shows higher peak ascribed for the peak for imperfect crystalline. Such variations in the crystal structure may affect mass transport resistance of the membrane (as discussed in Sections 3.2 and 3.3), as well as thickness, porosity, and pore size.

Among four membranes prepared at the temperature range of 40–100°C, PVDF-60 shows the lowest degree of crystallinity (Table 2). This can be explained by the variable rate of demixing across the cast film thickness. There are two major factors affecting liquid-liquid demixing and crystallization. High temperature leads to rapid solvent and nonsolvent exchanges from and to the cast film which also fasten the solidification of the top skin layer, resulting in a membrane with low crystallinity [21]. However, when the temperature is too high, skin layer formation occurs so fast that it inhibits further solvent and nonsolvent exchanges for the bottom layer of the cast film to allow crystallization. Those phenomena lead to an optimum condition at a temperature of about 60°C. Under this temperature, the solvent and nonsolvent exchange rates are fast enough to suppress crystallization but insufficient to hinder further solvent and nonsolvent exchanges from the lower part of the cast film.

The crystallinity behaviors of PVDF matrix in membranes prepared from different dope solution temperatures are also visible in the FTIR-ATR spectra (Figure 5). All membranes have identical peak locations ranging from 720 to 2915 $cm^{-1}$, but those peaks have distinct differences in intensities. There are strong absorption bands at 721, 839, and 880 $cm^{-1}$ corresponding to C-H rocking and C-F stretching vibration of PVDF, respectively. The presences of the absorption band at 1179, 1403, and 1467 $cm^{-1}$ may be attributed to C-C bending; and absorption band at 2845 and 2915 $cm^{-1}$ might be related to C-H stretching, as described elsewhere [18, 22].

The peak intensity of PVDF-60 is unique. The value is higher at C-C bending but weaker at C-H stretching. This may explain the presence of different crystal types most likely attributed by imperfect crystallization representing bending of PVDF back-bond not in an ideal crystal shape.

Results from XRD in Figure 6 show that all the membranes exhibit the same XRD diffraction pattern with minor difference in intensity at certain peaks. As reported earlier, time and temperature of crystallization are factors that determine the predominant phase of crystal [23]. Although the dope solution temperature was varied, the applied range (25 to 100°C) seems not wide enough to notice the obvious change of the PVDF crystal structure in the membrane matrix. However, Figure 6 shows that PVDF-60 has a less $\alpha$-phase structure at 20°. Since $\beta$-phase is most stable at 60°C, the transformation of $\alpha$- to $\beta$-phase occurs efficiently at this temperature compared to others [23, 24], thus explaining the low intensity of $\alpha$-phase obtained by PVDF-60. Considering its low total crystallinity (Table 2), the difference of peak intensity of $\beta$-phase could not be observed.

Figure 6 (inset) illustrates the XRD results of selected membranes (dope solutions of 40, 60, and 80°C) penetrated at the top surface layer. The findings support the result obtained in Figure 6 by showing that PVDF-60 has the highest peak intensity of $\beta$-phase [19, 25]. At 20.1° which indicate predominant formation of $\alpha$-phase, PVDF-80 shows the highest intensity. It is in line with a study by Gregorio and Cestari [23] that suggests that $\alpha$-phase is stable at the range of 80–160°C. The predominance of $\beta$-phase in PVDF-60 might be attributed to the formation of more imperfect crystal in the membrane matrix (Figure 4).

*3.2. Short-Term DCMD Performance.* Synthetic seawater (Figure 7) and brine (Figure 8) were used as feeds to assess the flux performance of the developed membranes. From the permeation test, the flux and rejection trends of the samples are almost the same for both feeds and both membrane orientations. However, a small decrease of flux is detected when brine is used as the feed instead of seawater for both membrane orientations. This is because the higher salinity in brine depresses the vapor pressure in the feed which acts as driving force in MD [20, 26].

Figure 7 shows that the maximum flux of 18 $l/m^2h$ was obtained by PVDF-60, when seawater was used as the feed. The value is considerably high for the given feed (65°C) and permeate side (25°C) temperatures for both top layer facing the feed (Figures 7(a) and 8(a)) and top layer facing permeate side (Figures 7(b) and 8(b)) orientations. The maximum flux obtained by PVDF-60 can be ascribed to its low thickness and low crystallinity due to formation of imperfect crystal and due to its desirable pore size distribution as discussed in Section 3.1.2. The presence of imperfect crystalline structures helps to improve MD flux, especially under the top membrane surface facing the hot feed orientation

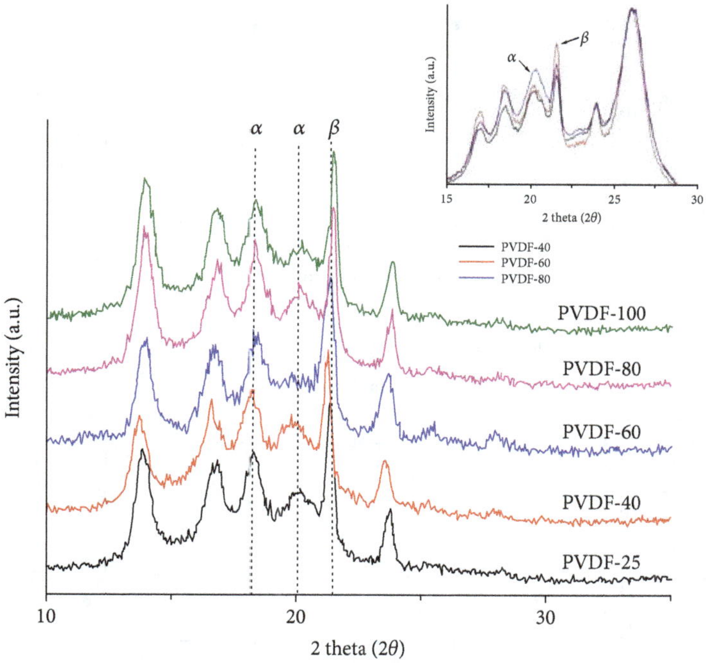


FIGURE 6: XRD (stacked) spectra of the PVDF membranes. The inset shows the spectra of selected membranes specifically obtained for a thin layer near the top surface.

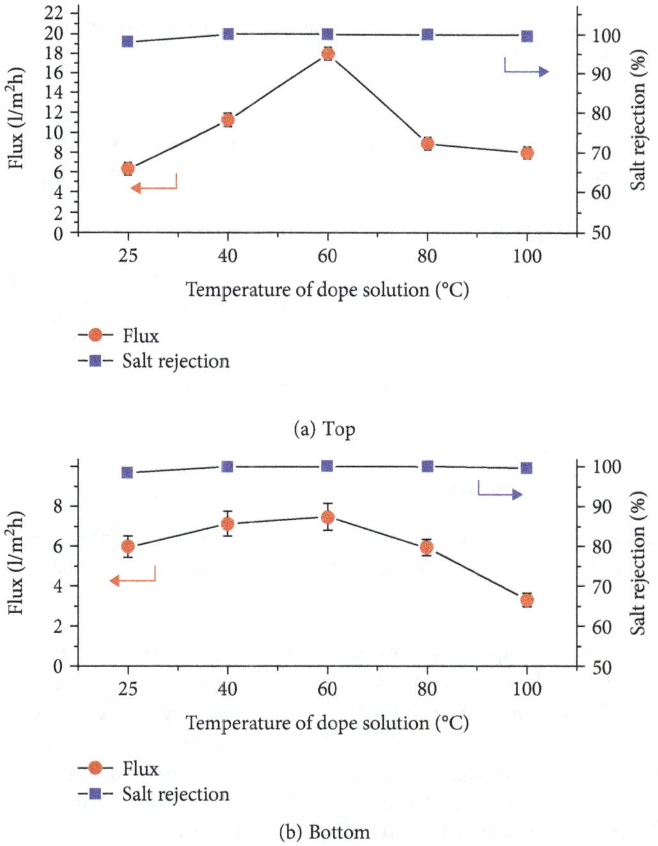


FIGURE 7: Flux and salt rejection of resulting membranes positioned under the (a) top layer facing feed and (b) top layer facing permeate. The tests were performed using seawater at 65°C as the feed and cold side temperature of 25°C.

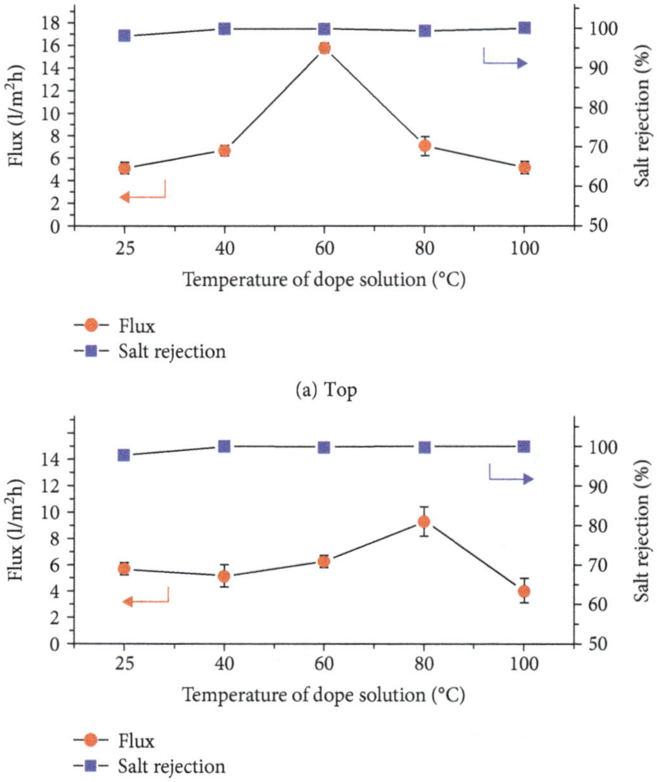


(a) Top

(b) Bottom

FIGURE 8: Flux performance and salt rejection of the resulting membranes using brine as the feed. The membranes were positioned under the (a) top layer facing feed and (b) top layer facing permeate.

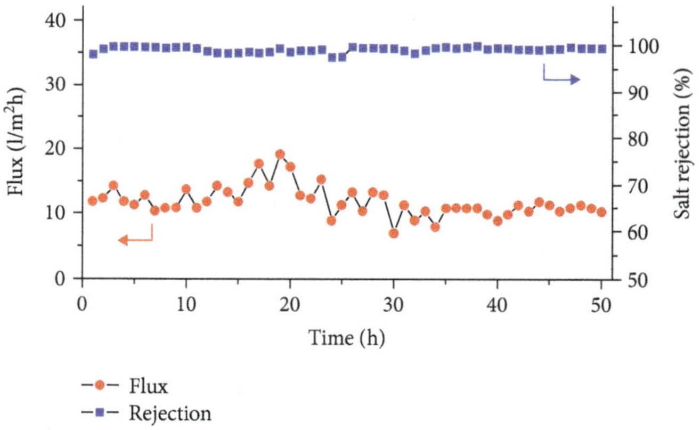


FIGURE 9: Long-term MD performance of PVDF-60 for treating brine as a feed flux and rejection. The feed and cold temperature was 65°C and 25°C, respectively. The membrane was positioned in the top surface facing feed configuration.

(Figure 7(a)). Low crystallinity favors free space for water vapor to pass by. It can then further decrease tortuosity property, thanks to a low crystallinity from an imperfect crystalline structure.

In addition, the orientation of the top surface layer also affects the flux performance of the resulting membranes. From the permeation results, the trend shows that the fluxes obtained are always lower when the top layer facing permeate orientation was applied instead of the top layer facing feed. Referring to prior discussions in Section 3.1, the rapid solidification process at the membrane top layer provides more time for crystallization to take place at the bottom layer, thus resulting in denser structures and responsible for flux decrement. Hence, the orientation of the membrane's top layer in DCMD is important because of the morphological difference between the top and bottom layers.

The lowed flux exhibited by other membranes can be ascribed by partial pore wetting, particularly for PVDF-25 ascribed from its poor salt rejection. The wetted pores become untenable for transport and inhibit the transport of water vapor and reduce the flux. When a complete pore wetting occurs, it facilitates salt bridging that deteriorates salt rejection. The PVDF-25 has the lowest salt rejection value than the rest. Pore wetting lowers water flux by decreasing the temperature gradient. This happens because the penetrant liquid in the pore is static inaccessible from external mixing. It is worth noting that these results were obtained during a short-term test and should be confirmed via a long-term test discussed in Section 3.3.

When compared with recent literature, the flux of our membranes (even for the most optimum one) is considered low [7, 27–33]. However, the flux value is not only a function of intrinsic membrane properties. Our low flux value is due to our low applied crossflow velocities, limited by our experimental set-up.

*3.3. Long-Term Performance (Wetting Resistance Test).* A long-term permeability test has been conducted to determine the capability of the developed membrane to retain their performance stability in term of flux and salt rejection. Considering the optimum flux in the short-term test (Section 3.2) was obtained by PVDF-60 under the top layer facing feed orientation, it has been selected for the long-term test. Figure 9 shows the trends of water flux and solute rejection of PVDF-60 during 50 h operation by using synthetic brine as the feed.

Both the flux and the salt rejection are quite stable during the entire testing period with flux average of $10.5\,l/m^2$ h and salt rejection of >99%, which proves that PVDF-60 possesses a high wetting resistance. The flux is also highly encouraging, 50% higher than the one obtained for the corrugated PVDF membrane reported elsewhere [31]. Relatively high MD flux offered by PVDF-60 is attributed from its lower crystallinity that increases the free volume (voids) in the polymer matrix and thus facilitates the transport of water vapor.

The flux obtained by this test, nevertheless, is still lower than the ones obtained in the short-term DCMD test (Figures 7 and 8). This might be due to the temperature polarization since DCMD configuration was used. The decrement of temperature differences across the membrane eventually reduces the rate of water vapor transportation.

## 4. Conclusion

Manipulating dope solution temperature was found effective to increase PVDF MD membrane performance, being optimum at 60°C. It offers considerably high steady state flux ($18.5\,l/m^2$ h) and high wetting resistance over long-term operation (50 h). The gain on MD performance for PVDF-60 can be ascribed from its low thickness, desirable pore size distribution, and most importantly, low crystallinity that offers high free space for water vapor to pass by. The combination of the aforementioned factors leads to improvements of the mass transfer across the membrane.

## Additional Points

*Highlights.* The role of dope solution temperature on PVDF crystallization was explored. An optimum temperature of 60°C helps to reduce PVDF crystallization. Crystallization was reduced via formation of imperfect crystal. The optimum dope solution temperature was 60°C in terms of improved MD performance.

## Conflicts of Interest

The authors declare that they have no conflicts of interest.

## Acknowledgments

We acknowledge Yayasan Universiti Teknologi Petronas-Malaysia for providing grants (Grant number: 0153AA-E96) and the Chemical Engineering Department for providing facilities for conducting the research activities.

## References

[1] A. Alkhudhiri, N. Darwish, and N. Hilal, "Membrane distillation: a comprehensive review," *Desalination*, vol. 287, pp. 2–18, 2012.

[2] M. C. García-Payo, M. A. Izquierdo-Gil, and C. Fernández-Pineda, "Wetting study of hydrophobic membranes via liquid entry pressure measurements with aqueous alcohol solutions," *Journal of Colloid and Interface Science*, vol. 230, no. 2, pp. 420–431, 2000.

[3] L. D. Tijing, Y. C. Woo, J.-S. Choi, S. Lee, S.-H. Kim, and H. K. Shon, "Fouling and its control in membrane distillation—a review," *Journal of Membrane Science*, vol. 475, pp. 215–244, 2015.

[4] R. W. Schofield, A. G. Fane, and C. J. D. Fell, "Heat and mass transfer in membrane distillation," *Journal of Membrane Science*, vol. 33, no. 3, pp. 299–313, 1987.

[5] N. N. Li, A. G. Fane, W. W. Ho, and T. Matsuura, *Advanced Membrane Technology and Applications*, John Wiley & Sons, 2011.

[6] G.-D. Kang and Y.-M. Cao, "Application and modification of poly(vinylidene fluoride) (PVDF) membranes—a review," *Journal of Membrane Science*, vol. 463, pp. 145–165, 2014.

[7] F. A. AlMarzooqi, M. R. Bilad, and H. A. Arafat, "Development of PVDF membranes for membrane distillation via vapour induced crystallisation," *European Polymer Journal*, vol. 77, pp. 164–173, 2016.

[8] M. Peng, H. Li, L. Wu, Q. Zheng, Y. Chen, and W. Gu, "Porous poly(vinylidene fluoride) membrane with highly hydrophobic surface," *Journal of Applied Polymer Science*, vol. 98, no. 3, pp. 1358–1363, 2005.

[9] H. Fan, Y. Peng, Z. Li, P. Chen, Q. Jiang, and S. Wang, "Preparation and characterization of hydrophobic PVDF membranes by vapor-induced phase separation and application in vacuum membrane distillation," *Journal of Polymer Research*, vol. 20, no. 6, p. 134, 2013.

[10] Y. Peng, H. Fan, Y. Dong, Y. Song, and H. Han, "Effects of exposure time on variations in the structure and hydrophobicity of polyvinylidene fluoride membranes prepared via vapor-induced phase separation," *Applied Surface Science*, vol. 258, no. 20, pp. 7872–7881, 2012.

[11] M. G. Buonomenna, P. Macchi, M. Davoli, and E. Drioli, "Poly(vinylidene fluoride) membranes by phase inversion: the role the casting and coagulation conditions play in their morphology, crystalline structure and properties," *European Polymer Journal*, vol. 43, no. 4, pp. 1557–1572, 2007.

[12] X. Wang, L. Zhang, D. Sun, Q. An, and H. Chen, "Formation mechanism and crystallization of poly(vinylidene fluoride) membrane via immersion precipitation method," *Desalination*, vol. 236, no. 1-3, pp. 170–178, 2009.

[13] N. I. M. Nawi, M. R. Bilad, and N. A. H. M. Nordin, "Effect of dope solution temperature on the membrane structure and membrane distillation performance," *IOP Conference Series: Earth and Environmental Science*, vol. 140, article 012032, 2018.

[14] E. Pramono, A. Simamora, C. Radiman, and D. Wahyuningrum, "Effects of PVDF concentration on the properties of PVDF membranes," *IOP Conference Series: Earth and Environmental Science*, vol. 75, article 012027, 2017.

[15] K. Nakagawa and Y. Ishida, "Annealing effects in poly(vinylidene fluoride) as revealed by specific volume measurements, differential scanning calorimetry, and electron microscopy," *Journal of Polymer Science Part A-2: Polymer Physics*, vol. 11, no. 11, pp. 2153–2171, 1973.

[16] D. Hou, H. Fan, Q. Jiang, J. Wang, and X. Zhang, "Preparation and characterization of PVDF flat-sheet membranes for direct contact membrane distillation," *Separation and Purification Technology*, vol. 135, pp. 211–222, 2014.

[17] Z. Zhenxin and T. Matsuura, "Discussions on the formation mechanism of surface pores in reverse osmosis, ultrafiltration, and microfiltration membranes prepared by phase inversion process," *Journal of Colloid and Interface Science*, vol. 147, no. 2, pp. 307–315, 1991.

[18] H. Bai, X. Wang, Y. Zhou, and L. Zhang, "Preparation and characterization of poly(vinylidene fluoride) composite membranes blended with nano-crystalline cellulose," *Progress in Natural Science: Materials International*, vol. 22, no. 3, pp. 250–257, 2012.

[19] J. Liu, X. Lu, and C. Wu, "Effect of preparation methods on crystallization behavior and tensile strength of poly(vinylidene fluoride) membranes," *Membranes*, vol. 3, no. 4, pp. 389–405, 2013.

[20] L. Eykens, I. Hitsov, K. De Sitter et al., "Influence of membrane thickness and process conditions on direct contact membrane distillation at different salinities," *Journal of Membrane Science*, vol. 498, pp. 353–364, 2016.

[21] M. Zhang, A.-Q. Zhang, B.-K. Zhu, C.-H. Du, and Y.-Y. Xu, "Polymorphism in porous poly(vinylidene fluoride) membranes formed via immersion precipitation process," *Journal of Membrane Science*, vol. 319, no. 1-2, pp. 169–175, 2008.

[22] A. Bottino, G. Camera-Roda, G. Capannelli, and S. Munari, "The formation of microporous polyvinylidene difluoride membranes by phase separation," *Journal of Membrane Science*, vol. 57, no. 1, pp. 1–20, 1991.

[23] J. R. Gregorio and M. Cestari, "Effect of crystallization temperature on the crystalline phase content and morphology of poly(vinylidene fluoride)," *Journal of Polymer Science Part B: Polymer Physics*, vol. 32, no. 5, pp. 859–870, 1994.

[24] R. Gregorio, "Determination of the $\alpha$, $\beta$, and $\gamma$ crystalline phases of poly(vinylidene fluoride) films prepared at different conditions," *Journal of Applied Polymer Science*, vol. 100, no. 4, pp. 3272–3279, 2006.

[25] S. Rajabzadeh, T. Maruyama, Y. Ohmukai, T. Sotani, and H. Matsuyama, "Preparation of PVDF/PMMA blend hollow fiber membrane via thermally induced phase separation (TIPS) method," *Separation and Purification Technology*, vol. 66, no. 1, pp. 76–83, 2009.

[26] I. Hitsov, L. Eykens, W. D. Schepper, K. D. Sitter, C. Dotremont, and I. Nopens, "Full-scale direct contact membrane distillation (DCMD) model including membrane compaction effects," *Journal of Membrane Science*, vol. 524, pp. 245–256, 2017.

[27] L. Zheng, J. Wang, J. Li, Y. Zhang, K. Li, and Y. Wei, "Preparation, evaluation and modification of PVDF-CTFE hydrophobic membrane for MD desalination application," *Desalination*, vol. 402, pp. 162–172, 2017.

[28] S. Munirasu, F. Banat, A. A. Durrani, and M. A. Haija, "Intrinsically superhydrophobic PVDF membrane by phase inversion for membrane distillation," *Desalination*, vol. 417, pp. 77–86, 2017.

[29] D. Singh and K. K. Sirkar, "Performance of PVDF flat membranes and hollow fibers in desalination by direct contact membrane distillation at high temperatures," *Separation and Purification Technology*, vol. 187, pp. 264–273, 2017.

[30] L. Eykens, K. De Sitter, C. Dotremont, L. Pinoy, and B. Van der Bruggen, "Characterization and performance evaluation of commercially available hydrophobic membranes for direct contact membrane distillation," *Desalination*, vol. 392, pp. 63–73, 2016.

[31] M. R. Bilad, E. Guillen-Burrieza, M. O. Mavukkandy, F. A. Al Marzooqi, and H. A. Arafat, "Shrinkage, defect and membrane distillation performance of composite PVDF membranes," *Desalination*, vol. 376, pp. 62–72, 2015.

[32] J. A. Kharraz, M. R. Bilad, and H. A. Arafat, "Flux stabilization in membrane distillation desalination of seawater and brine using corrugated PVDF membranes," *Journal of Membrane Science*, vol. 495, pp. 404–414, 2015.

[33] D. Sun, M.-Q. Liu, J.-H. Guo, J.-Y. Zhang, B.-B. Li, and D.-Y. Li, "Preparation and characterization of PDMS-PVDF hydrophobic microporous membrane for membrane distillation," *Desalination*, vol. 370, pp. 63–71, 2015.

7

# Fatigue Fracture Properties and Morphology of Polyoxymethylene (POM) Plates Produced under Moderate Processing Conditions

**M. Berer,[1] M. Halb,[1] M. Feuchter,[2] G. Pacher,[1] and G. Pinter[2]**

[1]*Polymer Competence Center Leoben GmbH, Roseggerstrasse 12, 8700 Leoben, Austria*
[2]*Material Science and Testing of Polymers, Montanuniversitaet Leoben, Otto Gloeckel-Strasse 2, 8700 Leoben, Austria*

Correspondence should be addressed to M. Berer; michael.berer@pccl.at

Academic Editor: Vito Speranza

The present study was inspired by different industry projects in which a strong dependence of the fatigue fracture performance of POM on the processing conditions was observed. To examine the relationships under more reproducible conditions, plates of two different POM homopolymer resins (one significantly nucleated) were produced by compression molding and by injection molding under moderate conditions. For the injection molding, three different mold temperatures were used. At specific locations, the plates were analyzed concerning their hierarchical structure on the micro- and nanolevel and concerning their fatigue fracture performance. For the fatigue fracture performance, the dependence on the processing conditions was rather small for the nucleated resin but much more significant for the other resins. This dependence could not be related to morphological effects only, and thus, a combined effect of morphology and residual stresses was assumed.

## 1. Introduction

POM is a semicrystalline, thermoplastic polymer with high degrees of crystallinity. It is frequently used for components which are exposed to cyclic loading [1–7], and hence, its fatigue properties are of technical interest. Generally, this material shows only small crack tip plasticity in fatigue crack growth tests, and thus, it can be investigated quite well using linear elastic fracture mechanics [8–16].

It is state of the art that changes in the morphology of polymers, which are introduced by varying processing parameters, additives, physical ageing, postcrystallization, or annealing, have an influence on the resulting crack growth behavior. In the literature, this is mainly attributed to changes in the tie molecule and entanglement densities [13, 17–21]. However, the determination of the latter is very complex and time-consuming [13, 20], and thus, easier accessible parameters are usually used to correlate changes in the crack growth behavior with the corresponding changes in the morphology. This strategy was also used in the present study, and the morphology was interpreted in terms of (mass and volumetric) degree of crystallinity and amorphous fraction, long period, lamellar thickness and thickness of the amorphous layer, contribution of the different X-ray diffraction peaks, spherulite size and size distribution, and the cross-sectional ("skin–core") layer structure.

Generally, the number of publications focusing on the interactions between processing conditions, resulting material morphology and mechanical properties of polymers, is not so small. However, this number becomes significantly smaller when only those are considered, which examine the full chain (starting with processing and ending with the mechanical properties) and not only parts of it. From a material's point of view, so far, most of the work was done for polypropylene [18, 22–35], which seems to be the best studied polymer in this field. Nevertheless, aside from studies done on other polymers [17, 36, 37], there are surprisingly many publications available for POM [13, 19, 23, 38–40] as well. These POM studies have quite different motivations and approaches. Runt and Gallagher [13] used a combination of quenching, annealing, and isothermal crystallization to obtain specimens with different morphologies, which were analyzed and compared for their fatigue fracture properties (crack growth kinetics). In [19, 38], the decrease of the fracture toughness ($K_{IC}$) in dependence on the crystallization temperature used in compression molding was modelled (via the loss in entanglement density) and experimentally verified. These three studies are of scientific interest mainly since the annealing and crystallization temperatures examined were rather high (100–160°C) and thus usually not relevant for conventional processing conditions. Zhao and Ye [40] studied the effect of drawing (different draw ratios) on morphology and mechanical properties of POM. Most relevant for our study are the publications of Wright et al. and Viana [23, 39] in which POM was injection-molded using conventional processing conditions. However, in both studies, the morphological analysis is incomprehensive and the mechanical characterization focuses more on monotonic properties (tensile and fracture toughness tests) than on the fatigue fracture performance.

The present study was inspired by different industry projects in which a strong dependence of the fatigue fracture performance of POM on the processing conditions was observed. To examine the relationships under more reproducible conditions, moderate processing conditions were selected. For this, two POM homopolymer resins (one significantly nucleated) were processed by compression molding and by injection molding under moderate conditions. The molded plates were analyzed concerning their fatigue crack growth behavior (cyclic fatigue fracture tests) and concerning their morphology. The material morphology was characterized using differential scanning calorimetry (DSC), polarized light microscopy (PLM), small-angle X-ray scattering (SAXS), and wide-angle X-ray diffraction (WAXD).

## 2. Materials and Methods

The basic work of this study was conducted within a diploma thesis at the Polymer Competence Center Leoben, which is recommended as reference for more specific information concerning the experimental details [41].

*2.1. Materials and Processing.* Two different POM homopolymer resins were examined in this study: Delrin 111PF from DuPont (DuPont, Wilmington, Delaware, USA) and Tenac 3010 from Asahi Kasei (Asahi Kasei, Tokyo, Japan). The Delrin resin is a special grade, which is based on Delrin 100 but with modifications in the additive contents. The most relevant modification for this study is that it contains a heterogeneous nucleation agent to improve the fatigue crack growth behavior. The materials will be called "Delrin" and "Tenac" in the following discussions. For both resins, the molecular weight and the polydispersity index are given in Table 1. For Tenac, this data was already published previously [15]. The corresponding data for the Delrin resin used in this study was also determined at that time.

Both materials were processed by the same processing techniques, i.e., compression molding and injection molding with three different mold temperatures. For compression molding, the molding press "Collin PCS II" (Dr. Collin GmbH, Ebersberg, Germany) was used. The corresponding compression molding parameters are listed in Table 2. The injection molding machine used was an "Arburg Allrounder

TABLE 1: Molecular characteristics of the POM resins used in this study.

| Material | Weight (average molecular weight) $M_w$ (g/mol) | Polydispersity index |
|---|---|---|
| Delrin | 149,000 | 2.6 |
| Tenac | 137,000 | 2.6 |

TABLE 2: Processing parameters used for compression molding (plate dimension: 150 × 150 × 4 mm).

| | Heating | Compression | Cooling |
|---|---|---|---|
| *Delrin* | | | |
| Pressure (bar) | 1 | 100 | 100 |
| Temperature (°C) | 210 | 210 | 30 |
| Time span (min) | 16 | 15 | 18 |
| *Tenac* | | | |
| Pressure (bar) | 1 | 100 | 100 |
| Temperature (°C) | 210 | 210 | 30 |
| Time span (min) | 18 | 15 | 18 |

470A 1000-400 Alldrive" (Arburg GmbH & Co. KG, Loßburg, Germany). The settings for the most important injection molding parameters are given in Table 3. The processing parameters for both techniques were based on the corresponding data sheets provided by the material producers. Nevertheless, they had to be adapted empirically to gain acceptable component quality. In order to guarantee comparability between the two resins, the processing conditions were kept as constant as possible. However, slight adaptations of some processing parameters could not be avoided in order to achieve similar component quality. In both techniques, plates with a side length of 150 × 150 mm were produced. Due to the different molds available, the thickness was not uniform; the compression-molded plates had a thickness of 4 mm, while the injection-molded ones were 5 mm thick. In all cases, the material was dried before processing. This was an important detail because without drying, an insufficient plate quality was obtained in compression molding as well as injection molding. For the compression molding, the dried granules were used directly (no previous melting). The amount required was uniformly distributed in a steel frame and subsequently heated and compressed in the molding press using the processing parameters in Table 2. In case of injection molding, filling studies for all mold temperatures and resins were conducted directly before the molding of the plates. They confirmed a symmetrical filling of the mold in all cases. This was especially important for the definition of the characterization spots.

*2.2. Analysis Techniques*

*2.2.1. SAXS and WAXD.* In order to facilitate the comparison between morphology and fracture mechanical properties, the SAXS and WAXD analysis was carried out directly on the compact tension (CT) specimens used for the subsequent fracture mechanical tests. The specimens were made from the plates as described below. In analogy to [15], the

Table 3: Processing parameters used for injection molding (plate dimension: 150 × 150 × 5 mm).

| | Delrin | | Tenac |
|---|---|---|---|
| Screw diameter (mm) | | 40 | |
| Metering volume ($cm^3$) | | 160 | |
| Back pressure (bar) | | 40 | |
| Melt temperature (°C) | 215 | | 210 |
| Injection flow rate ($cm^3$/s) | | 10 | |
| Max. injection pressure (bar) | 700 | | 630/700 (90°C mold temperature only) |
| Mold temperature (°C) | | 60/90/110 | |
| Switch-over volume ($cm^3$) | | 27 | |
| Holding pressure (bar) | | 600 | |
| Holding pressure time (s) | | 12 | |
| Remaining cushion ($cm^3$) | 12 | | 11 |
| Residual cooling time (s) | | 50 | |
| Total cycle time (s) | 82.8 | | 80.8 |

measurement position for the X-ray measurements was about 2.5 mm ahead of the premilled notch of the CT specimens. Based on our experience with fatigue fracture testing of POM, it was expected that this area is passed by the crack during fatigue crack growth. The measurements were conducted for both resins, all processing conditions, and all position variations except for those made from the injection-molded plates and rotated by 90° (crack orientation perpendicular to the flow direction). They were expected to show the same morphology as the specimens, which were not rotated (crack orientation in flow direction). For the details regarding the CT specimen alignment on the plates, please refer to the corresponding section below. For each condition examined, 5 CT specimens were analyzed using SAXS and WAXD before the fatigue fracture tests were conducted.

All SAXS and WAXD experiments were conducted on a Bruker NANOSTAR (Bruker Corporation, Massachusetts, USA). The X-ray source I$\mu$S (Incoatec GmbH, Geesthacht, Germany) provided monochromatic radiation with a wavelength of 1.54 Å. The X-ray beam was focused by a circular aperture to a diameter of 300 $\mu$m. SAXS and WAXD tests were performed simultaneously using the setup shown in Figure 1. The exact distances between samples and detectors were determined using a silver behenate standard for the SAXS detector and a silicon standard for the WAXD detector. For all measurements, an irradiation time of one hour was used. A software tool called "Fit2d" (European Synchrotron Radiation Facility, Grenoble, France) was used for the integration of the SAXS and WAXD patterns [42]. The integrated patterns will be termed as "profiles" in the subsequent discussion.

For SAXS, an automatic detector of the type "VÅNTEC-2000 2D MicroGap" (Bruker Corporation, Massachusetts, USA) was used. The radial integration of the recorded patterns was done up to a scattering angle of 1.5°. For the morphological interpretation of the SAXS profiles, two different methods were used: the so-called "1D correlation" [43, 44] and a fitting procedure, which uses a 3-component model developed especially for POM [21, 45, 46]. The background

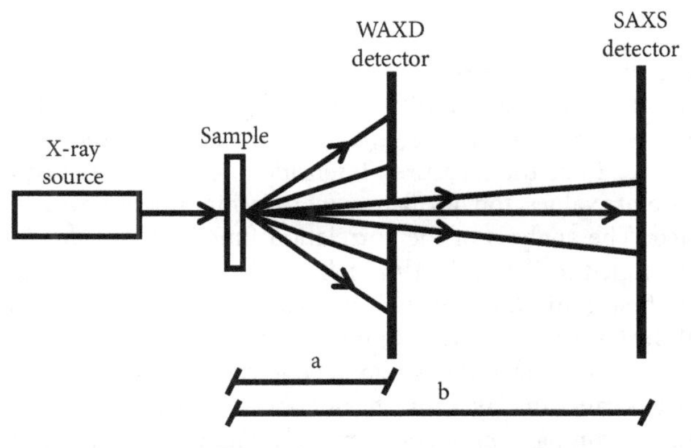


Figure 1: Setup for simultaneous SAXS and WAXD measurements (a was 49.9 mm for all measurements, b was 1142.5 mm for the plate scanning and 1097 mm for the CT specimen measurements).

for this proceeding was that for POM it is nowadays accepted that there are three different phases (two different types of crystalline lamellae and the amorphous phase) present in a conventionally crystallized material [1, 21, 45–47]. Since 1D correlation is based on a two-phase model, a part of the morphology information is lost, when using this analysis technique. This loss of information can be prevented when using the 3-component model developed by Hama and Tashiro [21, 46]. However, to the best knowledge of the authors, this method has not been applied on such high molecular weight resins as used in this study so far. Hence, its applicability for our purposes was unclear at the beginning. For both methods (1D correlation and 3-component model), a self-written MATLAB (MathWorks Inc., Natick, Massachusetts, USA) script was used for the analysis and parameter calculation.

In accordance with the literature [43, 44], the 1D correlation script allowed the extrapolation of the scattering profiles on both ends to overcome the experimental limitations and calculated the correlation function afterwards. For the extrapolation at high $q$, a modified Porod law assuming a

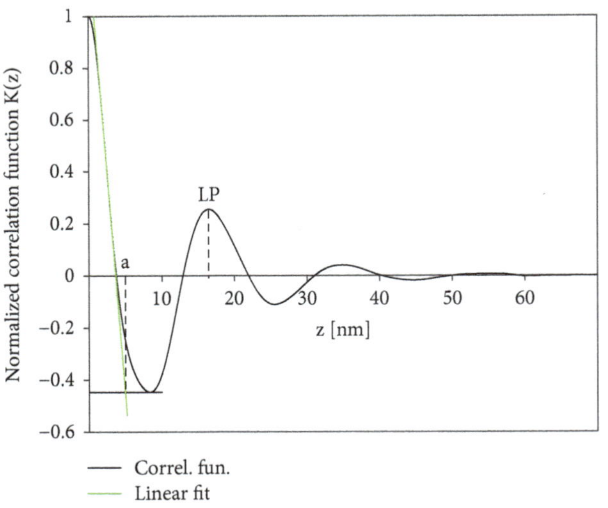


FIGURE 2: Example for the correlation functions obtained (Delrin, injection molded, 110°C mold temperature, CT specimen, "Delrin_IM_110°C-MT") and scheme of the analysis used.

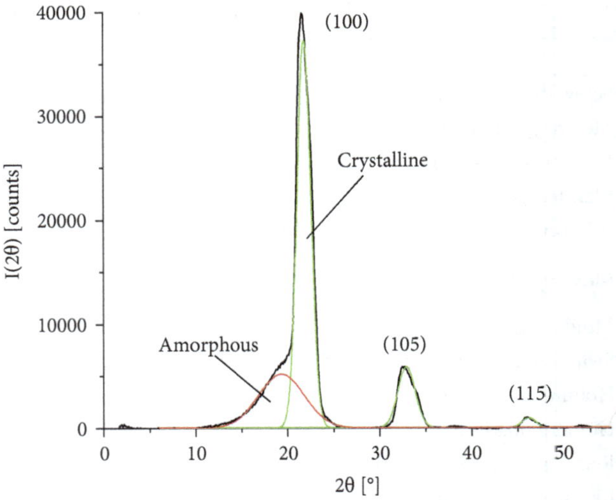


FIGURE 3: Schematic illustration of the analysis of the WAXD profiles including the peak used for the amorphous halo.

sigmoidal electron density transition [48, 49] was fit to the tail of the intensity profile. The background scatter, which was obtained as a fit result, was subtracted from the whole profile. Since the measured intensities were not obtained as absolute values, the normalized correlation function was created. The analysis of the correlation functions was done as illustrated in Figure 2. Although no plateau was observed at the first minimum, the latter was used for the determination of the layer thicknesses and linear crystallinity. It is expected that this simplification is a good engineering approach in our case because the layer thicknesses obtained were mainly used for the comparison of comparatively similar material conditions at one temperature (room temperature). Nevertheless, it has to be kept in mind that the values are only an estimation of the "real" layer thicknesses. The layer thicknesses and linear crystallinities were also calculated based on the first zero of the correlation function. However, according to our experience from an internal study (in which a layer-by-layer SAXS and WAXD analysis was conducted on some of the plates used in this study), these values are not very reliable for our materials especially for the Delrin resin. Hence, the plateau values were preferred. The script for the implementation of the 3-component model fitted this model to the measured SAXS data using a nonlinear least squares optimization. For the conversion between layer thicknesses, long period, and "linear crystallinity" (volumetric degree of crystallinity within the crystalline stack [50]), (1) (1D correlation) and (2) (3-component model) were used, respectively.

$$X_{\mathrm{C,1D,>50\%}} = 1 - \frac{a}{\mathrm{LP}}, \tag{1}$$

where $a$ is the amorphous layer thickness and LP is the long period (both obtained from 1D correlation). $X_{\mathrm{C,1D}}$ is the volumetric degree of crystallinity in the crystalline stack based on the 1D correlation results. The formula above is valid only for materials with a crystallinity of more than 50% [44], which was the case for the POM materials examined in this study.

$$X_{\mathrm{C,3comp}} = \frac{d_1 + d_2}{\mathrm{LP}}, \tag{2}$$

where $d_1$ and $d_2$ are the lamellar thicknesses of "folded chain lamellae" and "extended chain lamellae", respectively [21, 45, 46]. All parameters are fitting coefficients of the 3-component model. $X_{\mathrm{C,3comp}}$ is the volumetric degree of crystallinity in the crystalline stack based on the 3-component model.

For the WAXD measurements, an analogue detector was used. It was placed directly in the test chamber with a distance of 49.9 mm to the sample (Figure 1). The detector had a central hole so that the scattered SAXS beam could pass. After every measurement, the detector was removed from the chamber and digitized, and finally, the detected pattern was deleted before starting the next measurement. The WAXD profiles generated from the patterns were analyzed concerning the (mass) degree of crystallinity (3) and regarding the contribution of the different diffraction peaks to the total area. For the determination of the amorphous fraction, the amorphous halo was defined in a similar way as shown in [40] (a more regular peak shape was used in our case). The analysis procedure is illustrated in Figure 3.

$$X_{\mathrm{C,WAXD}} = \frac{a_{\mathrm{c}}}{a_{\mathrm{c}} + a_{\mathrm{a}}}, \tag{3}$$

where $a_{\mathrm{c}}$ and $a_{\mathrm{a}}$ are the total area under the crystalline peaks and the area of the amorphous halo, respectively. $X_{\mathrm{C,WAXD}}$ is the (mass) degree of crystallinity obtained from WAXD.

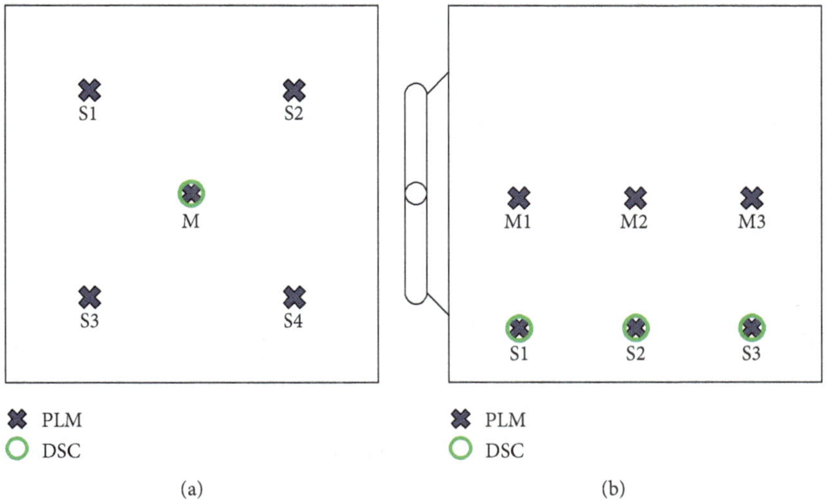


Figure 4: Sample positions used for the PLM studies and the DSC measurements. Compression-molded plate (a) and injection-molded plate (b).

*2.2.2. PLM.* PLM (polarized light microscopy) was used to qualitatively examine the cross section microstructure of the differently processed POM plates. This comprised the spherulite size and distribution as well as the layer structure introduced through injection molding.

The PLM analysis was conducted on the plates only. For each processing condition, one selected plate was analyzed. The samples used for the analysis were thin films, which were prepared in a way that the entire thickness of the plate could be examined. They were taken from multiple positions of the molded plates, which are marked in Figure 4 (S stands for side and M for middle). For the analysis, it was assumed that the compression-molded plates were homogenous and that the injection-molded plates were symmetric in the flow direction. The latter assumption was supported by the previously conducted filling studies. For the preparation of the films, the plates were first cut into stripes with a buzz saw to get close to the measuring points; the injection-molded plates were cut in filling direction, while the compression-molded plates had no preferred direction. Then, several layers of the cut face were removed with a microtome to exclude any influence of the saw cutting on the microstructure and to make the surface smooth and clean. Finally, the sample films, which had a thickness of 10 $\mu$m, were cut off with the microtome. They were placed on a microscope slide and fixed with baby oil and a thin glass plate.

The PLM images were recorded on an "Olympus SZX12 stereo microscope" (Olympus Corporation, Tokyo, Japan). For this, the samples were placed between two crossed polarizers (twisted by 90°) and the images were made using the transmitted light mode.

*2.2.3. DSC.* The samples used for the differential scanning calorimetry (DSC) were taken from the same plates as those for the PLM. However, as shown in Figure 4, only selected points were analyzed using DSC (compression-molded plates: position "M"; injection-molded plates: positions "S1," "S2," and "S3"). Based on our experiences from previous studies, the authors consider DSC as a rather rough method for the determination of the crystallinity of POM. Hence, the analysis was limited to the compression-molded plates and to the injection-molded plates processed with 90°C mold temperature. After preparing the samples for PLM, small pieces with 5–7 mg of weight were cut off the remaining parts of the plates with a cutter. Because each sample was intended to represent the entire thickness of the plate, thin rods with 4 mm (for the compression-molded plates) and 5 mm (for the injection-molded plates) in length, respectively, were obtained. The samples were put into 40 $\mu$l aluminum crucibles and sealed with caps with two holes for pressure compensation. For each measurement position, in total thirteen DSC measurements were conducted.

Table 4: Temperature profile used for the DSC measurements.

| 1$^{st}$ heating run | Hold | Cooling run | Hold | 2$^{nd}$ heating run |
|---|---|---|---|---|
| 25°C → 230°C | 230°C | 230°C → 25°C | 25°C | 25°C → 230°C |
| 10 K/min | 3 min | 10 K/min | 3 min | 10 K/min |

The DSC measurements were done on a "DSC 1 STAR$^e$ System" (Mettler-Toledo International Inc., Columbus, Ohio, USA). Before the measurements, the calibration of the machine was checked using a standard indium sample. All samples were tested using the same temperature profile given in Table 4. The degree of crystallinity ($X_{C,DSC}$) of the samples was calculated using (4).

$$X_{C,DSC} = \frac{\Delta H}{\Delta H_0} * 100, \quad (4)$$

where $\Delta H_0$ is the specific melt enthalpy of 100% crystalline POM, which is 250 J/g according to [38]. $\Delta H$ is the specific melt enthalpy measured in the DSC tests (1$^{st}$ heating run). $X_{C,DSC}$ is the mass degree of crystallinity obtained by DSC.

*2.2.4. Fatigue Fracture Tests.* Fatigue fracture tests were selected for the fracture mechanical characterization

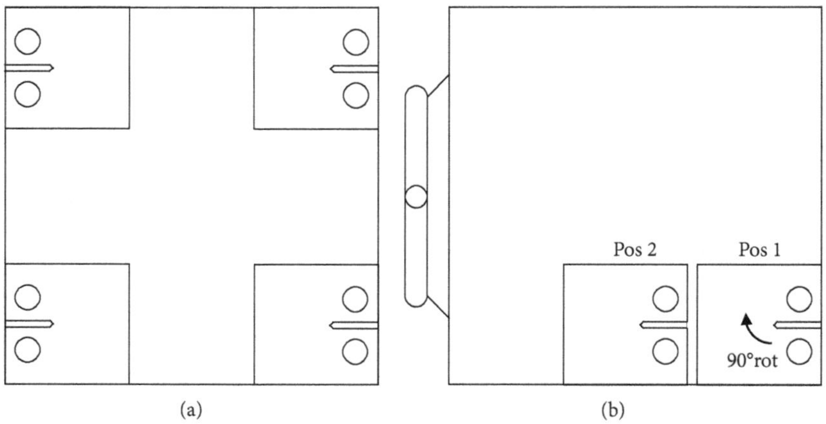


FIGURE 5: Alignment of the CT specimens on the compression-molded plates (a) and on the injection-molded plates (b).

TABLE 5: Load levels used for the fatigue fracture tests ($F_{max}$, $F_{min}$ are the maximum and minimum load of the sinusoidal signal, respectively).

| | Data point | 1 | 2 | 3 | 4 |
|---|---|---|---|---|---|
| Compression-molded CT specimens | $F_{max}$ (N) | 400 | 440 | 460 | 520 |
| | $F_{min}$ (N) | 40 | 44 | 46 | 52 |
| Injection-molded CT specimens | $F_{max}$ (N) | 500 | 550 | 575 | 600 |
| | $F_{min}$ (N) | 50 | 55 | 57.5 | 60 |

because they are known to be very useful for material ranking [15, 16]. Moreover, they were found to be very sensitive on morphological changes [13, 15, 17, 51]. In general, the method used in this study is based on the recently approved standard ISO 18489 [52]. However, CT specimens instead of the recommended cracked round bar (CRB) specimens were used in our case because they can be produced of thinner plates. The applicability of this specimen type for POM tested under similar testing conditions was shown in previous studies [14, 15]. As it is suggested in the attachment of ISO 18489 [52], the stress intensity factor was used to express the local load level inside the specimens (ahead of the crack tip). The aim of the fatigue fracture tests was to determine the number of cycles to final fracture in dependence on the initial stress intensity factor amplitude ($\Delta K_{init}$). The corresponding diagrams will be termed "fatigue fracture curves" in the following discussions.

The CT specimens were prepared from the molded plates using a mill cutter. The positions where they were taken from are given in Figure 5. For the injection-molded plates with 60°C and 110°C mold temperature, the specimens were taken from position 1 only. For the injection-molded plates with 90°C mold temperature, the specimens were taken from positions 1 and 2 in order to examine areas near and far from the sprue. Additionally, specimens from position 1 but rotated clockwise by 90° were made for this mold temperature in order to investigate the dependence of the fatigue fracture performance on the melt flow direction (in Results and Discussion, these samples will have the suffix "Pos1-90°rot" in their name). The CT specimens were made with a specimen width $W$ of 40 mm. Because of the different processing techniques used, they had two different thicknesses: 4 mm (compression-molded plates) and 5 mm (injection-molded plates). Directly before the measurements, the pre-milled notch was sharpened with a razor blade (geometry (approximate values): thickness 0.1 mm, tip radius 1–2 $\mu$m, opening angle 15°) to generate a defined and uniform initial crack geometry.

The tests were carried out on a servohydraulic testing machine of the type "MTS 858 horizontal" (MTS System Corporation, Minneapolis, Minnesota, USA). A load ratio $R$ (minimum load/maximum load) of 0.1 and a frequency of 10 Hz (sinusoidal signal) were used for all measurements. All tests were done in load-controlled mode. For both resins, all processing conditions and all position variations examined, one complete fatigue fracture curve was generated. Each of them consisted of four measurement points tested at four different load levels. The corresponding load values are shown in Table 5. Because of the different plate thicknesses and fatigue fracture resistances, these load values were different for compression- and injection-molded plates. Following ISO 18489 [52] and owed to the long test duration of fatigue tests, for each load level, only one single specimen was tested. Based on our experience with previous tests made on POM and confirmed by the present experiments, the scatter between the tests is sufficiently small to allow this proceeding.

## 3. Results and Discussion

From a currently running study, we know that POM can show significant post processing changes when stored at room temperature. This is important because several months were required to finish the characterization program of the plates. However, since in this work the materials were

TABLE 6: SAXS measurements of the CT specimens analyzed using 1D correlation (LP is the long period, *a* is the thickness of the amorphous layer, *l* is the lamellar thickness of the crystalline lamella).

| CT specimens | LP[1] (nm) | $a^1$ (nm) | $l^1$ (nm) | Crystallinity[2] (vol%) | Amorphous fraction[2] (vol%) |
|---|---|---|---|---|---|
| Delrin_CM | 17.2 ± 0.2 | 4.8 ± 0.1 | 12.4 ± 0.2 | 72.0 | 28.0 |
| Delrin_IM_60°C-MT | 14.6 ± 0.1 | 4.6 ± 0.1 | 10.0 ± 0.1 | 68.5 | 31.5 |
| Delrin_IM_90°C-MT_Pos1 | 15.5 ± 0.1 | 4.9 | 10.6 | 68.7 | 31.3 |
| Delrin_IM_90°C-MT_Pos2 | 15.4 ± 0.1 | 4.7 ± 0.1 | 10.7 ± 0.1 | 69.4 | 30.6 |
| Delrin_IM_110°C-MT | 16.4 ± 0.1 | 5.0 | 11.5 | 69.8 | 30.2 |
| Tenac_CM | 16.3 ± 0.2 | 4.8 ± 0.1 | 11.5 ± 0.1 | 70.7 | 29.3 |
| Tenac_IM_60°C-MT | 14.2 | 4.7 ± 0.1 | 9.5 ± 0.1 | 66.7 | 33.3 |
| Tenac_IM_90°C-MT_Pos1 | 15.1 | 4.9 ± 0.1 | 10.2 ± 0.1 | 67.6 | 32.4 |
| Tenac_IM_90°C-MT_Pos2 | 15.2 | 4.9 | 10.3 | 67.5 | 32.5 |
| Tenac_IM_110°C-MT | 15.9 | 5.1 ± 0.1 | 10.9 ± 0.1 | 68.2 | 31.8 |

[1]Average value ± standard deviation (if the latter is not given it is smaller than 0.05). [2]Average value.

TABLE 7: WAXD measurements of the CT specimens analyzed for the contribution of the different diffraction peaks to the total crystallinity.

| CT specimens | Crystallinity[1] (%) | (100)[2] (%) | (105)[2] (%) | (115)[2] (%) | (205)[2] (%) | Amorphous fraction[2] (%) |
|---|---|---|---|---|---|---|
| Delrin_CM | 75.8 ± 0.8 | 60.1 | 12.2 | 2.9 | 0.6 | 24.2 |
| Delrin_IM_60°C-MT | 72.6 ± 1.0 | 60.4 | 10.6 | 1.4 | 0.1 | 27.4 |
| Delrin_IM_90°C-MT_Pos1 | 75.1 ± 0.9 | 62.0 | 11.2 | 1.7 | 0.2 | 24.9 |
| Delrin_IM_90°C-MT_Pos2 | 75.9 ± 0.7 | 64.1 | 10.2 | 1.5 | 0.2 | 24.1 |
| Delrin_IM_110°C-MT | 74.3 ± 0.3 | 61.2 | 11.4 | 1.6 | 0.2 | 25.7 |
| Tenac_CM | 74.8 ± 0.5 | 58.6 | 12.6 | 3.0 | 0.7 | 25.2 |
| Tenac_IM_60°C-MT | 70.7 ± 0.7 | 57.0 | 11.8 | 1.7 | 0.2 | 29.3 |
| Tenac_IM_90°C-MT_Pos1 | 70.5 ± 0.8 | 56.4 | 12.1 | 1.7 | 0.3 | 29.5 |
| Tenac_IM_90°C-MT_Pos2 | 71.6 ± 0.6 | 58.0 | 11.7 | 1.7 | 0.2 | 28.4 |
| Tenac_IM_110°C-MT | 71.3 ± 0.4 | 56.9 | 12.3 | 1.8 | 0.3 | 28.7 |

[1]Average value ± standard deviation. [2]Average value.

processed under comparatively moderate conditions, the aging effect is expected to be of minor importance.

In order to facilitate the readability of this section, a uniform nomenclature is used for the different plates/samples/specimens examined. The following abbreviations are used in this context:

(i) "CM" for compression molding and "IM" for injection molding

(ii) "MT" is the mold temperature used in injection molding

(iii) "M," "M1," "M2," "M3," "S1," "S2," "S3," and "S4" are the position indicators for the PLM and DSC samples (in accordance with Figure 4)

(iv) "Pos1," "Pos2," and "Pos1-90°rot" give details about the CT specimen alignment on the injection-molded plates processed with 90°C mold temperature (in accordance with Figure 5)

### 3.1. Material Morphology

*3.1.1. SAXS and WAXD.* The results of the SAXS and WAXD measurements made on the CT specimens are summarized in Tables 6 and 7. The values given in these tables are averages of five measurements. In Table 6, the results of the SAXS measurements analyzed using 1D correlation are given. For both materials, the CT specimens made of the compression-molded plates have the biggest long periods and lamellar thicknesses. The amorphous layer thicknesses, on the other hand, are similar to those of the injection-molded samples. For the CT specimens made of the injection-molded plates, long period, lamellar thickness, and amorphous layer thickness increase with increasing mold temperature. However, even for the highest mold temperature used (110°C), the long period and lamellar thickness do not reach the levels of the compression-molded specimens. In comparison to Tenac, the Delrin specimens reveal slightly higher long periods and lamellar thicknesses. For the amorphous layer thicknesses, both materials show very similar values. The two different positions, which were examined for the injection-molded plates with 90°C mold temperature ("Pos1" and "Pos2" in Table 6), show no significant differences in long period, lamellar thickness, and amorphous layer thickness.

In addition to the long period and layer thicknesses, the (volumetric) degree of crystallinity was calculated using (1). The highest values for this parameter are observed for the

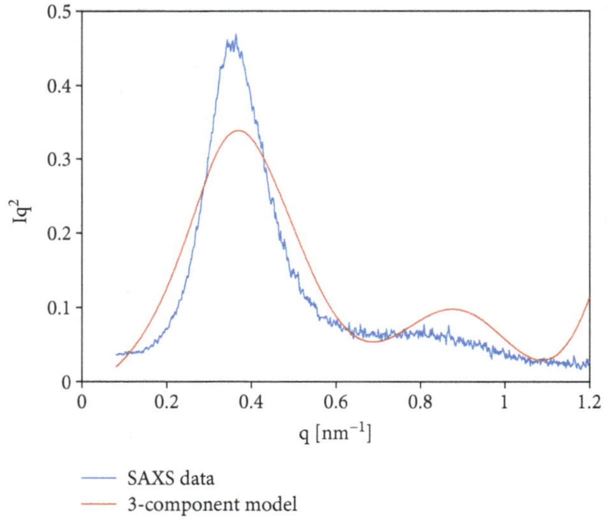


FIGURE 6: Representative result of the 3-component model fit (SAXS data belongs to the Delrin_IM_110°C-MT sample).

compression-molded CT specimens. For the specimens made of the injection-molded plates, only a small increase with increasing mold temperature is observed. For the two materials examined, Delrin shows the slightly higher degree of crystallinity in all processing conditions used. However, it is evident from Table 6 that the changes in the (volumetric) degree of crystallinity are less pronounced than those in long period and lamellar thickness.

As described above in Materials and Methods, a second analysis method for the SAXS data was used in this study: a fitting procedure, which uses a 3-component model developed especially for POM [21, 45, 46]. The background for this proceeding was, as mentioned before, that for POM it is nowadays accepted that there are three different phases (two different types of crystalline lamellae and the amorphous phase) present in a conventionally crystallized material [1, 21, 45–47]. Since 1D correlation is based on a two-phase model, a part of the morphology information is lost, when using this analysis technique. This loss of information can be prevented when using the 3-component model developed by Hama and Tashiro [21, 46]. However, for our SAXS data, no reasonable results were obtained with this analysis method. Although a lot of effort was made (many different starting points for the fit were tried, even ridiculous ones), the fitting quality achieved was not acceptable. During the analysis procedure, the 3-component model generally caused some doubts about its reliability. Figure 6 is a representative for all fitting results; it was generally observed that the first peak in the Lorentz-corrected SAXS profile was underestimated by the fit, whereas the second one was overestimated, and the peak maxima of the data and the fit did not coincide. Additionally, a further, rather significant peak was observed in the fit results, which started at a $q$-value of approximately 1.1 nm$^{-1}$. This additional peak was always obtained, and hence, the diagrams were truncated at 1.2 nm$^{-1}$ as it was done by Hama and Tashiro in their papers [21, 46]. From personal communications, we know that an additional peak at higher $q$ levels was also observed by other working groups. It is speculated if this peak is not an erroneous artefact of the model. The described issues were observed similarly for Delrin and Tenac. Based on the observations made and although Hama and Tashiro mentioned Delrin and Tenac in their work [21], the applicability of the 3-component model for these two materials is questionable at least for the high molecular weight resins used in our study.

For the analysis of the WAXD data given in Table 7, two main aspects are taken into account: the (mass degree of) crystallinity and the quality of the crystal packing. The latter is deduced from the contribution of the crystal peaks located at higher diffraction angles to the total area of the WAXD profile. The compression-molded samples show both, slightly higher crystallinity and higher quality of the crystal packing (higher contribution of (105), (115), and (205) diffraction peaks) than the injection-molded ones. For the injection-molded CT specimens, the quality of the crystal packing does not change a lot with the mold temperature. In addition, the increase in crystallinity is rather small when taking into account the standard deviation. However, the increase is more pronounced for Delrin than for Tenac. For both processing techniques used (compression molding and injection molding), Delrin shows a higher level of crystallinity. The comparison of the CT specimens obtained from plate positions 1 and 2 ("Pos1" and "Pos2" in Table 7) shows that at position 2 the total crystallinity is similar or slightly higher, but the quality of the crystal packing is lower (higher fraction of (100), smaller or similar fraction of (105), (115), and (205)). For the SAXS results, such a position dependence was not observed.

For the interpretation of the SAXS and WAXD results, it has to be kept in mind that the whole cross section was examined. Hence, the quantitative results for crystallinities, long periods, layer thicknesses, and diffraction peak contributions (to the overall crystallinity) represent average values along the thickness. To obtain this information on a local scale, a layer-by-layer SAXS/WAXD analysis was carried out as a follow-up to the present study. The analysis was made for selected conditions of the ones discussed in this study. However, the additional information obtained from this layer-by-layer analysis was rather limited. It was shown that on this local scale the influence of sample preparation and analysis parameters becomes dominant. Hence, the corresponding results are not implemented in the present study.

*3.1.2. PLM.* In the following, the cross section microstructure of the compression- and injection-molded plates will be discussed for both materials. In all cases, the PLM images selected are representative for the condition examined. In Figure 7, the images of the compression-molded plates are given. Due to the added heterogeneous nucleation agent, the compression-molded Delrin sample (Figures 7(a) and 7(b)) has a very fine structure over the entire thickness. A rough estimation of the spherulite diameters observed in Figure 7(b) gave an approximate diameter range of 25–50 $\mu$m. A skin region at the surface cannot be observed for this sample, which leads to the conclusion that Delrin does not form a skin layer when processed by compression

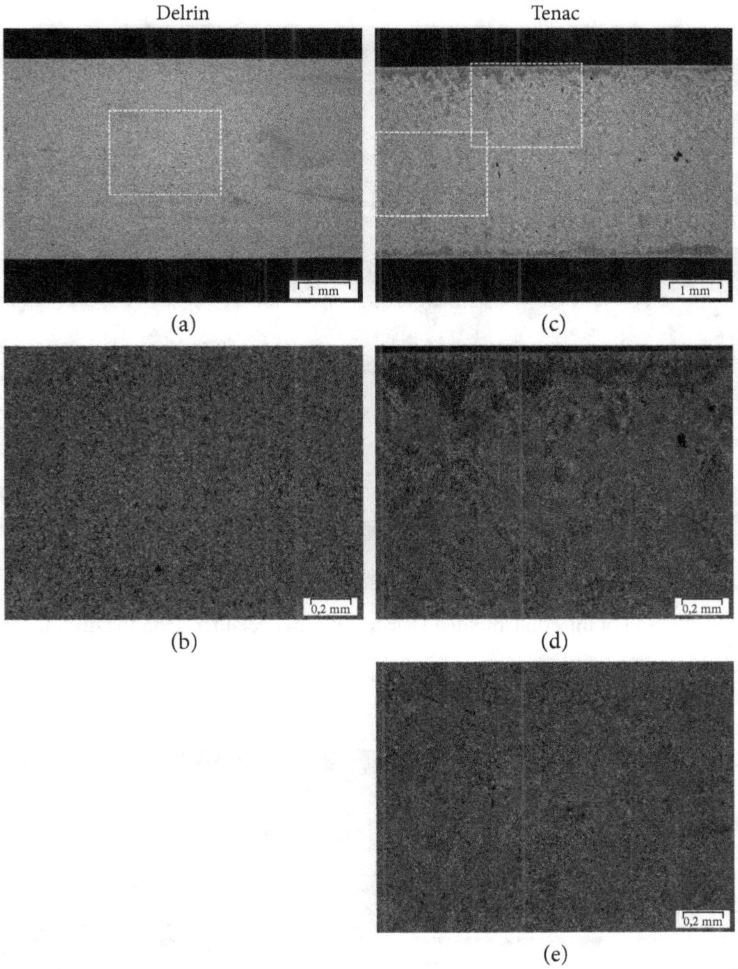


FIGURE 7: PLM images of the cross section of compression-molded plates. Overview and details for Delrin (a, b) and Tenac (c–e).

molding with the processing conditions used. There are "fuzzy" areas distributed irregularly over the plate, which were observed at all five sample positions shown in Figure 4. They seem to contain even finer spherulites than the rest of the material. Their origin has not been identified yet, but they are assumed to represent the transition zone between adjacent granulate grains of the resin.

Compared to Delrin, the compression-molded Tenac plate (Figures 7(c)–7(e)) has a much coarser structure and a distinct skin layer. A skin layer for the compression-molded plate was rather unexpected since the cooling conditions were very moderate. However, this skin layer was observed at all five sample positions investigated. It is shown in Figure 7(d) with a higher magnification. The area directly below the plate surface shows less pronounced crystal structures (no spherulites) than the rest of the sample. Below, irregular-shaped spherulites with longitudinal orientation towards the center of the plate can be observed (size approximately 100–250 $\mu$m). Below a depth of about 0.15 mm, the microstructure is dominated by spherulites with regular shape (rough diameter range is 100–250 $\mu$m). Towards the center of the plate, the spherulite sizes do not change further but the overall spherulitic structure becomes more and more compact.

As it is illustrated in Figures 8–11, the microstructure of the injection-molded plates was found to be more complex than that of the compression-molded ones. In Figures 8 and 9, the corresponding PLM images of Delrin and Tenac are shown for the different mold temperatures used in the injection molding process. They are aligned along the filling direction from near the sprue to far from the sprue (in accordance with the positions "S1," "S2," and "S3" in Figure 4). The "S" positions were selected because the CT specimens used for the fracture mechanical tests were also made of this area. The PLM images at the "M" positions were very similar to the corresponding "S" position images.

In the following, the layer structure observed will be discussed with respect to previous findings in the literature. Liparoti et al. determined a general structure for injection-molded polypropylene plates, which consists of the following three layers starting from the outside [26]:

(i) Skin layer with two subphases

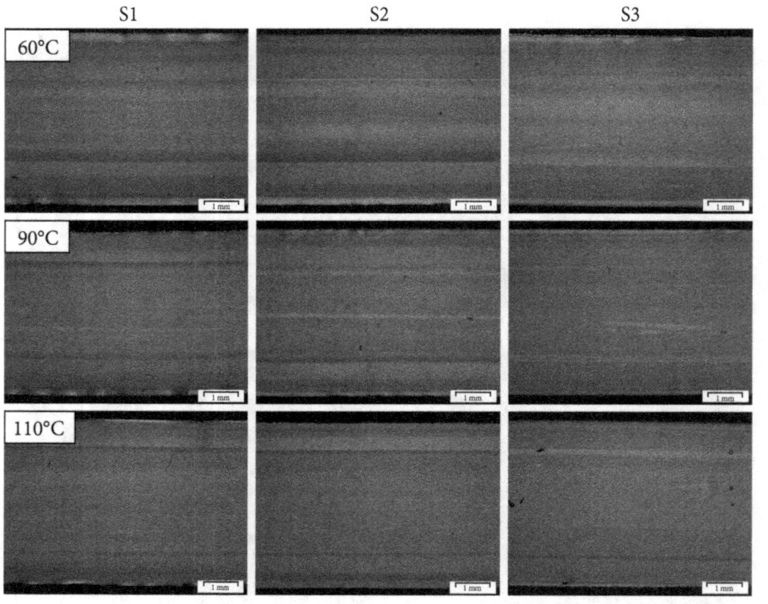


FIGURE 8: PLM images of the cross section of injection-molded Delrin plates in dependence on the distance from the sprue and on the mold temperature used.

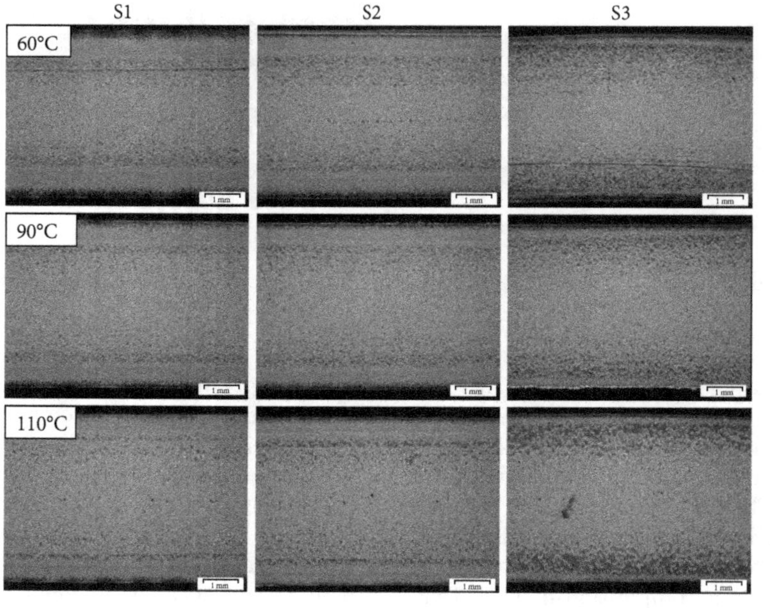


FIGURE 9: PLM images of the cross section of injection-molded Tenac plates in dependence on the distance from the sprue and on the mold temperature used.

(ii) Shear layer with structures aligned along the flow direction

(iii) Spherulitic layer in the central zone of the plate

The study of Liparoti et al. is the most elaborated one concerning the layer structure of unfilled and injection-molded semicrystalline thermoplastics. Although it was conducted for an alternative material (polypropylene), it will be used as reference in this work.

For Delrin, a quite distinct layer structure with numerous layers and very fine spherulites can be observed in Figure 8. This layer structure does not correlate well with what was observed by Liparoti et al. for polypropylene. As a rough trend, the number of layers decreases from low to high mold temperature and from near the sprue ("S1") to far from the sprue ("S3"). It seems that the lower the temperature difference between the mold and the melt, the less layers appear. It is interpreted that most of the additional layers represent

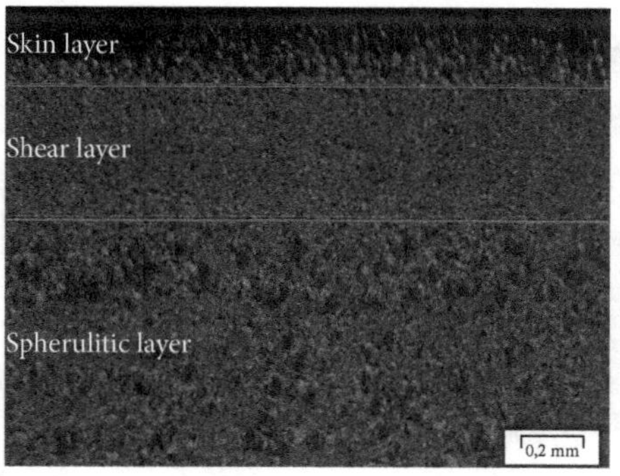


FIGURE 10: Layer structure of the injection-molded Tenac plates (PLM image of "Tenac_IM_90°C-MT_S1").

additional shear zones because of the quite fast solidification of the nucleated Delrin resin in the mold. Far from the sprue, the number of layers decreases because less material has to pass this area and thus less material shearing occurs. A classical skin layer with less pronounced crystal structures is only observed for the "harshest" conditions used ("Delrin_IM_60°C-MT-S1": 60°C mold temperature and close to the sprue). Even in this case, the thickness of this skin layer is very small (in the order of 50 $\mu$m). Besides the thinner layers, three main layers can be distinguished in Figure 8 (especially for the samples with high mold temperature and far from the sprue): two outer layers (bottom and top side) of about 1 mm thickness and a layer in the middle with a thickness of about 2 mm. In Figures 11(b) and 11(c), the surface and the middle layer for the sample "Delrin_IM_90°C-MT_S3" are shown with higher magnification. Differences in the spherulite size cannot be observed in these images. Because of their very small size, a quantitative estimation of the spherulite diameters was not possible in the light microscope. For the sake of completeness, it has to be mentioned that some white lines in Figure 8 like the ones in the middle of the samples "Delrin_IM_90°C-MT_S2" and "Delrin_IM_90°C-MT_S3" come from the sample preparation (microtome knife) and not from the processing. However, they can be identified rather easily because of their almost white color. No mentionable differences between the samples at the side positions ("S") and those in the middle of the plate ("M") were found.

The layer structure of the injection-molded Tenac plates, which is given in Figure 9, is quite different to that of Delrin. Similar to what was observed for polypropylene [26], a skin layer, a shear layer, and a spherulitic layer in the center of the plate can be observed for this material. The different layers are schematically illustrated in Figure 10. Interestingly, also the skin layer shows crystalline structures (refer to Figure 11(e) for higher magnification). For the shear layers, a slight decrease in thickness with increasing mold temperatures and between the "S1" and "S2" sample positions can be observed. However, far from the sprue (sample positon "S3"), the shear layer is much thinner for all three mold temperatures. The strong decrease in the layer thickness from "S2" to "S3" is accompanied by a less compact spherulite structure in the shear layer. The spherulitic layer in the center of the plate has a similar morphology as the compression-molded plate (Figures 7(c)–7(e)): comparably coarse spherulites, which are packed more and more compact from below the shear layer towards the center of the plate. The described microstructure is shown with higher magnification in Figures 11(d)–11(f). No influence of the mold temperature on the spherulitic layer can be observed except for the lowest one (60°C), where the spherulites in the center appear less compact. Far from the sprue (sample position "S3"), the spherulitic layer is significantly thicker than in the sample positions closer to the sprue ("S1" and "S2"). For Figures 11(e) and 11(f), the following spherulite diameters were estimated from the PLM images: shear layer around 40 $\mu$m, area in spherulitic layer directly below the shear layer around 50–70 $\mu$m, and spherulitic layer at some distance from shear layer around 70–130 $\mu$m. Also, for the Tenac samples, the influence of the sample preparation could not be avoided completely. Mainly for the 60°C mold temperature images, some scratches of the microtome knife are visible. For Tenac, the samples in the middle of the plate ("M") showed some additional effects compared to those at the side positions ("S"). In both, the "M1" and the "M2" positions, the shear layers clearly become smaller with increasing mold temperature. Moreover, double shear layers at the bottom and at the top of the plate were found for the "M1" positions (near the sprue) for all three mold temperatures. Similar to the multiple shear layers observed for Delrin, this is related to the solidification of the outermost shear layer during the filling process of the mold.

*3.1.3. DSC.* The results of the DSC examination are summarized in Table 8 as average and standard deviation of the thirteen measurements conducted per sample position. As already mentioned in Section 2.2.3, the DSC examination was limited to selected processing conditions and sample positions. The reason for this was that in our previous research, DSC was found to be very insensitive to morphology changes in POM. However, as shown in Table 8, it was sufficiently sensitive to detect some differences in the crystallinity of the samples investigated. For both materials, the compression-molded samples had the highest degree of crystallinity. The difference between compression-molded samples and injection-molded ones was bigger in the case of Tenac. Comparing both materials, Delrin has slightly higher degrees of crystallinity in both the compression-and injection-molded samples. At least for the injection-molded plates analyzed, no crystallinity trend from near the sprue to far from the sprue is indicated by the results. For the interpretation of the results, it has to be kept in mind that some of the crystallinity differences documented in Table 8 are in the same order as the standard deviation.

*3.2. Fatigue Fracture Tests.* The results of the fatigue fracture tests of the Delrin and Tenac CT specimens are summarized in Figure 12. While there is rather little influence of the processing conditions on the lifetime of Delrin, the Tenac results show a clear dependence. For better illustration and because the logarithmic scale is usually misleading, the lifetime ratios between best and worst performing curves are exemplarily given in Table 9 for both materials. The required cycles to fracture values were calculated based on the fitting equations

FIGURE 11: PLM images of the cross section of injection-molded plates. Overview and details for "Delrin_IM_90°C-MT_S3" (a–c) and "Tenac_IM_90°C-MT_S3" (d–f).

TABLE 8: DSC crystallinity of compression- and injection-molded POM plates.

| Sample position | Crystallinity (%) | Standard deviation (%) |
|---|---|---|
| Delrin_CM_M | 73.3 | 2.9 |
| Delrin_IM_90°C-MT_S1 | 70.5 | 2.2 |
| Delrin_IM_90°C-MT_S2 | 71.2 | 2.8 |
| Delrin_IM_90°C-MT_S3 | 69.0 | 2.8 |
| Tenac_CM_M | 71.5 | 2.6 |
| Tenac_IM_90°C-MT_S1 | 67.3 | 1.9 |
| Tenac_IM_90°C-MT_S2 | 67.7 | 1.8 |
| Tenac_IM_90°C-MT_S3 | 67.1 | 1.7 |

of the corresponding trend lines in Figure 12. The lifetime ratios were calculated for a specific $\Delta K_{init}$ value of 2.9 MPa $m^{0.5}$. This value was arbitrarily selected for comparison purposes. It is clear from Figure 12 that the lifetime ratios and even the best and worst performing curves are not uniform over the whole $\Delta K_{init}$ range examined. However, Table 9 shows that for the $\Delta K_{init}$ value selected, the difference in the best and worst performing curves is about 10% for Delrin and almost 50% for Tenac. Obviously, the nucleation agent contained in Delrin reduces the processing sensitivity of the fatigue fracture properties significantly. On the other side, the variation of almost 50% in the lifetime of Tenac is quite a lot, when considering the rather moderate injection molding conditions used.

In the case of Delrin, the best performing processing condition was "Delrin_IM_110°C-MT." It was slightly better than the other conditions, which did not differ significantly over a wide range of $\Delta K_{init}$ values. Nevertheless, "Delrin_IM_90°C-MT_Pos1-90°rot" showed a significantly reduced lifetime at the measurement point with the highest $\Delta K_{init}$ value. For the corresponding Tenac CT specimen of "Tenac_IM_90°C-MT_Pos1-90°rot," this effect was even more distinct and it broke after a few thousand cycles (far below the scaling minimum of Figure 12(b)). For Tenac, the best performing curve was that of the compression-molded material ("Tenac_CM"). For small $\Delta K_{init}$ values, the worst performing curve was "Tenac_IM_60°C-MT." This

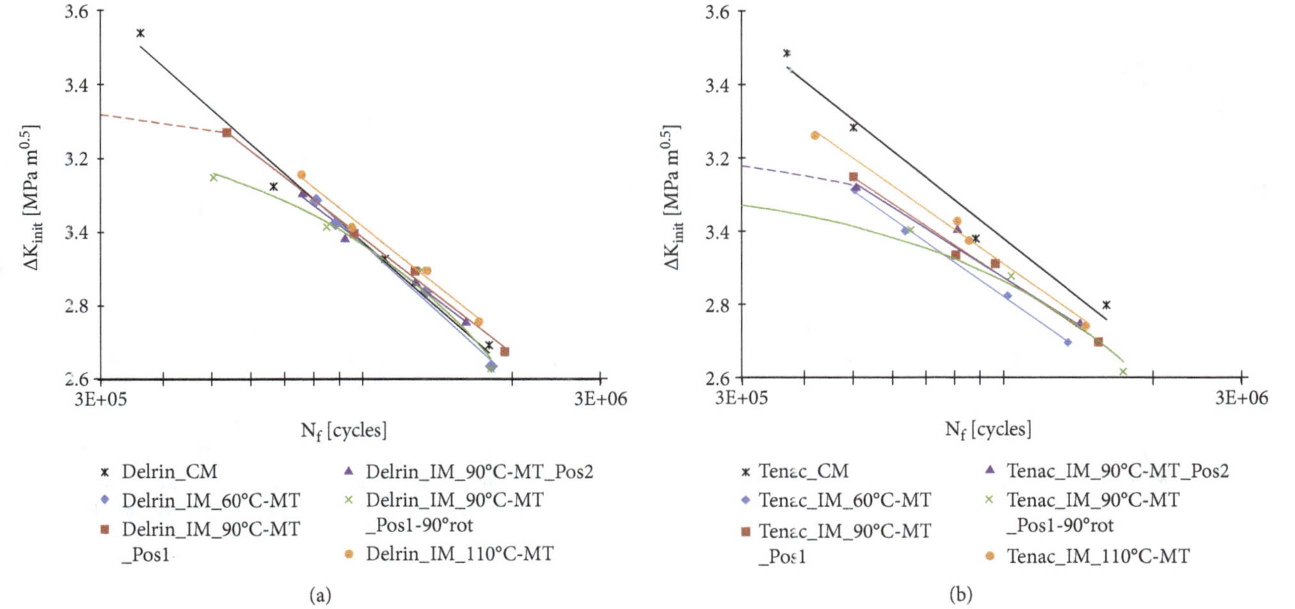


FIGURE 12: Fatigue fracture tests of CT specimens made of compression- and injection-molded plates. Results for Delrin (a) and Tenac (b).

TABLE 9: Lifetime ratios for Delrin and Tenac calculated for one specific $\Delta K_{init}$ value.

| Resin | $\Delta K_{init}$ (MPa m$^{0.5}$) | Worst performing trend line | Best performing trend line | Lifetime ratio (Best/Worst * 100%) |
|---|---|---|---|---|
| Delrin | 2.9 | Delrin_IM_60°C-MT | Delrin_IM_110°C-MT | 112% |
| Tenac | 2.9 | Tenac_IM_60°C-MT | Tenac_CM | 145% |

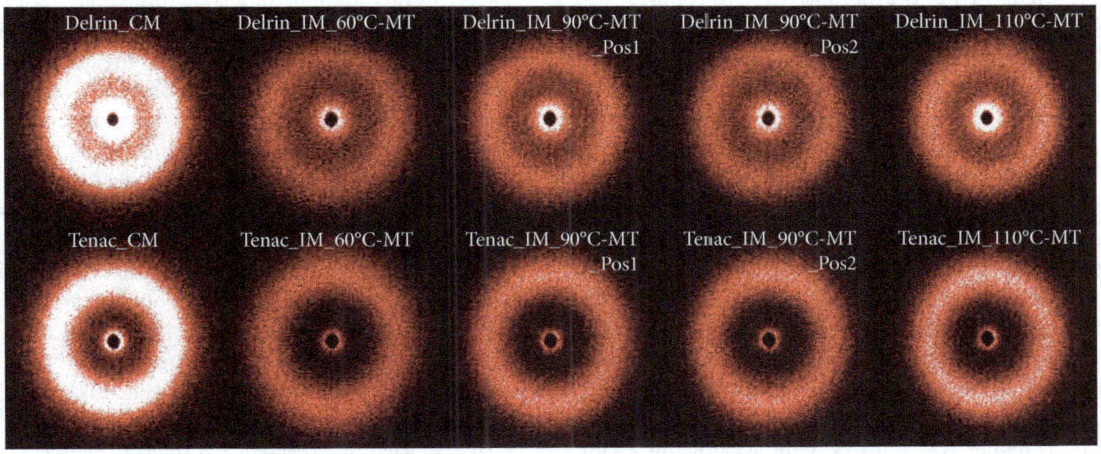


FIGURE 13: Representative SAXS patterns obtained for the different processing conditions.

changed abrupt with the kink in the "Tenac_IM_90°C-MT_Pos1-90°rot" curve at approximately 2.9–3.0 MPa m$^{0.5}$.

Interestingly, an abrupt decrease in the lifetime similar to that of the rotated specimens was also found for "Delrin_IM_90°C-MT_Pos1" and "Tenac_IM_90°C-MT_Pos2" during the conception phase of the fatigue fracture tests (the specimens broke after a few thousand cycles and thus far below the scaling minimums in Figure 12). This is indicated by the dashed lines in Figure 12, and it was the reason why the maximum $\Delta K_{init}$ values used for the tests of the injection-molded specimens were lower than those of the compression-molded ones. Thus, the kink in the fatigue fracture curve is not a phenomenon of the rotated samples only. For the $\Delta K_{init}$ values examined, the compression-molded samples did not show any abrupt lifetime decrease. For the injection-molded specimens with 60°C and 110°C mold temperature, this additional information is not available. For the sake of completeness, it has to be mentioned that although the different injection molding conditions were all tested with identical load levels (refer to Table 5), some

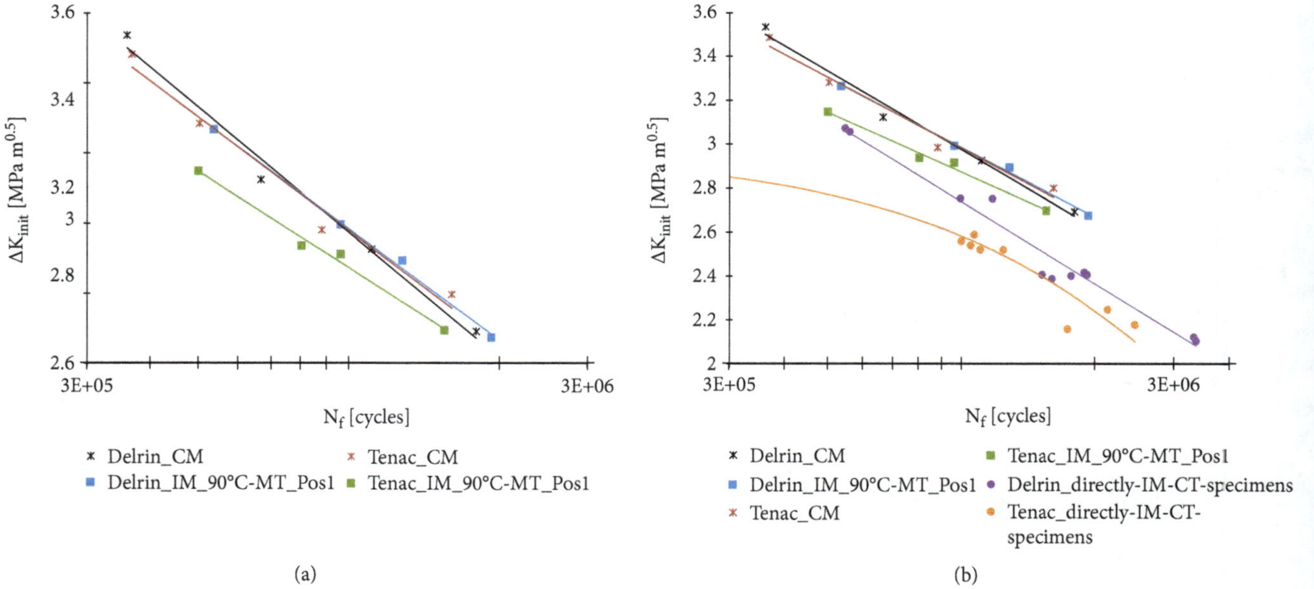


FIGURE 14: Comparison of fatigue fracture test results of Delrin and Tenac CT specimens. (a) Selected processing conditions used in this study. (b) Additional results obtained for directly injection-molded CT specimens produced under industrial conditions.

variation in the $\Delta K_{init}$ values can be observed in Figure 12. This variation was caused by different initial crack lengths introduced through the sharpening process of the notch using the razorblade.

Since no significant orientations could be detected using SAXS (Figure 13), the difference between the rotated and the not rotated (injection-molded) CT specimens was unexpected. However, despite the lack of (detectable) orientations, it is assumed that the difference was caused by the shear deformation in flow direction. It is speculated that during the filling process, the molecular orientation caused by this shear deformation was accompanied by partial molecular disentanglement through molecular slipping. While the orientations were obviously able to relax before the solidification (no trace in the SAXS images/profiles), these disentanglements were frozen in the crystallized polymer. Consequently, the molecules in flow direction were partially disentangled, which caused a decreased fatigue fracture performance for the rotated CT specimens with their crack plane orientation perpendicular to this direction.

The ability of the compression-molded samples to withstand even higher $\Delta K_{init}$ levels without showing a kink in the fatigue fracture curve is related to the higher quality of the crystal packing in comparison with the injection-molded samples. This higher crystal packing quality is deduced from the WAXD results; the compression-molded samples showed a higher contribution of higher diffraction peaks ((105), (115), and (205)). Additionally, in the SAXS 1D correlation results, the compression-molded samples had the biggest lamellar thicknesses, while the amorphous layer thicknesses were similar to the injection-molded samples.

For the injection-molded specimens, the lifetime differences at lower $\Delta K_{init}$ levels (Figure 12, below the kinks) are not reflected by the results of the morphology investigation. For Delrin, the results of SAXS 1D correlation (Table 6) and WAXD (Table 7) show some reaction on the changed processing conditions but the fatigue fracture performance (Figure 12) almost does not. For Tenac, on the other hand, the SAXS 1D correlation results and the fatigue fracture curves change with the processing conditions but for the WAXD results there is only a slight change. Based on these results, it is concluded that, at least for the injection-molded specimens, the influence of the processing conditions on the fatigue fracture performance (at lower $\Delta K_{init}$ levels) cannot be explained by the morphology changes only. Since no orientations are indicated by the SAXS patterns in Figure 13, their influence is also excluded and the additional effects are attributed to residual stresses inside the specimens.

Indeed, the fatigue fracture performance of the injection-molded specimens in Figure 12 can be explained quite well by residual stresses, when taking into account the following considerations. The PLM images of the injection-molded plates (Figures 8 and 9) show a very homogeneous spherulite structure for Delrin, whereas it is rather inhomogeneous for Tenac. Hence, it is expected that the (nucleated) Delrin resin crystallizes and solidifies in a more uniform manner than the Tenac resin does. As a result, the residual stresses in the injection-molded Delrin plates are expected to be smaller and less dependent on the mold temperature than those in the injection-molded Tenac plates.

Based on these assumptions, in the following, the fatigue fracture curves are reinterpreted by taking into account the expected residual stress effects. The significant mold temperature dependence of the Tenac fatigue fracture curves observed in Figure 12 (while the corresponding WAXD crystallinity is almost unaffected) is related to a strong change in the residual stress level. According to the literature, the lower the mold temperature, the higher is the expected residual stress level [53, 54] and for Tenac, the lower is the observed fatigue fracture performance. Hence, it is interpreted that there is an inverse proportionality between the fatigue fracture performance and the residual stress level, which is consistent with the effect of residual stresses on the tensile properties [55] and on the slow crack growth behavior of polymeric pipes [56, 57]. For Delrin, the insensitivity of the fatigue fracture curves to the mold temperature is, as described above, related to the nucleation of the material, which is expected to lead to a rather uniform residual stress level for the processing conditions examined. It has to be emphasized at this point that in terms of the morphology, the changes in dependence on the mold temperature were similar for Delrin and Tenac (SAXS 1D correlation, contribution of higher diffraction peaks to total crystallinity in WAXD) or even slightly higher for Delrin (WAXD total crystallinity).

To compare Delrin and Tenac directly, fatigue fracture curves of compression- and selected injection-molded CT specimens of both materials are plotted in Figure 14(a). Generally, the compression-molded samples are expected to reveal a very small (if at all) residual stress level, and the different morphologies in relation to the injection-molded samples were already discussed above. Interestingly, the compression-molded curves of both materials and the injection-molded one of Delrin practically coincide while the injection-molded one of Tenac has a worse fatigue fracture performance. The coincidence of the compression-molded curves is not surprising when considering the similar WAXD results, the mentioned small residual stress level expected, and the similar molecular weight and polydispersity of both resins (Section 2.1). Obviously, the differences in terms of the SAXS 1D correlation results do not significantly affect the fatigue fracture performance. Taking into account the completely different spherulite sizes and distributions shown by the PLM images in Figure 7, it is confirmed what has already been observed by other authors [13, 19]; at least for POM, there is no direct influence of the spherulite size and distribution on the fracture mechanical properties. This is plausible because in POM and other polymeric materials, the crack path was found to run through the spherulites [8, 11, 12, 18, 58–61] as long as the latter do not pass a distinct level of size and crystalline perfection [18, 58]. The additional coincidence of the injection-molded Delrin curve with the compression-molded ones is not surprising from the point of the total WAXD crystallinity. However, there are also morphological differences: the already mentioned lower contribution of the higher WAXD peaks to the total crystallinity (interpreted as lower quality in the crystal packing) and clear differences in the SAXS 1D correlation results. The level of the residual stresses in the injection-molded Delrin specimens is unknown especially since the nucleated Delrin resin seems to be less prone to the generation of residual stresses than the Tenac resin (details were discussed above). Hence, it is speculated that either the morphological differences between the compression- and the selected injection-molded Delrin specimens do not significantly affect the fatigue fracture curves and/or that they are compensated by residual stress effects. The worse fatigue fracture performance of the injection-molded Tenac specimens in comparison to the compression-molded curves is interpreted as a combined effect of different morphologies (smaller total crystallinity and lower contribution of the higher diffraction peaks to the total crystallinity in WAXD, clear differences in the SAXS 1D correlation results) and different residual stress levels.

In Figure 14(b), fatigue fracture curves of directly injection-molded CT specimens of Delrin and Tenac are added to the curves of Figure 14(a). The Tenac curve was already published previously together with information concerning processing conditions, specimen mold design (crack growth perpendicular to flow direction), and morphological details [15]. For the directly injection-molded Delrin specimens, the same processing conditions and mold design were used. The important point is that technical conditions with much smaller cycle times than in our study were used for the processing. It is obvious that this has a much stronger influence on the fatigue fracture performance than the comparatively small processing variations used in our study. The comparison with the directly injection-molded specimens is intended as outlook to demonstrate the wide field in which the fatigue fracture performance of unfilled POM can be varied (logarithmic scale on $x$-axis) only by modifying the processing conditions.

## 4. Conclusions

Quadratic plates of two POM homopolymer resins (Delrin and Tenac) were produced by compression molding and by injection molding under moderate conditions. The injection molding was conducted with three different mold temperatures. At specific locations, the plates were analyzed concerning their morphological structures on the micro- (spherulite size and distribution, processing layer structure/PLM) and nanolevel (long period, lamellar thickness, and amorphous layer thickness/SAXS), concerning their crystallinity (DSC, SAXS, and WAXD) and concerning their fatigue fracture properties. For the latter, CT specimens were cut out from the plates at defined positions. The PLM images showed distinct differences between the resins and between compression and injection molding, while for SAXS and WAXD, the differences were much smaller. Concerning the fatigue fracture performance, the Delrin resin showed only a small dependence on the processing conditions. For Tenac, on the other hand, this dependence was more pronounced. The differences in this dependence could not be explained by morphological considerations alone. Since no evidence for orientations was found in the SAXS patterns, it was concluded that residual stresses had to be responsible for them. Because the nucleated Delrin resin revealed less dependence of the fatigue fracture performance on the processing conditions than the Tenac resin (in combination with similar morphological changes), it was further concluded that the nucleation caused a reduced residual stress level in the plates. This was explained by a more uniform solidification in the mold because of the faster crystallization in the presence of the nucleation agent.

As a general conclusion, for future studies in which the influence of the processing conditions on the (fracture) mechanical behavior of polymers is examined, it is strongly suggested to combine the morphological analysis with residual stress measurements, especially when injection molding is involved. For the interpretation of the SAXS and WAXD results, it has to be kept in mind that the whole cross section was examined. Hence, the quantitative results for crystallinities, long periods, layer thicknesses, and diffraction peak contributions (to the overall crystallinity) represent average values along the thickness. A local layer-by-layer analysis could help in the interpretation of the morphology data. However, a follow-up study showed that such layer-by-layer analyses are error-prone, and significant method improvement is needed to obtain reliable results. The 3-component model fitting, which was used in this study as an alternative method to analyze the SAXS profiles of POM (and which was developed especially for POM [21, 45, 46]), did not provide acceptable fitting results. The authors doubt its applicability to high molecular weight resins of POM.

## Disclosure

A part of this paper was previously presented as a Poster at the 17th International Conference on Deformation, Yield and Fracture of Polymers (25th to 29th of March 2018 in Kerkrade/Netherlands).

## Conflicts of Interest

The authors declare that there is no conflict of interest regarding the publication of this paper.

## Acknowledgments

The research work was performed at the Polymer Competence Center Leoben GmbH (PCCL, Austria) within the framework of the COMET-program of the Federal Ministry for Transport, Innovation and Technology and Federal Ministry for Economy, Family and Youth with contributions by the Montanuniversitaet Leoben (Institute of Materials Science and Testing of Polymers). The PCCL is funded by the Austrian Government and the State Governments of Styria, Lower Austria and Upper Austria.

## References

[1] S. Lüftl, P. M. Visakh, and S. Chandran, *Polyoxymethylene Handbook: Structure, Properties, Applications and Its Nanocomposites*, John Wiley & Sons, Hoboken, NJ, USA, 2013.

[2] J. Rösler, *Zur Tragfähigkeitssteigerung Thermoplastischer Zahnräder mit Füllstoffen*, Technische Universität Berlin, 2005, Dissertation.

[3] B. Heym and W. Beitz, "Zur belastbarkeit von stirnzahnrädern aus dem hochtemperatur-thermoplast PEEK," *Konstruktion*, vol. 47, pp. 351–357, 1995.

[4] M. Berer and Z. Major, "Characterization of the global deformation behaviour of engineering plastics rolls," *International Journal of Mechanics and Materials in Design*, vol. 6, no. 1, pp. 1–9, 2010.

[5] M. Berer and Z. Major, "Characterisation of the local deformation behaviour of engineering plastics rolls," *Strain*, vol. 48, no. 3, pp. 225–234, 2012.

[6] M. Berer, I. Mitev, and G. Pinter, "Finite element study of mode I crack opening effects in compression-loaded cracked cylinders," *Engineering Fracture Mechanics*, vol. 175, pp. 1–14, 2017.

[7] B. Ramoa, M. Berer, M. Schwaiger, and G. Pinter, "Effect of cyclic fatigue on the fracture toughness of Polyoxymethylene," *Journal of Physics: Conference Series*, vol. 843, article 012052, 2017.

[8] P. E. Bretz, R. W. Hertzberg, and J. A. Manson, "The effect of molecular weight on fatigue crack propagation in nylon 66 and polyacetal," *Journal of Applied Polymer Science*, vol. 27, no. 5, pp. 1707–1717, 1982.

[9] S. Bandyopadhyay, J. R. Roseblade, and R. Muscat, "Fracture study of a polyacetal resin," *Journal of Materials Science*, vol. 28, no. 14, pp. 3777–3782, 1993.

[10] S. Hashemi and J. G. Williams, "Size and loading mode effects in fracture toughness testing of polymers," *Journal of Materials Science*, vol. 19, no. 11, pp. 3746–3759, 1984.

[11] R. W. Hertzberg, M. D. Skibo, and J. A. Manson, "Fatigue crack propagation in polyacetal," *Journal of Materials Science*, vol. 13, no. 5, pp. 1038–1044, 1978.

[12] A. Lazzeri, A. Marchetti, and G. Levita, "Fatigue and fracture in polyacetal resins," *Fatigue and Fracture of Engineering Materials and Structures*, vol. 20, no. 8, pp. 1207–1216, 1997.

[13] J. Runt and K. P. Gallagher, "The influence of microstructure on fatigue crack propagation in polyoxymethylene," *Journal of Materials Science*, vol. 26, no. 3, pp. 792–798, 1991.

[14] M. Berer and G. Pinter, "Determination of crack growth kinetics in non-reinforced semi-crystalline thermoplastics using the linear elastic fracture mechanics (LEFM) approach," *Polymer Testing*, vol. 32, no. 5, pp. 870–879, 2013.

[15] M. Berer, G. Pinter, and M. Feuchter, "Fracture mechanical analysis of two commercial polyoxymethylene homopolymer resins," *Journal of Applied Polymer Science*, vol. 131, no. 19, pp. 1–15, 2014.

[16] F. Arbeiter, B. Schrittesser, A. Frank, M. Berer, and G. Pinter, "Cyclic tests on cracked round bars as a quick tool to assess the long term behaviour of thermoplastics and elastomers," *Polymer Testing*, vol. 45, pp. 83–92, 2015.

[17] J. Runt and M. Jacq, "Effect of crystalline morphology on fatigue crack propagation in polyethylene," *Journal of Materials Science*, vol. 24, no. 4, pp. 1421–1428, 1989.

[18] R. Greco and G. Ragosta, "Isotactic polypropylenes of different molecular characteristics: influence of crystallization conditions and annealing on the fracture behaviour," *Journal of Materials Science*, vol. 23, no. 11, pp. 4171–4180, 1988.

[19] C. J. G. Plummer, P. Menu, N. Cudré-Mauroux, and H. H. Kausch, "The effect of crystallization conditions on the properties of polyoxymethylene," *Journal of Applied Polymer Science*, vol. 55, no. 3, pp. 489–500, 1995.

[20] R. Seguela, "Critical review of the molecular topology of semicrystalline polymers: the origin and assessment of intercrystalline tie molecules and chain entanglements," *Journal of Polymer Science Part B: Polymer Physics*, vol. 43, no. 14, pp. 1729–1748, 2005.

[21] H. Hama and K. Tashiro, "Structural changes in non-

isothermal crystallization process of melt-cooled polyoxymethylene [II] evolution of lamellar stacking structure derived from SAXS and WAXS data analysis," *Polymer*, vol. 44, no. 7, pp. 2159–2168, 2003.

[22] G. Kalay and M. J. Bevis, "Injection molding of isotactic polypropylene," in *Polypropylene*, Polymer Science and Technology Series, J. Karger-Kocsis, Ed., pp. 329–334, Springer, Dordrecht, 1999.

[23] D. G. M. Wright, R. Dunk, D. Bouvart, and M. Autran, "The effect of crystallinity on the properties of injection moulded polypropylene and polyacetal," *Polymer*, vol. 29, no. 5, pp. 793–796, 1988.

[24] R. Čermák, M. Obadal, P. Ponížil, M. Polášková, K. Stoklasa, and A. Lengálová, "Injection-moulded α- and β-polypropylenes: I. structure vs. processing parameters," *European Polymer Journal*, vol. 41, no. 8, pp. 1838–1845, 2005.

[25] G. Kalay and M. J. Bevis, "Processing and physical property relationships in injection-molded isotactic polypropylene. 2. Morphology and crystallinity," *Journal of Polymer Science Part B: Polymer Physics*, vol. 35, no. 2, pp. 265–291, 1997.

[26] S. Liparoti, A. Sorrentino, G. Guzman, M. Cakmak, and G. Titomanlio, "Fast mold surface temperature evolution: relevance of asymmetric surface heating for morphology of iPP molded samples," *RSC Advances*, vol. 5, no. 46, pp. 36434–36448, 2015.

[27] R. Pantani, I. Coccorullo, V. Speranza, and G. Titomanlio, "Modeling of morphology evolution in the injection molding process of thermoplastic polymers," *Progress in Polymer Science*, vol. 30, no. 12, pp. 1185–1222, 2005.

[28] R. Čermák, M. Obadal, P. Ponížil, M. Polášková, K. Stoklasa, and J. Hečková, "Injection-moulded α- and β-polypropylenes: II. Tensile properties vs. processing parameters," *European Polymer Journal*, vol. 42, no. 9, pp. 2185–2191, 2006.

[29] G. Kalay and M. J. Bevis, "Processing and physical property relationships in injection-molded isotactic polypropylene. 1. Mechanical properties," *Journal of Polymer Science Part B: Polymer Physics*, vol. 35, no. 2, pp. 241–263, 1997.

[30] J.-W. Housmans, M. Gahleitner, G. W. M. Peters, and H. E. H. Meijer, "Structure–property relations in molded, nucleated isotactic polypropylene," *Polymer*, vol. 50, no. 10, pp. 2304–2319, 2009.

[31] M. Fujiyama, T. Wakino, and Y. Kawasaki, "Structure of skin layer in injection-molded polypropylene," *Journal of Applied Polymer Science*, vol. 35, no. 1, pp. 29–49, 1988.

[32] J. C. Viana, A. M. Cunha, and N. Billon, "The tensile behaviour of an injection-moulded propylene-ethylene copolymer: the effect of the local thermomechanical processing conditions," *Polymer International*, vol. 43, no. 2, pp. 159–166, 1997.

[33] J. C. Viana, A. M. Cunha, and N. Billon, "The effect of the skin thickness and spherulite size on the mechanical properties of injection mouldings," *Journal of Materials Science*, vol. 36, no. 18, pp. 4411–4418, 2001.

[34] J. C. Viana, A. M. Cunha, and N. Billon, "The thermomechanical environment and the microstructure of an injection moulded polypropylene copolymer," *Polymer*, vol. 43, no. 15, pp. 4185–4196, 2002.

[35] J. C. Viana, N. Billon, and A. M. Cunha, "The thermomechanical environment and the mechanical properties of injection moldings," *Polymer Engineering & Science*, vol. 44, no. 8, pp. 1522–1533, 2004.

[36] W.-W. Tseng, *Relationships between Processing, Structure, Fatigue Crack Growth and J-Integral Fracture Toughness of Neat Polyetheretherketone*, University of Texas, 1986, Dissertation.

[37] B. Ozcelik, A. Ozbay, and E. Demirbas, "Influence of injection parameters and mold materials on mechanical properties of ABS in plastic injection molding," *International Communications in Heat and Mass Transfer*, vol. 37, no. 9, pp. 1359–1365, 2010.

[38] C. J. G. Plummer, P. Béguelin, and H.-H. Kausch, "The temperature and strain-rate dependence of mechanical properties in polyoxymethylene," *Polymer Engineering and Science*, vol. 35, no. 16, pp. 1300–1312, 1995.

[39] J. C. Viana, "The local thermomechanical conditions and the fracture behavior of an injection-molded poly(oxymethylene)," *Polymer Engineering and Science*, vol. 46, no. 2, pp. 181–187, 2006.

[40] X. Zhao and L. Ye, "Structure and properties of highly oriented polyoxymethylene produced by hot stretching," *Materials Science and Engineering: A*, vol. 528, no. 13-14, pp. 4585–4591, 2011.

[41] M. Halb, *Influence of the Processing Conditions on Morphology and Fracture Mechanical Properties of Unfilled Polyoxymethylene (POM), [Ph.D. Thesis]*, Montanuniversitaet Leoben, 2016.

[42] A. P. Hammersley, "FIT2D: a multi-purpose data reduction, analysis and visualization program," *Journal of Applied Crystallography*, vol. 49, no. 2, pp. 646–652, 2016.

[43] C. G. Vonk and G. Kortleve, "X-ray small-angle scattering of bulk polyethylene," *Kolloid-Zeitschrift & Zeitschrift für Polymere*, vol. 220, no. 1, pp. 19–24, 1967.

[44] G. R. Strobl and M. Schneider, "Direct evaluation of the electron density correlation function of partially crystalline polymers," *Journal of Polymer Science: Polymer Physics Edition*, vol. 18, no. 6, pp. 1343–1359, 1980.

[45] H. Hama and K. Tashiro, "Structural changes in non-isothermal crystallization process of melt-cooled polyoxymethylene. [I] Detection of infrared bands characteristic of folded and extended chain crystal morphologies and extraction of a lamellar stacking model," *Polymer*, vol. 44, no. 10, pp. 3107–3116, 2003.

[46] H. Hama and K. Tashiro, "Structural changes in isothermal crystallization process of polyoxymethylene investigated by time-resolved FTIR, SAXS and WAXS measurements," *Polymer*, vol. 44, no. 22, pp. 6973–6988, 2003.

[47] M. Shimomura, *Morphology Dependence of Vibrational Spectra of Polyoxymethylene and P(Ethylene Oxide) Crystals*, Osaka University, 1993, Dissertation.

[48] J. T. Koberstein, B. Morra, and R. S. Stein, "The determination of diffuse-boundary thicknesses of polymers by small-angle X-ray scattering," *Journal of Applied Crystallography*, vol. 13, no. 1, pp. 34–45, 1980.

[49] B. Goderis, H. Reynaers, M. H. J. Koch, and V. B. F. Mathot, "Use of SAXS and linear correlation functions for the determination of the crystallinity and morphology of semi-crystalline polymers. Application to linear polyethylene," *Journal of Polymer Science Part B: Polymer Physics*, vol. 37, no. 14, pp. 1715–1738, 1999.

[50] N. Stribeck, *X-Ray Scattering of Soft Matter*, Springer, Berlin, Heidelberg, 2007.

[51] J. Karger-Kocsis, R. Walter, and K. Friedrich, "Annealing effects on the fatigue crack propagation of injection moulded

PEEK and its short-fibre composites," *Journal of Polymer Engineering*, vol. 8, no. 3-4, pp. 221–255, 1988.

[52] International Standard, *Polyethylene (PE) Materials — Determination of Resistance to Slow Crack Growth under Cyclic Loading — Cracked Round Bar Test Method*, International Organization for Standardization, 2015, no. 18489.

[53] A. Siegmann, A. Buchman, and S. Kenig, "Residual stresses in polymers I: the effect of thermal history," *Polymer Engineering and Science*, vol. 22, no. 1, pp. 40–47, 1982.

[54] A. Siegmann, A. Buchman, and S. Kenig, "Residual stresses in polymers III: the influence of injection-molding process conditions," *Polymer Engineering and Science*, vol. 22, no. 9, pp. 560–568, 1982.

[55] A. Siegmann, A. Buchman, and S. Kenig, "Residual stresses in polymers. II. Their effect on mechanical behavior," *Polymer Engineering and Science*, vol. 21, no. 15, pp. 997–1002, 1981.

[56] P. Hutař, M. Ševčík, A. Frank, L. Náhlík, J. Kučera, and G. Pinter, "The effect of residual stress on polymer pipe lifetime," *Engineering Fracture Mechanics*, vol. 108, pp. 98–108, 2013.

[57] J. Poduška, P. Hutař, J. Kučera et al., "Residual stress in polyethylene pipes," *Polymer Testing*, vol. 54, pp. 288–295, 2016.

[58] C. J. G. Plummer, N. Cudré-Mauroux, and H.-H. Kausch, "Deformation and entanglement in semicrystalline polymers," *Polymer Engineering and Science*, vol. 34, no. 4, pp. 318–329, 1994.

[59] S. Bandyopadhyay and H. R. Brown, "Studies of environmental stress-crack propagation in low-density polyethylene," *Journal of Polymer Science: Polymer Physics Edition*, vol. 19, no. 5, pp. 749–761, 1981.

[60] P. J. Rae, E. N. Brown, and E. B. Orler, "The mechanical properties of poly(ether-ether-ketone) (PEEK) with emphasis on the large compressive strain response," *Polymer*, vol. 48, no. 2, pp. 598–615, 2007.

[61] N. A. Jones and A. J. Lesser, "Morphological study of fatigue-induced damage in isotactic polypropylene," *Journal of Polymer Science Part B: Polymer Physics*, vol. 36, no. 15, pp. 2751–2760, 1998.

8

# New Solvent for Polyamide 66 and Its use for Preparing a Single-Polymer Composite-Coated Fabric

**Mostafa Jabbari, Mikael Skrifvars, Dan Åkesson, and Mohammad J. Taherzadeh**

*Swedish Centre for Resource Recovery, University of Borås, Borås, Sweden*

Correspondence should be addressed to Mostafa Jabbari; mostafa.jabbari@hb.se

Academic Editor: Andrea Camposeo

Polyamides (PAs) are one of the most important engineering polymers; however, the difficulty in dissolving them hinders their applications. Formic acid (FA) is the most common solvent for PAs, but it has industrial limitations. In this contribution, we proposed a new solvent system for PAs by replacing a portion of the FA with urea and calcium chloride (FAUCa). Urea imparts the hydrogen bonding and calcium ion from the calcium chloride, as a Lewis acid was added to the system to compensate for the pH decrease due to the addition of urea. The results showed that the proposed solvent (FAUCa) could readily dissolve PAs, resulting in a less decrease in the mechanical properties during the dissolution. The composite prepared using the FAUCa has almost the same properties as the one prepared using the FA solution. The solution was applied on a polyamide 66 fabric to make an all-polyamide composite-coated fabric, which then was characterized. The FAUCa solution had a higher viscosity than the one prepared using the neat FA solvent, which can be an advantage in the applications which need higher viscosity like preparing the all-polyamide composite-coated fabric. A more viscous solution makes a denser coating which will increase the water /gas tightness. In conclusion, using the FAUCa solvent has two merits: (1) replacement of 40% of the FA with less harmful and environmentally friendly chemicals and (2) enabling for the preparation of more viscous solutions, which makes a denser coating.

## 1. Introduction

Aliphatic polyamides (PAs), also called nylons, are a class of semicrystalline polymers that contain amide groups which are intercalated along linear alkane chains [1]. Solution processing of aliphatic PAs is quite challenging due to the fact that only a few solvents, such as formic acid (FA) and cresol [2] or fluoric solvents [3], can dissolve them. All of the solvents being used for the dissolution of PAs have severe environmental challenges. A few attempts have been made [2] to replace the existing solvents or propose new solvents. Papadopoulou et al. [2] mixed FA with trifluoroacetic acid and acetone. Charlet et al. [4] studied the crystallization and dissolution behavior of polyamide 6-water systems under pressure. Nirmala et al. [5] used FA (85 wt%), acetic acid, dichloromethane, 1,1,1,3,3,3-hexafluoro-2-propanol, trifluoroacetic acid, and chlorophenol in their study. Basically, dissolving PAs is difficult due to two reasons: (a) polyamides are highly crystalline, whereas the above treatment holds for amorphous polymers, and (b) solvents for polyamides are believed to act by virtue of strong, highly specific polar forces [6].

Polyamide (PA) is a well-known high-performance engineering plastic (technical thermoplastic [7]) with high strength and good fatigue resistance [8] and excellent mechanical and physical properties, which is why it is increasingly used in industrial machinery [7]. Nylon is the common name of linear aliphatic PAs. Nylons are important commercial polymers, with uses ranging from fibers to cooking bags to coatings [9], carpets, upholstery, and apparel [10].

Coated fabrics are flexible composites, consisting of a fabric substrate and a polymeric coating [10]. The coating could be on one side or on both sides, with either the same or different polymeric coating per side [10]. Coated woven fabrics are used in a wide range of structural applications to provide lightweight, architecturally striking solutions [11]. The physical properties of a coated fabric depend on the specific properties of the substrate, coating formulation, coating technique, and processing conditions during coating [10]. There are two principal types of coated woven fabric: a glass fiber fabric with a PTFE (polytetrafluoroethylene) coating and a polyester fabric with PVC (polyvinyl chloride) coating [12]. Both PTFE- and PVC-coated fabrics are employed today in different types of tents and architectural membrane structures all over the world [13]. As the fabric is prepared from a high-crystalline polymer (spun fiber), it can resist environmental loads, as tensile stresses in the plane of the fabric [12]. These fabrics are popular mainly due to their affordable prices, high strength, durability, resistance to wear and tear, various colors, and soft texture [11]. They are often used for wide-span surfaces, membrane-cable structures, and pneumatic constructions [13]. One recent application of coated fabrics is their application as a textile bioreactor which is a vessel to conduct the biological processes (fermentations) [14]. The PVC-coated fabric was the first material proposed for this purpose. The recent promising candidate was the all-polyamide composite-coated fabric (APCCF) which showed superior properties over the former one [14]. However, using a high amount of formic acid as the solvent is still one industrial challenge for scaling-up the production.

The aim of this paper was to introduce a new solvent based on the replacement of FA with urea, calcium chloride (industrially abundant chemicals), and water to produce an all-polyamide composite-coated fabric from a PA by solvent casting of only one component to have strong adhesion between the coating and the fabric as well as enhanced recyclability.

Table 1: Ingredients for the solutions (in some cases, the sum is not 100%—due to the mathematical rounding of the numbers—the names are only for labeling.)

| Sample name (FA-U-Ca-W, %) | FA (%) | U (%) | Ca (%) | W (%) |
|---|---|---|---|---|
| 100-0-0-0 | 100.00 | 0.00 | 0.00 | 0.00 |
| 60-7-20-13 | 60.00 | 6.67 | 20.00 | 13.33 |
| 64-7-14-14 | 64.29 | 7.14 | 14.29 | 14.29 |
| 90-0-0-10 | 90.00 | 0.00 | 0.00 | 10.00 |
| 90-10-0-0 | 90.00 | 10.00 | 0.00 | 0.00 |
| 82-9-9-0 | 81.82 | 9.09 | 9.09 | 0.00 |
| 69-8-15-8 | 69.23 | 7.69 | 15.38 | 7.69 |
| 67-4-30-0 | 66.67 | 3.70 | 29.63 | 0.00 |
| 82-0-0-18 | 81.82 | 0.00 | 0.00 | 18.18 |
| 75-8-17-0 | 75.00 | 8.33 | 16.67 | 0.00 |
| 75-0-0-25 | 75.00 | 0.00 | 0.00 | 25.00 |
| 55-6-39-0 | 54.55 | 6.06 | 39.39 | 0.00 |
| 50-8-42-0 | 50.00 | 8.33 | 41.67 | 0.00 |
| 47-8-39-5 | 47.37 | 7.89 | 39.47 | 5.26 |
| 56-6-25-13 | 56.25 | 6.25 | 25.00 | 12.50 |
| 50-6-28-17 | 50.00 | 5.56 | 27.78 | 16.67 |
| 43-7-36-14 | 42.86 | 7.14 | 35.71 | 14.29 |
| 38-6-43-13 | 38.30 | 6.38 | 42.55 | 12.77 |
| 35-6-48-12 | 34.62 | 5.77 | 48.08 | 11.54 |

## 2. Material and Methods

*2.1. Materials.* The formic acid, urea, and calcium chloride used in this work were supplied by Sigma-Aldrich (ACS reagent grade, >98%). The PA66 plain woven fabric (70 grams per square meter (gsm)) was provided by FOV Fabrics AB (Borås, Sweden). As a polymer source to produce a solution, PA fiber production waste from the weaving process at FOV Fabrics was used.

*2.2. Solvent Preparation.* The solutions were prepared by adding 35 g of PA waste scraps into 100 g solvent mixtures, which, in turn, were obtained by mixing different amounts of formic acid (FA), urea (U), calcium chloride (Ca), and distilled water (W) at room temperature, according to the values tabulated in Table 1. For the cases that had more urea and calcium chloride, it took a while (1–3 minutes) to get a clear solution.

*2.3. Composite Preparation.* All-polyamide composites were prepared in the form of a flat laminate on the substrate fabric using an isothermal immersion-precipitation method. The solutions of the PA production waste in the formic acid were made by dissolving 35 g of PA in 100 g solvent at room temperature. In low concentrations (less than ≈30% *w/w*), PA66 readily dissolves in the formic acid containing the solvent mixture at room temperature, but for higher concentrations, the solution should be agitated for a longer time. In order to assure the completion of dissolution and have the same agitation condition for all the solutions, the sealed solution flasks were put in a shaker at a speed of 100 rpm for 20 hours at 55°C to obtain a homogeneous solution [11]. The solutions were cooled to room temperature, and after centrifugation for 10 minutes at a speed of 16,000 × $g$ (meaning 16,000 times Earth's gravitational force) to remove the bubbles, the dopes were cast on a PA fabric with a size of 18 × 24 cm fixed on a glass with adhesive tape, using a ZUA 2000 universal film applicator (Zehntner GmbH Testing Instruments, Sissach, Switzerland) with a gap of 175 $\mu$m. Once the casting process was done, after waiting 30 seconds, the glass plate (carrying the fabric and a layer of the PA solution on top of it) was immersed in a distilled water coagulation bath (10 L) at room temperature to induce polymer precipitation (phase separation). After one hour of coagulation (in the water bath), the composites obtained were first washed three times with distilled water and then held under light press between two sheets of filter papers and dried at 55°C in a vacuum oven (≈0.1 bar) for 2 hours. The samples were kept in a vacuum zipper storage bag for further analysis. The composites were named according to their corresponding solution ingredients (Table 1). The thickness of the composites varied between 150 and 200 $\mu$m. The reason of variation in the thickness could be the strength of the solvent. In other words, if the solvent is stronger, it will dissolve the surface of the fabric in a greater extent which leads to a more reduction of the composite thickness. The thickness of the fabric was 100 $\mu$m initially.

*2.4. Characterization Methods.* The tensile strength properties were evaluated in accordance with the standard method ISO 527 [15]. Dumbbell-shaped test bodies, 75 mm long (with a width of 4 mm), were tested on an MTS 20/M tensile strength tester (MTS Systems Corporation, Eden Prairie, MN, USA), fitted with a 5 kN load cell and a special grip for films, using a crosshead speed of 5 mm/min. The gauge length, the preload force, and the first approach speed were 33 mm, 0.5 N, and 2 mm/min, respectively. The thickness of the composites was measured using an Elastocon thickness meter (Elastocon, Sweden). A minimum of five test bodies was tested for each material. The specimens were all cut along the warp direction of the fabric.

To investigate the viscoelastic properties of the composites, dynamic mechanical thermal analysis (DMA Q800, TA Instruments, Waters LLC, USA) was performed on the prepared composites. The specimens were run with a film tension clamp using the temperature ramp procedure with a sample dimension of approximately 15 × 9 mm. The temperature ranged from room temperature to 180°C with a heating rate of 3°C/min; the frequency and the amplitude were 2 Hz and 15% elongation at max, respectively.

Scanning electron microscopy (SEM) was used to monitor the fracture surface morphology of the cross sections of the composites and to check the adhesion between the coating (the PA film) and the fabric. The specimens were obtained by quenching in liquid nitrogen and breaking by hand. As the samples contained fabric, using only quenching in liquid nitrogen did not take apart the sample completely; therefore, after breaking, the unbroken parts were cut by a sharp blade. The studied surface was sputtered with a layer of gold before the measurements. SEM analysis was performed using AIS2100 (Seron Technology, Korea) operated at an acceleration voltage of 18 kV.

The viscosity measurements were carried out with a Brookfield viscometer (MA, USA) at 40°C. The tests were done within 30 seconds for each sample.

Thermogravimetric analysis (TGA) was performed on the composites using the Q500 machine (TA Instruments, MA, USA). About 8 mg of the material was heated from room temperature to 700°C at a heating rate of 10°C/min in a nitrogen purge stream.

The pH measurements were carried out using a Jenway 3505 pH meter (Staffordshire, UK) at room temperature. The pH sensor was placed in the samples for 30 seconds while being stirred at 200 rpm to reach to an equilibrium.

## 3. Results and Discussion

The most common solvent for polyamides is formic acid (FA), which is neither safe nor easy to handle. To replace a portion of the FA to make it more appropriate, more economical, and safer to handle solvent mixtures, different amounts of urea, calcium chloride, and water were mixed and added to the FA. Thereafter, the polymer was added to the solvent mixtures to make homogeneous solutions. The solutions were used to make all-polyamide composite-coated fabrics by means of a universal film applicator and consecutive coagulation in a water bath as a nonsolvent in order to induce phase separation (phase inversion). As a result, a composite composed of a thin continuous PA layer (the coating) and a PA fabric out of the most common type of aliphatic PA (PA66) was obtained. The hypothesis in this work was that since each urea molecule contains four hydrogens and can establish strong intermolecular interactions (H-bonding) with the amide groups in the PA, it can disrupt the intermolecular interactions (hydrogen bonding) among the polymer chains, therefore resulting in an easier dissolution. In general, the dissolution of semicrystalline polymers such as polyamide comprises several steps including solvent penetration [16], decrystallization of crystalline domains, amorphous polymer swelling, and chain untangling [17]. Formic acid (as the main solvent) has enough penetration power to penetrate into the PA chains [18]. However, the main goal in this paper was to replace a portion of the FA, meaning that the replacement molecule needed to have enough interactions with the PA chains to start the penetration. As discussed earlier, urea has four hydrogen bonding per unit of molecule, and even if all the four hydrogen cannot engage in hydrogen bonding, urea still has high capability of hydrogen bonding interactions. On the other hand, PA chains contain amide groups which are the strong hydrogen bonding donor/acceptor groups [19], while FA has only one hydrogen bonding per unit of molecule. Therefore, the mixture of FA/urea will have a higher tendency to penetrate into the PA chains. The other steps of dissolution (decrystallization of crystalline domains, amorphous polymer swelling, and chain untangling) would be done faster and easier if the first step (solvent penetration) occurs faster/easier [20, 21]. This process will be enhanced by using calcium chloride as a Lewis acid (to nullify the increased pH due to the addition of urea—an organic base) as well as localizing the paired electrons on the amide groups of PA chains. Different properties were analyzed, and the process parameters were optimized. The selection of different percentages of the solvent's components (FA, urea, calcium chloride, and water) was done by trying to keep the percentage of the FA as low as possible. Water was added to the solvent mixture, both for imparting hydrogen bonding [22] and also for its well-known effect on the dissolution of polyamides [4, 23] when it is used in lower percentages.

*3.1. Mechanical Properties.* The maximum force divided by the cross-sectional surface area of the samples needed to break the composite in the tensile strength mode; also, the elongation at max of the composites is shown in Figure 1. The thickness of the composites varied between 150 and 200 $\mu$m.

The results from the tensile strength testing (Figure 1) implied an irregular trend. However, a slight trend is evident in terms of the FA content. Four composites had the tensile strength values higher than the one obtained from a solvent out of pure FA. It could be related to the fact that FA partially decomposes the PA chains during the dissolution process [11]. In other composites, which have an FA content that is even less than the FA content of those four good composites,

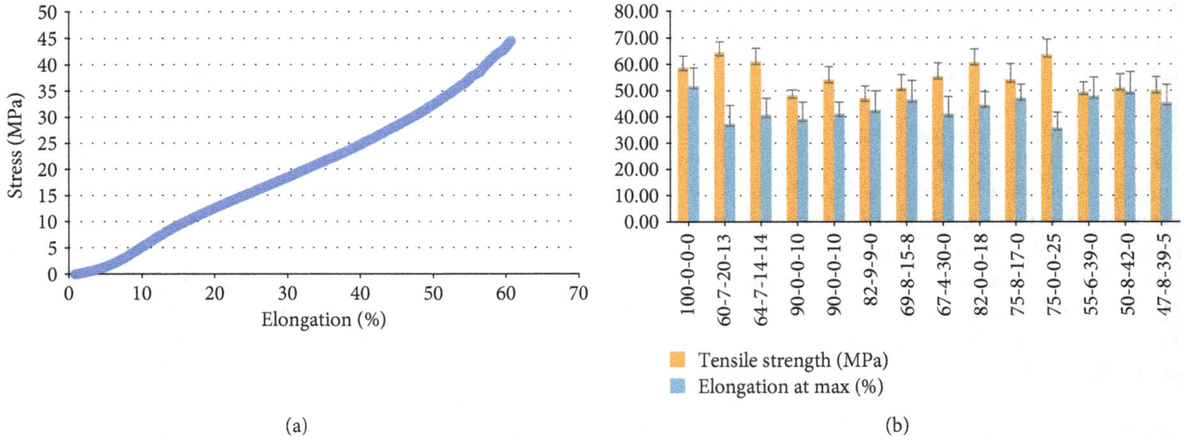


FIGURE 1: An exemplary stress-elongation at max curve for the sample 60-7-20-13 (a), and the tensile strength and the elongation at max values for the composites (b).

the tensile strength values are not higher than those of the composite 100-0-0-0 (with pure FA). Although it might seem contradictory with the above proposal for increasing the tensile strength value, it could be the effect of the urea/calcium chloride content (and even water content) on the crystallization of the PA chains in the phase inversion process. Regarding the elongation at max, all the composites prepared from the new proposed solvent mixtures in this work have lower values compared to the 100-0-0-0 composite. This might be related to the effect of the localization of the paired electron on the amide groups of PA chains, which plays a role like a spacer [24] during the phase inversion process. We can conclude that the optimized FA content for dissolving PA with the least possible decomposition/degradation is 60% (the sample 60-7-20-13).

*3.2. Viscoelastic Properties.* The loss modulus is a measure of energy dissipation. In the region of the glass transition, molecular segmental motions are activated; however, motions occur with difficulty, described as the molecular friction that dissipates much of the force. Therefore, though the material is less stiff, more force is dissipated as heat, increasing the loss modulus. However, after that region, as the chains are free to move (at the temperature higher than the glass transition temperature), much less energy is stored since the molecules can move with the force, resulting in a rapid decline in the storage modulus. As evident in Figure 2, in the sample prepared from pure FA, the maximum in the curve of the loss modulus decreases with the reduction in the amount of FA in the solution. The solvent 60-7-20-13 contains water, and there might be some water molecules left. As the water molecules can act as a plasticizer for polyamide [25], it could plasticise the polymer. More so, it might be related to the effect of higher density of hydrogen bonding on a urea-containing solvent (in 60-7-20-13) as well as the localization of the paired electrons of the amide groups from the PA chains. One of the main characteristics of PAs is the strong interchain interaction that arises from the hydrogen bonding between the amide groups on adjacent chains [26]. In other words, in the solvation process, the amide groups from the PA chains meet a higher density of hydrogen bonding (offering by urea), leading to weaker intrainteractions among the PA chains, which, in return, decreases the interaction among them. Therefore, they have much more freedom to move in the solvent. This freedom increases the Brownian motion of the chains, which increases their distances from each other [27]. Finally, when they are about to change their phase from solution to solid (coagulation process in the coagulant), they come together in a relatively less-packed structure. Moreover, when they are less-packed in the solid form (having lower crystallinity), their Tg is slightly lower (Figure 2 for tan delta), and the peak of the loss modulus is seen at a lower temperature for a solution with a higher amount of urea. This proposed reason is intensified with the presence of calcium ion as a Lewis acid in a different way. Calcium ion has free orbital to localize the paired electrons of the nitrogen from the amide group of the PA chains; hence, the hydrogen bonding ability of the amide groups of the PA chains also decreases, and finally the same effect happens. The same trend is seen in the storage modulus as well as in the tan delta.

*3.3. Viscosity.* The viscosity values are different for various solutions (Figure 3). However, in all solutions, the viscosity value is higher than the one for the solution using only formic acid as a solvent (100-0-0-0). Regarding the merit of the increase in viscosity, during the process of making the APCCF, it was observed several times that when the viscosity of the solution is low, the formed film (coating) does not have enough coherency. It makes sense as it might be due to the less aggregation of polymer chains on top of the fabric and also letting the solution go through more into the fabric which is not favorable. Because, for the application of making the APCCF, only the surface of the fabric must be dissolved partially, not the whole fabric which will decrease the whole tensile strength of the composite, a solution which has higher viscosity is more favorable for making APCCF as it makes a well-formed coating on top of the PA fabric. When

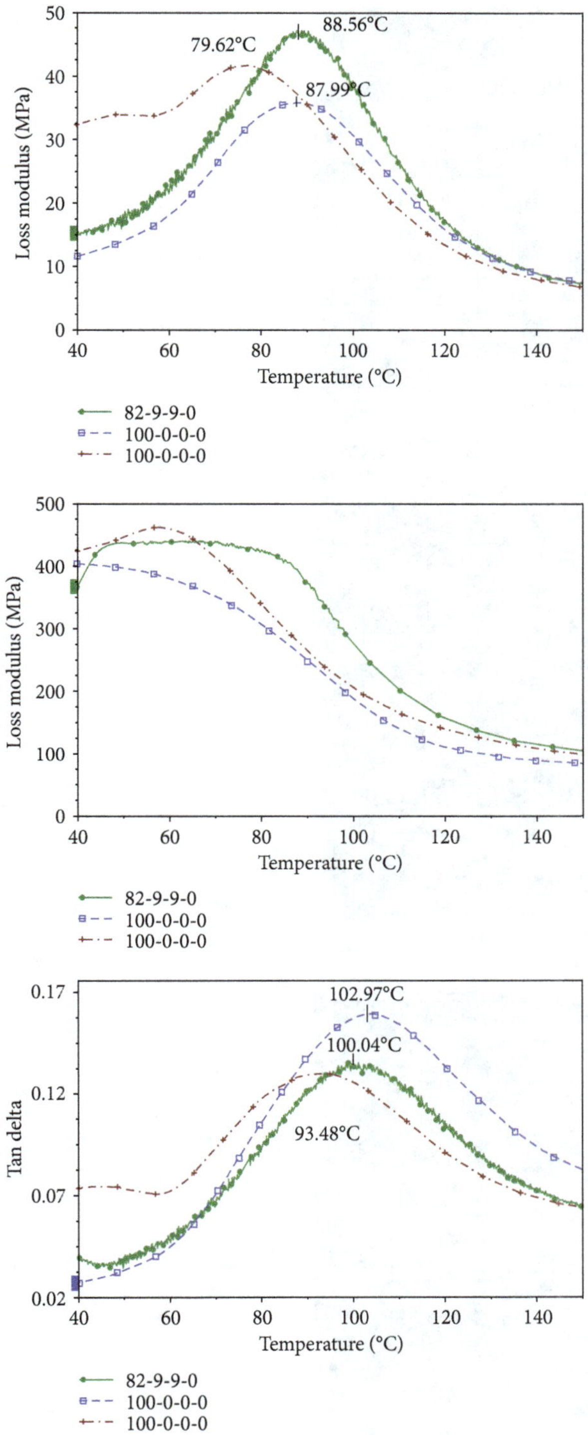


FIGURE 2: Loss modulus, storage modulus, and tan delta for the composites prepared from the new solvents.

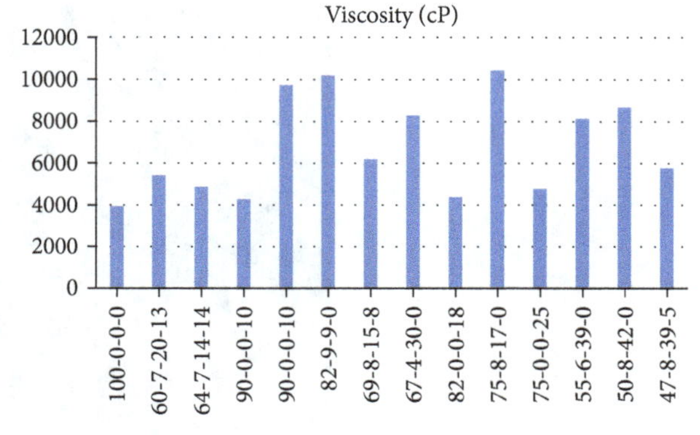


FIGURE 3: The viscosity values for the composites. The solution made from the solvent mixture 75-8-17-0 is the most viscous solution due to the intermolecular interactions between the ingredients of the solvent mixture and the polymer.

the viscosity of the solution is low, the formed film does not have enough coherency. Comparatively, when the same percentage of PA solutions in different solvents gives a different viscosity and, on the other hand, the viscosity of around 4000 cP (for the solution 100-0-0-0) is enough to make a proper coating, the solution 75-8-17-0 (which has the highest viscosity value, around 10,500 cP) will use less polymer. Additionally, as the APCCF must be waterproof and be as gas tight as possible, the coating being formed from a higher viscosity solution will be denser and will increase the APCCF's gas tightness.

The reason for the difference in the viscosity values in different solutions could be related to the difference in the intermolecular interactions between the solvent's molecules and the PA's molecules. The ones that have a higher amount of urea have higher values in viscosity. This is due to the fact that urea can establish hydrogen bonding in large numbers (up to four units) from two sides, which can make a bridge between two PA chains (two bonds from the one side and two bonds from the other side). By establishing these four bidirectional hydrogen bonds, the fluidity of the PA solution decreases, which, in turn, increases the viscosity.

*3.4. Morphological Properties.* According to the scanning electron microscopic images of the cross section (Figure 4), in almost all the APCCF specimens, there is an excellent adhesion between the coating (formed film) and the fabric. In other words, the boundaries between the fiber of the fabric and the coating are not clear because they have faded due to the adhesion of the two components.

It is obvious that a large part of the cross-sectional area of the fabric is dissolved, and the polymer chains are interdiffused with each other, enabling the APCCF to have a good adhesion between the coating and the fabric, both of which are made out of PA. Although higher surface dissolution helps to create a better adhesion between the fabric and the coating, it also disintegrates the fabric structure and changes the fabric, from being a fibrous form to an amorphous film. Fibers are spun and pose a high crystallinity and thus a good strength while the films are amorphous as they do not have enough time to rearrange their chains in a nice ordered way (fast coagulation in water). Therefore, converting PA from a highly crystalline form to a less crystalline (highly amorphous) form is not favorable from a mechanical point of view. The amount of the surface dissolution should be as low as possible to impart a good adhesion between the fabric

FIGURE 4: Selected cross-sectional SEM micrographs of the composite 60-7-20-13. Strong adhesion is seen between the coating (the top film) and the fibers from the fabric. The fibrils of the fiber closer to the coating are fused to each other.

and the coating, but a higher dissolution is not favorable [11]. Using this new solvent helps the fabric's surface to be partially dissolved more quickly, due to the higher hydrogen bond intensity (due to the presence of urea); hence, the adhesion between the fabric and the coating—the coating (formed film) on top of the fabric—would be stronger.

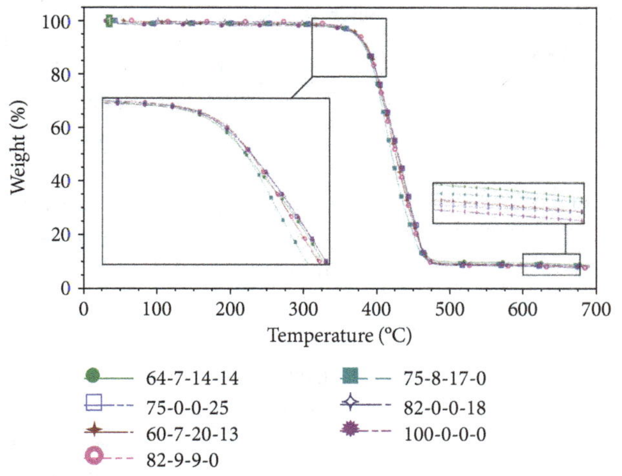


FIGURE 5: TGA curves of the composites prepared via the new solvent.

*3.5. Thermogravimetric Analysis (TGA).* The TGA curves showed almost the same values and shape (Figure 5). The only composite that deviates more from the rest is the 75-8-18-0, which tends to decompose slightly earlier than the others, though the difference is not significant. However, it has higher residue percentage compared to the rest of the composites (except the 64-7-14-14). Although PA66 is partially crystalline, all good solvents penetrate the crystallites, disrupting them completely [6]. Calcium ion (which is the acidic ion) is smaller than the urea molecule, resulting in an easier entrance to the crystallites. The final swollen gel is thus amorphous, even though the initial polymer is crystalline [6]. In the composite 75-8-17-0, the ratio of urea to calcium chloride (the source of calcium ion) is 0.5 (Table 2) and the ratio of U/FA is 0.11. These ratios are the same as those of the composites 69-8-15-8 and 64-7-14-14; however, their Ca/(FA + U + W) and U/(FA + Ca + W) are different. From the last two ratios and the TGA curves, it could be concluded that out of the four ingredients of the new solvent mixture (U, FA, Ca, and W), the ratio of U and Ca to the other ingredients should be kept as low as possible to have the least disruption of PA crystallites in order not to sacrifice the thermal stability. It is noteworthy that the composite 82-9-9-0, which has the same Ca/(FA + U + W) and U/(FA + Ca + W) ratios (equal to 10), has the lowest residue in the TGA curve, meaning that the closer the two ratios, the less the thermal stability of the composite.

*3.6. pH Values and Observations.* From the pH measurements, it is obvious and reasonable that the solutions that have a higher amount of FA have lower pH. The amine group (-NH2) in urea can accept a hydrogen ion, making it a basic substance. However, the carbonyl group (-C=O) offers a significant opportunity for resonance, which will stabilise the amine group, meaning that the paired electrons on N is in resonance. This means that urea is very slightly basic (and very close to neutral). However, in the presence of an acid, the basicity of urea increases (as there is plenty of H+ in the solution and the -NH2 is converted to -NH3+). Hence, in the solution, urea is a base, so the reason for adding calcium chloride (which acts as a Lewis base) was to decrease the pH to make the solution as similar as possible to the pure FA solution. The highest pH is attributed to the solution 75-0-0-25 in which 25% of the water has increased the pH from −1.74 to −0.30. Basically, polyamides are polymers with a relatively high density of hydrogen bonds [1]. As PA is a weak acid, a solvent with acidic properties can protonate the amide group; consequently, the amide group will have less possibility to establish hydrogen bonding with other chains. Therefore, it will be easier for the solvent to swell into the polymer and dissolve it. In the first six solutions of the solutions tabulated in Table 2, it was observed that the solution 90-0-0-10 was excellent in terms of homogeneity and dissolution power. This might be related to the lower pH they have compared to the others, which have higher pH. However, the pH by itself is not a good criterion to judge the dissolution power of these solutions. For example, in the solution 35-6-48-12, which has a pH value very close to the pH of FA (−1.66 vs. −1.74), the observation showed that the solution does not have a high power of dissolution. This might be related to the calcium chloride content (100*Ca/(FA + urea + W)), which is 92.59%, while the value is less than 25% in those showing "excellent" dissolution. The reason for this might be due to the effect of the counterion, chloride. Chloride is a relatively big ion. In lower numbers, they can help the dissolution; however, when they are increased, they most probably promote some interferences with the interactions of other ingredients (FA, urea, and water) with polyamide chains. Therefore, it could be concluded that the calcium chloride content should be less than 25%.

## 4. Conclusions

The preparation of all-polyamide composite-coated fabric (APCCF), a specific form of single-polymer composites, made through a phase inversion method using a new solvent is described. The solvent was obtained by replacing 40% of the FA with less harmful, cost-effective, more environmentally friendly, safer-to-handle, and industrially available chemicals: urea, calcium chloride, and water. The APCCF prepared in this work using the new solvent mixture does not have inferior properties over the one prepared through FA, such as mechanical and thermal properties. The prepared composite has a strong adhesion between the two constituents due to the use of the same polymer (PA66) in the constituents. The composite is fully recyclable since it contains no other materials except PA, which can be melted or dissolved and reused as a PA source. The prepared composite has a broad range of semistructural applications for providing lightweight, architecturally striking solutions as well as wide-span surfaces, membrane-cable structures, hanging roofs, pneumatic constructions, water-/gas-proof fabric reactors, temporary houses and tents, facade coverings, container linings, and tarpaulins and as material of construction of textile bioreactors.

TABLE 2: Values for the pH and the relative components and also the observations.

| Sample name | pH | U/Ca | U/FA | 100*U/(FA + Ca + W) | 100*Ca/(FA + U + W) | Observation* |
|---|---|---|---|---|---|---|
| 100-0-0-0 | −1.74 | — | 0.00 | 0.00 | 0.00 | Excellent |
| 60-7-20-13 | −0.74 | 0.33 | 0.11 | 7.14 | 25.00 | Excellent+ |
| 64-7-14-14 | −0.54 | 0.50 | 0.11 | 7.69 | 16.67 | Excellent |
| 90-0-0-10 | −1.10 | — | 0.00 | 0.00 | 0.00 | Excellent |
| 90-10-0-0 | −0.18 | — | 0.11 | 11.11 | 0.00 | Excellent |
| 82-9-9-0 | −0.75 | 1.00 | 0.11 | 10.00 | 10.00 | Excellent |
| 69-8-15-8 | −0.74 | 0.50 | 0.11 | 8.33 | 18.18 | Good |
| 67-4-30-0 | −1.64 | 0.13 | 0.06 | 3.85 | 42.11 | Good |
| 82-0-0-18 | −0.56 | — | 0.00 | 0.00 | 0.00 | Good |
| 75-8-17-0 | −0.95 | 0.50 | 0.11 | 9.09 | 20.00 | Good |
| 75-0-0-25 | −0.30 | — | 0.00 | 0.00 | 0.00 | Fairly good |
| 55-6-39-0 | −1.69 | 0.15 | 0.11 | 6.45 | 65.00 | Fairly poor |
| 50-8-42-0 | −1.61 | 0.20 | 0.17 | 9.09 | 71.43 | Fairly poor |
| 47-8-39-5 | −1.50 | 0.20 | 0.17 | 8.57 | 65.22 | Fairly poor |
| 56-6-25-13 | −0.91 | 0.25 | 0.11 | 6.67 | 33.33 | Poor |
| 50-6-28-17 | −0.90 | 0.20 | 0.11 | 5.88 | 38.46 | Poor |
| 43-7-36-14 | −1.08 | 0.20 | 0.17 | 7.69 | 55.56 | Bad |
| 38-6-43-13 | −1.42 | 0.15 | 0.17 | 6.82 | 74.07 | Bad |
| 35-6-48-12 | −1.66 | 0.12 | 0.17 | 6.12 | 92.59 | Bad |

*The markings were assigned based on the homogeneity of the solutions and the uniformity of the cast film on the glass.

## Conflicts of Interest

The authors declare that they have no conflicts of interest.

## Acknowledgments

The authors are grateful to FOV Fabrics AB (Sweden) for providing the PA fabric and PA residue and Prof. Dr. Parviz Rashidi Ranjbar for his help.

## References

[1] M. Laurati, A. Arbe, A. Rios de Anda, L. A. Fillot, and P. Sotta, "Effect of polar solvents on the crystalline phase of polyamides," *Polymer*, vol. 55, no. 12, pp. 2867–2881, 2014.

[2] E. L. Papadopoulou, F. Pignatelli, S. Marras et al., "Nylon 6, 6/ graphene nanoplatelet composite films obtained from a new solvent," *RSC Advances*, vol. 6, no. 8, pp. 6823–6831, 2016.

[3] K. Behler, M. Havel, and Y. Gogotsi, "New solvent for polyamides and its application to the electrospinning of polyamides 11 and 12," *Polymer*, vol. 48, no. 22, pp. 6617–6621, 2007.

[4] K. Charlet, V. Mathot, and J. Devaux, "Crystallization and dissolution behaviour of polyamide 6-water systems under pressure," *Polymer International*, vol. 60, no. 1, pp. 119–125, 2011.

[5] R. Nirmala, H. R. Panth, C. Yi et al., "Effect of solvents on high aspect ratio polyamide-6 nanofibers via electrospinning," *Macromolecular Research*, vol. 18, no. 8, pp. 759–765, 2010.

[6] L. Valentine, "Interaction of polyamides with solvents. I. A preliminary survey of the swelling of crosslinked nylon 66 in various types of solvents," *Journal of Polymer Science*, vol. 23, no. 103, pp. 297–314, 1957.

[7] G. Vasile, C. Fetecau, and A. Serban, "Experimental research on the roughness of surfaces processed through milling polyamide composites," *Materiale Plastice*, vol. 51, pp. 205–212, 2014.

[8] Y. Gong and G. Yang, "Manufacturing and physical properties of all-polyamide composites," *Journal of Materials Science*, vol. 44, no. 17, pp. 4639–4644, 2009.

[9] C. G. Johnson and L. J. Mathias, "In synthesis and solid state NMR of 15n-labeled nylon 12," in *1989 Boston, Massachusetts Meeting of ACS*, pp. 523-524, Boston, MA, USA, 1990.

[10] A. K. Sen, *Coated Textiles: Principles and Applications*, CRC Press, Boca Raton, FL, USA, 2nd edition, 2007.

[11] M. Jabbari, M. Skrifvars, D. Åkesson, and M. J. Taherzadeh, "Introducing all-polyamide composite coated fabrics: a method to produce fully recyclable single-polymer composite coated fabrics," *Journal of Applied Polymer Science*, vol. 133, no. 7, 2016.

[12] B. N. Bridgens and P. D. Gosling, "Direct stress–strain representation for coated woven fabrics," *Computers & Structures*, vol. 82, no. 23-26, pp. 1913–1927, 2004.

[13] A. Ambroziak and P. Kłosowski, "Mechanical properties for preliminary design of structures made from PVC coated fabric," *Construction and Building Materials*, vol. 50, pp. 74–81, 2014.

[14] M. Jabbari, O. Osadolor, R. Nair, and M. Taherzadeh, "All-polyamide composite coated-fabric as an alternative material of construction for textile-bioreactors (TBRs)," *Energies*, vol. 10, no. 11, p. 1928, 2017.

[15] M. Jabbari, D. Åkesson, M. Skrifvars, and M. J. Taherzadeh, "Novel lightweight and highly thermally insulative silica aerogel-doped poly (vinyl chloride)-coated fabric composite," *Journal of Reinforced Plastics and Composites*, vol. 34, no. 19, pp. 1581–1592, 2015.

[16] B. A. Miller-Chou and J. L. Koenig, "A review of polymer dissolution," *Progress in Polymer Science*, vol. 28, no. 8, pp. 1223–1270, 2003.

[17] M. Ghasemi, M. Tsianou, and P. Alexandridis, "Dissolution of semicrystalline polymers: solvent-induced decrystallization and chain untangling," in *2015 AIChE Annual Meeting*, p. 29, Salt Lake City, UT,USA, 2015.

[18] K. V. S. N. Raju and M. Yaseen, "Influence of nonsolvents on dissolution characteristics of nylon-6," *Journal of Applied Polymer Science*, vol. 43, no. 8, pp. 1533–1538, 1991.

[19] E. Vinken, A. E. Terry, O. van Asselen, A. B. Spoelstra, R. Graf, and S. Rastogi, "Role of superheated water in the dissolution and perturbation of hydrogen bonding in the crystalline lattice of polyamide 4, 6," *Langmuir*, vol. 24, no. 12, pp. 6313–6326, 2008.

[20] C. H. R. M. Wilsens, Y. S. Deshmukh, B. A. J. Noordover, and S. Rastogi, "Influence of the 2, 5-furandicarboxamide moiety on hydrogen bonding in aliphatic-aromatic poly (ester amide)s," *Macromolecules*, vol. 47, no. 18, pp. 6196–6206, 2014.

[21] A. N. Derbyshire, E. D. Harvey, and D. Parr, "Solvent-assisted dyeing of nylon 6.6 and polyester fibres," *Journal of the Society of Dyers and Colourists*, vol. 91, no. 4, pp. 106–111, 1975.

[22] M. G. M. Wevers, V. B. F. Mathot, T. F. J. Pijpers, B. Goderis, and G. Groeninckx, "Full dissolution and crystallization of polyamide 6 and polyamide 4.6 in water and ethanol," in *Lecture Notes in Physics*, G. Reiter and G. R. Strobl, Eds., vol. 714, pp. 151–168, 2007.

[23] M. Pegoraro, A. Penati, M. Zocchi, and G. Albertini, "Influence of nylon-6 film solvent treatments on water and water solutions permeability," *Annali di Chimica*, vol. 74, pp. 589–605, 1984.

[24] Z. L. Wang, J. L. Xu, L. J. Wu et al., "Dissolution, hydrolysis and crystallization behavior of polyamide 6 in superheated water," *Chinese Journal of Polymer Science*, vol. 33, no. 9, pp. 1334–1343, 2015.

[25] P. Y. Le Gac, M. Arhant, M. Le Gall, and P. Davies, "Yield stress changes induced by water in polyamide 6: characterization and modeling," *Polymer Degradation and Stability*, vol. 137, pp. 272–280, 2017.

[26] Q. Zhou, J. Zhang, J. Fang, and W. Li, "The influence of nanofillers migration on the mechanical property of pa 6/chitosan nanocomposites," *RSC Advances*, vol. 5, no. 22, pp. 16631–16639, 2015.

[27] S.-i. Manabe and R. Fujioka, "Solvent concentration in a specified region of regenerated cellulose solid evaluated from dynamic viscoelasticity in a hydrophilic solvent," *Carbohydrate Polymers*, vol. 41, no. 1, pp. 75–82, 2000.

9

# Numerical Study on Seismic Behavior of Underwater Bridge Columns Strengthened with Prestressed Precast Concrete Panels and Fiber-Reinforced Polymer Reinforcements

**Yu Tang,[1] Gang Wu,[1,2] and Zeyang Sun[1]**

[1]*Key Laboratory of Concrete and Prestressed Concrete Structures of the Ministry of Education, Southeast University, Nanjing 210096, China*
[2]*Laboratory of Industrialized Structural and Bridge Engineering of Jiangsu Province, Nanjing 210096, China*

Correspondence should be addressed to Gang Wu; g.wu@seu.edu.cn and Zeyang Sun; sunzeyang@seu.edu.cn

Academic Editor: Filippo Berto

The seismic performance of the bridge column, such as pier or pile, is a time-dependent property which may decrease in resistance to the deterioration or natural hazards along the structure's service life. The most effective strengthened method for degraded bridge columns is the jacketing method, which has been widely developed and investigated through numerous studies since the 1980s. This paper presented a modeling method, as well as a comprehensive parametric study, on seismic performance of bridge columns strengthened by a newly developed strengthening method with prestressed precast concrete panels and fiber-reinforced polymer reinforcements (PPCP-FRP). A modeling method of bridge columns strengthened with PPCP-FRP was first presented and validated with test results. The influence of design parameters, such as axial load ratio, equivalent FRP reinforcement ratio rate (EQFRR), expansion ratio, and shear span ratio of strengthened columns, were then further evaluated in terms of lateral load capacity, ductility, energy dissipation, lateral stiffness, and residual displacement of strengthened columns. The peak load of strengthened columns increases with the increasing of EQFRR due to the unique failure model of strengthened columns characterized by the fracture of FRP bars. The initial stiffness of strengthened columns increased by 300% with the increasing of expansion ratio by 45%, and a stable postyield stiffness stage was obtained by most strengthened columns in analysis. The residual displacement of strengthened columns decreases rapidly with the increasing of EQFRR, which indicated that a better repairability could be achieved by the strengthened column with a relatively high EQFRR.

## 1. Introduction

The seismic performance degradation objectively exists on underwater bridge columns, as piers and piles, during their long-term service periods due to unpredictable factors, such as extreme scouring or corrosion, which have been identified as the most common causes of bridge failures in the United States [1, 2]. The most effective retrofitting method for the degraded bridge columns is jacking; their forms, as well as their seismic effects, have been widely developed and investigated through numerous experimental studies since the 1980s.

Concrete jacketing was the earliest developed method used to enhance the lateral stiffness and strength of the columns. Priestley et al. [3] reported that adding a relatively thick layer of reinforced concrete (RC) in the form of a jacket around the column could enhance the flexural strength and shear strength of columns. Bett et al. [4] reported a repairing study with concrete jacket subject to a shear deficient RC column, showing that the columns strengthened by concrete jacket became much stiffer and stronger laterally than the original column. Meanwhile, the drawbacks of concrete jacketing were also obvious such as high cost, low efficiency, and limited effect on flexural strength and ductility improvement; therefore, steel jacketing was developed in the 1990s. Chai et al. [5] reported a retrofitting investigation on a circular column, which was designed with an insufficient lap splice length of the longitudinal bars, with a bonded steel jacket at its plastic hinge region, and showed that steel jacket retrofitting could effectively restore the flexural strength and ductility of the column and improve the lateral stiffness by 10 to 15 percent. Studies on failure mode investigation conducted by Priestley et al. [6] proved that steel jacket retrofitting could also change the inelastic deformation pattern of the columns, from the predominantly shear deformation for the as-built columns to the predominately flexural deformation for the retrofitted columns, and the elastic stiffness and ductility of the columns could also be significantly improved simultaneously.

From the 21st century, new materials and methods have been developed to keep up with the gradually increasing demand for retrofitting, such as high durability and low vulnerability, in which the development and use of FRP composites could be the most representative example in structural retrofitting projects. One of the most attractive applications of FRP composites is their use as confining devices for concrete columns, which may result in remarkable increases in strength and ductility [7, 8]. It is proved by many researches that the FRP composites, especially with high deformability and low elastic modulus, such as basalt fiber composites [9], polyethylene terephthalate (PET) fiber-reinforced polymer composites [8, 10], and FRP ropes [11, 12], have a positive effect to the axial compressive ability of FRP-confined concrete. The FRP composites could also be used as the longitudinal reinforcement to improve the seismic performance of bridge columns such as flexural capacity, yet have little impact on, or might even lower, the lateral stiffness of the strengthened columns, especially when the FPR bars were embedded into the columns [13–16]. Other researchers concentrated on the combination of the different jacketing methods. The prestressed steel jacketing (PSJ) was developed by Fakharifar et al. [17] for rapid and cost-effective repair of severely damaged circular RC columns. The results showed that the PSJ successfully restored the ultimate stress and displacement ductility of the columns, but only limitedly restored the initial stiffness of the column by 84% of the as-built column. Meanwhile, the anchored dowel bar showed its essential role in resisting lateral force; by linking the column and footing with anchored dowel bars, the lateral performance of the column such as ultimate stress, initial stiffness, and displacement ductility were all enhanced by at least 20%. Li et al. [18] also reported an experimental investigation on the corrosion-damaged RC columns strengthened with a combined CFRP sheet and steel jacket, which indicated a better effectiveness than the columns strengthened with only CFRP sheets or steel jacket in terms of strength and ductility. Conventional jacketing methods could enhance the flexural strength, shear strength, or even ductility of columns but with unneglectable drawbacks, such as high cost, low efficiency, and poor durability. With the recognition of life-cycle maintenance requirement and the advancement of retrofitting methods for the structures, new efficient jacketing methods were demanded to keep up with the gradually increased demand on retrofitting, such as high durability and low vulnerability.

This paper presented a modeling method, as well as a comprehensive parametric study, on seismic performance of bridge columns strengthened by a newly developed strengthening method with PPCP-FRP. The PPCP-FRP method is mainly developed for the rapid retrofitting of the underwater bridge columns, for which conventional retrofitting usually requires a cofferdam construction ahead of the repairing to build a relatively easy-to-work construction environment, which is a far less ideal one due to its low cost-efficiency and applicability. The PPCP-FRP method enhanced the cost-efficiency and applicability by using the prestressed precast concrete panel (PPCP) as the formwork (cofferdam). A modeling method of bridge columns strengthened with PPCP-FRP was first presented and validated with test results. The influence of design parameters, such as axial load ratio, EQFRR, expansion ratio, elastic modulus of FRP reinforcement, compressive strength of concrete, and shear span ratio of column, were then further evaluated in terms of the lateral load capacity, ductility, energy dissipation, lateral stiffness, and residual displacement of strengthened columns.

## 2. Modeling Method of Bridge Columns Strengthened with PPCP-FRP Method

*2.1. Design of PPCP-FRP Method.* The PPCP-FRP retrofitting is mainly composed of precast concrete panels, steel strands, longitudinal FRP bars, FRP spiral stirrup, and mortar, as in Figure 1(a). The principle of strengthening with reinforced concrete is adopted in PPCP-FRP retrofitting to achieve a significant improvement on the lateral stiffness of the column; however, to enlarge the section of the column, instead of using a common steel mold, a newly developed quick-assembled concrete mold, composed of precast concrete panels (PCP), is applied to facilitate the construction process. The longitudinal FRP bars, which are embedded into the column base, as well as the FRP spiral stirrup were adopted as the strengthening reinforcement to enhance the overall flexural capacity and repairability of the column. Prestressed steel strands were used to integrate PCPs into a fixed ring; however, due to the fact that prestressing was relatively low (less than 10% of the ultimate tensile strength of steel strands), its confinement effect (to concrete materials) was mostly neglected in this paper considering the stress lagging. The mortar was used for filling up the space between the original column and PCP to join the original column, FRP reinforcement, and PCP together. The retrofitting processes could be summarized as follows: (a) embedding of FRP bars into the column base, (b) installation of the FRP spiral stirrup, (c) assembly of PCP, (d) prestress steel strands, and (e) grouting, as shown in Figure 1(b). The detailing of embedding FRP bars into the column base was shown in Figure 1(c), and an embedding depth of 350 mm was applied for all kinds of FRP bars in this paper.

*2.2. Modeling.* The modeling of bridge columns strengthened with the PPCP-FRP method was based on the fiber analysis

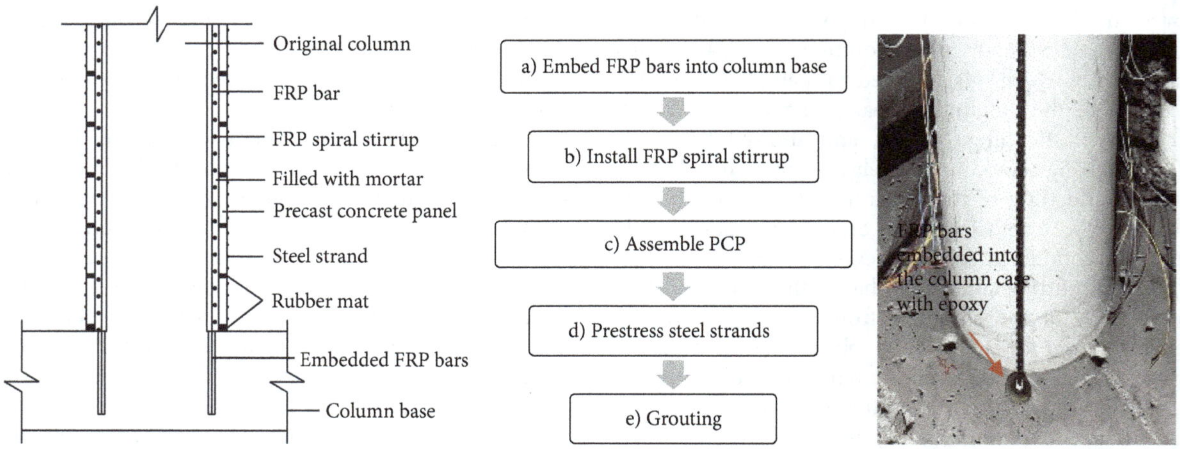


(a) Design of the PPCP-FRP method (b) Retrofitting processes (c) Embedding FRP bars

Figure 1: Design and retrofitting processes of the PPCP-FRP method.

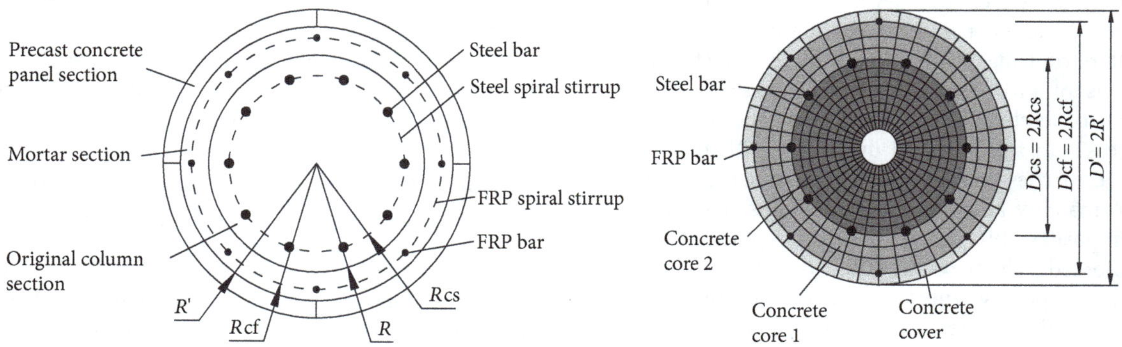


(a) Cross section in experiment (b) Cross section in OS

Figure 2: Cross section of the strengthened column.

concept, in which the flexural member was represented by unidirectional material fibers, and was realized in OpenSees (OS) [19].

The real cross section and the modeling cross section of strengthened specimens are shown in Figure 2. The real cross section is composed of the PCP section, the mortar section, and the original column section, as shown in Figure 2(a). The PCP section was excluded from the modeling cross section due to its minor contribution on vertical load bearing and transverse confinement to the column. The mortar and original column sections were divided into three sections in modeling, which were concrete core 1, concrete core 2, and concrete cover, as shown in Figure 2(b), based on the different confinement conditions applied by steel spiral stirrup and FRP spiral stirrup. Particularly, the cover of the original column and the mortar confined by the FRP spiral stirrup were considered with only one section (concrete core 1) in modeling, which was based on the fact that the concrete cover of the original column usually had deteriorated and would be removed anyway before the retrofitting, as in the experiment.

The real column in the experiment and the column model in the modeling are shown in Figure 3. In the experiment, three 1/4-scale circular-section bridge columns were initially designed and manufactured. The diameter and height of the initial column were 300 mm and 1275 mm, respectively, and the shear span ratio equaled to 4.25, a little larger than those in other common columns, which was to ensure that the shear span ratios of the specimens would be larger than 3 after strengthening; therefore, the flexure failure mode could be achieved. The volumetric transverse reinforcement ratio was equal to 0.09%. One of the columns was regarded as the control column, while the other two were further strengthened with PPCP-FRP, as shown in Figure 3(a). Different types of FRP reinforcement, which are referred to as basalt fiber-reinforced polymer (BFRP) reinforcement and carbon fiber-reinforced polymer (CFRP) reinforcement, were used as strengthening reinforcements for different specimens; however, the strengthening reinforcement ratio, calculated based on the equivalent strength principle, remained almost the same for both specimens, as shown in Table 1, in which $\rho_{orl}$ refers to the longitudinal reinforcement ratio of the original column (i.e., control column) before strengthening, as in (1); $n_s$ refers to the amount of longitudinal steel bars; $A_s$ refers to the cross-sectional area of longitudinal steel bars; $A_{cr}$ refers to the

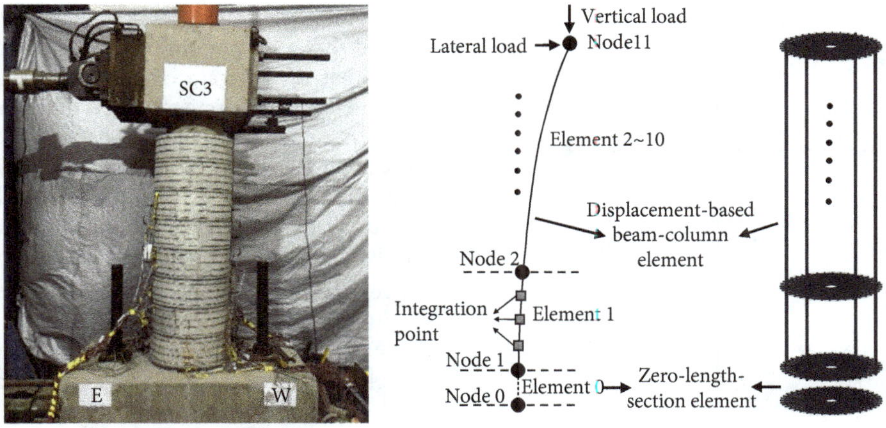


(a) Real column under testing (b) Column model in OS

FIGURE 3: Real column and column model.

TABLE 1: Details of all the specimens.

| Specimen label | $\rho_{orl}$ (%) | Strengthening reinforcement Type | FRP bar/$\rho_{eqfl}$ (%) | FRP spiral stirrup/$\rho_{eqft}$ (%) | $\rho_{eql}$ (%) |
|---|---|---|---|---|---|
| CC | 1.1 | \ | \ | \ | 1.1 |
| SCB | 1.1 | BFRP | 8Φ7.8/1.1 | Φ5.9@50/1.8 | 1.8 |
| SCC | 1.1 | CFRP | 8Φ6/1.2 | Φ4@40/1.9 | 1.9 |

cross-sectional area of the original column; $\rho_{eqfl}$ and $\rho_{eqft}$ refer to the additional equivalent longitudinal and transverse reinforcement ratio of strengthened columns, respectively, as in (2) and (3); $n_f$ refers to the amount of longitudinal FRP bars; $A_{cr}{}'$ and $d_{cr}{}'$ refer to the cross-sectional area and diameter of the strengthened columns, respectively; $d_{cr}{}'$ equals to 370 mm in this paper; $s$ refers to the spacing of the FRP spiral stirrup; $A_{eqfl}$ and $A_{eqft}$ refer to the equivalent longitudinal and transverse cross-section areas of the FRP bars, which were calculated based on the equivalent strength princeple, as shown in (4) and (5); $f_{ufl}$ and $f_{uft}$ refer to the ultimate stress of longidufinal and transverse FRP bars, respectively; $A_{fl}$ and $A_{ft}$ refer to the actual cross-sectional areas of longitudinal and transverse FRP bars, respectively; and $\rho_{eql}$ refers to the equivalent longitudinal reinforcement ratio of the strengthened column, as in (6).

$$\rho_{orl} = \frac{n_s \cdot A_s}{A_{cr}}, \tag{1}$$

$$\rho_{eqfl} = \frac{n_f \cdot A_{eqfl}}{A_{cr}{}'}, \tag{2}$$

$$\rho_{eqft} = \frac{4A_{eqft}}{s \cdot d_{cr}{}'}, \tag{3}$$

$$A_{eqfl} = \frac{f_{ufl}A_{fl}}{f_s}, \tag{4}$$

$$A_{eqft} = \frac{f_{uft}A_{ft}}{f_s}, \tag{5}$$

$$\rho_{eql} = \frac{n_f A_{eqfl} + n_s A_s}{A_{cr}{}'}. \tag{6}$$

The tests were performed using a loading system, as shown in Figure 3(a). All of the specimens were subjected to a combination of constant axial load and lateral cyclic loads. A constant axial load of 287 kN (normalized axial load ratio equals to 0.16) was first applied to each specimen with a hydraulic jack. Then, the lateral cyclic load was first controlled by a force which started at ±10 kN and then increased by ±10 kN at each load step until the column yielded. After the column yielded, the cyclic loading pattern was converted to a displacement-based control and each load step was repeated three times. Each specimen was instrumented with the following devices to measure its lateral load and corresponding displacement, deformations, and strains: (1) load sensor installed at the head of the actuator, (2) LVDT installed on the top of the specimen that works with the load sensor to measure displacement, and (3) strain gauges bonded onto the reinforcing bars.

A column model, which is composed of displacement-based beam-column (DBBC) elements (element 1 ~ 10) and zero-length-section element (element 0), was built in modeling, as shown in Figure 3(b). Each DBBC element contained three integration points. The zero-length-section element was located at the bottom of the model to consider the strain penetration of longitudinal bars. Node 0 was fully fixed during the test, while node 1 was constrained only at dof 1 and 2. In addition, the vertical and lateral loads were applied to node 11 and the loading patterns were exactly the same as in the experiment.

*2.3. Material Parameters.* Two different kinds of concrete were used to build the original column and PCP, respectively. The 28-day compressive strengths ($f_c{}'$) of these concretes were equal to 25.1 MPa and 40.5 MPa, respectively, and the corresponding elastic moduli ($E_o$) were equal to 24.9 GPa and 36.3 GPa. The 28-day compressive strength

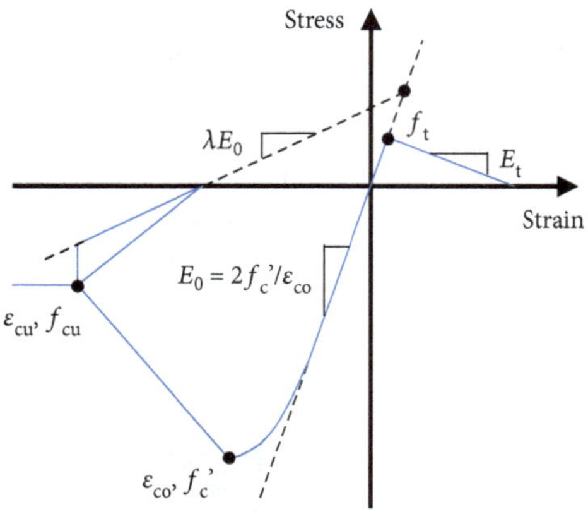


FIGURE 4: Model of concrete in OS.

and elastic moduli of mortar were equal to 32.7 MPa and 22.6 GPa, respectively.

*Concrete 02* in OS was used for simulating the concrete material, which was realized by defining the initial elastic tangent ($E_o$), compressive strength ($f_c'$), compressive strain ($\varepsilon_{co} = 4f_c'/E_o$), ultimate strength (or crushing strength for unconfined concrete) ($f_{cu}$), ultimate strain (or crushing strain for unconfined concrete) ($\varepsilon_{cu}$), tensile strength ($f_t$), and ratio ($\lambda$) between unloading slopes at $\varepsilon_{cu}$ and $E_o$, as shown in Figure 4. The modified Kent-Park model [20] was used as the compressive skeleton curve, which maintained a balance between simplicity and accuracy. For unconfined concrete, the crushing strength was usually set to "0" (or a quiet small value), which represents that no strength was considered for unconfined concrete after it was crushed. However, for concrete cover in this paper, the crushing strength should be considered due to the fact that the outside PCPs kept in a good shape until the end of the experiment, which means that a relatively strong confinement could be continuously applied to the "concrete cover" even after it has crashed. Due to a lack of experimental studies on axial compressive behavior of concrete confined with PCPs, 0.5 times of compressive strength was assumed for the crushing strength of concrete cover, and the crushing strain equals to −0.004. For concrete core 1 and concrete core 2, the ultimate strengths and strains would be enhanced by the single or dual confinements provided by the FRP spiral stirrup and steel spiral stirrup. However, due to a lack of experimental studies, a series of existing FRP-confined concrete models [21–23] were adopted to calculate the ultimate strengths and strains for the concrete materials. For concrete core 1, the ultimate strength and strain were calculated by using the FRP sheet confined concrete model proposed by Wu et al. [21]; the ultimate strength was taken as 1.8 times of its compressive strength, and the ultimate strain equals to −0.02. For concrete core 2, the ultimate strength and strain were expected to be larger than those of concrete core 1 due to an extra confinement provided by the steel spiral stirrup [22]. However, due to the potential negative effects such as stress lagging of the FRP spiral stirrup or a weak old-new concrete interface, which could possibly lower the ultimate strength, a conservative ultimate strength, which equals to 2 times of the compressive strength [23], was taken for concrete core 2, and the corresponding crushing strain remains the same with that of concrete core 1. The detailed setting of the parameters of concrete materials in OS are shown in Table 2.

The dimensions and mechanical properties of all reinforcements used in the experiment were experimentally measured, as listed in Table 3, in which $D$ and $E$ refer to the diameter and elastic modulus of reinforcement, respectively, $f_y$ and $f_u$ refer to the yield stress and ultimate stress of reinforcement, respectively, and $\varepsilon_u$ refers to the fracture strain of reinforcement.

*Steel 02 material* in OS was adopted for modeling the longitudinal steel bars in modeling, by defining parameters such as elastic modulus of steel bar ($E_s$), yield stress, strain-hardening ratio ($b = 0.01$) between postyield tangent and initial elastic modulus, and transition parameters from elastic to plastic branches ($R_0 = 18$, $cR_1 = 0.9$, and $cR_2 = 0.15$) [19]. *Hysteretic material* in OS was applied for modeling the longitudinal FRP bars by defining stress and strain values of feature points on the envelope curve [19]. The compressive strength of FRP bars was assumed equal to 50% of its tensile strength, and the compressive elastic modulus was assumed the same as the tensile elastic modulus based on the experimental studies conducted by Deitz et al. [24]

*2.4. Comparison between Test Results and Modeling Results.* In the experiment, a classic flexural failure mode including concrete crushing and longitudinal steel bar buckling (some fractured in the end) was observed for the CC specimen, as shown in Figure 5(a). The test phenomena of SCB and SCC were basically the same. After the test, the concrete panels of SCB and SCC were removed and the fracture of longitudinal FRP bar was observed, as in Figure 5(b).

The hysteresis load-displacement (L-D) curves of all specimens showed that the overall seismic performances of all strengthened specimens, including the lateral loading capacity, stiffness, and ductility, were significantly enhanced through the strengthening process with PPCP-FRP, as shown in Figure 6. Due to the relatively big ultimate strain of BFRP reinforcements, the force level of BFPR reinforcements in SCB would be smaller than that of CFPR reinforcements in SCC when similar strains were applied to these FRP reinforcements, as illustrated in (7), in which $F_f$ refers to the force level of the FPR reinforcement, $E_f$ refers to the elastic modulus of the FRP reinforcement, and $A_f$ and $A_{eqf}$ refer to the actual and equivalent cross-section area of the FPR reinforcement. This could be further illustrated as follows: The increasing speed of the force level would be relatively slow for the FRP reinforcement with a relatively high ultimate strain in the strengthened column, which could be beneficial to the adequate utilization of the material strength (especially to the steel reinforcement in original column) and therefore enhance the stability and energy dissipation capacity of the strengthened column.

$$F_f = \varepsilon E_f A_f = \varepsilon f_s A_{eqf} \frac{E_f}{f_u} = \varepsilon f_s A_{eqf} \cdot \frac{1}{\varepsilon_u}. \quad (7)$$

TABLE 2: Detailed setting of the parameters of concrete materials in OS.

| Concrete material parameters | Setting value in OS | | |
|---|---|---|---|
| | Concrete core 2 | Concrete core 1 | Concrete cover |
| $f'_c$ compressive strength | −25.1 MPa | −32.7 MPa | −32.7 MPa |
| $E_o$ elastic modulus | 24.9 GPa | 22.6 GPa | 22.6 GPa |
| $\varepsilon_{co}$ compressive strain | $4 * f'_c/E_o$ | $4 * f'_c/E_o$ | $4 * f'_c/E_o$ |
| $f_{cu}/f'_c$ ratio of ultimate strength (or crushing strength) to compressive strength | 2.0 | 1.78 | 0.5 |
| $\varepsilon_{cu}$ ultimate strain (or crushing strength) | −0.02 | −0.02 | −0.004 |
| $f_t$ tensile strength | $-0.14 * f'_c$ | $-0.14 * f'_c$ | $-0.14 * f'_c$ |
| $E_t$ tension softening stiffness | $f_t/0.002$ | $f_t/0.002$ | $f_t/0.002$ |
| $\lambda$ ratio of unloading slope at $\varepsilon_{cu}$ to $E_o$ | 0.1 | 0.1 | 0.1 |

TABLE 3: Dimensions and mechanical properties of all reinforcements.

| Reinforcement type | D/mm | E/GPa | $f_y$/MPa | $f_u$/MPa | $\varepsilon_u$/% |
|---|---|---|---|---|---|
| Steel bar | 10 | 210 | 446 | 615 | 16.8 |
| Steel spiral stirrup | 6 | 210 | 446 | 615 | 16.8 |
| Steel strand | 3 | 161 | \ | 1653 | 2.4 |
| BFRP bar | 7.8 | 63 | \ | 1382 | 2.2 |
| BFRP bar/BFRP spiral stirrup | 5.9 | 63 | \ | 1382 | 2.2 |
| CFRP bar | 6 | 170 | \ | 2560 | 1.5 |
| CFRP spiral stirrup | 4 | 170 | \ | 2560 | 1.5 |

The L-D curves of all specimens were consistent with the experimental curves, as shown in Figure 6. The hysteretic curves of all specimens agreed well with those in experiments. The characteristic values of L-D curves, as shown in Figure 7, as well as their errors with the characteristic values of experimental curves, were calculated and are listed in Table 4. The yield load ($F_y$) and peak load ($F_p$), as well as their corresponding displacements ($\Delta_y$, $\Delta_p$), agreed well with the test results with the maximal errors of 3.6% and 12.1%, respectively. In addition, the postyield stiffness of the L-D curves was calculated, as shown in Table 4, which also agreed well with those of test results with a maximal error of 11.1%. Moreover, the residual displacement of the L-D curves, which equals to the mean value of the positive and negative residual displacements of the L-D curve of each cycle, was compared with that of experimental curves in Figure 8, in which the residual displacement values were basically consistent with the experimental residual displacement values especially within the residual displacement limit, which was 1% of the column height according to the Japanese code [25], as shown in Figure 8(a). Furthermore, the development of residual displacement with the drift ratio of L-D curves agreed well with those of experimental curves, as shown in Figure 8(b), all of which indicated a high accuracy of the modeling method on predicting the seismic behavior of columns strengthened by the PPCP-FRP method.

## 3. Parametric Study

*3.1. Parameter Setting.* A comprehensive parametric study was conducted to further evaluate the seismic performance of the strengthened columns with different parameters, such as axial load ratio ($n$), EQFRR ($\alpha_p$), expansion ratio ($\kappa_s$), elastic modulus of FRP reinforcement ($E_f$), compressive strength of concrete ($f_{co}'$), and shear span ratio of the column ($\lambda$). The set value of each parameter, corresponding to the contrivable range for each parameter, are listed in Table 5.

The axial load ratio in this paper particularly equals to the axial load divided by the compressive strength of concrete and the cross-sectional area of the original column. A common range of axial load ratio for bridge columns, which was 0.05 ~ 0.2, was adopted.

The EQFRR equals to the ratio of the additional equivalent longitudinal reinforcement ratio of the strengthened column to the longitudinal reinforcement ratio of the original column, as in (8), and a range of 0.05 ~ 0.43 was adopted for EQFRR.

$$\alpha_p = \frac{\rho_{\mathrm{eqfl}}}{\rho_{\mathrm{orl}}}. \tag{8}$$

The expansion ratio equals to the ratio of $D_{cf}$ to $D_{cs}$, as shown in Figures 2(b) and (9). It should be noted that two major factors were coupled in $\kappa_s$, which were the expansion of the cross-section area and the variation of the distribution of FRP bars, considering the irrationality of the strengthening design by simply enlarging the cross-sectional area without rearranging the FRP bars. In this paper, the thickness of the concrete cover was fixed at 20 mm for all possible set values of the expansion ratio, which means that when a bigger $\kappa_s$ was applied in the analysis, a farther distance would be achieved by FRP bars away from the neutral axis of the cross section.

$$\kappa_s = \frac{D_{cf}}{D_{cs}}. \tag{9}$$

The elastic modulus of FRP reinforcement was designed for considering the different types of FRP reinforcements manufactured with different fibers, such as glass fiber, aramid fiber, basalt fiber, and carbon fiber. The ultimate

(a) CC (b) SCB

FIGURE 5: Failure mode of the specimens.

strength of these FRP reinforcements usually differs from each other but within a certain range. In this paper, a conservative value (1400 MPa) for the ultimate strength of FRP reinforcements was set for all kinds of FRP reinforcement to simplify the modeling approach. Therefore, the ultimate strain of FRP reinforcement would vary with $E_f$ due to the inverse relationship between these two factors; basically, when a bigger $E_f$ was applied, a smaller ultimate strain would be achieved by FRP reinforcement in the parametric study.

The compressive strength of the concrete and the shear span ratio of the column were also considered in a parametric study according to the design and analysis basics for the bridge column [3, 26]. The shear span ratio equals to the column height divided by the cross-sectional diameter of the original column.

A default value for each parameter was proposed, as in Table 5, for conducting the qualitative study on seismic behavior of strengthened columns. When a specific parameter was being investigated, other parameters could fix at the default value and would not impose any effect to the results.

### 3.2. Typical Load-Drift Curves with the Variation of Specific Parameter

*3.2.1. Axial Load Ratio.* The typical load-drift curves with the variation of axial load ratio are shown in Figure 9(a), in which the yield load and peak load of the curves increase with the increase in axial load ratio. The increase in yield load was due to the fact that the yield of steel bars was postponed because of the relatively slow development of stress in steel bars when a relatively high axial load ratio was applied. In addition, increasing the axial load ratio would potentially increase the flexural bearing capacity of the column by increasing the depth of the compression zone of the cross section of the column, in which case the peak force of the column would be enhanced, as displayed on the typical load-drift curves.

*3.2.2. EQFRR.* The typical load-drift curves with the variation of EQFRR are shown in Figure 9(b), in which the postyield stiffness, peak force, and corresponding drift ratio increase with the increase in EQFRR. A relatively high postyield stiffness could be expected for the specimens with a relatively large EQFRR due to the fact that a relatively high lateral load would be achieved by a strengthened column during the postyield stiffness stage when a relatively large EQFRR was applied under a certain drift ratio. In addition, since the cross-section failure of the column was characterized by the fracture of FRP bars in analysis, the peak moment of the cross section would increase, while the depth of the compression zone of the cross section would decrease with the increase in EQFRR; therefore, the peak force and corresponding drift ratio would increase with the increase in EQFRR.

*3.2.3. Expansion Ratio.* The typical load-drift curves with the variation of expansion ratio are shown in Figure 9(c), in which the initial stiffness, postyield stiffness, and peak force increase, while the drift ratio of the peak force decreases with the increase in expansion ratio. Since the lateral stiffness of the circular column was proportional to the fourth power of the cross-sectional diameter, a significant enhancement of initial stiffness and postyield stiffness should be expected by simply increasing the expansion ratio. In addition, enlarging the cross-sectional area of the column fundamentally equals to a decrease in shear span ratio of the column and reduction in the depth of the compression zone of the cross section, which would further lead to an increase in peak force and a decrease in its corresponding drift ratio.

*3.2.4. Elastic Modulus of FRP Reinforcement.* The typical load-drift curves with the variation of elastic modulus of FRP reinforcement are shown in Figure 9(d), in which the postyield stiffness increases, while the drift ratio of the peak force decreases with the increase in elastic modulus of FRP reinforcement. The increase in the postyield stiffness was due to the same reason as in EQFRR. The decrease in drift ratio of the peak force was due to the fact that a relatively high

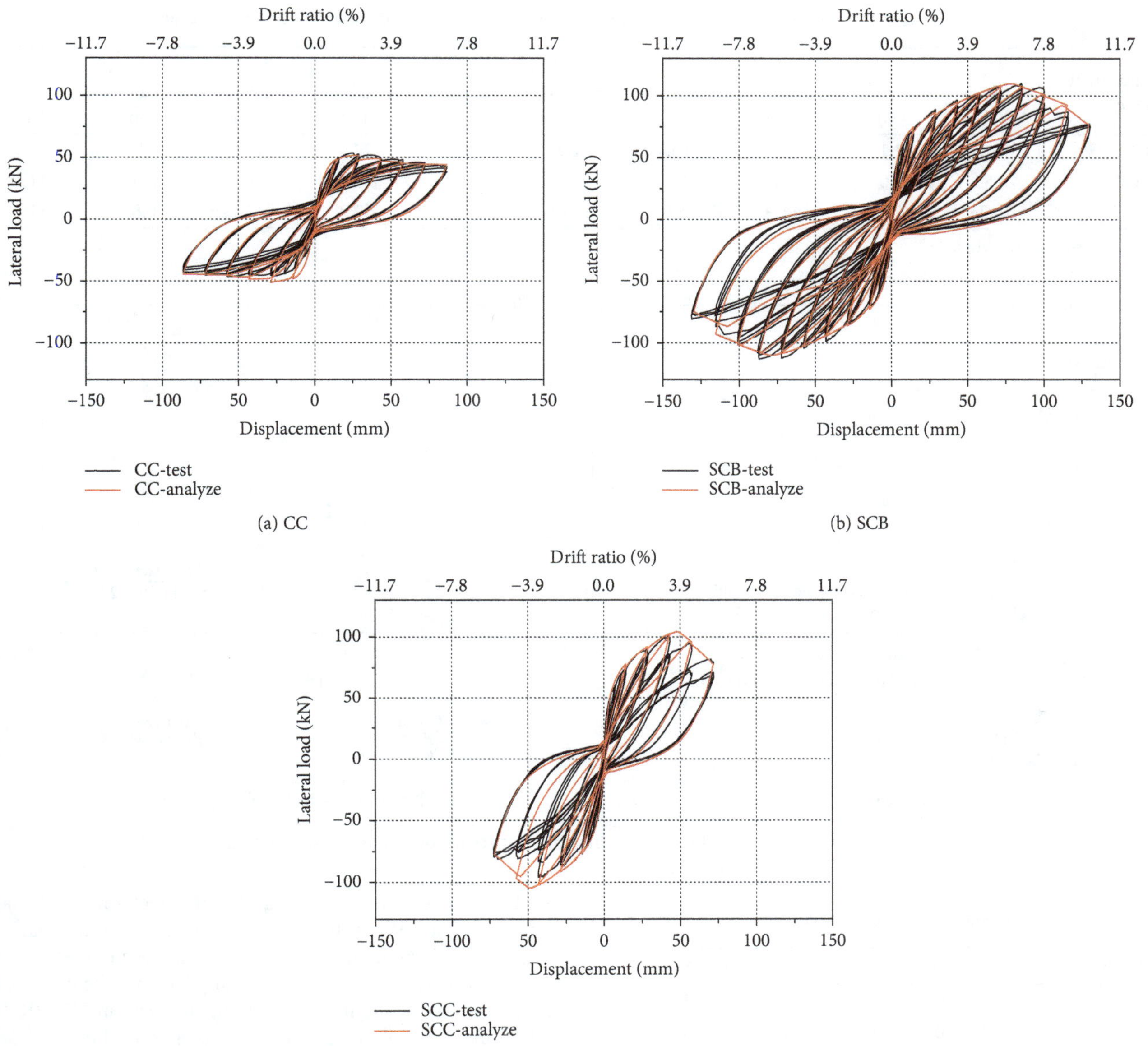


FIGURE 6: L-D hysteresis curves of all specimens.

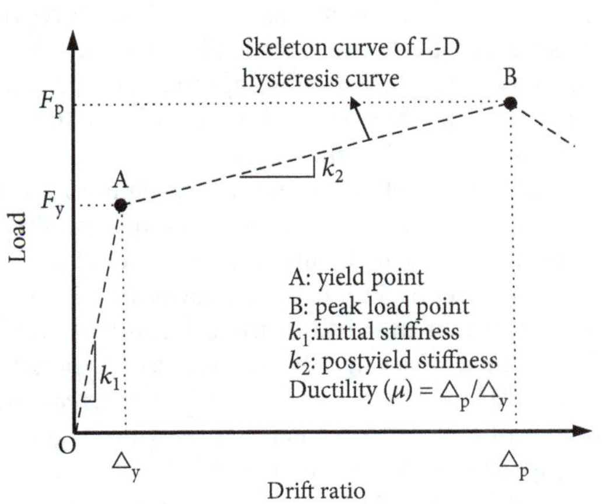


FIGURE 7: Schematic diagram of characteristic values of L-D hysteresis curves.

elastic modulus implied a relatively low ultimate strain for FRP reinforcement when the ultimate strength of FRP reinforcement was fixed in a parametric study.

*3.2.5. Compressive Strength of Concrete and Shear Span Ratio of Column.* The typical load-drift curves with the variation of the compressive strength of concrete are shown in Figure 9(e), in which the drift ratio of the peak force decreases with the increase in compressive strength of the concrete, which was due to the indirect effect imposed by the decrease in depth of the compression zone of the cross section. The typical load-drift curves with the variation of the shear span ratio of the column are shown in Figure 9(f). Since the effect of the shear span ratio of the column was basically overlapped within the effect of the expansion ratio, a similar tendency was displayed on these two sets of curves.

Table 4: The characteristic values of calculated L-D hysteresis curves and the error to the test results.

| Specimen label | $F_y$ Analysis (kN) | $F_y$ Error (%) | $\Delta_y$ Analysis (mm) | $\Delta_y$ Error (%) | $F_p$ Analysis (kN) | $F_p$ Error (%) | $\Delta_p$ Analysis (mm) | $\Delta_p$ Error (%) | $k_2$ Analysis (mm) | $k_2$ Error (%) |
|---|---|---|---|---|---|---|---|---|---|---|
| CC | 47.3 | 2.1 | 12.9 | 10.4 | 51.97 | 3.0 | 27.6 | 12.2 | \ | \ |
| SCB | 74.53 | 1.0 | 14 | 9.1 | 109.8 | 0.1 | 78.8 | 8.8 | 0.54 | 11.1 |
| SCC | 74.2 | 2.9 | 11.9 | 4.0 | 104.1 | 3.6 | 47.3 | 12.1 | 0.84 | 11.1 |

Error = |test − analyse|/test.

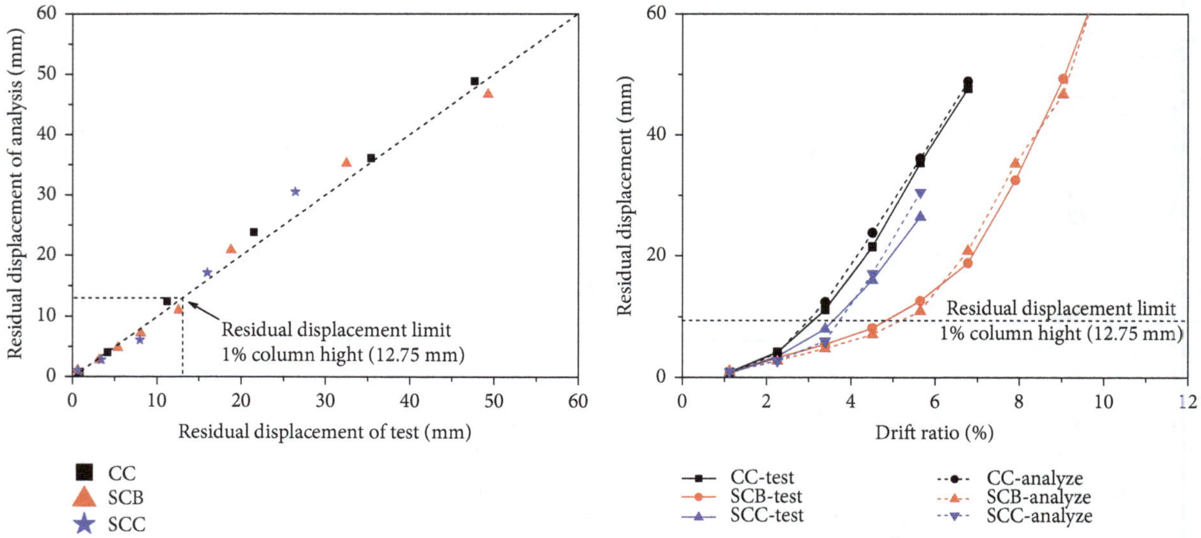


(a) Residual displacement value

(b) Drift ratio-residual displacement curve

Figure 8: Comparison between modeling and test results on residual displacement.

Table 5: Parameter value setting.

| Parameters | $n$ | $\alpha_\rho$ | $\kappa_s$ | $E_f$/GPa | $f_{co}{}'$/MPa | $\lambda$ |
|---|---|---|---|---|---|---|
| Set value | 0.05 | 0.05 | 1.1 | 25 | 15 | 3.5 |
| | 0.1 | 0.11 | 1.2 | 50 | 25 | 4.5 |
| | 0.15 | 0.18 | 1.4 | 100 | 35 | 5.5 |
| | 0.2 | 0.25 | 1.6 | 150 | 45 | 6.5 |
| | | 0.33 | | 200 | | |
| | | 0.43 | | | | |
| Default value | 0.1 | 0.11 | 1.4 | 50 | 25 | 4.5 |

The increase in initial stiffness and postyield stiffness, with the decrease in shear span ratio, was due to the fact that the lateral stiffness of the circular column was inversely proportional to the square of the column height. The increase in peak force and the decrease in corresponding drift ratio, with the decrease in shear span ratio, were basically due to the same reason as analyzed in the expansion ratio.

### 3.3. *Effect of the Parameters to the Seismic Performance*

*3.3.1. Lateral Load Capacity and Ductility.* The yield load and peak load are two representative characteristics of the lateral load capacity of the column. Among all parameters, EQFRR showed the most significant effect on yield load and peak load. When the steel bars of the column began to yield, a relatively high stress would be achieved by FRP bars when a relatively high EQFRR was applied, and the yield load of the column would increase with the increase in EQFRR. In addition, since column failure was characterized by the fracture of FRP bars, when a higher EQFRR was applied for strengthened columns, a higher peak load would be obtained. As shown in Figure 10(a), the peak load of specimens significantly increased with the increase in EQFRR. Moreover, the shear span ratio affects the yield load and peak load of the column as well, by presenting an approximate inverse relationship with both yield load and peak load, due to the same reason as common RC columns [27].

The ductility ($\mu$) of the column, as in Figure 7, is mostly influenced by EQFRR and elastic modulus of FRP reinforcement, as shown in Figure 11. The ductility decreases with the increase in EQFRR, as shown in Figure 11(a), due to the fact that the yield of the column had been postponed with the increase in EQFRR; the reason is presented in Figure 10(a). The ductility decrease with the increase in elastic modulus of FRP reinforcement, as shown in Figure 11(b), was simply due to the fact that the drift ratio of the peak load had significantly decreased with the increase in elastic modulus of FRP reinforcement; the reason is illustrated in Figure 9(d).

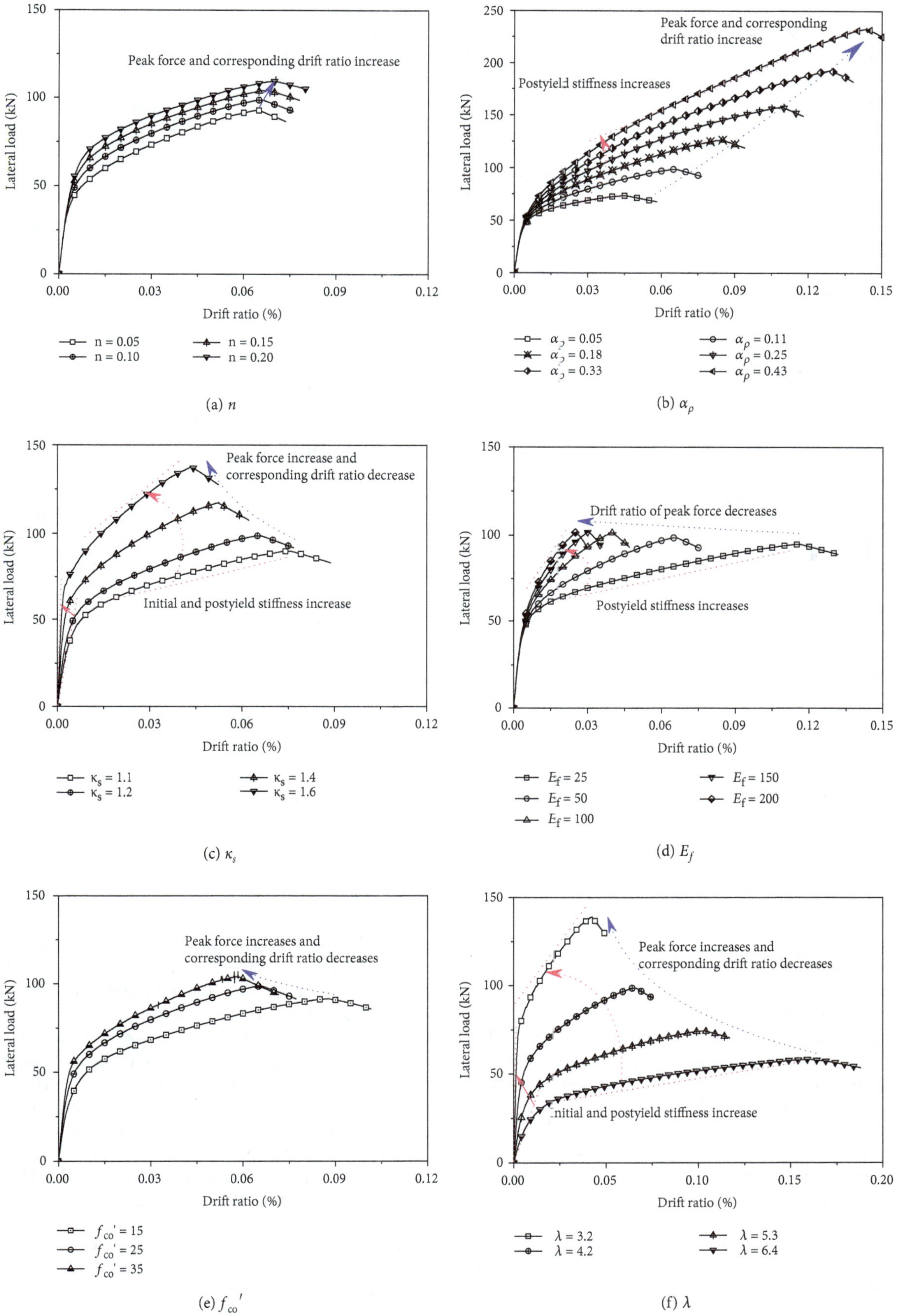


FIGURE 9: Typical load-drift curves with the variation of different parameters.

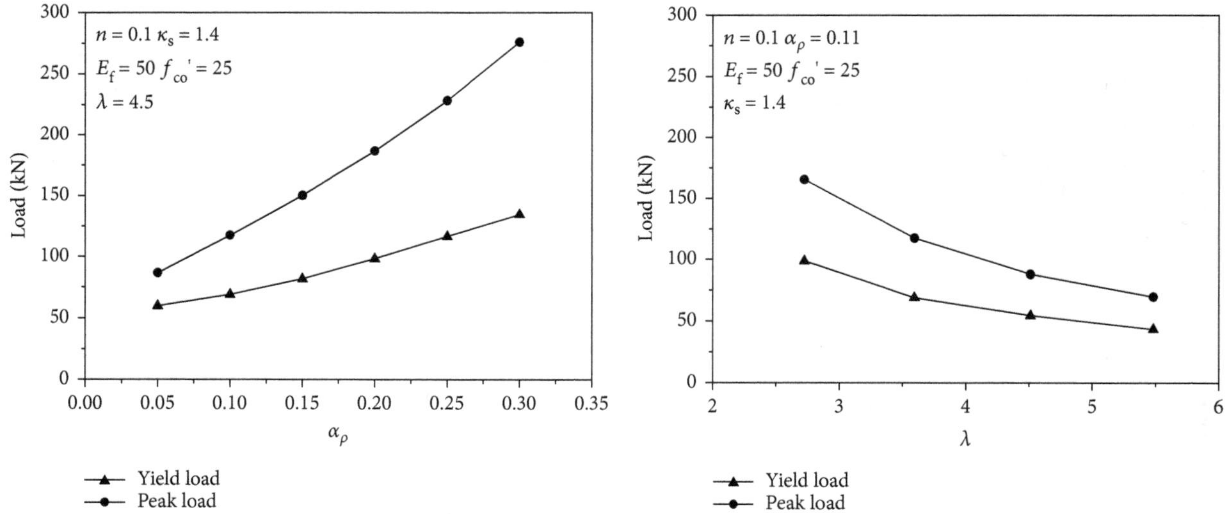


(a) $\alpha_p$ (b) $\lambda$

FIGURE 10: Variation of the characteristic load with different parameters.

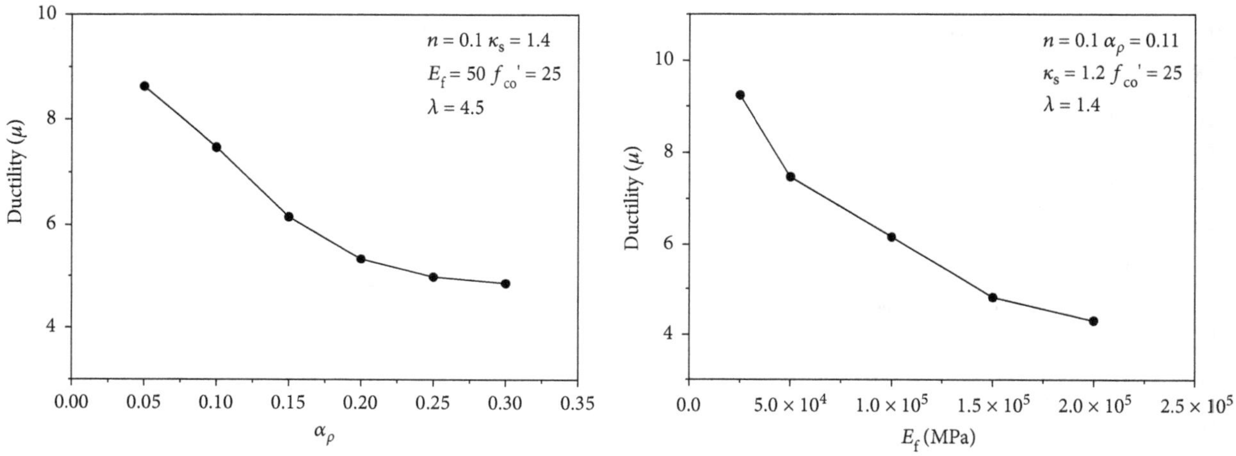


(a) $\alpha_p$ (b) $E_f$

FIGURE 11: Variation of ductility with different parameters.

*3.3.2. Energy Dissipation.* To evaluate the energy dissipation capacity of the strengthened columns, a cyclic loading analysis was conducted in OS with a similar loading pattern as in experiments for all specimens in analysis with different parameters. For each specimen, three major factors [3], including energy dissipation ($E_h$), elastic strain energy ($E_e$), and equivalent viscous damping ratio ($\xi_{eq}$), were calculated for each hysteric loop with MATLAB.

Among all parameters, the shear span ratio showed the most significant effect on energy dissipation capacity. Both of the energy dissipation and elastic strain energy decrease rapidly with the increase in shear span ratio, as shown in Figure 12(a), due to the stress level decrease in columns. In addition, the equivalent viscous damping ratio, which reflects the plastic strain energy, was relatively small for a column with a relatively big shear span ratio, which was due to the fact that the plastic strain energy was mainly contributed by concrete cracking, which would be relatively less to a column with a relatively big shear span ratio, especially when the drift ratio was relatively small. As shown in Figure 12(b), when the drift ratio equals to 0.01, the equivalent viscous damping ratio of the column (=0.048) when $\lambda = 8.85$ was only 32% that of the column (=0.148) when $\lambda = 3.18$, and this difference would gradually diminish with the increase in drift ratio. After the drift ratio reached 0.09, the development of the equivalent viscous damping ratio was almost coincident for all specimens with the increase in the drift ratio.

Due to the elastic property of FRP reinforcement, which contributes no energy dissipation to the column, the energy dissipation of columns with different EQFRR would remain constant, which has been proved in analysis, as shown in Figure 13(a); the energy dissipation curves with different EQFRR were basically coincident with each other. In addition, increasing EQFRR would downgrade the stress level of steel bars by "transforming" more stress from the steel bars to the FRP bars; the elastic strain energy of each cycle of

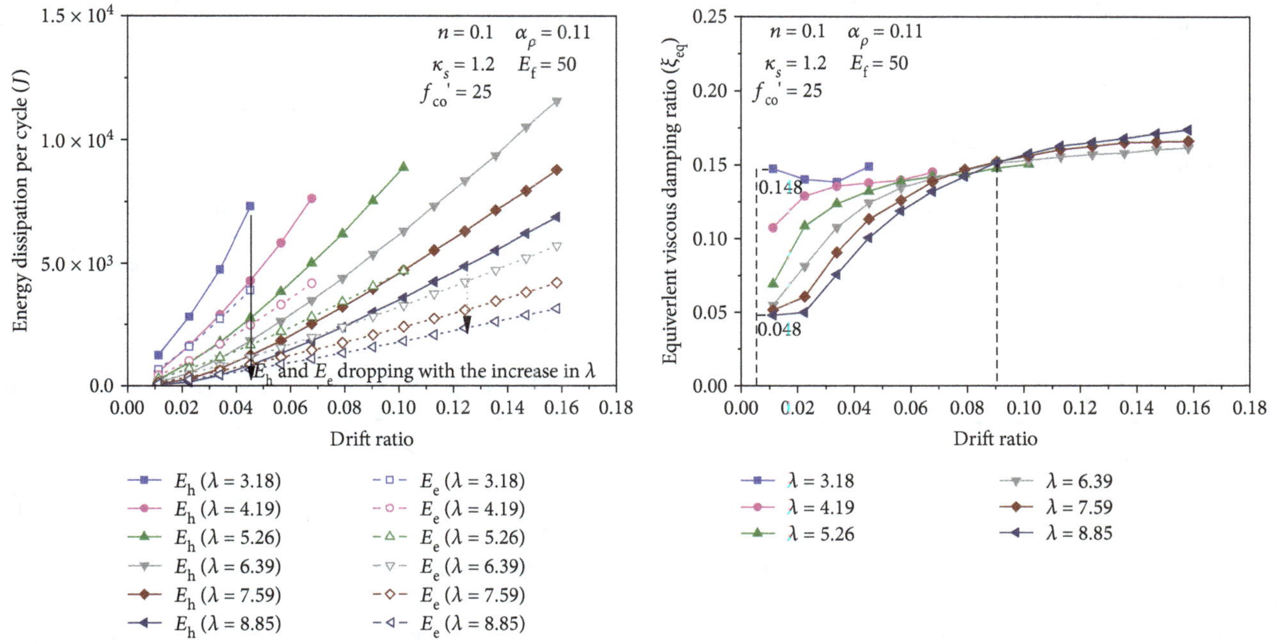


(a) Energy dissipation (b) Equivalent viscous damping ratio

FIGURE 12: Variation of energy dissipation indicators with drift ratio along with $\lambda$.

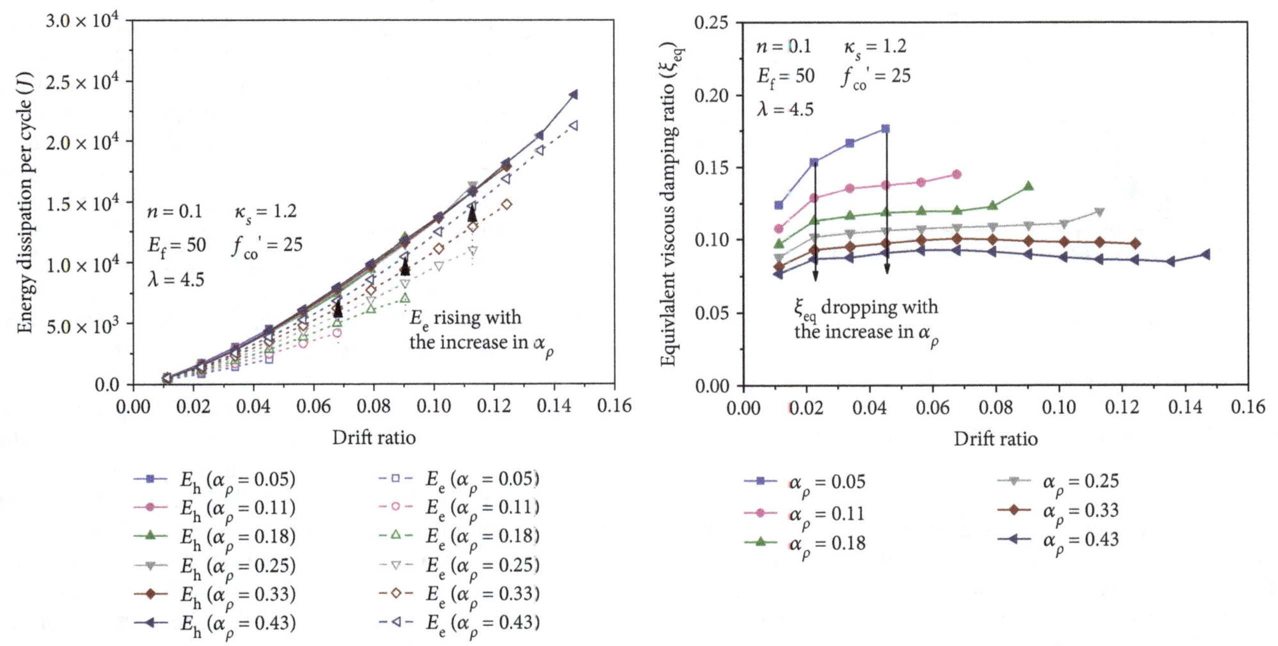


(a) Energy dissipation (b) Equivalent viscous damping ratio

FIGURE 13: Variation of energy dissipation indicators with drift ratio along with $\alpha_\rho$.

hysteresis loops may increase with the increase in EQFRR under a certain drift ratio, which has also been proved in analysis. The elastic strain energy and equivalent viscous damping ratio, which refers to plastic strain energy, gradually increases and decreases with the increase in EQFRR, as shown in Figures 13(a) and 13(b), respectively. Moreover, the equivalent viscous damping ratio increased relatively slowly with the increase in drift ratio when a relatively high EQFRR was applied, as shown in Figure 13(b), which indicated that the increase in EQFRR could slow down the development of plastic strain energy in columns.

*3.3.3. Lateral Stiffness.* The lateral stiffness of the bridge column is a time-dependent property which may decrease in

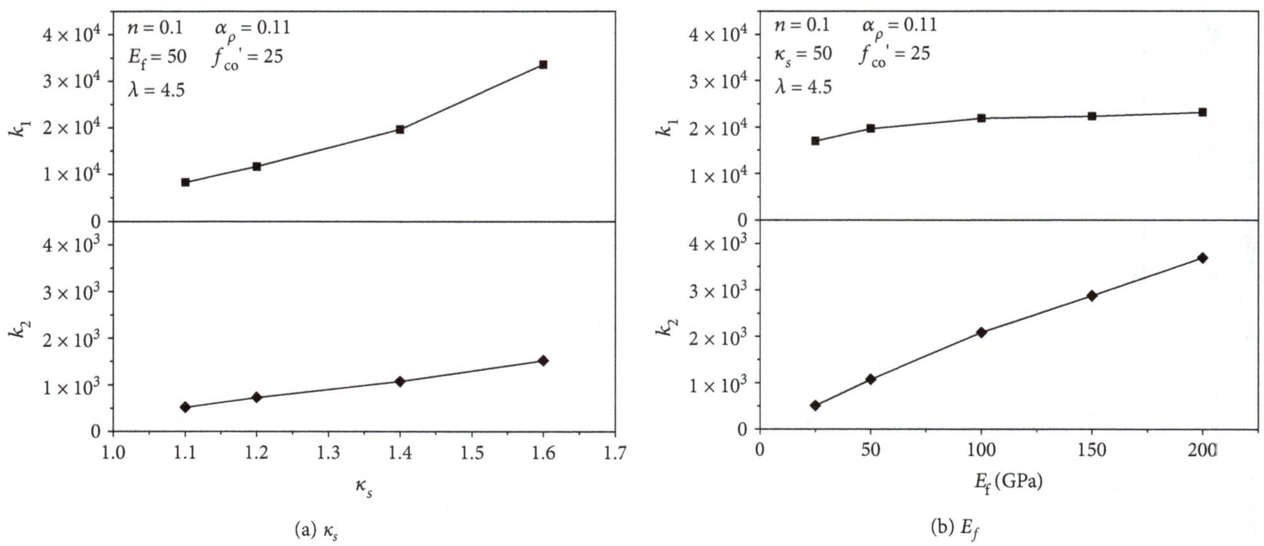


(a) $\kappa_s$

(b) $E_f$

FIGURE 14: Variation of $k_1$ and $k_2$ with different parameters.

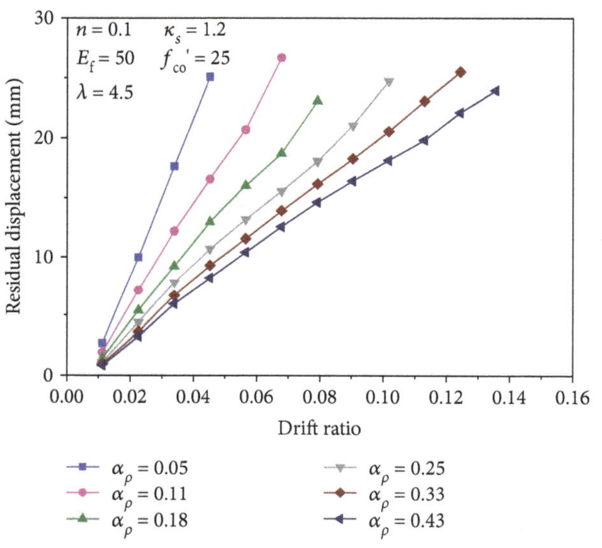


FIGURE 15: Variation of residual displacement with drift ratio along with $\alpha_p$.

resistance of the gradually weakening anchoring condition affected especially by scouring along the structure's service life [28]. Therefore, extra stiffness demands should be considered in the strengthening process. Since the stiffness of the column is proportional to the flexural stiffness of the cross section (EI) and inversely proportional to the second power of the column height ($h$), the most effective way to enhance the stiffness is to change the dimensions of the column, such as enlarging the cross-sectional area or "shortening" the height of the column, which had been directly and indirectly considered in analysis through the expansion ratio and shear span ratio of the column, respectively. Take expansion ratio as an example; the initial stiffness ($k_1$) of the specimen increased by 300% with the expansion ratio increased by 45%, as shown in Figure 14(a), which suggested a high efficiency of expansion ratio on improving the initial stiffness. Other parameters, such as elastic modulus of FRP reinforcement, could also contribute to the initial stiffness yet not as remarkable as the expansion ratio and shear span ratio of the column, as shown in Figure 14(b).

A stable postyield stiffness stage was obtained by most specimens in the analysis. The column with a higher postyield stiffness generally represents less damage and better repairability when the earthquake strikes [29, 30], and the variation of postyield stiffness with different parameters was also evaluated in the analysis. Since the FRP reinforcement mainly contributes to the postyield stiffness of the column, the postyield stiffness increases rapidly and almost linearly with the increase in the elastic modulus of FRP reinforcement, as shown in Figure 14(b). Other parameters such as expansion ratio could also contribute to the postyield stiffness yet not as remarkable as the elastic modulus of FRP reinforcement, as shown in Figure 14(a).

*3.3.4. Residual Displacement.* The residual displacement is a critical parameter for evaluating the repairability of the structure after earthquakes [25], and it basically increases linearly with the drift ratio regardless of any parameters applied to the model, as shown in Figure 15. Residual displacement was mostly affected by EQFRR which showed a rapid decrease with the increase in EQFRR, as shown in Figure 15. This was due to the fact that the plastic deformation of the column developed rather slowly when a relatively high EQFRR was applied, which further indicated that a better repairability could be achieved with a relatively high EQFRR for the strengthened column.

## 4. Conclusions

A modeling method based on the fiber analysis concept was realized in OS, in which the cross-section model of the strengthened column was simplified based on different confinement conditions applied by a steel spiral stirrup and an FRP spiral stirrup, and the strain penetration of longitudinal bars was also considered in the column model by using a zero-length-section element. A comprehensive parametric study was then further conducted to evaluate the influence of design parameters, such as axial load ratio, equivalent FRP reinforcement ratio rate (EQFRR), expansion ratio, elastic modulus of FRP reinforcement, compressive strength of concrete, and shear span ratio of the column, in terms of lateral load capacity, ductility, energy dissipation, lateral stiffness, and residual displacement of the strengthened column. The conclusions could be summarized as follows:

(1) The hysteric curves in the modeling were consistent with those in experiments. The yield load and peak load in modeling agree well with those in experiment with a maximal error of 3.6%. The postyield stiffness of hysteric curves is also closed to that of experimental hysteric curves with a maximal error of 11.1%. In addition, the residual displacement of each hysteric loop in modeling is basically consistent with that in the experiment, especially within the residual displacement limit, all of which indicates the high accuracy of the modeling method on predicting the seismic behavior of columns strengthened by the PPCP-FRP method.

(2) The lateral load capacity and ductility of strengthened columns were mostly affected by EQFRR. The yield load of strengthened columns increases with the increase in EQFRR due to the relatively high stress achieved by FRP bars when the strengthened column started yielding. The peak load of strengthened columns increases as well with the increase in EQFRR but due to the different fact that the column failure is characterized by the fracture of FRP bars; therefore, when a higher EQFRR was applied, the higher peak load would be obtained by strengthened columns. The ductility of the strengthened columns decreases with the increase in EQFRR due to the fact that the yield of the column was postponed with the increase in EQFRR.

(3) The shear span ratio of the column showed the most significant effect on the energy dissipation capacity of strengthened columns. With the increase in shear span ratio, both energy dissipation and elastic strain energy decrease rapidly due to the stress level decrease in columns. In addition, the equivalent viscous damping ratio, which reflects the plastic strain energy, was relatively small for a column with a relatively big shear span ratio, which was due to the fact that the plastic strain energy was mainly contributed by concrete cracking, which would be relatively less to a column with a relatively big shear span ratio, especially when the drift ratio was relatively small.

(4) The initial stiffness of strengthened columns was extremely sensitive to the expansion ratio, which increased by 300% with the expansion ratio increased by 45%. A stable postyield stiffness stage was obtained by most strengthened columns in a parametric study, and the value of postyield stiffness increases almost linearly and rapidly with the increase in elastic modulus of FRP reinforcement. The residual displacement of strengthened columns was mostly influenced by EQFRR, which rapidly decreases with the increase in EQFRR due to the fact that the plastic deformation of the column developed relatively slow when a relatively high EQFRR was applied, which further indicated that a better repairability could be achieved by applying a relatively high EQFRR in the strengthening process.

## Conflicts of Interest

The authors declare that there are no conflicts of interest regarding the publication of this paper.

## Acknowledgments

This research was financially supported by the National Key Research and Development Program of China (Grant no. 2016YFC0701100), the National Natural Science Foundation of China (Grant nos. 51525801 and 51778130), and the Fundamental Research Funds for the Central Universities (Grant no. 2242017k30002).

## References

[1] J. Kattell and M. Eriksson, "Bridge scour evaluation: screening, analysis, & countermeasures," Tech. Rep. No. 9877 1207, SDTDC, 1998.

[2] P. F. Lagasse, *Countermeasures to Protect Bridge Piers from Scour*, Transportation Research Board, 2007.

[3] M. J. N. Priestley, F. Seible, and G. M. Calvi, *Seismic Design and Retrofit of Bridges*, John Wiley & Sons, 1996.

[4] B. J. Bett, R. E. Klingner, and J. O. Jirsa, "Lateral load response of strengthened and repaired reinforced concrete columns," *ACI Structural Journal*, vol. 85, no. 5, pp. 499–508, 1988.

[5] Y. H. Chai, M. N. Priestley, and F. Seible, "Seismic retrofit of circular bridge columns for enhanced flexural performance," *ACI Structural Journal*, vol. 88, no. 5, pp. 572–584, 1991.

[6] M. N. Priestley, F. Seible, Y. Xiao, and R. Verma, "Steel jacket retrofitting of reinforced concrete bridge columns for enhanced shear strength-part 1: theoretical considerations and test design," *ACI Structural Journal*, vol. 91, no. 4, pp. 394–405, 1994.

[7] A. Mirmiran and M. Shahawy, "Behavior of concrete columns confined by fiber composites," *Journal of Structural Engineering*, vol. 123, no. 5, pp. 583–590, 1997.

[8] Y.-L. Bai, J. G. Dai, and J. G. Teng, "Cyclic compressive behavior of concrete confined with large rupture strain FRP composites," *Journal of Composites for Construction*, vol. 18, no. 1, article 04013025, 2014.

[9] L.-Z. Yao and G. Wu, "Fiber-element modeling for seismic performance of square RC bridge columns retrofitted with

NSM BFRP bars and/or BFRP sheet confinement," *Journal of Composites for Construction*, vol. 20, no. 4, article 04016001, 2016.

[10] J.-G. Dai, L. Lam, and T. Ueda, "Seismic retrofit of square RC columns with polyethylene terephthalate (PET) fibre reinforced polymer composites," *Construction and Building Materials*, vol. 27, no. 1, pp. 206–217, 2012.

[11] T. C. Rousakis, "Reusable and recyclable nonbonded composite tapes and ropes for concrete columns confinement," *Composites Part B: Engineering*, vol. 103, pp. 15–22, 2016.

[12] T. C. Rousakis, "Inherent seismic resilience of RC columns externally confined with nonbonded composite ropes," *Composites Part B: Engineering*, vol. 135, pp. 142–148, 2018.

[13] H. Saadatmanesh, M. R. Ehsani, and L. Jin, "Repair of earthquake-damaged RC columns with FRP wraps," *ACI Structural Journal*, vol. 94, pp. 206–215, 1997.

[14] M. S. Saiidi and Z. Cheng, "Effectiveness of composites in earthquake damage repair of reinforced concrete flared columns," *Journal of Composites for Construction*, vol. 8, no. 4, pp. 306–314, 2004.

[15] A. Vosooghi and M. S. Saiidi, "Design guidelines for rapid repair of earthquake-damaged circular RC bridge columns using CFRP," *Journal of Bridge Engineering*, vol. 18, no. 9, pp. 827–836, 2013.

[16] A. Vosooghi and M. S. Saiidi, "Shake-table studies of repaired reinforced concrete bridge columns using carbon fiber-reinforced polymer fabrics," *ACI Structural Journal*, vol. 110, no. 1, p. 105, 2013.

[17] M. Fakharifar, G. Chen, C. Wu, A. Shamsabadi, M. A. ElGawady, and A. Dalvand, "Rapid repair of earthquake-damaged RC columns with prestressed steel jackets," *Journal of Bridge Engineering*, vol. 21, no. 4, article 04015075, 2016.

[18] J. Li, J. Gong, and L. Wang, "Seismic behavior of corrosion-damaged reinforced concrete columns strengthened using combined carbon fiber-reinforced polymer and steel jacket," *Construction and Building Materials*, vol. 23, no. 7, pp. 2653–2663, 2009.

[19] S. Mazzoni, F. Mckenne, M. Scott, and G. L. Fenves, *Open System for Earthquake Engineering Simulation User Manual, Version 2.0*, University of California, Pacific Earthquake Engineering Center, Berkeley, CA, USA, 2009.

[20] N. G. Bunni, B. Scott, R. Park, and M. Priestley, "Stress-strain behavior of concrete confined by overlapping hoops at low and high-strain rates," *Journal of the American Concrete Institute*, vol. 79, no. 6, pp. 496–498, 1982.

[21] G. Wu, Z. T. Lü, and Z. S. Wu, "Strength and ductility of concrete cylinders confined with FRP composites," *Construction and Building Materials*, vol. 20, no. 3, pp. 134–148, 2006.

[22] T. C. Rousakis and I. S. Tourtouras, "Modeling of passive and active external confinement of RC columns with elastic material," *ZAMM - Journal of Applied Mathematics and Mechanics/ Zeitschrift für Angewandte Mathematik und Mechanik*, vol. 95, no. 10, pp. 1046–1057, 2015.

[23] J. Teng, J.-F. Chen, S. T. Smith, and L. Lam, *FRP: Strengthened RC Structures*, Wiley, 2002.

[24] D. H. Deitz, I. E. Harik, and H. Gesund, "Physical properties of glass fiber reinforced polymer rebars in compression," *Journal of Composites for Construction*, vol. 7, no. 4, pp. 363–366, 2003.

[25] Committee J E E, *Earthquake Resistant Design Codes in Japan*, Japan Society of Civil Engineers (JSCE), 2000.

[26] L. Aashto, *Bridge Design Specifications*, American Association of State Highway and Transportation Officials, Washington, DC, USA, 1998.

[27] R. Park and T. Paulay, *Reinforced Concrete Structures*, John Wiley & Sons, 1975.

[28] A. Fioklou and A. Alipour, "Seismic behavior of bridges with deep foundations under effects of scouring," in *Structures Congress 2014*, pp. 313–323, Boston, MA, USA, 2014, ASCE.

[29] Z.-Y. Sun, G. Wu, Z.-S. Wu, and M. Zhang, "Seismic behavior of concrete columns reinforced by steel-FRP composite bars," *Journal of Composites for Construction*, vol. 15, no. 5, pp. 696–706, 2011.

[30] M. F. M. Fahmy, Z. Wu, G. Wu, and Z. Sun, "Post-yield stiffnesses and residual deformations of RC bridge columns reinforced with ordinary rebars and steel fiber composite bars," *Engineering Structures*, vol. 32, no. 9, pp. 2969–2983, 2010.

# 10

# Side-Chain Polyimides as Binder Polymers for Photolithographic Patterning of a Black Pixel Define Layer for Organic Light Emitting Diode

**Genggongwo Shi,[1] Sung Hoon Park,[2] Jeseob Kim,[1] Minji Kim,[2] and Lee Soon Park[2]**

[1]*CCTech Inc., 72, Sicheng-Ro, 721Beon-Gil, Paltan-Myeon, Haseong-Si, Gyeonggi-Do 18524, Republic of Korea*
[2]*School of Material Science and Engineering, Ulsan National Institute of Science and Technology (UNIST), Ulsan 44919, Republic of Korea*

Correspondence should be addressed to Lee Soon Park; parkls@unist.ac.kr

Academic Editor: Renbo Wei

A pixel define layer (PDL) in an organic light emitting diode (OLED) is patterned using a photolithographic process before the deposition of organic layers on top of ITO anode. If the patterning of PDL on OLED panels can be achieved using a black photoresist, the patterning of black matrix (BM) on top of PDL patterns can be omitted by reducing the reflection of ambient light from OLED panels. In this study, we synthesized a series of side-chain-type polyimides as binder polymers of black photoresists and investigated the potential of using the black photoresist for the fine patterning of black PDL on OLED panels.

## 1. Introduction

Organic light emitting diodes (OLEDs) have become one of the major trends in the fabrication of flat panel displays because of their advantages, such as light weight, slim panels, fast response time, wide viewing angle, high resolution, and low power consumption [1-4]. The light emitting area of an OLED device is disposed at the luminous region of the pixel array on the OLED panel connected to a thin film transistor unit. Before the deposition of OLED organic layers, a pixel define layer (PDL) is patterned via a photolithographic process, as shown in Figure 1 [5-8].

Currently, positive-tone photosensitive polyimide [9, 10] is used to fabricate PDL microstructures. The corresponding PDL is yellowish brown similar to that of polyimide films. Since the yellowish-brown polyimide layer of PDL reflects the incident light, the visibility of the OLED display is reduced, particularly when used outdoors. Therefore, in the current positive-tone photolithographic process, a black matrix (BM) layer is patterned on top of the PDL pattern to improve the visibility of the OLED device (Figure 1).

If the patterning of PDL on OLED panels can be achieved using a negative-tone black photoresist [11] containing black pigment instead of the positive-tone photoresist based on photosensitive polyimide, the photolithographic process of patterning BM on top of the PDL pattern would be omitted. The required components for a typical negative-tone black photoresist include the photoinitiator, photosensitizer, multifunctional monomer, black pigment millbase, and binder polymer developable with aqueous alkaline solution. The binder polymer is crucial in the patterning of black PDL on OLED panels. It is not only one of the major components in terms of weight but also affects the shape of PDL patterns during its development in the photolithographic process. The process temperature of OLED panel fabrication reaches close to 300 °C, including the postcure step after photolithographic patterning of PDL; therefore, the thermal stability of the binder polymer

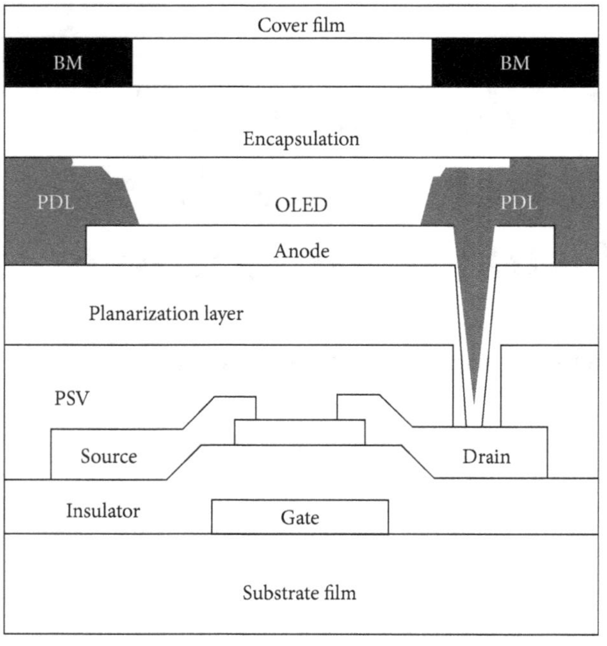


FIGURE 1: Typical cross-sectional view of an OLED panel.

should also be considered. Herein, new binder polymers with thermally stable imide linkages were synthesized and applied to the black photoresist formulation to pattern black PDL on OLED panels.

## 2. Experimental Methods

*2.1. Materials and Characterization.* Chemicals and solvents were reagent grades and used without further purification: poly(styrene-*co*-maleic anhydride), cumene terminated, $M_n \sim 1600$ (PSMA); 5-aminoisophthalic acid (AIPA); 3,5-bis(trifluoromethyl) aniline (6FAL); 4-aminostyrene (AS); 2-hydroxylethylacrylate (2-HEA); glycidyl methacrylate (GMA); dimethyl acetamide (DMAc); 3,5-di-*tert*-4-butylhydroxytoluene (BHT); tetrabutylphosphonium bromide (TBPB); propylene glycol monomethyl ether acetate (PGMEA); and pentaerythritol triacrylate (PETA).

Thermogravimetric analysis (TGA) was performed on a TA Instruments Q500 thermogravimetric analyzer with a programmed temperature at 10°C/min. Fourier-transform infrared spectroscopy (FT-IR) was performed on a Varian 670/620 spectrometer using the KBr pellet method. $^1$H NMR spectroscopy was taken on a Bruker AVANCE III HD 400 MHz spectrometer.

*2.2. Synthesis of Styrene-Type Side-Chain Polyimides.* PSMA (20 mmol based on maleic anhydride units) was dissolved in 20 mL of DMAc. Varying molarities of AS, AIPA, and 6FAL were added sequentially once every hour and stirred at room temperature. After the formation of amic acid linkages between maleic anhydride units of PSMA and three aromatic amines (AS, AIPA, and 6FAL), a free radical polymerization inhibitor (BHT) and a chemical imidization agent (acetic anhydride; 10 mL) were added and the reaction mixture was reacted for 4.5 h at 100°C to afford side-chain polyimides (PSI-x) [12] (Figure 2). After cooling down, the reaction mixture was poured into excess water (>10× volume) and the precipitates were collected by centrifugation. EtOAc was added to the precipitates, and the insoluble was removed via filtration. The organic solution was washed twice using brine, dried with anhydrous $Na_2SO_4$, and evaporated to dryness. The solid product was washed using diethyl ether multiple times to afford PSI series side-chain polyimide as brown powder.

*2.3. Synthesis of Acrylate-Type Side-Chain Polyimides.* PSMA (20 mmol based on maleic anhydride units) was dissolved in 20 mL of DMAc solvent. First, 2-HEA (10 mmol) and pyridine (1 mmol) were added to the PSMA solution and the reaction mixture was stirred at room temperature for 16 h. Second, 6FAL (10 mmol) was added and stirred at room temperature for 1 h. Acetic anhydride (10 mL) was added, and the mixture was stirred at 100°C for 4.5 h for the imidization reaction to occur (Figure 3). After the reaction, the mixture was poured into excess amount of water. The subsequent procedure was the same as that of PSI series in Section 2.2 to afford PSEI series.

*2.4. Synthesis of Methacrylate-Type Side-Chain Polyimides.* PSMA (20 mmol based on maleic anhydride unit) was also dissolved in 20 mL of DMAc. AIPA (6 mmol) and 6FAL (14 mmol) were added sequentially once an hour and stirred at room temperature. The reaction mixture was heated to 130°C and stirred for 16 h [13]. After cooling down, GMA (4 mmol), TBPB (0.4 mmol), and BHT (0.04 mmol) were added and stirred at 90°C for another 3 h (Figure 4). The reaction mixture was poured into excess water. The subsequent procedure was the same as PSI series to afford PSEA-1.

For the PSEA-2 synthesis, PSMA was dissolved in PGMEA (net. 30 wt%) instead of DMAc, and the reaction was proceeded as above with GMA.

## 3. Results and Discussions

*3.1. Synthesis of Polyimide-Based Binder Polymers.* The use of negative-tone black photoresist, instead of positive-tone photoresist based on photosensitive polyimide could remove both the BM patterning and the 1/4$\lambda$ polarizing film by reducing the reflection of ambient light from the OLED panel. Positive-tone photoresist based on photosensitive polyimide is used herein because the resulting PDL made of polyimide thin film has a high thermal stability and can withstand temperatures up to 300°C during the OLED panel fabrication process. However, the positive-tone photoresist based on photosensitive polyimide cannot be used to develop a black photoresist due to limited penetration depth of the UV light in the photolithographic patterning of black PDL on the OLED panel. The negative-tone black photoresist has deeper penetration depth than positive-tone photoresist since the photolithographic patterning is performed via a crosslinking reaction mechanism involving free

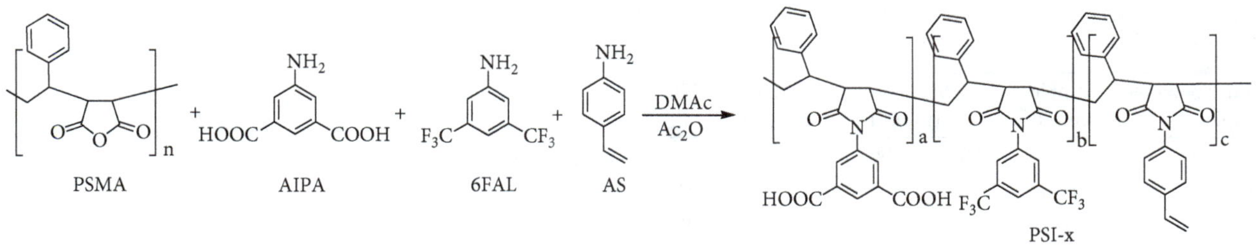


FIGURE 2: Synthetic scheme of styrene-type side-chain polyimides.

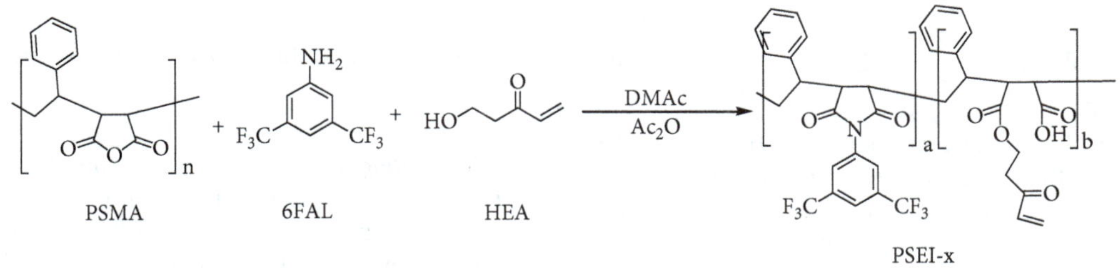


FIGURE 3: Synthetic scheme of acrylate-type side-chain polyimides.

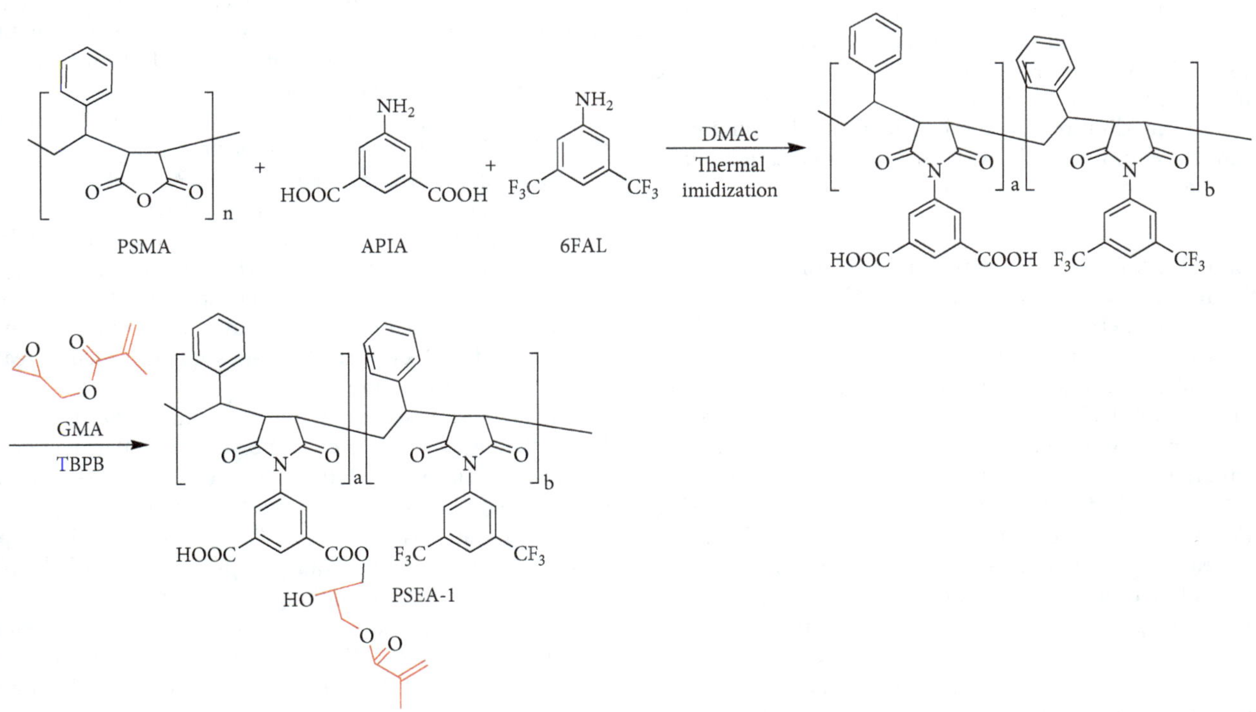


FIGURE 4: Synthetic scheme of methacrylate-type side-chain polyimides.

radical chain polymerization among multifunctional monomers and the binder polymer.

The binder polymer should have carboxylic groups to be developed easily with aqueous alkaline developer. Moreover, the binder polymer should also be crosslinked tightly with the multifuncional monomers upon UV light irradiation in order not to be soluble in the aqueous alkaline developer. Further, a high thermal stability of the binder polymer that corresponds to the photosensitive polyimide in the positive-tone photoresist is required. According to this guideline, three different binder polymers with thermally stable imide linkages have been synthesized.

*3.1.1. PSI and PSEI Series.* In the synthesis of PSI series binder polymers, three aromatic amines reacted with the maleic anhydride units of PSMA. The three amines (AIPA, 6FAL, and AS) were used to afford carboxyl groups required in the development step of photolithography. The solubility of the binder polymer in the PGMEA, a common solvent in the photoresist, was also examined. Besides, double bonds

Table 1: Synthesis of side-chain polyimides for binder polymers of black photoresists.

| Binder polymer samples | Sample codes | Molarities (mmol) | | | Yield (%) | GPC $M_w$ |
|---|---|---|---|---|---|---|
| | | AS | AIPA | 6FAL | | |
| PSI series | PSI-1 | 4 | 6 | 10 | 52 | 4716 |
| | PSI-2 | 4 | 8 | 8 | 44 | 4863 |
| | | 2-HEA | | 6FAL | | |
| PSEI series | PSEI-1 | 10 | 10 | 53 | 4637 | |
| | | AIPA | 6FAL | GMA | | |
| PSEA series | PSEA-1 | 6 | 14 | 4 | 56 | 4735 |
| | PSEA-2 | 6 | 14 | 6 | — | 4819 |

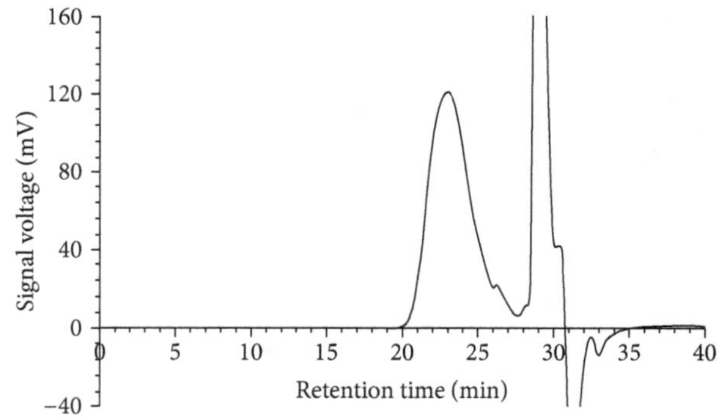


Figure 5: GPC chromatogram of PSEA-2 polyimide.

responsible for the photocrosslinking reaction were also included through the AS monomer. In the synthesis of PSI series, the chemical imidization was carried out utilizing acetic anhydride under mild condition of 100 °C to avoid self-polymerization of AS via a thermal initiation mechanism. The PSI product was recovered by precipitation in excess water to remove the excess acetic anhydride followed by filtration, drying, and then tested in the black photoresist.

In PSEI series, 2-HEA was reacted with maleic anhydride units of PSMA to generate both carboxyl groups and double bonds. Then, aromatic amine (6FAL) was imidized using chemical imidization method under mild conditions. Because acetic anhydride was used for imidization, PSEI has to be worked-up in the same way as PSI series.

*3.1.2. PSEA Series.* PSEA series were synthesized using two different processes. First, APIA and 6FAL amines were reacted via thermal imidization with maleic anhydride groups of PSMA in DMAc solvent at 130 °C. Then side-chain polyimide intermediate was reacted with GMA to afford a PSEA-1 binder polymer, in which carboxyl groups were partially converted to epoxymethacrylates. The PSEA-1 sample was recovered in the same manner as PSI series. In the second process, PSEA-2 was synthesized using PGMEA solvent instead of DMAc. However, PSEA-2 was obtained using a one-pot solution method without work-up and was tested directly as the binder polymer in the photolithographic patterning of black PDL in the OLED panel.

Table 1 shows that the yields of polyimides were not high (44%–56%) because the polyimide products had relatively high carboxyl groups and underwent weight loss during precipitation in excess water. GPC, FT-IR, and $^1$H NMR analyses of the PSEA-2 sample are shown in Figures 5, 6, and 7, respectively. The molecular weight of PSEA-2 ($M_w = 4819$ g/mol) was higher than that of PSMA ($M_n = 1600$ g/mol) due to the expansion of hydrodynamic volume after an imidization step that introduced bulky substituents in side chains of a relatively flexible PSMA polymer. According to FT-IR spectra, anhydrides were converted to imides and methacrylate esters. Moreover, $^1$H NMR spectrum indicated that electron-poor aromatic amines ($\delta$ 8.18–8.45), epoxy ($\delta$ 4.16), and methacrylate groups ($\delta$ 1.87–1.96, 5.66, and 6.07) were introduced to PSMA chains.

*3.2. Photolithography and Thermal Stability of Black PDL Patterns.* The black photoresist for the patterning of the black PDL of OLED panel is composed of the photoinitiator, photosensitizer, multifunctional monomer, polyimide as a thermal stabilizer, black millbase, and binder polymer. The binder polymer and multifunctional monomer are the main components of the photoresist, which affect the exact patterning of PDL via the photocrosslinking reaction in the UV-irradiated area and development of UV-unirradiated area by aqueous alkaline solution.

In this study, a commercial binder polymer SR-6300 (SMS Co., Korea) known as cresol novolac epoxyacrylate, dissolved in PGMEA solvent at 30 wt%, was used as a reference binder polymer as shown in Table 2. A commercial sample PETA was used as multifunctional monomer, which has the functionality of 3. The thermal stability of the reference black photoresist (PT-0), determined by TGA, was not high enough (up to 300 °C) to endure the photolithographic patterning of PDL even after the subsequent postcure treatment at 250 °C for 30 minutes.

Figure 8 shows the photolithographic patterns of the black PDL using the side-chain polyimide binder polymers. The black photoresists PT-0 and PT-3 exhibited fine patterns of black PDL. However, PT-1 and PT-2 showed irregular or wavy PDL patterns. In case of PT-1 photoresist PSI type binder polymer had styrenic groups as source of double bonds for the photocrosslinking reaction, so that the adhesion to the silicon wafer was not as good as the reference binder polymer (SR-6300), which has epoxy acrylate as a source of double bonds. In the case of the PT-2 black photoresist, the wavy PDL pattern was related to the high acid value of the binder polymer PSEI-1 in which acrylate groups were introduced via the reaction of 2-HEA with maleic anhydride units of PSMA. High amount of 2-HEA monomer had to be used for the complete solubility of the resulting PSEI-1 in PGMEA. The PT-3 black photoresist showed fine PDL patterns due to the balance of good adhesion to the silicon wafer and the optimum acid value from benzoic acid-type carboxyl groups.

The effect of side-chain polyimide binder polymers on the thermal stability of the resulting PDL patterns was evaluated via TGA analysis. The black PDL patterns formed on the silicon wafer were scratched off after photolithographic

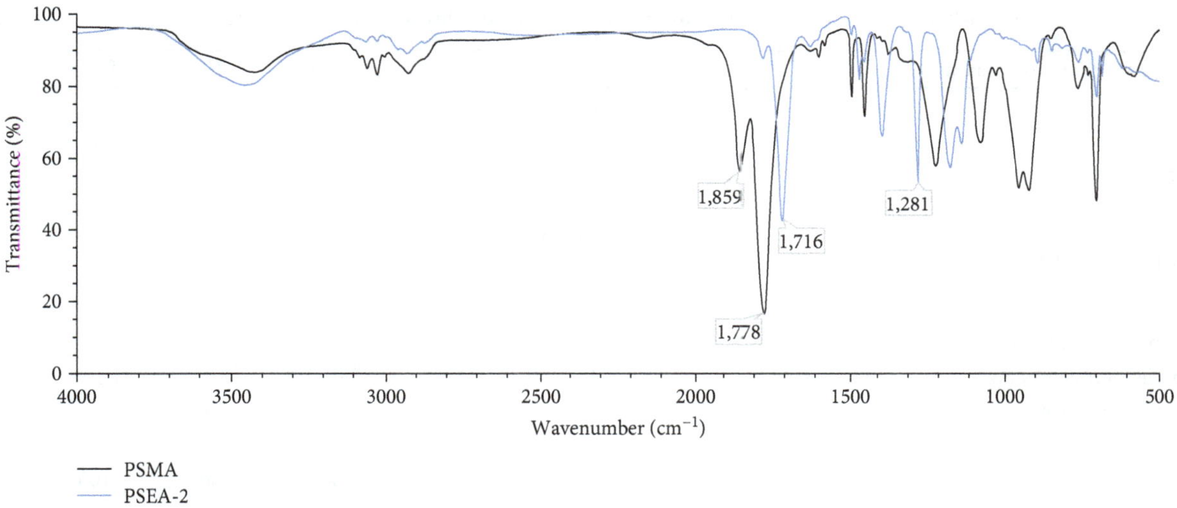


FIGURE 6: FT-IR charts of PSEA-2 and its starting material PSMA.

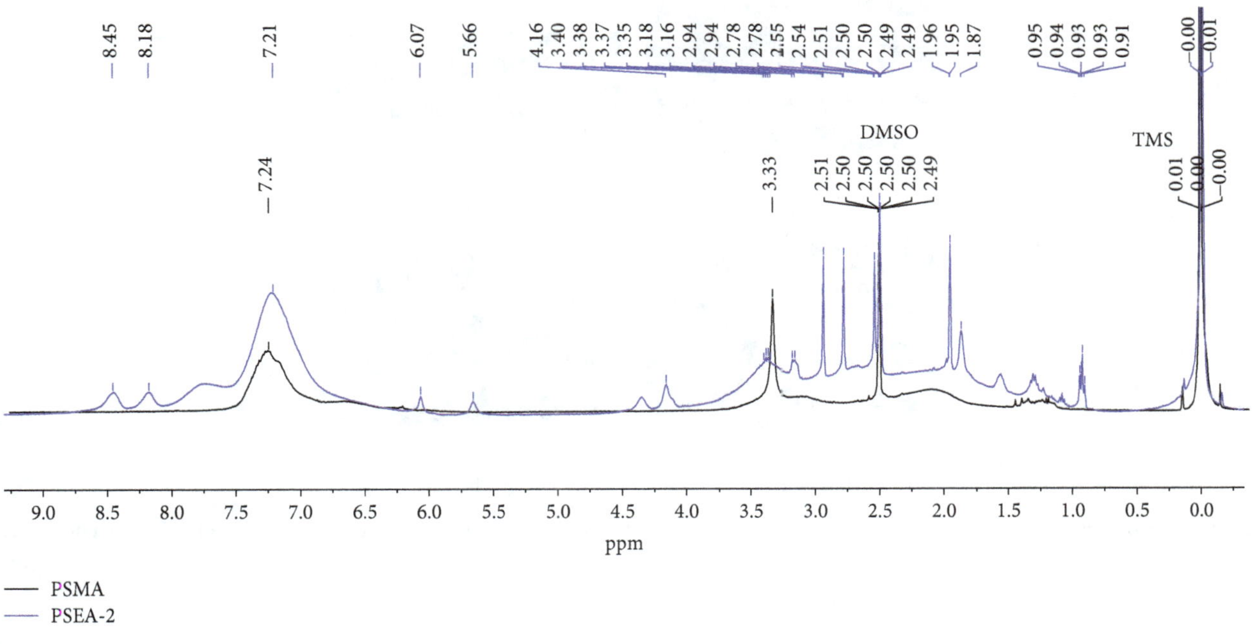


FIGURE 7: $^1$H NMR diagrams of PSEA-2 and its starting material PSMA in $d_6$-DMSO.

patterning using a sharp razor blade, and the fine black powder was subjected to TGA analysis. As shown in Table 2 and Figure 9, the PT-3 photoresist having thermally stable side-chain polyimide PSEA-2 as a binder polymer exhibited higher 1 wt% loss temperature than the PT-0 reference photoresist which had commercial SR-6300 as binder polymer.

## 4. Conclusion

In this study, a series of new binder polymers having thermally stable side-chain imide linkages were synthesized and used in the black photoresist formulation to pattern black PDL on the OLED panel. For the synthesis of binder polymers, poly(styrene-*co*-maleic anhydride) was used as the starting material and three different types of side-chain polyimides were obtained. The epoxy methacrylate-type side-chain polyimide afforded a fine black PDL pattern with high thermal stability, comparable to that of the currently used positive-tone photosensitive polyimide.

## Disclosure

Part of this work's content will be used on "The 18$^{th}$ International Meeting on Information Display (IMID)" to be held on Aug. 29, 2018.

TABLE 2: Formulation of black photoresists and thermal stability of a black PDL pattern [14].

| Components/photoresist samples | | PT-0 | PT-1 | PT-2 | PT-3 |
|---|---|---|---|---|---|
| Photoinitiators | Irgacure 754 | 1 | 1 | 1 | 1 |
| | Irgacure TPO | 6 | 6 | 6 | 6 |
| Photosensitizer | Darocure ITX | 2 | 2 | 2 | 2 |
| Binder polymer (30 wt% in PGMEA) | SR-6300 | 35 | | | |
| | PSI-2 | | 35 | | |
| | PSEI-1 | | | 35 | |
| | PSEA-2 | | | | 35 |
| Multifunctional monomer | PETA | 8 | 8 | 8 | 8 |
| Polyimide thermal stabilizer (30 wt% in PGMEA) | MY-10 | 10 | 10 | 10 | 10 |
| Black millbase (23 wt% in PGMEA) | LT-1 (SKC htm Co., Korea) | 35 | 35 | 35 | 35 |
| Solvent | PGMEA | 5 | 5 | 5 | 5 |
| Total | wt% | 100 | 100 | 100 | 100 |
| Thermal stability | 1 wt% loss temp. (°C, TGA) | 282 | — | — | 303 |

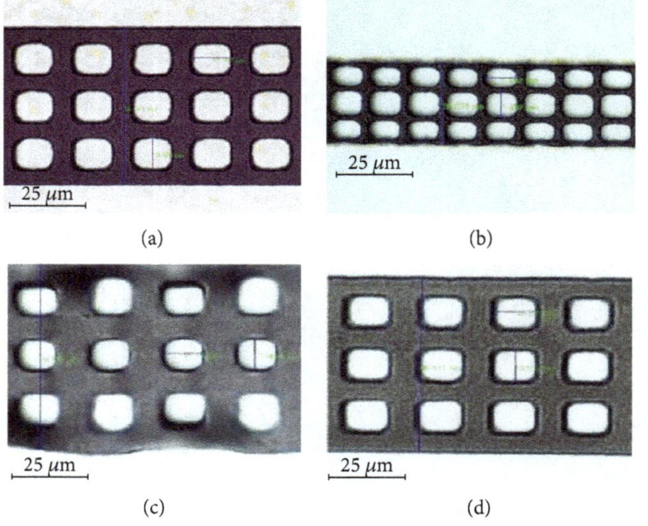


FIGURE 8: The optical microscopic images of black PDL patterns obtained with (a) PT-0, (b) PT-1, (c) PT-2, and (d) PT-3 black photoresists.

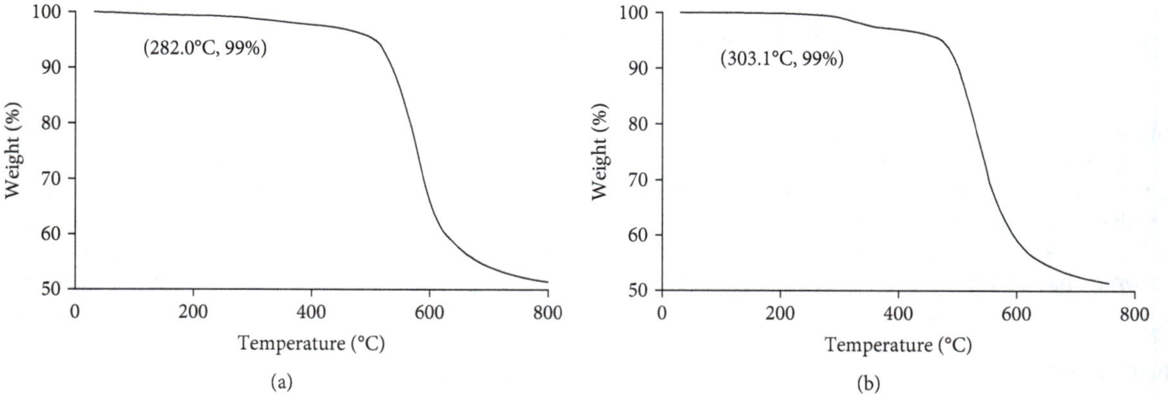


FIGURE 9: TGA thermograms of black powders obtained from the black PDL patterns. (a) PT-0 and (b) PT-3.

## Conflicts of Interest

The authors declare that they have no competing interests.

## Acknowledgments

This study was supported by the Technology Innovation Program (or Industrial Strategic Technology Development Program) (10063289, Development of High Temperature Negative Tone Photosensitive Black Resin and Fabrication Process for Pol-less AMOLED Devices) funded by the Ministry of Trade, Industry & Energy (MOTIE, Korea).

## References

[1] J. K. Lee, S. Cho, and D. W. Kang, "Analysis of light leakage between the adjacent pixels in a color-filter stacked white OLED display," *Displays*, vol. 45, pp. 6–13, 2016.

[2] M. Kim and G. H. Jin, "ITO/AlNdN/Al contact process for active matrix OLED displays," *Electronics Letters*, vol. 45, no. 8, p. 421, 2009.

[3] M. Noda, N. Kobayashi, M. Katsuhara et al., "47.3: a rollable AM-OLED display driven by OTFTs," *SID Symposium Digest of Technical Papers*, vol. 41, no. 1, p. 710, 2010.

[4] W. Liu, C. Jiang, D. Wang, and Y. Li, "Organic light-emitting diode (OLED) display panel, pixel define layer (PDL) and preparation method thereof," 2016, U.S. Patent 9373814 B2.

[5] D. J. Liaw, K. L. Wang, Y. C. Huang, K. R. Lee, J. Y. Lai, and C. S. Ha, "Advanced polyimide materials: syntheses, physical properties and applications," *Progress in Polymer Science*, vol. 37, no. 7, pp. 907–974, 2012.

[6] S. Tamai, A. Yamaguchi, and M. Ohta, "Melt processible polyimides and their chemical structures," *Polymer*, vol. 37, no. 16, pp. 3683–3692, 1996.

[7] J. A. Defranco, B. S. Schmidt, M. Lipson, and G. G. Malliaras, "Photolithographic patterning of organic electronic materials," *Organic Electronics*, vol. 7, no. 1, pp. 22–28, 2006.

[8] M. Aleksandrova, "Specifics and challenges to flexible organic light-emitting devices," *Advances in Materials Science and Engineering*, vol. 2016, Article ID 4081697, 8 pages, 2016.

[9] K.-i. Fukukawa and M. Ueda, "Recent progress of photosensitive polyimides," *Polymer Journal*, vol. 40, no. 4, pp. 281–296, 2008.

[10] Y. Tsuda, Y. Matsuda, and T. Matsuda, "Soluble polyimides bearing long-chain alkyl groups on their side chainviapolymer reaction," *International Journal of Polymer Science*, vol. 2012, Article ID 972541, 10 pages, 2012.

[11] C.-K. Lee, T.-M. Don, D.-J. Lin, C.-C. Chen, and L.-P. Cheng, "Characterization of acrylic copolymers applied in negative-type photoresist via a ternary composition diagram," *Journal of Applied Polymer Science*, vol. 109, no. 1, pp. 467–474, 2008.

[12] H.-C. Yu, J. W. Jung, J.-Y. Choi, and C.-M. Chung, "Kinetic study of low-temperature imidization of poly(amic acid)s and preparation of colorless, transparent polyimide films," *Journal of Polymer Science: Part A*, vol. 54, no. 11, pp. 1593–1602, 2016.

[13] E. Pyun, R. J. Mathisen, and C. S. P. Sung, "Kinetics and mechanisms of thermal imidization of a polyamic acid studied by ultraviolet-visible spectroscopy," *Macromolecules*, vol. 22, no. 3, pp. 1174–1183, 1989.

[14] S. H. Park, G. Shi, D. M. Seo, M. Kim, J. Kim, and L. S. Park, "Synthesis of multifunctional monomers for patterning pixel define layer of organic light emitting diode," *Molecular Crystals and Liquid Crystals*, vol. 663, no. 1, pp. 168–173, 2018.

# 11

# Alginate Biocomposite Films Incorporated with Cinnamon Essential Oil Nanoemulsions: Physical, Mechanical, and Antibacterial Properties

**Kporwodu Frank,[1] Coralia V. Garcia,[1] Gye Hwa Shin,[2] and Jun Tae Kim[1]**

[1]*Department of Food Science and Technology, Keimyung University, Daegu 42601, Republic of Korea*
[2]*Department of Food and Nutrition, Kunsan National University, Gunsan 54150, Republic of Korea*

Correspondence should be addressed to Jun Tae Kim; jtkim92@kmu.ac.kr

Academic Editor: Cornelia Vasile

Alginate-based antibacterial biocomposite films were prepared by incorporating with cinnamon essential oil nanoemulsions (CEO-NE). CEO-NE was first prepared by mixing and homogenizing the oil phase and aqueous phase containing polysorbate 80 (Tween 80), using a probe-type ultrasonication equipment. The biocomposite films were then prepared by incorporating the CEO-NE to an aqueous solution of alginate and glycerol, homogenizing, casting, and drying. The mean droplet size, zeta potential, and polydispersity index of the CEO-NE were 92.2 nm, −15.58 mV, and 0.25, respectively. Young's modulus of the CEO-NE/alginate biocomposite films was significantly increased with an increasing concentration of CEO-NE, while their elongation at break was significantly decreased. However, the tensile strength of the CEO-NE/alginate biocomposite films was not significantly changed. The maximum tensile strength and elongation at break of the biocomposite films were 15.63 MPa and 23.67%, respectively, corresponding to the biocomposite films containing 20% CEO-NE. The biocomposite films containing 20% CEO-NE also showed strong antibacterial effects against *Salmonella typhimurium*, *Bacillus cereus*, *Escherichia coli*, and *Staphylococcus aureus*, achieving inhibition zones from 29.7 to 53.0 mm. These results show the potential of the CEO-NE/alginate biocomposite films as antibacterial packaging for extending the shelf life of fresh foods.

## 1. Introduction

Active packaging has been defined as films or containers incorporating certain active compounds to maintain and extend the shelf-life of a food product [1]. Examples of active packaging include polysaccharide-based biodegradable and edible films and composites that could be used as carriers of active compounds such as antimicrobials, antioxidants, and texture enhancers [2–4]. In addition, these films not only protect foods from dehydration but also act as a gas barrier against the surrounding media [1, 2]. Natural polysaccharides have been used in both medical and food processing applications because of their biodegradability, biocompatibility, and interactions with drugs and foods [4]. Nevertheless, although biopolymer-based films showed excellent gas barrier properties, they exhibited poor water resistance because of their hydrophilic nature [5, 6].

Alginates are natural polymers extracted from brown seaweed and are composed of 1-4-$\beta$-D-mannuronic acid and $\alpha$-L-guluronic acid. It has been commonly known that alginates can be cross-linked with a tridimensional network by a reaction with polyvalent metal cations, specifically calcium ions [5, 6]. Since alginate films have a hydrophilic feature, cross-linking with calcium ions has been performed to improve their mechanical strength and water barrier properties [6]. Essential oils have also been incorporated into the alginate film to improve its water barrier properties and antibacterial activities [5, 7].

Essential oils (EOs) have widespread use as flavoring materials and are also used widely in the nutritional, pharmaceutical, and agricultural fields for their reported antibacterial and antioxidant properties [8, 9]. Thus, they could be incorporated into the packaging material or used as edible coatings to extend the shelf-life of foods. Cinnamon essential oil (CEO) is one of the most effective antibacterial oils; its major component is cinnamaldehyde, having minimum inhibitory concentrations (MIC) of 0.05–5 $\mu$L/mL *in vitro* [8]. Cinnamon is not harmful when consumed in food products, and it is proven to inhibit the growth of molds, yeast, and bacteria [9]. However, when preparing biocomposite films, direct addition of the hydrophobic cinnamon oil to the hydrophilic alginate may result in a film-forming solution that is not uniform and exhibits some phase separation; hence, in this study, cinnamon oil was first incorporated into a nanoemulsion system.

Nanoemulsions have been defined as emulsions with a droplet size in the range of 20–200 nm [10, 11]. Recent investigations suggest that EO-loaded nanoemulsions exhibit better physical and antibacterial properties than do conventional emulsions [12–14]. The small particle size of nanoemulsions has been reported to increase the surface area per unit of mass of lipophilic compounds, thereby enhancing their physicochemical properties, stability, and biological activities [14, 15]. Therefore, the current study is aimed at developing biocomposite alginate-based films incorporated with CEO-NE as a potential food packaging material. The physical, mechanical, and antibacterial properties of the films were also evaluated.

## 2. Materials and Methods

*2.1. Materials.* Sodium alginate, agar, Tween 80, and glycerol were purchased from Duksan Pure Chemicals (Ansan, Korea). Cinnamon essential oil (CEO; *Cinnamomum zeylanicum*) was purchased from NOW Foods Company (Bloomingdale, IL, USA). Nutrient broth was obtained from Becton Dickinson and Company (Franklin Lakes, NJ, USA). *Staphylococcus aureus* (ATCC 6538P), *Streptococcus mutans* (ATCC 25175), *Bacillus cereus* (ATCC11778), *Escherichia coli* (ATCC 8739), *Listeria monocytogenes* (ATCC 19115), and *Salmonella typhimurium* (KCCM 11862) were obtained from Sigma-Aldrich (St. Louis, MO, USA).

*2.2. GC-MS Analysis of CEO.* Gas chromatography-mass spectrometry (GC-MS) analysis of the CEO was performed on an Agilent 7890B-5977B system (Palo Alto, CA, USA) fitted with an Agilent J&W DB-5 column (30 m × 0.25 mm, 0.25 $\mu$m). The temperature program started at 60°C for 1 min, increased at a rate of 4°C/min to 280°C, and held at this temperature for 5 min. The injector and ion source were kept at 250 and 230°C, respectively. The carrier gas was helium, pumped at a flow rate of 1 mL/min. The sample (1 $\mu$L) was injected in split mode at a 50 : 1 split ratio. Mass spectra were obtained by electron impact ionization (EI) at 70 eV. The total ion current (TIC) in the range 30–600 $m/z$ was collected. Essential oil components were tentatively identified by comparison of their mass spectra with those stored in the mass spectral library of the GC-MS system and semiquantified as percentage of the total peak area.

*2.3. Preparation of CEO Nanoemulsions (CEO-NE).* CEO-loaded nanoemulsions were prepared as follows. Coarse emulsions were first prepared by mixing CEO and an aqueous phase containing Tween 80. CEO concentrations of 1, 2, 3, 4, and 5% ($v/v$) were used. Tween 80 was used at the same concentration as CEO. The mixture was vortexed for 2 min and further mixed under high-speed homogenization (WiseTis® HG-15D, DAIHAN Scientific, Seoul, Korea)) at 5000 rpm for 10 min. Then, the CEO nanoemulsions (CEO-NE) were obtained from the coarse emulsions by ultrasonication at 750 watts, 40% amplitude for 10 min using a Sonics Vibra-Cell™ (VC 750, Newtown, CT, USA).

*2.4. Droplet Size, Polydispersity Index (PDI), and Zeta Potential of CEO-NE.* The mean droplet size, polydispersity index (PDI), and zeta potential of the CEO-NE were measured by dynamic light scattering (DLS), using a zeta potential and particle size analyzer (ELSZ-1000, Otsuka Electronics, Tokyo, Japan). A 1 mL sample was loaded onto a polystyrene latex cell, and the mean droplet size and PDI were measured at 25°C, detector angle of 90°, and wavelength of 633 nm. Each sample was measured at least 3 times, and the average values were used.

*2.5. Preparation of CEO-NE/Alginate Biocomposite Films.* Sodium alginate (4 g) was dispersed in 200 mL distilled water, and 20% ($w/w$, based on alginate) glycerol was added as a plasticizer to avoid the brittleness of the films. To achieve complete dispersion, the mixture was stirred constantly at room temperature overnight using a magnetic stirrer at 300 rpm. To prepare the films, the nanoemulsion containing 4% ($v/v$) CEO was added to the alginate suspension in ratios from 20 to 40% ($v/w$ alginate solution). The mixtures were subjected to high-speed homogenization at 8000 rpm for 10 min to obtain the film-forming solution. An alginate film-forming solution without CEO was used as a control. The film-forming solutions were degassed and cast on a Teflon-coated glass plate. Then, the solutions were dried at 35°C in an oven for 24 h. The dried films were further conditioned at 25°C and 50% RH for 48 h before evaluation.

*2.6. Color of the Biocomposite Films.* The color of the biocomposite films was determined with a chroma meter (CR-400, Minolta, Tokyo, Japan). Film specimens were placed on a white standard plate ($L^* = 96.85$, $a^* = 0.01$, and $b^* = 1.64$), and the lightness ($L$) and chromaticity parameters $a$ (red-green) and $b$ (yellow-blue) were measured. Three readings on different sites of each film were recorded, and the average values were used. The total color difference ($\Delta E$) was determined as follows [16]:

$$\Delta E = \sqrt{(L^*-L)^2 + (a^*-a)^2 + (b^*-b)^2}, \qquad (1)$$

where $L^*$, $a^*$, and $b^*$ are the color parameter values of the standard and $L$, $a$, and $b$ are the color parameter values of the film sample.

*2.7. Moisture Content.* The moisture content of the biocomposite films was determined through the weight loss undergone by the film after 24 h of oven drying at 90°C until a constant dry weight was obtained. This temperature was chosen to avoid loss of CEO and plasticizer [16].

*2.8. Mechanical Properties.* The film thickness was measured using a manual digital micrometer (MDC-25MJ, Mitutoyo Co., Kanagawa, Japan) to the nearest 0.001 mm. Measurements were made in 10 random locations for each film, and an average value was calculated. The mechanical properties of the biocomposite films were tested using a universal testing machine (UTM; Zwick Z010TN, Zwick GmbH & Co. KG, Ulm, Germany) according to the ASTM D882 standard method [17] with some modifications. First, films were cut in rectangular strips of $10 \times 80$ mm and preconditioned for 48 h at 25°C and 50% RH before evaluation. The films were clamped between grips, and then force and deformation were recorded during extension. The grip distance was 50 mm, and the strain rate was 50 m/min. Tensile strength, elongation at break, and Young's modulus of the biocomposite films were measured according to ASTM D882 [17].

*2.9. Scanning Electron Microscopy (SEM).* The microstructure of the biocomposite film's fractured surface obtained during the mechanical analysis was examined using a scanning electron microscope (Hitachi S-4800 FE-SEM, Hitachi Ltd., Tokyo, Japan) with an acceleration voltage of 3.0 kV. The films were mounted on the specimen holder using a carbon tape and sputtered with a thin layer of osmium coating for 1 s using a plasma osmium coater (OPC-60A, West Chester, PA, USA).

*2.10. Antibacterial Activity of CEO-NEs and Biocomposite Films.* The disc diffusion method was used to investigate the antibacterial activity of the CEO-NE according to published methods with some modifications [18, 19]. Bacterial suspension at $10^8$ CFU/mL was spread over the plates containing nutrient agar medium. The paper discs were aseptically placed on the inoculated medium surface; 30 $\mu$L of CEO-NE was added and left for 30 min at room temperature for sufficient oil diffusion before incubation at 37°C for 24 h. The inhibition zone was then measured.

Antibacterial activity testing of the CEO-NE/alginate biocomposite films was also carried out using the disc diffusion (inhibition zone) method with some modifications [20]. The films were cut into 18 mm circular shapes and then aseptically placed on nutrient broth-agar plates previously inoculated with 0.15 mL of test bacteria ($10^8$ CFU/mL). The plates were then incubated at 37°C for 24 h. Observations on the diagonal of the inhibitory zone surrounding the films and contact area between the film and medium surface were made.

*2.11. Statistical Analysis.* Measurements of particle characterization, color, mechanical properties, and antibacterial activity were performed in triplicate, and data are expressed as mean ± standard deviation (SD). Statistical analysis was performed using SPSS version 20.0 (SPSS Inc., Chicago, IL, USA). Significant differences among treatments were determined by analysis of variance (ANOVA) and Duncan's multiple range test. The level of significance was set at $p < 0.05$.

## 3. Results and Discussion

*3.1. Composition of the CEO.* GC-MS analysis revealed that the main component in the CEO was (*E*)-cinnamaldehyde, representing 60.99% of the total peak area. This result was expected as cinnamaldehyde is known to be the major component of cinnamon oil. Other major compounds in the CEO were cinnamaldehyde dimethyl acetal (10.48%), eugenol (5.36%), which is also known for its antibacterial effects, and caryophyllene (4.26%). The composition of the CEO is summarized in Table 1.

*3.2. Characterization of the CEO-NE.* The effects of the CEO concentration on the mean droplet size, PDI, and zeta potential of the CEO-NE are displayed in Figure 1. These parameters have been shown to have a crucial effect on the color and mechanical properties of edible films [15]. The mean droplet size of the CEO-NE significantly ($p < 0.05$) increased with increasing oil concentration, with the smallest droplet size (55.8 nm) obtained at 1% ($v/v$) essential oil (Figure 1(a)). Small nanoparticles < 200 nm are desirable because of their enhanced properties and also result in smooth and stable films [21, 22]. The PDI value has been defined as the measure of heterogeneity in the droplet size distribution. PDI values close to 0 indicate homogeneous size distributions, whereas PDI values close to 1 indicate heterogeneous size distributions [15]. The PDI obtained in this study ranged between 0.16 and 0.26, showing that the CEO-NE obtained was homogeneous.

The zeta potential of the CEO-NE is shown in Figure 1(b). The zeta potentials of the CEO-NE ranged between −15.16 and −28.74 mV. Even though the nonionic surfactant Tween 80 was used, the CEO-NE droplets exhibited a highly negative zeta potential, which can be attributed to the adsorption of anionic fatty acid molecules of the oil phase [15]. Furthermore, the CEO-NE hindered the growth of both *S. aureus* and *E. coli* bacteria, being particularly effective against the latter as observed in the size of the inhibition zones (13 and 20 mm, respectively, for a concentration of 4% $v/v$) (data not shown).

*3.3. Mechanical Properties of the CEO-NE/Alginate Biocomposite Films.* The mechanical properties of the CEO-NE/alginate biocomposite films were characterized by measuring the tensile strength (TS), percent elongation at break (EB), and Young's modulus (YM), which represent the stretch capacity, flexibility, and rigidity of the films, respectively. The results obtained are shown in Figure 2. The TS, EB, and YM of the control film were 15.6 MPa, 23.7%, and 265.0 MPa, respectively. No significant ($p < 0.05$) differences between the TS of the CEO-NE/alginate biocomposite films and control alginate films were observed. Although several studies have reported that essential oil addition to polymer-based film formulations tends to weaken the films by decreasing the cohesive

Table 1: Composition of the cinnamon essential oil. Compounds were tentatively identified by their mass spectra.

| Peak no. | Compound | Retention time (min) | Peak area (%) |
|---|---|---|---|
| 1 | 3-Thujene | 6.6 | 0.14 |
| 2 | 2-Pinene | 6.8 | 0.63 |
| 3 | Camphene | 7.3 | 0.25 |
| 4 | Benzaldehyde | 7.6 | 0.36 |
| 5 | $\beta$-Pinene | 8.1 | 0.25 |
| 6 | $\alpha$-Phellandrene | 8.9 | 1.28 |
| 7 | $\alpha$-Terpinene | 9.2 | 0.46 |
| 8 | $p$-Cymene | 9.5 | 1.46 |
| 9 | D-Limonene | 9.6 | 0.45 |
| 10 | $\beta$-Phellandrene | 9.7 | 1.78 |
| 11 | Terpinolene | 11.4 | 0.13 |
| 12 | Linalool | 11.8 | 2.64 |
| 13 | Benzenepropanal | 14.0 | 0.23 |
| 14 | 3-Cyclohexen-1-ol | 14.6 | 0.18 |
| 15 | 3-Cyclohexen-1-methanol | 15.1 | 0.43 |
| 16 | ($E$)-Cinnamaldehyde | 18.1 | 60.99 |
| 17 | 3-Phenylacrylaldehyde | 18.3 | 0.36 |
| 18 | Eugenol | 20.3 | 5.36 |
| 19 | $\alpha$-Copaene | 21.1 | 0.60 |
| 20 | Cinnamaldehyde dimethyl acetal | 21.8 | 10.48 |
| 21 | Caryophyllene | 22.6 | 4.26 |
| 22 | ($E$)-Cinnamyl acetate | 23.3 | 3.57 |
| 23 | $\alpha$-Humulene | 23.6 | 0.87 |
| 24 | Phenol | 25.3 | 0.17 |
| 25 | 2-Methoxycinnamaldehyde | 25.7 | 0.18 |
| 26 | Caryophyllene oxide | 27.4 | 0.62 |
| 27 | ($E$)-2-Dodecen-1-ol | 28.2 | 0.47 |
| 28 | Benzoic acid | 32.4 | 1.12 |

forces within the microstructure [15, 20, 23], here we obtained resistant CEO-NE/alginate composite films. This result can be attributed to the plasticizing capacity of the oil [24]. The EB of the CEO-NE-alginate film decreased significantly ($p < 0.05$) as the amount of incorporated CEO-NE increased, whereas YM increased for low concentrations of CEO-NE but decreased when more than 30% CEO-NE was incorporated into the film. The decrease in YM may be explained by the decrease in the density and rigidity of the film matrix as more CEO-NE was incorporated [24, 25]. The decrease in EB at high oil concentrations has been mentioned by some researchers [20], although other studies have reported the opposite [15, 23]. This lack of consistency in the results reported for E B may be due to the different interactions between the polymers and essential oils used to prepare the various films. Here, it is possible that the formation of discontinuities at high CEO-NE concentrations created stressed regions in the film matrix, leading to rupture, and thus decreased EB [26, 27].

*3.4. Scanning Electron Microscopy (SEM).* Micrographs of the fracture surface of the prepared films after break (postmechanical test) were taken to examine the effect of the nanodroplet organization along the biopolymer matrix on the film properties. The SEM images in Figure 3 correspond to the fracture surface of the alginate (control) film and those containing different concentrations of CEO-NE. The incorporation of CEO-NE at higher concentrations induced notable changes in the microstructure of the films as shown in Figures 3(d) and 3(e). The biocomposite films incorporated with CEO-NE exhibited a coarse appearance and large discontinuities or cavity-like structures, which can be attributed to the migration of oil droplets towards the surface of the film when in contact with air and further volatilization when water evaporates during film processing [15, 28]. Moreover, at high CEO-NE concentrations, the surface area of the oil droplets increases, resulting in large pores or holes and leading to a heterogeneous biopolymer network formed between the lipophilic compounds and alginate, thereby decreasing chain-to-chain interactions [15].

*3.5. Moisture Content and Color of the Biocomposite Films.* Table 2 shows the moisture content and color parameters lightness ($L$), red ($a$), yellow ($b$), and total color difference ($\Delta E$) for the control (pure alginate) films and the biocomposite films containing different concentrations of CEO-NE. The control and biocomposite films with high CEO-NE content exhibited higher moisture contents than those with an intermediate amount of incorporated CEO-NE, although differences were small and the moisture content of the films remained at around 10%–12%. Jouki et al. [16] attributed the increase in moisture content to the glycerol content in the film, whereas Ghasemlou et al. [23] considered that the moisture increase may be due to the hydrophobic nature of essential oils, which has a direct effect on the water retention of the films.

The control films appeared clear and transparent, whereas the addition of CEO-NE showed significant ($p < 0.05$) effects on the color and transparency of the films. The increase in the $b$ value reflects the increase in yellowness with higher CEO-NE concentrations. This result is similar to that reported for alginate-based films incorporating garlic oil [20]. The total color difference ($\Delta E$) also increased as the amount of CEO-NE incorporated into the film increased. The decrease in $L$ values indicates that the films darkened with increasing CEO-NE concentration. This result was consistent with the previous reports where emulsified carrageenan and starch films became less transparent with the addition of essential oils [21, 23]. This darkening effect has been attributed to the increase in diffuse reflectance caused by light scattering in the lipid droplets, which lowers both the light scattering intensity and whiteness index of the film [23]. Hence, the manufactured CEO-NE/alginate biocomposite films became slightly yellow and tended to darken as more CEO-NE was incorporated.

*3.6. Antibacterial Activity.* CEO-NE incorporation into alginate resulted in biocomposite films with antibacterial activity. The inhibition zones increased significantly

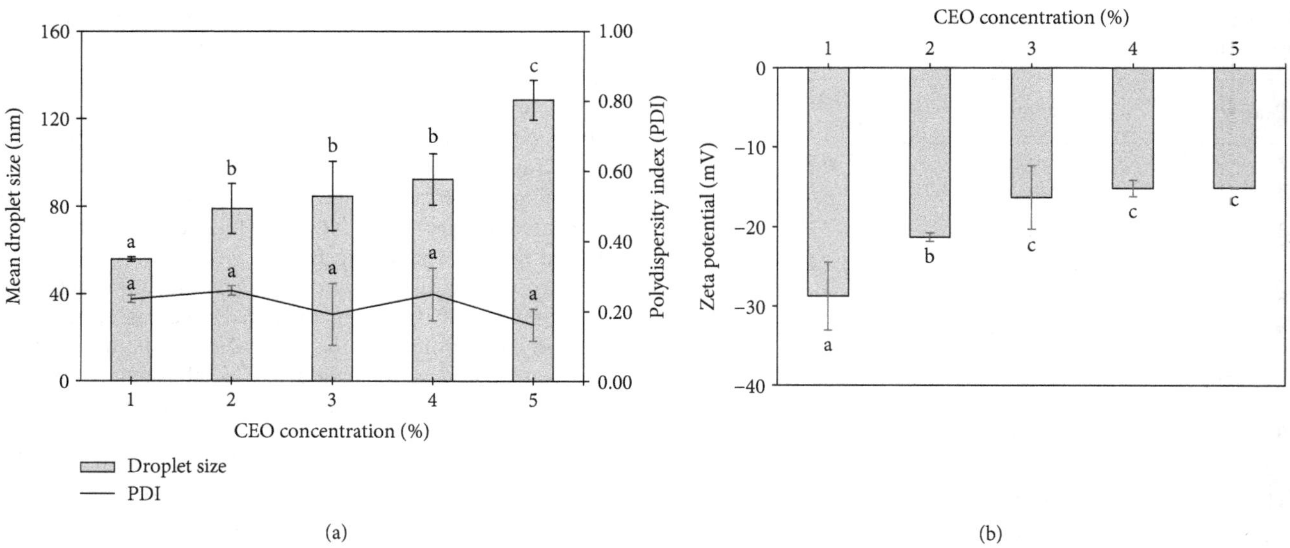


Figure 1: Effect of cinnamon essential oil (CEO) concentration on the (a) mean droplet size and PDI and (b) zeta potential of nanoemulsions (NEs). Different letters indicate a significant difference at $p < 0.05$ by the ANOVA test. Error bars represent standard deviation ($n = 3$).

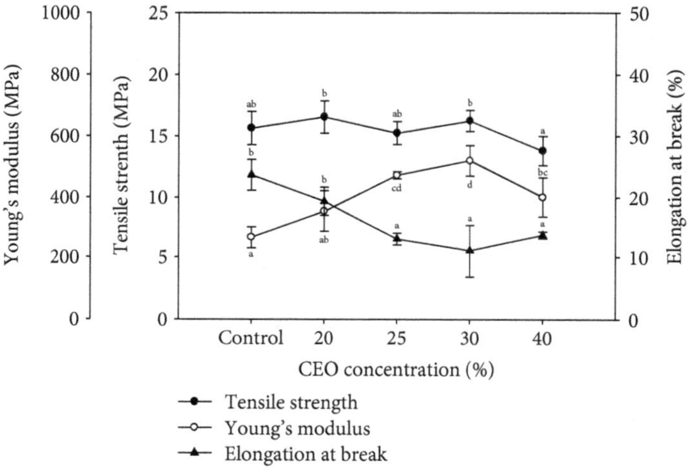


Figure 2: Tensile strength, elongation at break, and Young's modulus of pure alginate film and alginate films incorporating various CEO-NE concentrations. Different letters indicate a significant difference at $p < 0.05$ by the ANOVA test. Error bars represent standard deviation ($n = 3$).

($p < 0.05$) with increasing oil concentrations for the tested microorganisms (Table 3 and Figure 4). The CEO-NE/alginate biocomposite films were particularly effective against the Gram-negative *S. typhimurium*, which exhibited the largest inhibition halo among tested microorganisms, followed by the Gram-positive *B. cereus* and *S. aureus*. No antibacterial effect was observed for the control films. Several studies have shown that cinnamon has strong and consistent inhibitory effects against various food spoilage microorganisms [8]. Earlier studies have suggested that the antibacterial activity of cinnamon oil is mainly due to its major components cinnamaldehyde and cinnamic acid as well as to the presence of other compounds including carvacrol and eugenol [8, 9]. Here, GC-MS analysis of the CEO used for the formulation revealed that cinnamaldehyde was the major compound in it and eugenol was also a major component. In addition, several terpenoids were also present in the CEO and could have also contributed to its antibacterial properties. The hydroxyl group of eugenol may react with proteins thereby inhibiting the action of enzymes, and hydrophobic compounds could also cause damage to the outer membrane of the Gram-negative bacterial cell wall, releasing lipopolysaccharides [9, 19, 29]. Reduction of intracellular ATP may be another explanation for the antibacterial activity [7]. Furthermore, the CEO mechanism of action against *S. aureus* may be due to the leakage of the intracellular potassium ions from the cells of the bacteria [29].

The inhibition zones obtained when using the films were also larger than those obtained with the CEO-NEs, which may be explained by the protective action of the polymer matrix against the rapid evaporation of the essential oil as

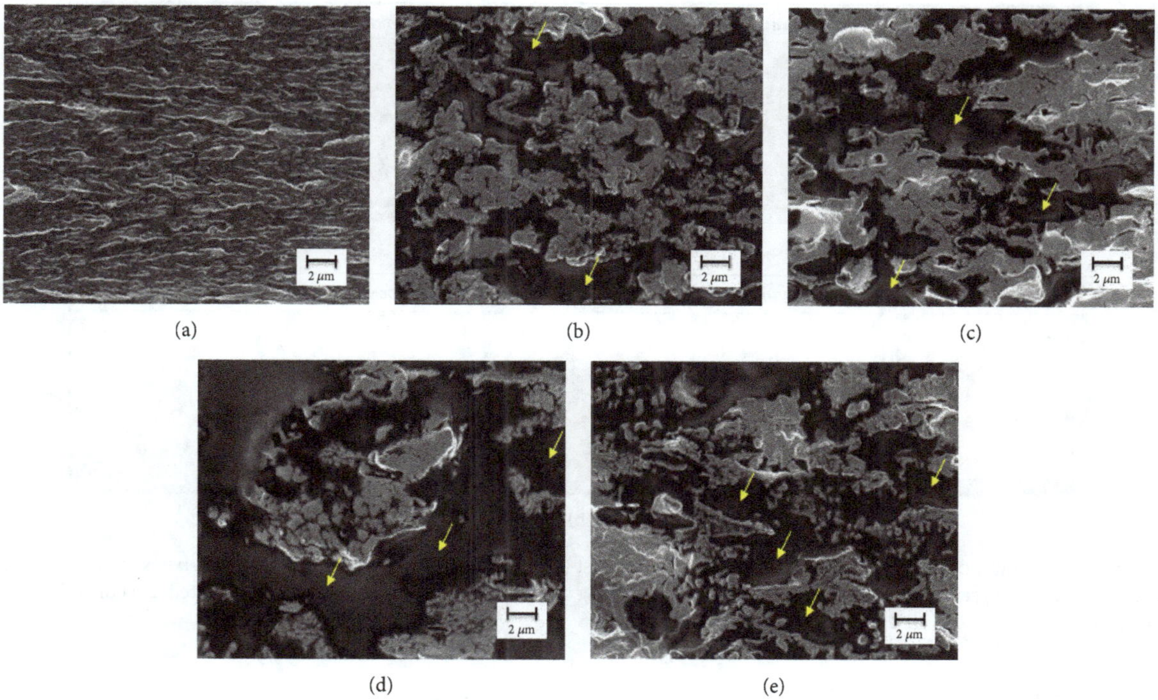


FIGURE 3: Scanning electron microscopy (SEM) images of the fracture surface of alginate films incorporating various CEO-NE concentrations viewed at a magnification of 5000x: (a) control, (b) CEO-NE 20%, (c) CEO-NE 25%, (d) CEO-NE 30%, and (e) CEO-NE 40%. Arrows show cavity-like structures in the film matrix.

TABLE 2: Optical properties ($L$, $a$, $b$, and $\Delta E$) and moisture content of the control and CEO-NE-alginate films. Different letters in the same column indicate significant differences at $p < 0.05$ by the ANOVA test. *The percentages indicate the amount of nanoemulsion incorporated into the films (%, $v/w$), based on total film-forming solution. Control = pure alginate film.

| Films | Color | | | | Moisture content (%) |
|---|---|---|---|---|---|
| | $L$ | $a$ | $b$ | $\Delta E$ | |
| Control | $40.58 \pm 0.33^{d}$ | $0.46 \pm 0.06^{d}$ | $0.54 \pm 0.06^{a}$ | $13.59 \pm 0.09^{a}$ | $12.11 \pm 0.44^{b}$ |
| CEO-NE 20%* | $39.18 \pm 0.78^{c}$ | $0.07 \pm 0.04^{c}$ | $1.48 \pm 0.34^{a,b}$ | $14.28 \pm 0.23^{a,b}$ | $10.18 \pm 0.19^{a}$ |
| CEO-NE 25% | $38.26 \pm 0.59^{b}$ | $-0.30 \pm 0.17^{b}$ | $2.97 \pm 0.66^{b,c}$ | $15.66 \pm 0.69^{b,c}$ | $10.69 \pm 0.26^{a}$ |
| CEO-NE 30% | $36.76 \pm 0.17^{a}$ | $-0.47 \pm 0.21^{a,b}$ | $3.43 \pm 1.63^{c}$ | $16.13 \pm 1.64^{c,d}$ | $12.27 \pm 1.54^{b}$ |
| CEO-NE 40% | $36.49 \pm 0.43^{a}$ | $-0.72 \pm 0.14^{a}$ | $4.75 \pm 0.53^{c,d}$ | $17.30 \pm 0.52^{d}$ | $12.95 \pm 0.12^{b}$ |

TABLE 3: Antibacterial activities of the control and CEO-NE-alginate films. Different letters in the same row indicate significant differences at $p < 0.05$ by the ANOVA test. *The percentages indicate the amount of nanoemulsion incorporated into the films (%, $v/w$), based on total film-forming solution. Control = pure alginate film.

| | Inhibition zone (mm) | | | | |
|---|---|---|---|---|---|
| | Control | CEO-NE 20%* | CEO-NE 25% | CEO-NE 30% | CEO-NE 40% |
| Gram negative | | | | | |
| *S. typhimurium* | $00.00 \pm 0.00^{a}$ | $53.00 \pm 1.00^{b}$ | $56.33 \pm 3.79^{b}$ | $64.67 \pm 1.15^{c}$ | $65.67 \pm 2.08^{c}$ |
| *E. coli* | $00.00 \pm 0.00^{a}$ | $29.67 \pm 1.53^{b}$ | $33.00 \pm 2.00^{c}$ | $38.33 \pm 2.08^{d}$ | $45.67 \pm 2.08^{e}$ |
| *L. monocytogenes* | $00.00 \pm 0.00^{a}$ | $18.00 \pm 0.00^{b}$ | $19.33 \pm 0.58^{c}$ | $26.33 \pm 0.58^{d}$ | $29.00 \pm 1.00^{e}$ |
| Gram positive | | | | | |
| *B. cereus* | $00.00 \pm 0.00^{a}$ | $31.00 \pm 2.65^{b}$ | $42.33 \pm 2.08^{c}$ | $55.67 \pm 2.52^{d}$ | $62.67 \pm 2.08^{e}$ |
| *S. aureus* | $00.00 \pm 0.00^{a}$ | $34.67 \pm 2.08^{b}$ | $36.00 \pm 2.00^{b}$ | $44.67 \pm 2.52^{c}$ | $48.67 \pm 1.53^{d}$ |
| *S. mutans* | $00.00 \pm 0.00^{a}$ | $19.67 \pm 1.15^{b}$ | $24.67 \pm 0.58^{c}$ | $31.00 \pm 2.00^{d}$ | $37.33 \pm 2.52^{e}$ |

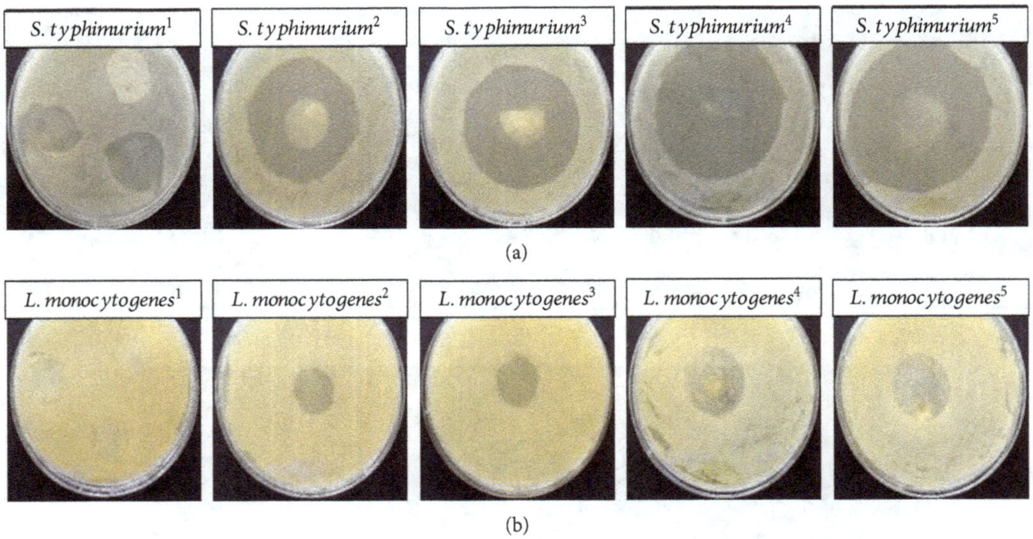


FIGURE 4: Images of inhibition zones of CEO-NE-alginate films against (a) *S. typhimurium* and (b) *L. monocytogenes*, showing the largest and smallest halos, respectively, among tested bacteria. Films incorporating various CEO-NE concentrations were tested: (1) control, (2) CEO-NE 20%, (3) CEO-NE 25%, (4) CEO-NE 30%, and (5) CEO-NE 40%.

well as the more sustained release of the volatile compounds embedded in the polymer.

The antibacterial activity observed against *E. coli* agrees with the previous study [7] in which cinnamon oil and cinnamaldehyde were incorporated into alginate-based edible coats. Nevertheless, the application of the edible coats loaded with essential oil to cut apple resulted in changes to the sensory properties of the fruit, particularly when the oil was added at high concentrations; thus, the evaluation of the developed films using actual foods would also be necessary.

## 4. Conclusions

The manufactured CEO-NE/alginate biocomposite films containing cinnamon essential oil nanoemulsions (CEO-NEs) demonstrated strong antibacterial activity against both Gram-positive and Gram-negative bacteria. Increasing the oil concentration did not significantly affect the physical and mechanical properties of the biocomposite films, which is advantageous because high concentrations of essential oils are required to achieve substantial antibacterial effects in food applications. Therefore, the biocomposite films developed here have the potential to be used as active films in several food packaging applications.

## Conflicts of Interest

The authors declare that there is no conflict of interest regarding the publication of this paper.

## Acknowledgments

This research was supported by the Keimyung University Research Grant of 2016.

## References

[1] J. Kerry and P. Butler, Eds., *Smart Packaging Technologies for Fast Moving Consumer Goods*, John Wiley & Sons, Ltd, 2008.

[2] S. Pashova and V. Zhivkova, "Edible films and coatings used for prolonging the term of storage of foods," *Forum Ware International*, vol. 1, pp. 23–31, 2010.

[3] M. E. Embuscado and K. C. Huber, Eds., *Edible Films and Coatings for Food Applications*, Springer, 2009.

[4] B. D. Ulery, L. S. Nair, and C. T. Laurencin, "Biomedical applications of biodegradable polymers," *Journal of Polymer Science Part B: Polymer Physics*, vol. 49, no. 12, pp. 832–864, 2011.

[5] M. Maizura, A. Fazilah, M. H. Norziah, and A. A. Karim, "Antibacterial activity and mechanical properties of partially hydrolyzed sago starch–alginate edible film containing lemongrass oil," *Journal of Food Science*, vol. 72, no. 6, pp. C324–C330, 2007.

[6] J. W. Rhim, "Physical and mechanical properties of water resistant sodium alginate films," *LWT - Food Science and Technology*, vol. 37, no. 3, pp. 323–330, 2004.

[7] R. M. Raybaudi-Massilia, M. A. Rojas-Graü, J. Mosqueda-Melgar, and O. Martín-Belloso, "Comparative study on essential oils incorporated into an alginate-based edible coating to assure the safety and quality of fresh-cut Fuji apples," *Journal of Food Protection*, vol. 71, no. 6, pp. 1150–1161, 2008.

[8] S. Burt, "Essential oils: their antibacterial properties and potential applications in foods—a review," *International Journal of Food Microbiology*, vol. 94, no. 3, pp. 223–253, 2004.

[9] S. Prabuseenivasan, M. Jayakumar, and S. Ignacimuthu, "In vitro antibacterial activity of some plant essential oils," *BMC Complementary and Alternative Medicine*, vol. 6, no. 1, p. 39, 2006.

[10] C. Solans, P. Izquierdo, J. Nolla, N. Azemar, and M. Garcia-Celma, "Nano-emulsions," *Current Opinion in Colloid & Interface Science*, vol. 10, no. 3-4, pp. 102–110, 2005.

[11] T. Tadros, P. Izquierdo, J. Esquena, and C. Solans, "Formation and stability of nano-emulsions," *Advances in Colloid and Interface Science*, vol. 108-109, pp. 303–318, 2004.

[12] L. Salvia-Trujillo, M. A. Rojas-Graü, R. Soliva-Fortuny, and O. Martín-Belloso, "Effect of processing parameters on physicochemical characteristics of microfluidized lemongrass essential oil-alginate nanoemulsions," *Food Hydrocolloids*, vol. 30, no. 1, pp. 401–407, 2013.

[13] U. Buranasuksombat, Y. J. Kwon, M. Turner, and B. Bhandari, "Influence of emulsion droplet size on antimicrobial properties," *Food Science and Biotechnology*, vol. 20, no. 3, pp. 793-800, 2011.

[14] R. Severino, K. D. Vu, F. Donsì, S. Salmieri, G. Ferrari, and M. Lacroix, "Antimicrobial effects of different combined non-thermal treatments against *Listeria monocytogenes* in broccoli florets," *Journal of Food Engineering*, vol. 124, pp. 1–10, 2014.

[15] A. Acevedo-Fani, L. Salvia-Trujillo, M. A. Rojas-Graü, and O. Martín-Belloso, "Edible films from essential-oil-loaded nanoemulsions: physicochemical characterization and antimicrobial properties," *Food Hydrocolloids*, vol. 47, pp. 168–177, 2015.

[16] M. Jouki, N. Khazaei, M. Ghasemlou, and M. HadiNezhad, "Effect of glycerol concentration on edible film production from cress seed carbohydrate gum," *Carbohydrate Polymers*, vol. 96, no. 1, pp. 39–46, 2013.

[17] ASTM, *Standard Test Method for Tensile Properties of Thin Plastic Sheeting. Standard D882*, American Society for Testing and Materials, Philadelphia, PA, USA, 2001.

[18] B. C. Nzeako, Z. S. Al-Kharousi, and Z. Al-Mahrooqui, "Antimicrobial activities of clove and thyme extracts," *Sultan Qaboos University Medical Journal*, vol. 6, no. 1, pp. 33–39, 2006.

[19] K. Kon and M. Rai, "Antibacterial activity of *Thymus vulgaris* essential oil alone and in combination with other essential oils," *Nusantara Bioscience*, vol. 4, pp. 50–56, 2012.

[20] Y. Pranoto, V. M. Salokhe, and S. K. Rakshit, "Physical and antibacterial properties of alginate-based edible film incorporated with garlic oil," *Food Research International*, vol. 38, no. 3, pp. 267–272, 2005.

[21] S. Shojaee-Aliabadi, H. Hosseini, M. A. Mohammadifar et al., "Characterization of κ-carrageenan films incorporated plant essential oils with improved antimicrobial activity," *Carbohydrate Polymers*, vol. 101, pp. 582–591, 2014.

[22] M. J. Fabra, R. Pérez-Masiá, P. Talens, and A. Chiralt, "Influence of the homogenization conditions and lipid self-association on properties of sodium caseinate based films containing oleic and stearic acids," *Food Hydrocolloids*, vol. 25, no. 5, pp. 1112–1121, 2011.

[23] M. Ghasemlou, N. Aliheidari, R. Fahmi et al., "Physical, mechanical and barrier properties of corn starch films incorporated with plant essential oils," *Carbohydrate Polymers*, vol. 98, no. 1, pp. 1117–1126, 2013.

[24] F. M. Pelissari, M. V. E. Grossmann, F. Yamashita, and E. A. G. Pineda, "Antimicrobial, mechanical, and barrier properties of cassava starch–chitosan films incorporated with oregano essential oil," *Journal of Agricultural and Food Chemistry*, vol. 57, no. 16, pp. 7499–7504, 2009.

[25] A. Aguirre, R. Borneo, and A. E. Leon, "Antimicrobial, mechanical and barrier properties of triticale protein films incorporated with oregano essential oil," *Food Bioscience*, vol. 1, pp. 2–9, 2013.

[26] P. R. Salgado, M. E. López-Caballero, M. C. Gómez-Guillén, A. N. Mauri, and M. P. Montero, "Sunflower protein films incorporated with clove essential oil have potential application for the preservation of fish patties," *Food Hydrocolloids*, vol. 33, no. 1, pp. 74–84, 2013.

[27] H. Tian, W. Wu, G. Guo, B. Gaolun, Q. Jia, and A. Xiang, "Microstructure and properties of glycerol plasticized soy protein plastics containing castor oil," *Journal of Food Engineering*, vol. 109, no. 3, pp. 496–500, 2012.

[28] L. Sánchez-González, A. Chiralt, C. González-Martínez, and M. Cháfer, "Effect of essential oils on properties of film forming emulsions and films based on hydroxypropylmethylcellulose and chitosan," *Journal of Food Engineering*, vol. 105, no. 2, pp. 246–253, 2011.

[29] S. Bouhdid, J. Abrini, M. Amensour, A. Zhiri, M. J. Espuny, and A. Manresa, "Functional and ultrastructural changes in *Pseudomonas aeruginosa* and *Staphylococcus aureus* cells induced by *Cinnamomum verum* essential oil," *Journal of Applied Microbiology*, vol. 109, no. 4, pp. 1139–1149, 2010.

# 12

# Preparation and Application in Drug Storage and Delivery of Agarose Nanoparticles

**Pablo G. Argudo,[1] Eduardo Guzmán,[1,2] Alejandro Lucia,[3] Ramón G. Rubio,[1,2] and Francisco Ortega[1,2]**

[1]*Departamento de Química Física I, Universidad Complutense de Madrid, Ciudad Universitaria s/n, 28040 Madrid, Spain*
[2]*Instituto Pluridisciplinar, Universidad Complutense de Madrid, Paseo Juan XXIII 1, 28040 Madrid, Spain*
[3]*Instituto de Ecología y Desarrollo Sustentable (INEDES, CONICET-UNLu), Ruta 5 y Avenida Constitución, 6700 Luján, Buenos Aires, Argentina*

Correspondence should be addressed to Eduardo Guzmán; eduardogs@quim.ucm.es and Francisco Ortega; fortega@quim.ucm.es

Academic Editor: Angel Concheiro

In this work, monodisperse agarose gel nanoparticles were prepared using a W/O microemulsion as a template to control the size of the obtained particles. The combination of this template method with a temperature-induced gelling and a solvent exchange methodology has allowed preparing stable aqueous dispersions of monodisperse agarose gel nanoparticles in water. The average size, measured as an apparent hydrodynamic diameter, of the obtained particles was around 150 nm. The ability of the obtained hydrogel particles for the encapsulation and release of a synthetic insecticide (azamethiphos) was tested. The results evidence that the insecticide molecules encapsulated in the fabricated nanoparticles are released following a diffusion-controlled mechanism. These results combined with the biodegradability of the agarose provide the bases for the design of a new vector with application in the control of parasites in water reservoirs.

## 1. Introduction

Encapsulation of active ingredients has become a multidisciplinary challenge with interest in several industries, e.g., pharmaceutical, medical, cosmetics, chemical, food, and agricultural [1–9]. This is mainly because encapsulation creates a barrier to protect encapsulated compounds (liquid, solid, and gas) that allows isolating them from the surrounding environment, thus avoiding their degradation and oxidation during its storage and limiting their interactions and reactions with other compounds. Therefore, encapsulation provides the bases for increasing the shelf life of the encapsulated compounds, especially those extremely labile. This enhances its stability and allows one to control the release process without modification of the encapsulated compound activity [10, 11]. Even though the development of encapsulation methodologies started in the last '50s of the last century [12], it was not until the last two decades when the progresses on the nanoscience and nanoengineering have allowed the development of efficient and reproducible encapsulation methodologies, lowering costs and enabling the industrial production [13, 14].

Many different nano- and microstructured supramolecular systems, e.g., colloidal particles, liposomes, and micelles, have been used on the development of efficient encapsulation platforms. Among such systems, biopolymer matrices are accounted as one of the most successful due to their availability, low cost, and reduced impact on the environment and human health [15, 16]. Furthermore, many different biopolymers, either proteins or polysaccharides, are frequently used on the fabrication of particles, e.g., whey and soy proteins, casein, gelatin, zein, starch, cellulose, chitosan, and alginate [17]. These particles are classified into three main groups depending on their structure: hydrogels, inclusion complexes, and interpolyelectrolyte complexes. This leads to the existence of particles with a broad range of physicochemical and functional attributes. It is also important that particles can be obtained following different methodologies (molding techniques, spray drying, solvent desorption, injection and microfluidic and shearing methods, and emulsion-templating methods) [18]. The choice of the most adequate system for a certain purpose and the methodology used for its preparation depends on different factors, e.g., properties of desired particles and biopolymers and the nature of the material to be encapsulated. In addition, the methodology used for the fabrication of the encapsulation platforms must be easy and cost effective [19].

A successful approach for fabricating biopolymer particles is the use of the emulsion templating method which relies on the use of drops of water in oil (W/O) emulsions as a template for the fabrication of the particles [18]. For this purpose, an aqueous solution of the biopolymer is homogenized with an oil solution containing a surfactant, thus leading to the formation of a W/O emulsion. The size of the drop can be controlled by tuning the mixture composition or the homogenization velocity. Once the emulsion is obtained, the water drops are gelled following a specific mechanism, either physical (temperature change) or chemical (pH change, addition of a cross-linking agent), depending on the nature of the biopolymer. The last step of the preparation process involves the solvent exchange process to obtain an aqueous dispersion of biopolymer particles [18]. This methodology has been successfully applied for preparing alginate nanoparticles and mixed nanoparticles of alginate and whey protein [20, 21]. The latter two were successfully used in the encapsulation of riboflavin [21]. Milk protein particles used for probiotic bacteria encapsulation have been also prepared using this methodology [22].

This work is focused on the preparation of agarose nanoparticles. Agarose is a polysaccharide frequently used in different biotechnological applications due to its natural, biodegradable, and nontoxic character which enables the formation of inert nanoparticles. In recent years, an extensive research on polysaccharide nanoparticles for several applications has been developed [23, 24]. Among such polysaccharides, the use of agarose for the fabrication of nanoparticles is widely extended [25, 26]. Wang and Wu [26] described the fabrication of agarose hydrogel nanoparticles using an emulsion-converted-to-suspension method and the application of the obtained particles for encapsulating proteins and peptides. Similar particles were used as batch reactors in dairy industry for the hydrolysis of lactose [27]. Furthermore, many works deal with the fabrication of hybrid metal-agarose particles for combining the electronic properties of metals and the biocompatibility of the agarose to obtain particles with application as biosensors [28, 29].

This work the fabrication of agarose nanoparticles using W/O microemulsions as a template instead of conventional emulsions. The use of microemulsions as a template is preferred due to their thermodynamic stability. Furthermore, drops with sizes in the nanometre scale that provide the bases for fabricating nanosized particles [30]. The aforementioned gelling step takes advantage of the thermal behavior of agarose which presents a gelling temperature of 35–52°C and melting temperature of 85–90°C [31, 32]. Thus, the preparation of W/O with solubilized agarose in the vicinity of the agarose melting followed by a cooling down of the mixture to room temperature, leads to the formation of agarose hydrogel nanoparticles, which after solvent evaporation can be transferred to water and used as a platform for encapsulation of a synthetic insecticide (azamethiphos). Azamethiphos is an insecticide for urban, domestic, or agricultural use, and their encapsulation and controlled release can provide the bases for controlling external parasites in water reservoirs [33].

## 2. Material and Methods

*2.1. Chemicals.* Dioctyl sulfosuccinate sodium salt (aerosol-OT, purity ≥ 99.0%), KCl BioXtra (purity ≥ 99.0%), and azamethiphos (S-[(6-chloro-2-oxo[1,3]oxazolo[4,5-b]pyridin-3(2H)-yl)methyl] O,O-dimethyl phosphorothioate) were purchased from Sigma-Aldrich (Germany). n-Heptane (purity > 99%, extra dry) and ultrapure agarose were purchased from Acros Organics (Belgium) and Invitrogen (United States of America), respectively. All the reagents were used as received without further purification.

All the used glassware was cleaned with a saturated solution of KOH in ethanol, followed by rinsing with abundant water. Milli-Q grade water obtained using a multi-cartridge purification system aquaMAX™ Ultra 370 Series (YoungLin Instrument Co. Ltd., South Korea) was used for rinsing the glassware and for solution preparation. This water obtained presents a resistivity at around 18 MΩ·cm$^{-1}$ and a total organic content lower than 6 ppb.

*2.2. Preparation of Agarose Nanoparticles.* A three-step method was followed for preparing agarose nanoparticles. The first step involves the preparation of water in oil (W/O) microemulsions that are used as a template to control the size distribution of the particles. The microemulsions were prepared using the ternary system n-heptane/aerosol-OT/water because it is a well-known system and its phase diagram has been extensively studied in the literature [34, 35]. The chosen template microemulsions were composed by 80 wt% of n-heptane, 10 wt% of aerosol-OT, and 10 wt% of water [36]. Once the composition is chosen, the required amount water phase containing an agarose volume fraction of 1% and a KCl concentration of 50 mM is dropped into the organic phase containing n-heptane and aerosol-OT under continuous stirring at 2500 rpm. It is worth mentioning that the aqueous solution of agarose needs to be preheated at 95°C (above the agarose melting temperature) to ensure agarose solubilization. The second step involves the gelling of the microemulsion drops to obtain agarose gel particles dispersed in n-heptane. For this step, the obtained W/O is cooling down at room temperature maintaining the stirring. To obtain a dispersion of agarose gel particles in water, the n-heptane was evaporated by freeze drying and the obtained powder was stored overnight at 4°C. This powder was redispersed in water leading to a stable aqueous dispersion of agarose gel particles. It is worth mentioning that during the solvent exchange process, some aggregates are obtained. To remove such aggregates and possible residues of aerosol-OT, the dispersion of gel particles was filtered using nylon membranes with a pore diameter of 0.45 $\mu$m (Millipore, United States of America) followed by dialysis during 8 hours of the filtered samples using membranes with a molecular weight cut-off of 12 kDa. The result is a stable aqueous dispersion of monodisperse agarose gel nanoparticles.

Once the process of gel particles was optimized, they were used for azamethiphos encapsulation. For this purpose, a concentration of 0.01 mM of the drug was added to the initial aqueous phase used for the microemulsion preparation.

*2.3. Experimental Techniques.* The apparent size of the gel particles was estimated as the apparent hydrodynamic diameter $d_h^{app}$, using dynamic light scattering (DLS) measurements. For this purpose, a Zetasizer Nano ZS instrument (Malvern Instruments Ltd., Worcestershire, UK) was used. DLS experiments were performed in quasibackscattering configuration (scattering angle, $\theta = 173°$) using radiation of the red line of a He-Ne laser (wavelength, $\lambda = 633$ nm). Before the measurements, samples were filtered, in a clean room, using nylon membranes with a pore diameter of 0.45 $\mu$m (Millipore, United States of America). Thus, it is possible to remove dust particles from the samples. The measurements were carried out using optical glass cell (Hellma®6030-OG Model; Hellma, Jena, Germany). In DLS experiments, the normalized intensity autocorrelation function, $g^{(2)}(q,t)$, follows an exponential decay with time. In the simplest case that considers the Brownian motion of monodisperse scatterers, $g^{(2)}(q,t)$ is defined as follows:

$$g^{(2)}(q,t) - 1 = \beta e^{-t/\tau} = \beta e^{-D_{app}q^2 t}, \quad (1)$$

where $t$ represents the time, $\tau$ is the characteristic relaxation time which is directly related to the apparent diffusion coefficient $D_{app}$, $q = (4\pi n/\lambda)\sin(\theta/2)$ is the wavevector, $n$ is the solution refractive index which was assumed to be close to that of the continuous phase ($n = 1.33$ and $n = 1.39$ for water and n-heptane, respectively), and $\beta$, which is close to 1, is the coherence factor. For spherical particles diffusing in a continuous Newtonian medium, $D_{app} = 1/\tau q^2$ can be assumed to follow the Stoke-Einstein relationship allowing for determining the apparent hydrodynamic diameter:

$$D_{app} = \frac{k_B T}{3\pi\eta d_h^{app}}, \quad (2)$$

where $k_B$ and $T$ are the Boltzmann constant and the absolute temperature, respectively, and $\eta$ is the viscosity of the continuous phase.

Information on the surface charge of the dispersed particles was also obtained using the Zetasizer Nano ZS instrument (Malvern Instruments Ltd., Worcestershire, UK). For this purpose, electrophoretic mobility ($\mu_e$) measurements were carried out. The measured $\mu_e$ values were transformed into $\zeta$-potential using Smoluchowski's relationship.

Studies about the encapsulation and release of azamethiphos from nanoparticles were performed by spectroscopic measurements. UV-visible spectroscopy measurements were performed using an UV-visible diode spectrophotometer model HP-8452a (Hewlett-Packard, United States of America), and fluorescence measurements were carried out using a fluorescence spectrophotometer FP-6500 (JASCO Inc., United States of America).

## 3. Results and Discussion

*3.1. Agarose Nanoparticle Preparation.* The preparation of agarose gel nanoparticles involves the use of W/O microemulsions as a template; thus, the first step of our study involves the study of the size of the water drops of the bare microemulsion stabilized with aerosol-OT (composition: 80 wt% n-heptane, 10 wt% aerosol-OT, and 10 wt% water). Figure 1(a) shows the intensity autocorrelation function obtained for the bare microemulsion measurements.

The autocorrelation function for water drops presents a monomodal character which is clearer from the $d_h^{app}$ distribution (Figure 1(b)). The $d_h^{app}$ distribution for the water drops in n-heptane is relatively narrow (polydispersity index, PDI < 0.1), appearing centered at around 15 nm. When the results obtained for the W/O microemulsions are compared with those obtained for the agarose nanoparticles after gelling, no significant changes were found neither on the intensity autocorrelation function (Figure 1(a)) nor on the $d_h^{app}$ distribution (Figure 1(b)). These results evidence that the use of W/O microemulsions as a template for preparing agarose gel nanoparticles allows controlling the size of the obtained particles; i.e., the size of the gel nanoparticles is similar to that of the water drops of the O/W microemulsion used as a template. The nanoparticles obtained in n-heptane had an average $d_h^{app}$ at around 12 nm.

In spite of the success on the preparation of monodisperse agarose gel particles in n-heptane, the application of such particles in technological relevant applications requires transferring the obtained particles to an eco-friendly solvent such as water. This process was carried out by freeze drying evaporation of the n-heptane followed by redispersion of the obtained powder in water. Figure 2 shows the intensity autocorrelation function and the $d_h^{app}$ distribution obtained for the particles after redispersion in water.

The results show that the monomodal character of the intensity autocorrelation function (Figure 2(a)) is lost during the evaporation of the n-heptane and the subsequent redispersion in water of the agarose particle and at least three well-separated components can be identified in the sample. The characteristic diffusion times for such processes are around 1 $\mu$s, 10–100 $\mu$s, and 0.1 s. This evidences that the preparation process of the nanoparticles leads to a rather polydisperse dispersion of nanoparticles. The $d_h^{app}$ distribution obtained from the analysis of the intensity autocorrelation function confirms the existence of three different

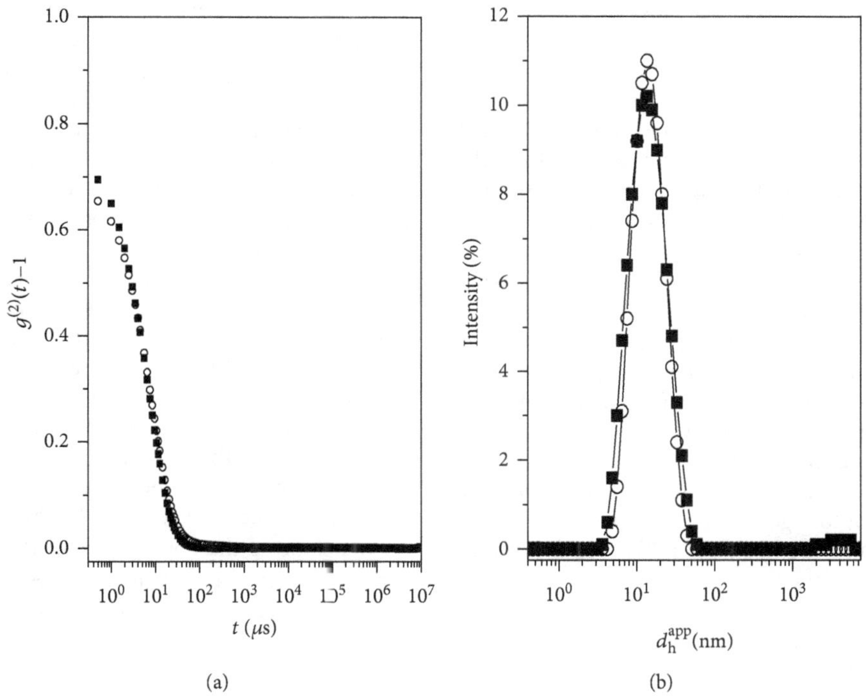


FIGURE 1: (a) Intensity autocorrelation functions as were obtained from DLS experiments for water drops in the template W/O microemulsion (○) and for agarose particles as were obtained in n-heptane (■). (b) $d_h^{app}$ distributions obtained from the analysis of the intensity autocorrelation function in (a) for water drops in the template W/O microemulsion (○) and for agarose particles as were obtained in n-heptane (■).

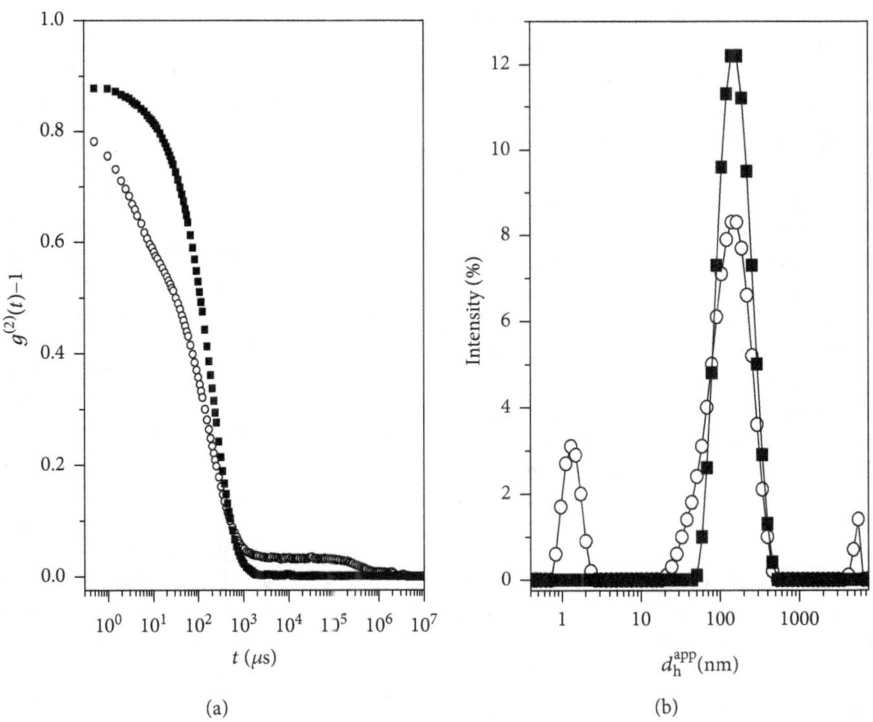


FIGURE 2: (a) Intensity autocorrelation functions as were obtained from DLS experiments for agarose gel particles in water after resuspension (○) and for agarose gel particles in water after the cleaning process (■). (b) $d_h^{app}$ distributions obtained from the analysis of the intensity autocorrelation function in (a) for agarose gel particles in water after resuspension (○) and for agarose gel particles in water after the cleaning process (■).

populations afterward the solvent exchange and subsequent redispersion of the agarose particles in water. The first population can be ascribed to micelles of the aerosol-OT used in the stabilization of the microemulsion. Aerosol-OT is also soluble in water; thus, it can be expected that a part of it can remain as free surfactant and above the critical micelle concentration (cmc) as micelles upon redispersion [37]. The second population is associated with the obtained agarose particles which present an average $d_h^{app}$ at around 150 nm. This evidences that the average size of the particles in water is tenfold of the one previously found in n-heptane. This swelling process occurring during the solvent exchange process can be explained assuming that n-heptane is not a good solvent for agarose; thus, nanoparticles remain in a collapsed state to minimize the number of contacts with the nonpolar phase. However, the dispersion of nanoparticles in water leads to the swelling of the agarose matrix because water is a good solvent for agarose. This justifies the increase of the particle size to its final value [38, 39]. The last population found is related to the formation of particle aggregates during the exchange solvent process. However, it is worth noting that the importance of this last population can be considered negligible in comparison to the contributions of the population centered at around 150 nm.

The technological application of the obtained agarose particles requires removing from aqueous suspension both free surfactant and aggregate molecules. The latter two (aggregates with sizes larger than 1 $\mu$m) are easily removed by filtration of the dispersions through a nylon membrane with a pore diameter of around 0.45 $\mu$m. In order to remove the excess of free aerosol-OT, the agarose particle dispersions were subjected to dialysis against pure water during 8 hours. This time was found to be enough to obtain a dispersion of particles in which there are neither free surfactant molecules nor particle aggregates. The study of such dispersion using DLS (see Figure 2) evidences the presence of rather monodisperse gel particles with an average $d_h^{app}$ centered at around 140 nm (PDI < 0.1). It is worth mentioning that the obtained gel particles present a good stability in aqueous medium due to its high negative charge as was found by $\zeta$-potential measurements. In the particular case of the here-studied particles, its $\zeta$-potential was $(-74 \pm 3)$ mV. This negative charge is explained considering that during the particle preparation, a shell of aerosol-OT remains attached onto the surface of the particles, providing electrostatic stability to the nanoparticles and minimizing the aggregation phenomena. Table 1 summarizes the main characteristics of the obtained particles.

*3.2. Thermoresponsiveness of Gel Nanoparticles.* It is well known that agarose presents a sol-gel transition around human physiological temperature (37°C). The sol-gel transition is strongly dependent on the length of agarose chains and the cross-linking degree of the polymer network [32, 40] Thus, it is expected that the increase of the temperature beyond the transition temperature can lead to a significant change of the nanoparticles size which can be an advantage for applications in which the load and release of active compounds from the gel matrix are required. However, most important than the gel swelling upon the increase of the temperature is the reversibility of such process, i.e., the reversible swelling-shrinking of the gel as result of its subsequent heating and cooling. This section analyses the response of the obtained agarose nanoparticles when they are subjected to an increase of temperature from room temperature to 65°C and after they were cooled down again to room temperature. Figure 3 shows the temperature dependence of the apparent hydrodynamic diameter during the thermal treatment of the nanoparticles.

The results point out that the increase of temperature leads to the swelling of agarose gel nanoparticles until their size reaches a value at 60°C that is twice the original size of nanoparticles in water at room temperature (25°C). The subsequent decrease of temperature down to 25°C leads to the shrinking of the particles till recovering the initial size. This result points out that the agarose gel nanoparticles obtained present a reversible response against temperature changes, without significant hysteresis. The results also show a change on the slope of the dependence of the $d_h^{app}$ in the 35–40°C range which can be explained as a result of the sol-gel transition of agarose. This behavior can be considered an advantage to trigger the release of active compounds from nanoparticles to human body [40].

The transition found in the nanoparticles may be explained assuming that as the temperature is approached at the melting temperature of agarose, water becomes a better solvent for nanoparticles. Thus, the here-obtained nanoparticles can be considered a physically cross-linked hydrogel with a lower critical solution temperature (LCST) at around 37°C.

*3.3. Encapsulation of Azamethiphos in Gel Nanoparticles.* Gel nanoparticles can be used for encapsulation and release of active compounds. Here, the encapsulation and release of a synthetic insecticide (azamethiphos) inside the agarose matrix have been explored. Table 1 summarised the apparent hydrodynamic diameter of agarose nanoparticles without and with encapsulated azamethiphos. The comparison of the results evidences clearly that the inclusion of azamethiphos does not change significantly the particle characteristics (size or surface charge). Thus, it is possible to use agarose nanoparticles as platforms for azamethiphos encapsulation without compromising significantly the physicochemical characteristics of the obtained particles. The evaluation of the degree of retention of azamethiphos during the preparation of the loaded nanoparticles was performed using UV-visible spectroscopy. Figure 4 shows the UV-visible spectrum of azamethiphos in water.

The UV-visible spectrum of azamethiphos in water shows three bands centered at 183, 211, and 274 nm. For the evaluation of the retention degree, we have followed the band at 274 nm in order to avoid overlapping with the absorbance of any other component of the sample. The inserted plot in Figure 4 represents the dependence of the absorbance ($A$) at 274 on the concentration for azamethiphos solutions in water. The results show a linear dependence of such absorbance on the concentration of the insecticide ($c_{azamethiphos}$) in the studied range (Lambert-Beer law). The molar absorptivity of the azamethiphos obtained

TABLE 1: Main characteristics of nanoparticles as were obtained dispersed in n-heptane and in water after the solvent exchange process.

| | Bare agarose nanoparticles | | Agarose nanoparticles loaded with azamethiphos | |
|---|---|---|---|---|
| | $d_h^{app}$ (nm) | $\zeta$-potential (mV) | $d_h^{app}$ (nm) | $\zeta$-potential (mV) |
| Dispersion in n-heptane | $12 \pm 2$ | — | $14 \pm 2$ | — |
| Dispersion in water | $140 \pm 10$ | $-74 \pm 4$ | $150 \pm 20$ | $-70 \pm 5$ |

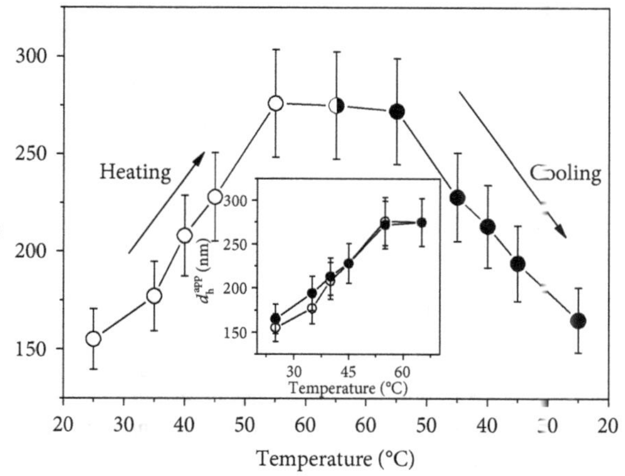


FIGURE 3: Dependence of the $d_h^{app}$ of agarose gel nanoparticles on the temperature during the heating (○)–cooling (●) cycle. The inserted figure represents the same results of the main panel showing the reversibility of the heating–cooling cycle. The heating–cooling steps were carried out slowly enough (0.25°C/min) to ensure the quasistatic character of the temperature change.

was $1.010 \pm 0.003 \times 10^7\,\mathrm{cm^2 \cdot mol^{-1}}$. From the concentration, dependence of the absorbance it is possible to determine the amount of azamethiphos remaining after the process of preparation of the loaded nanoparticles. The comparison of the UV-visible spectra of agarose solution with azamethiphos before the preparation process with that obtained after the preparation process of nanoparticles with encapsulated azamethiphos allows one to determine the efficiency of the encapsulation process. Figure 5 shows the UV-visible spectra of an agarose aqueous solution including a concentration 0.01 mM of azamethiphos and that of the nanoparticles with encapsulated azamethiphos obtained from the same initial solution. Note that the water volume used for the preparation of the template microemulsion is the same in which the final particles were dispersed. Therefore, in the case that all azamethiphos were encapsulated, the concentrations would be the same and both spectra should overlap.

The decrease of the absorbance evidences that during the preparation process of agarose-loaded nanoparticles, a part of the initial azamethiphos is lost. Thus, the encapsulation process yields efficiency in the 60–70% range of the initial amount of insecticide. This decrease of the azamethiphos concentration during the encapsulation may be probably associated with the amount of azamethiphos that remains trapped in the aggregates of big size that are removed during the solvent exchange process. It is important to note that the solubility of azamethiphos is n-heptane is negligible; thus, the transference of the insecticide to the oil phase during particle preparation can be ruled out.

Once the encapsulation of azamethiphos in the agarose particle is confirmed, the release of the insecticide from particles to the solvent at 25°C was evaluated taking advantage of the fluorescent emission of azamethiphos at 339 nm after excitation at 274 nm. For this purpose, a fresh dispersion of agarose nanoparticles with encapsulated azamethiphos was placed into dialysis tubes and the intensity of fluorescence was evaluated in aliquot of the surrounding water at several times after the preparation of the loaded nanoparticles. For the evaluation of the azamethiphos concentration released to water, a calibration curve was carried out measuring the fluorescence intensity of azamethiphos solutions of different concentrations (see Figure 6). From this calibration curve, it is possible to infer with the existence of a linear dependence of the fluorescence intensity ($I$) on the azamethiphos concentration within the studied concentration range ($I = Kc_{\text{azamethiphos}}$, where $K$ is an optical constant dependent on the studied process and assumes a value of $6.80 \pm 0.04 \times 10^7$ l/mol). Using the calibration curve, it is possible to evaluate the released amount of azamethiphos from the nanoparticles.

The azamethiphos release kinetics obtained from the fluorescence intensity of the water surrounding our particle dispersion inside the dialysis is represented in terms of the fraction of released azamethiphos in Figure 7. It is worth mentioning that to ensure the homogenization of the concentration of azamethiphos in water during the release process, the aqueous medium was subjected to continuous agitation at 500 rpm.

The release of azamethiphos from agarose gel nanoparticles occurs mainly in the early stages, with a released amount of around the 70% of the initial encapsulated azamethiphos after 15 minutes. Then a sustained release during almost 6 hours occurs till reaching a complete release of the encapsulated azamethiphos. On the basis of the study performed for the release in which temperature was kept constant at room temperature, and assuming that the physicochemical properties of particles do not change significantly with the azamethiphos encapsulation, it is possible to assume that the release of azamethiphos occurs by a diffusion-controlled mechanism through the porous matrix of agarose nanoparticles. An empirical model to describe a diffusion-controlled release of an encapsulated compound from a hydrogel matrix was developed by Peppas et al. [41, 42]. This model described the time

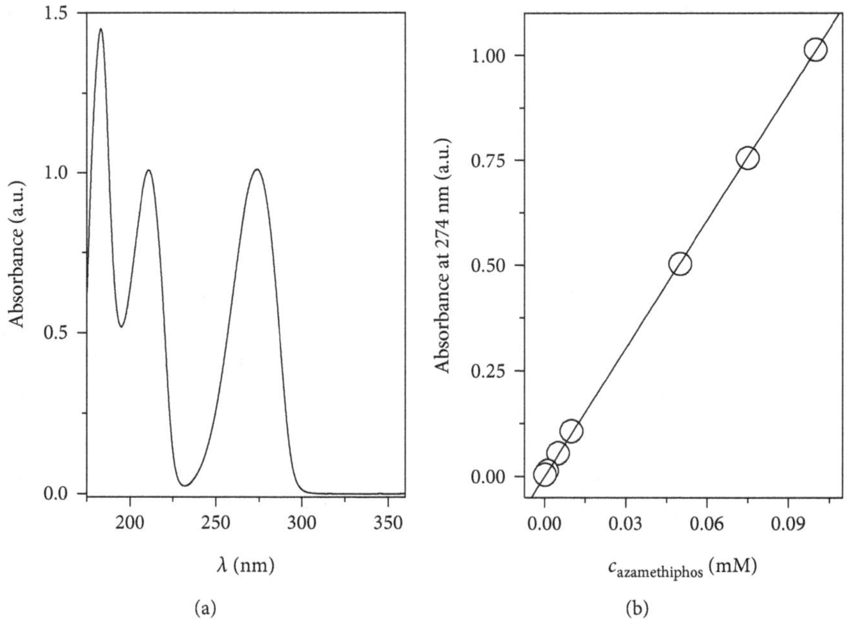


FIGURE 4: (a) UV-visible spectra of azamethiphos solution in water with 0.1 mM concentration. (b) Experimental dependence of the absorbance at 274 nm on the azamethiphos concentration for solutions of the insecticide in water (symbols) and the fit to the Lambert-Beer law (line).

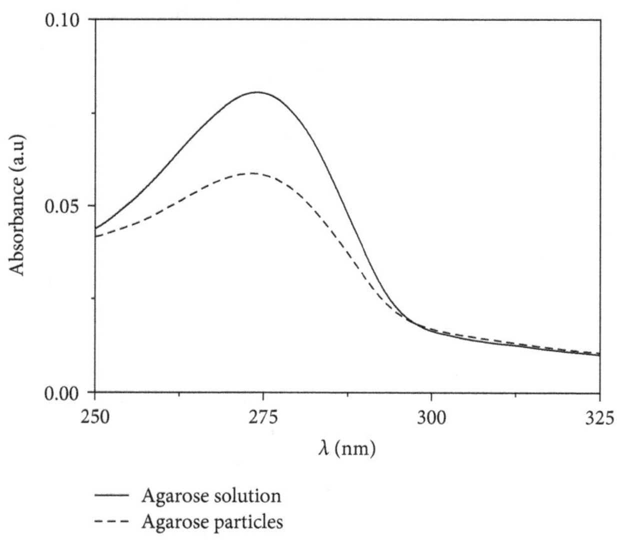


FIGURE 5: UV-visible spectra of azamethiphos solution in water with 0.01 mM concentration and agarose particles loaded with azamethiphos obtained using the same solution for its preparation.

dependence of the fraction of released compound $\phi$ as a power law defined as follows:

$$\phi = kt^n, \tag{3}$$

where $k$ is a structural/geometrical constant depending on the particular system and $n$ is the release exponent describing the release mechanism. The experimental data obtained for the release of the encapsulated azamethiphos agrees well with the description provided by the empirical law with the parameters assuming the following values: $k = 0.54 \pm 0.01$ and $n = 0.109 \pm 0.005$. Thus, the results confirm the diffusion-controlled release of azamethiphos from the agarose nanoparticles. Similar diffusion-controlled mechanisms were found by Wang and Wu [26] for the release of ovoalbumin from agarose nanoparticles and by Rekha and Sharma [43] for the release of insulin from chitosan-modified particles. It is worth mentioning that the diffusion-controlled mechanism has been described in the literature as the most extended mechanism for the release of molecules from polysaccharide particles [44].

## 4. Conclusions

This work presents a new route for preparation of aqueous dispersions of stable and monodisperse nanosized agarose gel nanoparticles. This was possible by the temperature-triggered gelation of the aqueous drops of the W/O microemulsions used as a template for controlling the size and morphology of the nanoparticles. Such nanoparticles can be transferred to aqueous medium to obtain a dispersion of highly charged nanoparticles which present a physicochemical behavior reminiscent of a physically cross-linked gel which can undergo a reversible swelling-shrinking process enabling for a controlled triggering of encapsulated compounds.

The obtained particles can be used for encapsulation of active compounds such as synthetic insecticides, e.g., azamethiphos, without any noticeable alteration of its physicochemical characteristics. The study of the release of azamethiphos encapsulated has evidenced a diffusion-controlled profile in agreement with previous studies using

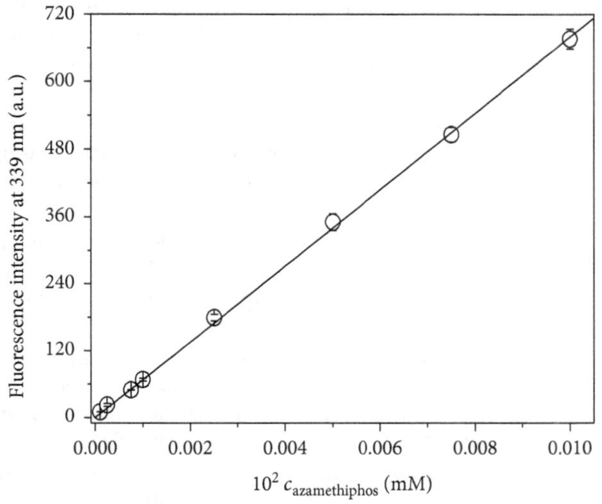


FIGURE 6: Fluorescence intensity for azamethiphos solutions of different concentrations. The symbols represent the experimental data and the line fitting to the relationship $I = Kc_{azamethiphos}$.

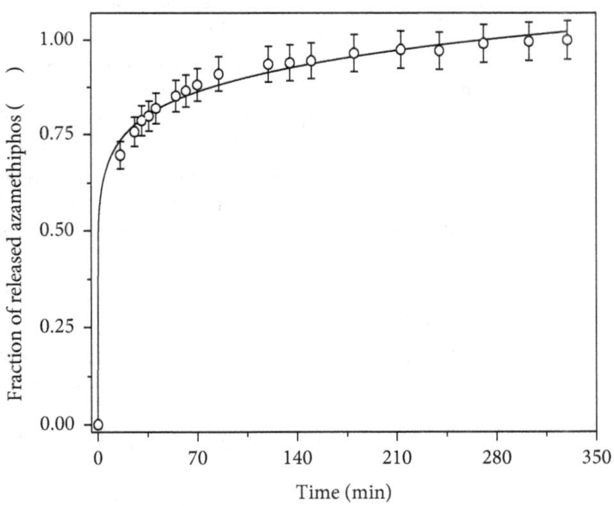


FIGURE 7: Release kinetics for azamethiphos from agarose nanoparticles represented as the time dependence of the fraction of azamethiphos released from the particles $\phi$. The symbols represent the experimental data and the solid lines the fit to the experimental model described by equation (3).

agarose particles. This work offers a new possibility for designing sketched vectors for controlling parasites in water reservoirs. However, further developments of this work require designing methodologies for a better control of the release profile of the encapsulated compounds to provide the bases for a sustained release.

## Disclosure

Pablo G. Argudo's current address is Departamento de Química Física, Universidad de Córdoba, Campus Rabanales, Ed. Marie Curie, 14014 Córdoba, Spain.

## Conflicts of Interest

The authors declare that they have no conflicts of interest.

## Acknowledgments

PGA, EG, RGR, and FO were funded by MINECO under grant CTQ2016-78895-R, and AL was supported by Agencia Nacional de Promoción Científica y Tecnológica (ANPCyT) PICT 2016-1111. We are grateful to C.A.I. Espectroscopia from the UCM for the use of their facilities.

## References

[1] M. A. Augustin and Y. Hemar, "Nano- and micro-structured assemblies for encapsulation of food ingredients," *Chemical Society Reviews*, vol. 38, no. 4, pp. 902–912, 2009.

[2] A. Lucia, A. C. Toloza, E. Guzmán, F. Ortega, and R. G. Rubio, "Novel polymeric micelles for insect pest control: encapsulation of essential oil monoterpenes inside a triblock copolymer shell for head lice control," *PeerJ*, vol. 5, article e3171, 2017.

[3] E. Guzmán, A. Mateos-Maroto, M. Ruano, F. Ortega, and R. G. Rubio, "Layer-by-layer polyelectrolyte assemblies for encapsulation and release of active compounds," *Advances in Colloid and Interface Science*, vol. 249, pp. 290–307, 2017.

[4] V. Nedovic, A. Kalusevic, V. Manojlovic, S. Levic, and B. Bugarski, "An overview of encapsulation technologies for food applications," *Procedia Food Science*, vol. 1, pp. 1806–1815, 2011.

[5] E. Guzman, R. Chulia-Jordan, F. Ortega, and R. G. Rubio, "Influence of the percentage of acetylation on the assembly of LbL multilayers of poly(acrylic acid) and chitosan," *Physical Chemistry Chemical Physics*, vol. 13, no. 40, pp. 18200–18207, 2011.

[6] B. F. Gibbs, S. Kermasha, I. Alli, and C. N. Mulligan, "Encapsulation in the food industry: a review," *International Journal of Food Sciences and Nutrition*, vol. 50, no. 3, pp. 213–224, 1999.

[7] E. Guzmán, J. A. Cavallo, R. Chuliá-Jordán et al., "pH-induced changes in the fabrication of multilayers of poly(acrylic acid) and chitosan: fabrication, properties, and tests as a drug storage and delivery system," *Langmuir*, vol. 27, no. 11, pp. 6836–6845, 2011.

[8] A. Madene, M. Jacquot, J. Scher, and S. Desobry, "Flavour encapsulation and controlled release–a review," *International Journal of Food Science & Technology*, vol. 41, pp. 1–21, 2006.

[9] A. Lucia, P. G. Argudo, E. Guzmán, R. G. Rubio, and F. Ortega, "Formation of surfactant free microemulsions in the ternary system water/eugenol/ethanol," *Colloids and Surfaces A: Physicochemical and Engineering Aspects*, vol. 521, pp. 133–140, 2017.

[10] G. A. Reineccius, "Role of carbohydrates in flavor encapsulation," *Journal of Dairy Science*, vol. 45, pp. 144–146, 1991.

[11] T. A. Tari and R. S. Singhal, "Starch based spherical aggregates: reconfirmation of the role of amylose on the stability of a model flavouring compound, vanillin," *Carbohydrate Polymers*, vol. 50, no. 3, pp. 279–282, 2002.

[12] B. K. Green and L. Scheicher, "Pressure sensitive record materials," US Patent 2,712,507, 1955, NCR Corp USA.

[13] R.-S. Liu, Ed., *Controlled Nanofabrication: Advances and Applications*, Pan Stanford Publishing, Boca Raton, FL, USA, 2013.

[14] Y. Wang, P. Li, T. T.-D. Tran, J. Zhang, and L. Kong, "Manufacturing techniques and surface engineering of polymer based nanoparticles for targeted drug delivery to cancer," *Nanomaterials*, vol. 6, no. 2, p. 26, 2016.

[15] J. P. Rao and K. E. Geckeler, "Polymer nanoparticles: preparation techniques and size-control parameters," *Progress in Polymer Science*, vol. 36, no. 7, pp. 887–913, 2011.

[16] S. Nitta and K. Numata, "Biopolymer-based nanoparticles for drug/gene delivery and tissue engineering," *International Journal of Molecular Sciences*, vol. 14, no. 1, pp. 1629–1654, 2013.

[17] O. G. Jones and D. J. McClements, "Functional biopolymer particles: design, fabrication, and applications," *Comprehensive Reviews in Food Science and Food Safety*, vol. 9, no. 4, pp. 374–397, 2010.

[18] A. Matalanis, O. G. Jones, and D. J. McClements, "Structured biopolymer-based delivery systems for encapsulation, protection, and release of lipophilic compounds," *Food Hydrocolloids*, vol. 25, no. 8, pp. 1865–1880, 2011.

[19] I. J. Joye and D. J. McClements, "Biopolymer-based nanoparticles and microparticles: fabrication, characterization, and application," *Current Opinion in Colloid & Interface Science*, vol. 19, no. 5, pp. 417–427, 2014.

[20] C. P. Reis, R. J. Neufeld, S. Vilela, A. J. Ribeiro, and F. Veiga, "Review and current status of emulsion/dispersion technology using an internal gelation process for the design of alginate particles," *Journal of Microencapsulation*, vol. 23, no. 3, pp. 245–257, 2006.

[21] L. Chen and M. Subirade, "Alginate–whey protein granular microspheres as oral delivery vehicles for bioactive compounds," *Biomaterials*, vol. 27, no. 26, pp. 4646–4654, 2006.

[22] T. Heidebach, P. Först, and U. Kulozik, "Microencapsulation of probiotic cells by means of rennet-gelation of milk proteins," *Food Hydrocolloids*, vol. 23, no. 7, pp. 1670–1677, 2009.

[23] M. R. Saboktakin, R. M. Tabatabaee, A. Maharramov, and M. A. Ramazanov, "Design and characterization of chitosan nanoparticles as delivery systems for paclitaxel," *Carbohydrate Polymers*, vol. 82, no. 2, pp. 466–471, 2010.

[24] S. Salatin, J. Barar, M. Barzegar-Jalali, K. Adibkia, M. A. Milani, and M. Jelvehgari, "Hydrogel nanoparticles and nanocomposites for nasal drug/vaccine delivery," *Archives of Pharmacal Research*, vol. 39, no. 9, pp. 1181–1192, 2016.

[25] R. Satar, S. A. Iizhar, M. Rasool, P. N. Pushparaj, and S. A. Ansari, "Investigating the antibacterial potential of agarose nanoparticles synthesized by nanoprecipitation technology," *Polish Journal of Chemical Technology*, vol. 18, no. 2, pp. 9–12, 2016.

[26] N. Wang and X. S. Wu, "Preparation and characterization of agarose hydrogel nanoparticles for protein and peptide drug delivery," *Pharmaceutical Development and Technology*, vol. 2, no. 2, pp. 135–142, 1997.

[27] S. Ansari, S. Ahmad, M. Jafri, M. Naseer, and R. Satar, "Utility of functionalized agarose nanoparticles in hydrolyzing lactose in batch reactors for dairy industries," *Química Nova*, vol. 41, pp. 429–433, 2018.

[28] V. Kattumuri, M. Chandrasekhar, S. Guha et al., "Agarose-stabilized gold nanoparticles for surface-enhanced Raman spectroscopic detection of DNA nucleosides," *Applied Physics Letters*, vol. 88, no. 15, article 153114, 2006.

[29] H. Hassanvand and P. Hashemi, "Synthesis of silver nanoparticles–agarose composite and its application to the optical detection of cyanide ion," *Analytical Sciences*, vol. 34, no. 5, pp. 567–570, 2018.

[30] S. Talegaonkar, A. Azeem, F. Ahmad, R. Khar, S. Pathan, and Z. Khan, "Microemulsions: a novel approach to enhanced drug delivery," *Recent Patents on Drug Delivery & Formulation*, vol. 2, no. 3, pp. 238–257, 2008.

[31] Z. Wang, K. Yang, H. Li et al., "In situ observation of sol-gel transition of agarose aqueous solution by fluorescence measurement," *International Journal of Biological Macromolecules*, vol. 112, pp. 803–808, 2018.

[32] R. Medina-Esquivel, Y. Freile-Pelegrin, P. Quintana-Owen, J. M. Yáñez-Limón, and J. J. Alvarado-Gil, "Measurement of the sol–gel transition temperature in agar," *International Journal of Thermophysics*, vol. 29, no. 6, pp. 2036–2045, 2008.

[33] B. T. Hargrave, Ed., *Environmental Effects of Marine Finfish Aquaculture*, Springer, Berlin, Heidelberg, 2005.

[34] G. Giammona, F. Goffredi, V. Turco Liveri, and G. Vassallo, "Water structure in water/AOT/*n*-heptane microemulsions by FT-IR spectroscopy," *Journal of Colloid and Interface Science*, vol. 154, no. 2, pp. 411–415, 1992.

[35] M. Goffredi, V. T. Liveri, and G. Vassallo, "Refractive index of water-AOT-*n*-heptane microemulsions," *Journal of Solution Chemistry*, vol. 22, no. 10, pp. 941–949, 1993.

[36] E. Negro, R. Latsuzbaia, A. H. de Vries, and G. J. M. Koper, "Experimental and molecular dynamics characterization of dense microemulsion systems: morphology, conductivity and SAXS," *Soft Matter*, vol. 10, no. 43, pp. 8685–8697, 2014.

[37] E. I. Franses and T. J. Hart, "Phase behavior and molecular motion of aerosol OT in liquid-crystalline phases with water," *Journal of Colloid and Interface Science*, vol. 94, no. 1, pp. 1–13, 1983.

[38] P. Serwer, "Agarose gels: properties and use for electrophoresis," *Electrophoresis*, vol. 4, no. 6, pp. 375–382, 1983.

[39] C. Rochas, A. Hecht, and E. Geissler, "Swelling properties of agarose gels," *Journal de Chimie Physique*, vol. 93, pp. 850–857, 1996.

[40] M. J. Taylor, P. Tomlins, and T. S. Sahota, "Thermoresponsive gels," *Gels*, vol. 3, no. 1, p. 4, 2017.

[41] N. A. Peppas, P. Bures, W. Leobandung, and H. Ichikawa, "Hydrogels in pharmaceutical formulations," *European Journal of Pharmaceutics and Biopharmaceutics*, vol. 50, no. 1, pp. 27–46, 2000.

[42] J. Siepmann and N. A. Peppas, "Modeling of drug release from delivery systems based on hydroxypropyl methylcellulose (HPMC)," *Advanced Drug Delivery Reviews*, vol. 48, no. 2-3, pp. 139–157, 2001.

[43] M. R. Rekha and C. P. Sharma, "Synthesis and evaluation of lauryl succinyl chitosan particles towards oral insulin delivery and absorption," *Journal of Controlled Release*, vol. 135, no. 2, pp. 144–151, 2009.

[44] A. Sosnik, "Alginate particles as platform for drug delivery by the oral route: state-of-the-art," *ISRN Pharmaceutics*, vol. 2014, Article ID 926157, 17 pages, 2014.

# 13

# Synthesis and Polymerization of Naphthoxazines Containing Furan Groups: An Approach to Novel Biobased and Flame-Resistant Thermosets

**Davi Rabelo de Oliveira, Selma Elaine Mazzetto, and Diego Lomonaco**

*Department of Organic and Inorganic Chemistry, Federal University of Ceara, 60440-900 Fortaleza CE, Brazil*

Correspondence should be addressed to Diego Lomonaco; lomonaco@ufc.br

Academic Editor: Atsushi Sudo

Naphthoxazines are a class of compounds with potential application in obtaining high-performance polymeric materials. Such application of these compounds, however, is still scarcely explored in the literature. Combined with the search for new high-performance materials, the development of biobased polymers has gained a lot of attention. In this sense, the inclusion of furan groups in polymers has been explored as a strategy that combines the search for high-performance materials with the search for the development of biobased materials. In this work, novel naphthoxazine monomers containing furan groups were synthetized. The syntheses were carried out in a single step, without the use of solvents and catalysts, obtaining the products in satisfactory yields and high purity. The naphthoxazines had their chemical structures completely characterized by FTIR, $^1$H NMR, and $^{13}$C NMR techniques. The thermal analyses (DSC and TGA) showed that all naphthoxazines exhibit exothermic typical polymerization events, making these compounds suitable for obtention of poly(naphthoxazines) resins, but suffer significant mass losses at temperatures below the onset polymerization temperature. In this way, a catalyst (1 mol% $MgCl_2$) was used in order to allow the polymerization of the compounds before the mass loss events. The FTIR analysis showed strong evidences of the formation of poly(naphthoxazines), and TGA analyses showed that the resins have high thermal stability, with high flame resistance and self-extinguishing properties (LOI > 28), which makes these compounds attractive in the development of biobased and high-performance materials.

## 1. Introduction

The search for high-performance polymeric materials has attracted the attention of several research groups, which seek to develop new compounds with attractive characteristics, such as high thermal and mechanical resistance, especially for application in the microelectronics and aerospace industries [1].

In this context, polymers derived from compounds containing 1,3-oxazinic rings (six-membered heterocyclic containing nitrogen and oxygen heteroatoms) arise as a relatively new class of high-performance thermosetting polymers. The oxazinic polymers or polyoxazines, can be considered as substituents for the traditional phenolic polymers because of their structural similarity in the main chain [2].

In addition to the advantages of traditional phenolic polymers, such as high thermal resistance, dimensional stability, good electrical properties, flame retardancy, and low smoke generation, polyoxazines exhibit their own attractive characteristics, such as low moisture absorption, good chemical resistance, high glass transition temperature ($T_g$), low shrinkage, and improved mechanical and thermal properties compared to traditional phenolic polymers. In addition, they offer great flexibility of molecular design and do not generate by-products during their polymerization [2, 3].

Within this group of compounds, benzoxazines are the most well-known and studied. Benzoxazines are a class of compounds in which an oxazinic ring is fused to a benzene ring. The main field of the study and application of benzoxazines is in the development of polymeric materials [4], and in the last years, several studies have been carried out to synthesize new compounds and to diversify their field of applications. When exposed to high temperatures, this class of molecules begins to polymerize through a process of ring-opening polymerization (ROP), generating poly(benzoxazines). These compounds have also been investigated because of their potential application in a number of biological activities [5–8].

Naphthoxazines are compounds analogous to benzoxazines, but in this class of compounds, the oxazinic ring is fusing to a naphthalene ring. Their potential application in a number of biological activities, especially their fungicide, bactericide, and principally in the treatment of Parkinson's disease [9, 10], has also attracted the attention of a number of research groups. Similarly to benzoxazines, the naphthoxazines also suffer ROP when exposed to high temperatures, generating poly(naphthoxazines). Its application in the field of materials is, however, still little explored, with only few papers reported in the literature describing the polymerization of these compounds to obtain high-performance polymeric materials. These limited activities are related in large part to the easy evaporation that some monomers of naphthoxazines exhibit during the polymerization [11], preventing their processing. However, the use of these compounds in order to obtain high-performance polymeric materials is very attractive, since polymers possessing the naphthalene group are expected to exhibit excellent mechanical property and thermal stability, resulting from high backbone rigidity and aromatic content provided by the multiaromatic ring structure [12].

Various approaches have been developed in order to obtain improvements in the properties of the polyoxazines. Especially designed molecules incorporated with naphthalene, biphenyl, or benzophenone groups were synthesized for this purpose [2, 13, 14]. These compounds showed high thermal stability but, nevertheless, presented poor processability properties, high melting, and polymerization temperatures, as also low solubility in organic solvents.

Another approach is the preparation of monomers containing other polymerizable groups, such as ethynyl, phenylethynyl, nitrile, propargyl, and allyl [12, 15–17]. The polyoxazines obtained from these compounds showed high glass transition temperature and good thermal stability. However, complicated polymerization reactions and high cure temperatures occur in these compounds. In addition, cross-linked structures formed from the first polymerization of these additional polymerizable groups may form barriers to subsequent polymerization reactions, reducing the total conversion of the polymerization reactions and increasing their temperature [1].

It is known that, in addition to the chemical structures, the hydrogen bonds present in the polyoxazines also play an important role in a wide variety of properties of these compounds. A number of studies have observed the existence of bonds of the type -OH$\cdots$O, -OH$\cdots$N, and -OH$\cdots\pi$ in poly(oxazines) and demonstrated that such bonds contribute to the improvement of their thermal properties, such as elevated glass transition temperatures [18–20]. Thus, the insertion of groups that promote the formation of these bonds in polyoxazine compounds is also a strategy for obtaining materials with improved properties.

Associated with the search for high-performance materials, there is a growing interest in green and sustainable chemistry. In this sense, the use of renewable raw materials is one of the principles of green chemistry [21] and one of the main goals of those working on the development of new compounds. Thus, the use of renewable raw materials in the synthesis of benzoxazines has been widely explored. Several studies have reported the replacement of petrochemicals with materials from renewable sources, such as resorcinol and hydroquinone [22], cardanol [23], and furfurylamine [24]. However, to the best of our knowledge, no works are reported in the literature regarding biobased naphthoxazines.

The insertion of furan groups into polymers has been explored as a strategy that combines the search for high-performance materials with the search for the development of biobased materials, since these groups have shown to be able to confer good thermal and mechanical properties to the materials and can be obtained from renewable raw materials such as hemicellulose. Works have been developed to synthesize benzoxazines containing these groups [1, 22], but this strategy has not yet been addressed in order to obtain high-performance and biobased naphthoxazines and poly(naphthoxazines).

Based on the above, naphthoxazines have potential to form polymers with high thermal resistance and the incorporation of a furan ring to its structures tends to provide new polymerizable groups and promote the formation of hydrogen bonds. Therefore, the main objective of this work was to synthesize and polymerize three naphthoxazines containing furan groups (Figure 1), starting from two lignocellulose-based bioproducts, furfurylamine, and furfural. The monomers synthesized were characterized by spectroscopic techniques ($^1$H and $^{13}$C NMR, FTIR), and their thermal properties were evaluated by DSC and TGA techniques. After spectral and thermal characterization, their polymers were analyzed by FTIR and TGA.

## 2. Materials and Methods

*2.1. Materials.* 2-Naphthol (99%) and furfurylamine (99%) were used as received by Sigma-Aldrich (USA). Formaldehyde (37%) was used as received by Dinâmica (Brazil). Furfural (99%, Sigma-Aldrich) was previously vacuum distilled prior to use in the syntheses. Ammonium acetate was produced from the reaction between ammonium hydroxide (28–30%, Dinâmica) and glacial acetic acid (Dinâmica). Methylene chloride, methanol, sodium hydroxide, and anhydrous sodium sulfate were used as received by Labsynth (Brazil).

### *2.2. Methods*

*2.2.1. Microwave-Assisted Synthesis.* The microwave-assisted synthesis was carried out in a Milestone microwave reactor, model Start SYNTH, under atmospheric pressure. The

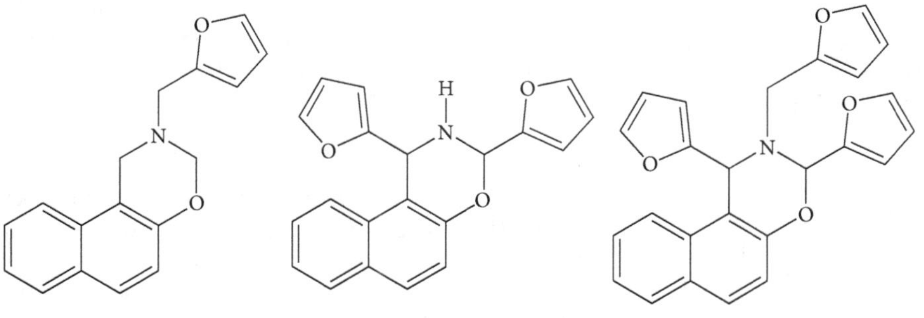

Naft-Fu-H Naft-H-Fu Naft-Fu-Fu

FIGURE 1: Naphthoxazines containing furan groups studied in this work.

temperature of each reaction was controlled by an infrared sensor, and the applied power was limited by 500 W.

*2.2.2. $^{13}C$ and $^{1}H$ Nuclear Magnetic Resonance Spectroscopy ($^{1}H$ NMR and $^{13}C$ NMR).* The $^{1}H$ NMR and $^{13}C$ NMR spectra were recorded on a Bruker spectrometer, model Avance DPX, operating at 300 MHz for $^{1}H$ and at 75 MHz for $^{13}C$. Deuterated chloroform ($CDCl_3$) and deuterated dimethyl sulfoxide (DMSO-$d_6$) were used as the solvent to solubilize the samples, and tetramethylsilane (TMS) was used as an internal standard.

*2.2.3. Fourier Transform Infrared Spectroscopy (FTIR).* The FTIR spectra were obtained on a PerkinElmer spectrometer model FTIR/NIR Frontier, using an attenuated total reflectance (ATR) accessory with zinc selenide (ZnSe) crystal surface, and resolution of 4 $cm^{-1}$ using the arithmetic average of four scans in the wavenumber range of 4000–550 $cm^{-1}$.

*2.2.4. Differential Scanning Calorimetry (DSC).* DSC analyses were made using Mettler Toledo DSC 823e, using 10 mg of samples in aluminum pans. Nonisothermal measurements were made with a heating rate of 10°C/min, in a temperature range of 30 to 400°C, under nitrogen ($N_2$) purge at a flow rate of 50 mL/min.

*2.2.5. Thermogravimetric Analyses (TGA).* TGA analyses were performed on a Mettler Toledo TGA/SDTA 851e, conducted under $N_2$ atmosphere (flow rate of 50 mL/min) with a heating rate of 10°C/min in a temperature range of 30°C to 800°C. Alumina crucibles were used with approximately 10 mg of samples.

### *2.3. Synthesis of Naphthoxazines*

*2.3.1. Synthesis of 3-(Furan-2-ylmethyl)-3,4-dihydro-2H-naphtho[1,2-e][1,3]oxazine (Abbreviated as Naft-Fu-H).* In a round bottom flask (25 mL) containing formaldehyde (13.86 mmol), furfurylamine (6.93 mmol) was slowly added under magnetic stirring. Then, 2-naphthol (6.93 mmol) was added, and the reaction mixture was homogenized before being brought to the microwave reactor. The flask was placed in the microwave cavity, equipped with a 50 cm Vigreux condenser, and distillation head. The microwave was set to a maximum power of 500 W and programmed to increase from ambient temperature to 70°C in a period of 2 minutes until the temperature stabilized and held for a further 5 minutes, totaling 7 minutes of irradiation. At the end of the reaction, the mixture was cooled to room temperature and methylene chloride (20 mL) was added to the flask. The organic solution was transferred to a separatory funnel (250 mL) and treated with 2 mol $L^{-1}$ aqueous NaOH solution (2 × 20 mL) and then with a saturated aqueous sodium chloride solution (2 × 20 mL). The organic phase was collected, dried over anhydrous sodium sulfate, and filtered, and the solvent was removed under reduced pressure to give an oily orange-colored product. The product was purified from that oil by recrystallization from methanol (20 mL) to give the pure product as light beige crystals in 60% yield. The product was characterized by FTIR, $^{1}H$ NMR, and $^{13}C$ NMR.

(i) $^{1}H$ NMR (300 MHz, $CDCl_3$): $\delta$ 7.76 (d, J = 8.0 Hz, 1H), 7.65 (d, J = 8.9 Hz, 1H), 7.56 (d, J = 8.4 Hz, 1H), 7.47 (dd, J = 6.8, 1.3 Hz, 1H), 7.43 (dd, J = 1.8, 0.8 Hz, 1H), 7.35 (ddd, J = 8.0, 6.8, 1.2 Hz, 1H), 7.05 (d, J = 8.9 Hz, 1H), 6.33 (dd, J = 3.1, 1.9 Hz, 1H), 6.24 (dd, J = 3.1, 0.5 Hz, 1H), 4.95 (s, 1H), 4.35 (s, 1H), 3.97 (s, 1H)

(ii) $^{13}C$ NMR (75 MHz, $CDCl_3$): $\delta$ 151.87, 151.84, 142.82, 132.06, 129.25, 128.83, 128.31, 126.75, 123.74, 121.22, 118.72, 111.52, 110.42, 109.23, 81.80, 48.89, 47.32

(iii) FTIR ($cm^{-1}$): 3016, 2948, 1633, 1599, 1515, 1470, 1431, 1403, 1357, 1285, 1224, 1126, 1058, 1011, 943, 899, 854, 807, 776, 733

*2.3.2. Synthesis of 2,4-Di(furan-2-yl)-3,4-dihydro-1H-naphtho[1,2-e][1,3]oxazine and 2,4-Di(furan-2-yl)-3-(furan-2-ylmethyl)-3,4,-dihydro-1H-naphtho[1,2-e][1,3]oxazine (Abbreviated as Naft-H-Fu and Naft-Fu-Fu, Respectively).* In a round bottom flask (25 mL) containing furfural (13.86 mmol), ammonium acetate (6.93 mmol) or furfurylamine (6.93 mmol) was slowly added. Then, 2-naphthol (6.93 mmol) was added, and the reaction mixture was left under magnetic stirring for 24 h. At the end of the reaction, the products were treated and purified similarly to the procedure described for the compound Naft-Fu-H (Section 2.2.1). The Naft-H-Fu compound was obtained as a white material with cotton-like fibrous appearance (62% yield), and the Naft-Fu-Fu compound was obtained as a light brown powder (yield 44%). Both compounds were also characterized by FTIR, $^1$H NMR, and $^{13}$C NMR.

(i) Naft-H-Fu

(1) $^1$H NMR (300 MHz, $CDCl_3$): $\delta$ 11.59 (s, 1H), 8.41 (s, 1H), 7.89–7.61 (m, 5H), 7.61–7.00 (m, 11H), 6.94 (d, J = 2.8 Hz, 1H), 6.44 (d, J = 35.0 Hz, 3H), 6.38–6.34 (m, 1H), 6.23 (d, J = 12.1 Hz, 2H), 5.83 (d, J = 59.1 Hz, 3H), 5.70 (s, 1H), 3.02 (s, 1H)

(2) $^1$H NMR (300 MHz, DMSO-$d_6$): $\delta$ 10.95 (s, 4H), 8.53 (s, 1H), 8.10 (s, 2H), 8.08–7.69 (m, 97H), 7.55 (dd, J = 33.3, 17.0 Hz, 41H), 7.43–6.91 (m, 103H), 6.62 (dd, J = 37.6, 20.8 Hz, 57H), 6.35 (s, 30H), 6.07 (s, 2H), 6.07–5.74 (m, 40H), 5.66 (s, 13H), 4.57 (d, J = 9.4 Hz, 13H), 2.50 (s, 11H)

(3) $^{13}$C NMR (75 MHz, $CDCl_3$): $\delta$ 156.09, 154.66, 152.81, 152.12, 151.02, 150.41, 146.46, 143.12, 142.72, 132.04, 131.47, 130.27, 129.84, 129.31, 128.55, 126.89, 123.72, 122.88, 122.52, 121.46, 120.70, 119.26, 117.13, 113.40, 112.43, 110.97, 110.49, 108.64, 108.31, 67.64, 48.72

(4) $^{13}$C NMR (75 MHz, DMSO-$d_6$): $\delta$ 155.60, 151.69, 151.14, 150.38, 146.67, 142.89, 142.57, 131.97, 130.99, 129.70, 129.15, 128.64, 128.14, 126.64, 126.06, 123.25, 122.53, 122.12, 118.95, 117.53, 112.96, 112.37, 110.50, 109.27, 107.99, 106.98, 77.65, 64.10, 63.84, 47.24

(5) FTIR ($cm^{-1}$): 1622, 1598, 1504, 1467, 1434, 1400, 1355, 1329, 1309, 1265, 1230, 1201, 1172, 1154, 1125, 1070, 1042, 1009, 987, 932, 909, 821, 742

(ii) Naft-Fu-Fu

(1) $^1$H NMR (300 MHz, $CDCl_3$): $\delta$ 7.85 (d, J = 8.9 Hz, 2H), 7.55 (s, 1H), 7.52 (s, 1H), 7.47 (s, 2H), 7.42 (s, 1H), 7.40 (d, J = 4.0 Hz, 1H), 7.30 (d, J = 8.9 Hz, 1H), 6.75 (d, J = 2.6 Hz, 1H), 6.50 (s, 1H), 6.38 (m, 2H), 6.30 (s, 2H), 5.98 (d, J = 2.7 Hz, 1H), 5.63 (s, 1H), 4.19 (d, J = 15.2 Hz, 1H), 3.69 (d, J = 15.1 Hz, 1H)

(2) $^{13}$C NMR (75 MHz, $CDCl_3$): $\delta$ 154.59, 152.14, 152.00, 149.71, 142.95, 142.81, 142.48, 132.55, 129.64, 129.45, 128.63, 126.80, 123.72, 122.51, 118.62, 111.18, 110.46, 110.38, 110.29, 110.18, 109.32, 108.48, 82.17, 53.38, 43.63

(3) FTIR ($cm^{-1}$): 3016, 2948, 1623, 1505, 1467, 1436, 1404, 1331, 1267, 1224, 1140, 1069, 1015, 936, 883, 821, 739

*2.4. Polymerization of Naphthoxazine Monomers.* Monomer powders, in an aluminum pan, were placed in a muffle oven and heated for a total of 4 h: 1 h at 150°C, 1 h at 170°C, 1 h at 190°C, and 1 h at 210°C in air. Dark brown cross-linked materials, poly(naphthoxazines), were obtained for all cases after the polymerization processes.

## 3. Results and Discussion

*3.1. Synthesis of Naphthoxazines Containing Furan Groups.* The naphthoxazines studied in this work were obtained by a one-pot reaction between 2-naphthol, an aldehyde (formaldehyde or furfural), and a nitrogenated compound (furfurylamine or ammonium acetate) under solvent-free conditions and without the use of catalysts, as shown in Figure 2. As far as it can be ascertained, the synthesis of the Naft-Fu-H and Naft-Fu-Fu compounds is not reported in the literature and even the Naft-H-Fu compound is reported in only one single work [25], where the thermal properties or polymerization behavior of this compound was not explored.

The synthesis of Naft-Fu-H was conducted under microwave heating, with satisfactory yield (60%), in a few minutes (7 min), especially if compared to the syntheses of oxazinic compounds conducted under traditional heating described in literature, which normally require several hours of reaction, thus demonstrating the efficiency of this method in the synthesis of this type of compound, as already reported in previous studies [3].

For the synthesis of the Naft-H-Fu and Naft-Fu-Fu compounds, it was chosen to conduct the syntheses at room temperature to avoid the polymerization of furfural, which would consequently reduce the yields of the reactions. The Naft-H-Fu compound was obtained in 62% yield, and the Naft-Fu-Fu compound was obtained in 44% yield. The lower yield of the Naft-Fu-Fu compound may be related to the greater steric hindrance caused by the three substituent groups present on the oxazinic ring, which would make it difficult to close the ring and consequently the formation of that compound.

It is worth mentioning that oxazinic compounds have been synthesized traditionally through Mannich condensation, between a phenolic compound, a primary amine, and formaldehyde, in a single step [4]. The synthesis of oxazinic compounds with aldehydes other than formaldehyde leading to the formation of compounds with substituents on the oxazinic ring carbons (2,4-substituted) is, however, uncommon and little reported in the literature; their methodologies for obtaining, reaction mechanisms and thermal properties are poorly exploited. In this sense, the synthesis and studies of the properties of Naft-H-Fu and Naft-Fu-Fu therefore represent an advance in the study of oxazinic compounds containing aldehydes other than formaldehyde.

*3.2. Structural Characterization of Naphthoxazines Containing Furan Groups*

*3.2.1. Fourier Transform Infrared Spectroscopy (FTIR).* In Figure 3, the FTIR spectra of 2-naphthol and naphthoxazines are compared. It is possible to observe that the broad band attributed to the stretching of the O-H bond present on the 2-naphthol spectrum (3248 $cm^{-1}$, indicated by the black arrow) is completely absent in the spectra of the naphthoxazines, indicating the transformation of the

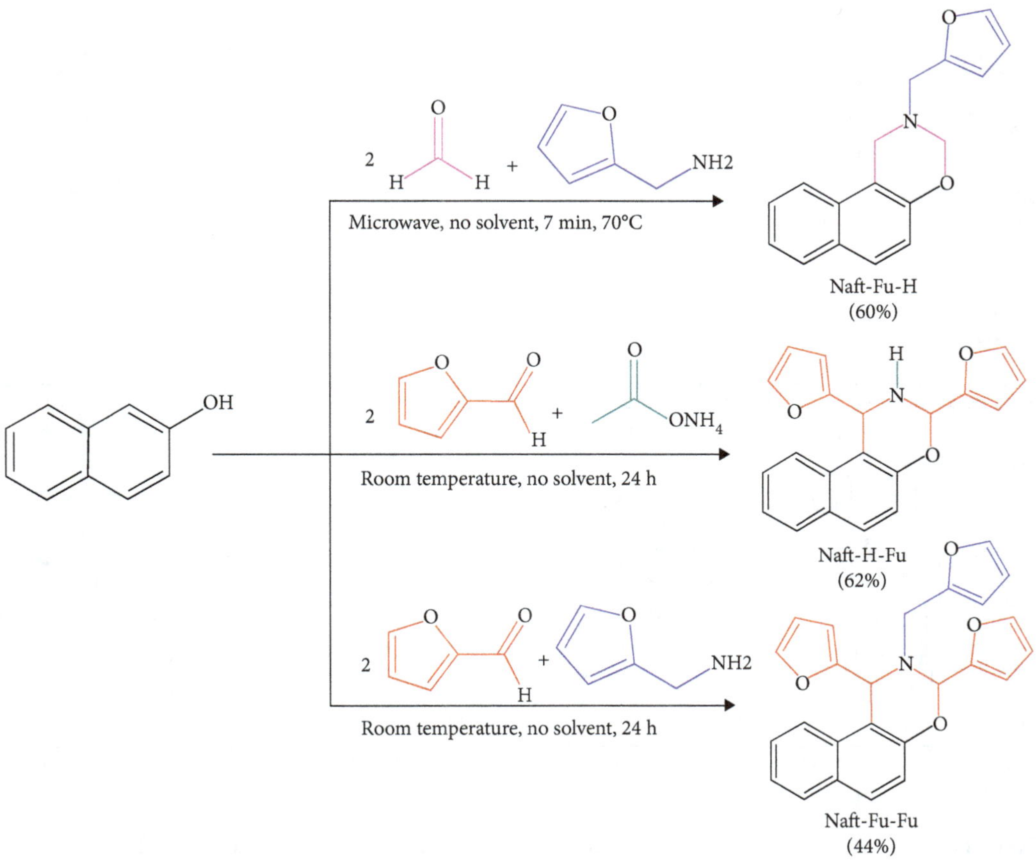


FIGURE 2: Synthetic scheme for the naphthoxazines containing furan groups.

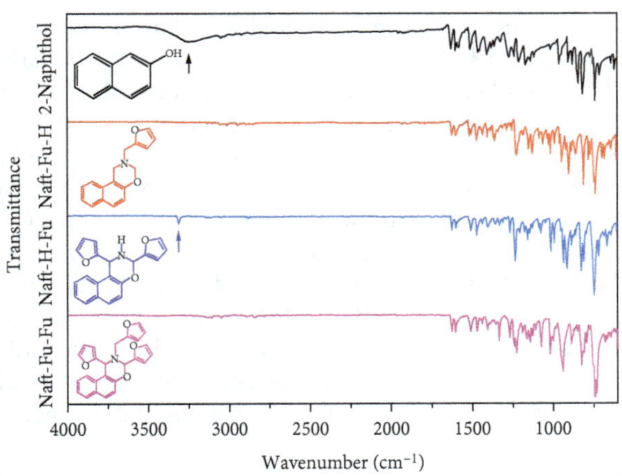


FIGURE 3: FTIR spectra of the naphthoxazines containing furan groups.

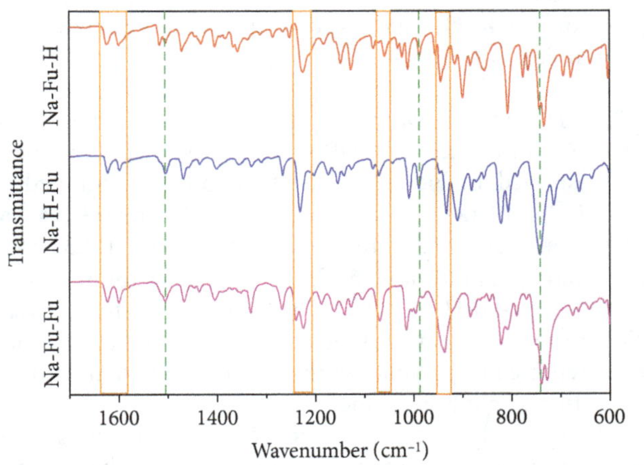


FIGURE 4: FTIR spectra of the naphthoxazines in the fingerprint region.

hydroxyl group, which can be attributed to the formation of the oxazinic ring. It is also possible to observe in the spectrum of the Naft-H-Fu a single sharp band attributed to the stretching of the N-H bond (3317 $cm^{-1}$, indicated by the blue arrow), which helps to confirm the proposed structure for that compound.

In the fingerprint region of the naphthoxazine spectra (Figure 4), the bands observed at 1624 and 1598 $cm^{-1}$ are attributed to the stretching of the C=C bonds of the naphthalene ring. The presence of characteristic bands around 930–940 $cm^{-1}$, attributed to the out-of-plane deformation mode of the $\alpha,\beta$-disubstituted naphthalene ring, indicates the formation of the oxazinic ring in all compounds. The bands observed at 1223–1229 $cm^{-1}$ and 1058–1070 $cm^{-1}$, assigned, respectively, to the asymmetric and symmetrical stretches of =C-O-C bonds, help to confirm the proposed structures for naphthoxazines containing furan groups [26]. The presence of the furan groups in all compounds was

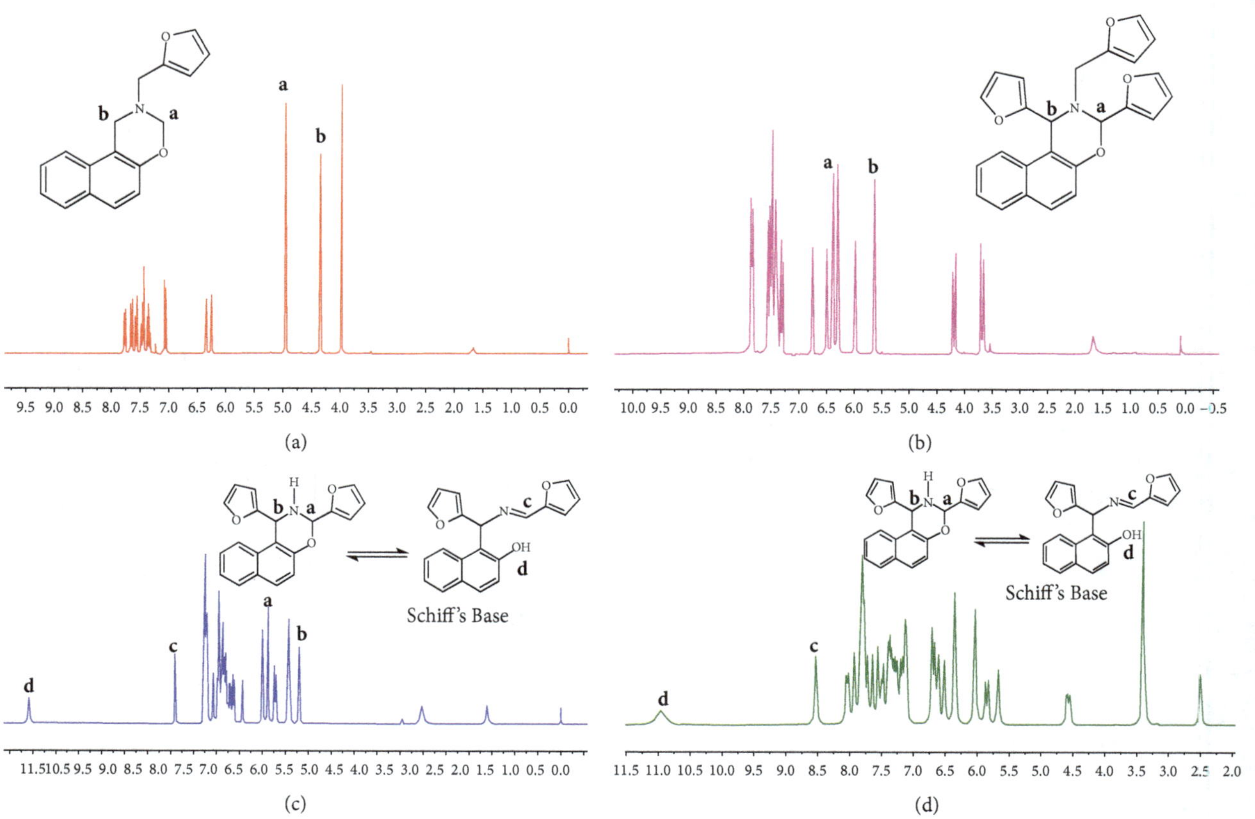

FIGURE 5: $^1$H NMR of naphthoxazines containing furan groups. (a) Naft-Fu-H, (b) Naft-Fu-Fu, (c) Naft-H-Fu, and (d) Naft-H-Fu (in DMSO-$d_6$).

confirmed by the observation of the characteristic peaks of this group in 1504, 987, and 742 cm$^{-1}$ [1], indicated by dashed lines in Figure 4.

*3.2.2. $^1$H and $^{13}$C Nuclear Magnetic Resonance ($^1$H NMR and $^{13}$C NMR).* The $^1$H NMR spectra of the naphthoxazines are displayed in Figure 5. The two singlets attributed to O-$\mathbf{CH_n}$-N (marked as **a**) and Ar-$\mathbf{CH_n}$-N (marked as **b**) of the oxazinic ring were observed at 4.94 and 4.34 ppm, 6.38 and 5.40 ppm, and 6.38 and 5.63 ppm for the compounds Naft-Fu-H, Naft-H-Fu, and Naft-Fu-Fu, respectively. These specific resonances of naphthoxazines have already been reported in previous works and are consistent with the formation of naphthoxazine rings [26–28].

Interestingly, the compound Naft-H-Fu exhibits ring-open-chain tautomerism in solution. This fact has already been observed in a similar compound by Karthikeyan et al. [29] and was attributed to a residual acidity of $CDCl_3$. In the $^1$H NMR spectra of this compound, it is possible to observe the signals attributed to Schiff's base at 8.40 ppm (-**CH**=N, marked as **c**) and 11.58 ppm (O-**H**, marked as **d**). The $^1$H NMR spectrum of Naft-H-Fu was also performed in DMSO-$d_6$ (Figure 5(d)), a solvent less acidic than $CDCl_3$, in order to avoid the ring-open-chain tautomerism, but even in this solvent, it was possible to observe the signals attributed to Schiff's base.

The $^{13}$C NMR spectra of naphthoxazines are displayed in Figure 6 and, in the same way of $^1$H NMR spectra, it is useful to verify the characteristic signals of naphthoxazines. To the compounds Naft-Fu-H and Naft-Fu-Fu, the signals attributed to O-$\mathbf{CH_n}$-N (marked as **a**) and Ar-$\mathbf{CH_n}$-N (marked as **b**) were observed at 81.8 and 48.8 ppm and 82.17 and 53.38 ppm, respectively. To the compound Naft-H-Fu, in the spectra performed in $CDCl_3$, it was possible to observe just the signal attributed to O-$\mathbf{CH_n}$-N at 67.64 ppm. The signals attributed to Schiff's base were observed at 67.64 ppm (C-N, marked as **d**) and at 156.09 ppm (C=N, marked as **c**). It was possible to observe the signal attributed to O-$CH_n$-N only in the spectra performed in DMSO-$d_6$, at 77.78 ppm. This fact can be attributed to the lower acidity of DMSO-$d_6$ in relation to $CDCl_3$, which causes the equilibrium displacement towards the naphthoxazine, allowing the observation of the signal.

### *3.3. Thermal Characterization of Naphthoxazines Containing Furan Groups*

*3.3.1. Differential Scanning Calorimetry (DSC).* The polymerization behavior of naphthoxazines containing furan groups was studied by differential scanning calorimetry (DSC). The nonisothermal DSC thermograms for each compound are shown in Figure 7. Thermal properties, including melting temperature ($T_m$), polymerization enthalpy ($\Delta H$), and onset polymerization temperature ($T_{onset}$) of the naphthoxazines are summarized in Table 1.

In all DSC curves, it was observed an endothermic event corresponding to the fusion of the compounds. Such events

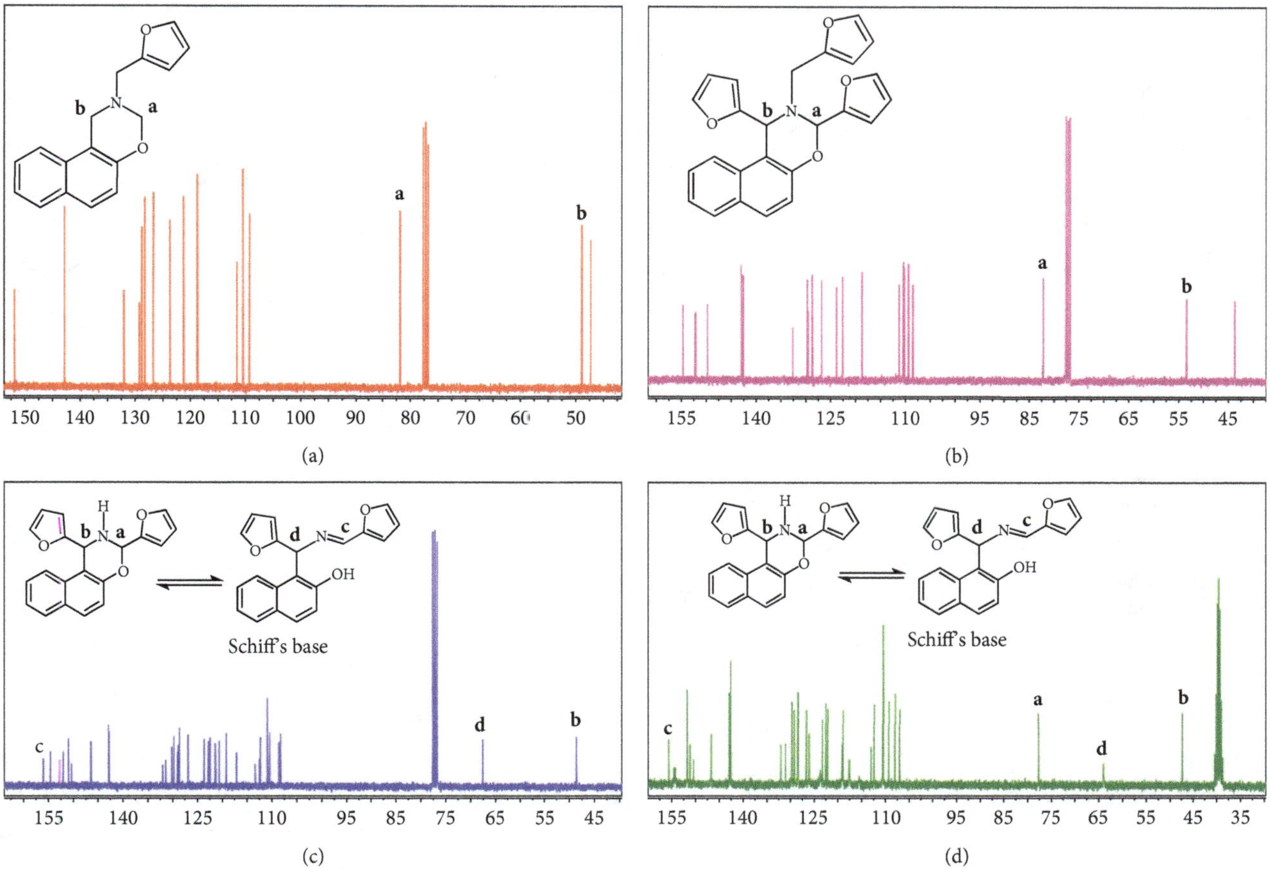


FIGURE 6: $^{13}$C NMR of naphthoxazines containing furan groups. (a) Naft-Fu-H, (b) Naft-Fu-Fu, (c) Naft-H-Fu, and (d) Naft-H-Fu (in DMSO-$d_6$).

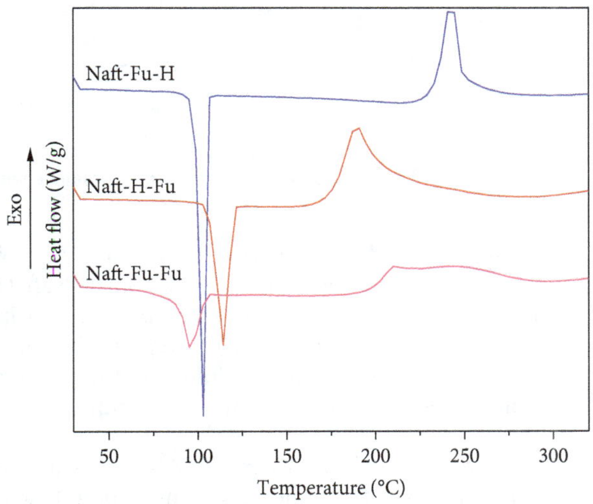


FIGURE 7: DSC thermograms of naphthoxazines containing furan groups.

are centered at 114°C, 100°C, and 95°C for Naft-H-Fu, Naft-Fu-H, and Naft-Fu-Fu, respectively.

The exothermic events observed at higher temperatures are associated with the opening of the oxazinic ring, initiating the process of the polymerization of the naphthoxazines.

TABLE 1: Summary of the values of the nonisothermal DSC thermograms of the naphthoxazines.

| Monomer | $T_m$ (°C) | $T_{onset}$ (°C) | $\Delta H$ (J g$^{-1}$) |
|---|---|---|---|
| Naft-H-Fu | 114 | 176 | −132.3 |
| Naft-Fu-H | 100 | 236 | −95.0 |
| Naft-Fu-Fu | 95 | 198 | −119.3 |

Figure 8 shows a proposed mechanism for the polymerization of Naft-Fu-Fu. The mechanism of polymerization of oxazinic compounds is divided into two main steps: the first is the opening of the oxazinic ring after the absorption of heat and the second occurs through electrophilic substitutions from the iminium ion formed in the ring opening. Depending on the conditions of this second step, two main polymeric structures can be formed in this mechanism: (i) a phenolic type, where phenolic hydroxyl groups are formed, and (ii) a phenoxy type, where is formed an ether bonding [4] (Figure 8(a)).

The onset polymerization temperature of Naft-Fu-H ($T_{onset}$ = 236°C) is comparable to the temperatures of other naphthoxazines found in the literature [12]. The Naft-H-Fu and Naft-Fu-Fu compounds, however, exhibit considerably lower starting polymerization temperatures ($T_{onset}$ = 176°C and 198°C, respectively) than the naphthoxazines reported

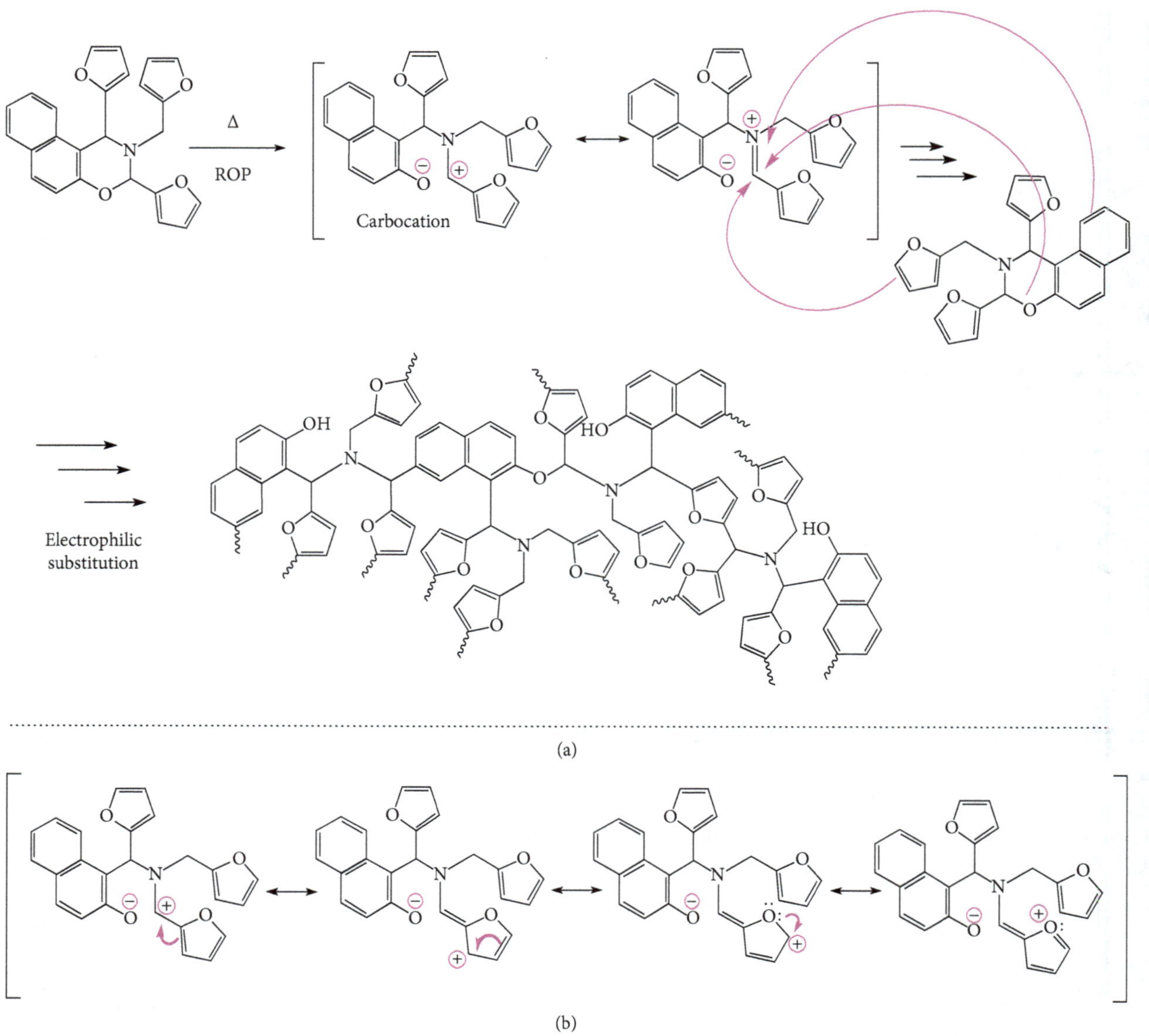


FIGURE 8: (a) Proposed mechanism for the polymerization of the compound Naft-Fu-Fu. (b) Stabilization of carbocation by resonance with the furan group.

in the literature. This effect may be associated with the fact that these compounds have substituents at the 2-position of the oxazinic ring, which facilitate the ring-opening process, since furan groups help to stabilize the carbocation formed at the opening of the oxazinic ring by resonance (Figure 8(b)) [30].

Furan groups may also participate in the polymerization mechanism through electrophilic substitution reactions, as reported in previous work involving the polymerization of benzoxazines containing furan groups [22, 24]. It is assumed, therefore, that the exothermic events observed in the thermograms are due to the simultaneous occurrence of cross-links of the furan groups and the opening of the oxazinic ring. The participation of the furan groups in the polymerization of these compounds is even more evident due to the high values of enthalpy of polymerization ($\Delta H$) and to the fact that the exothermic events associated with the polymerization process are enlarged.

The temperature difference between $T_m$ and $T_{onset}$ is defined as the processing window of a monomer. Naft-H-Fu, Naft-Fu-H, and Naft-Fu-Fu presented processing windows of 62.4°C, 135.5°C, and 102.2°C, respectively. These values are considered satisfactory, reinforcing the potential application of these compounds in material chemistry.

*3.3.2. Thermogravimetric Analyses (TGA).* The study of the thermal stability of the monomers is an important factor for obtaining polymers since, if the monomers undergo evaporation and/or degradation at temperatures below the polymerization temperature, this process will be impaired or even impossible.

The literature reports that one of the main limiting factors for obtaining poly(naphthoxazines) is the evaporation process that some monomers undergo before the polymerization process. In this way, the thermogravimetric analysis (TGA) of the synthesized naphthoxazines was carried

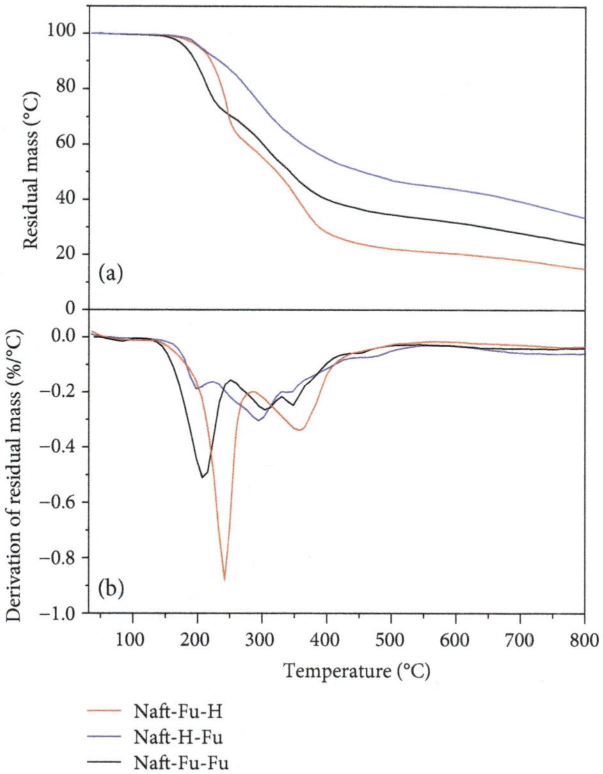


FIGURE 9: Thermograms of naphthoxazines containing furan groups. (a) TGA, (b) DTGA.

TABLE 2: Summary of the values of TGA thermograms of the naphthoxazines.

| Monomer | $T_{5\%}$ (°C) | $T_{10\%}$ (°C) | Char yield* (%) |
|---|---|---|---|
| Naft-Fu-H | 186 | 204 | 14.8 |
| Naft-H-Fu | 187 | 222 | 33.3 |
| Naft-Fu-Fu | 168 | 186 | 23.9 |

*Measured as the residual mass at 800°C.

out in order to evaluate their thermal stabilities before the polymerization process. TGA analysis was performed under conditions similar to the DSC analysis so that a suitable comparison could be made.

In Figure 9, the TGA and DTGA curves of the naphthoxazines containing furan groups are shown. The temperature of 5% and 10% of weight loss ($T_{5\%}$ and $T_{10\%}$, respectively), as well as the char yield (%) are summarized in Table 2.

It is possible to observe that all the compounds undergo a significative mass loss at temperatures lower than the onset polymerization temperatures ($T_{onset}$) observed by the DSC analysis. In this way, the use of catalysts was evaluated as a way of lowering the polymerization temperature of the compounds, making it possible to obtain the polymers without higher losses, as will be discussed later.

Analyzing the DTGA curves, it is possible to observe two main events of weight loss for Naft-Fu-H and three events for the Naft-H-Fu and Naft-Fu-Fu. Such events may be associated with the evaporation of the compounds or degradation. The formation of imines as by-products of the benzoxazine polymerization process has been hypothesized by several authors [31, 32]. The formation of this by-product appears to be favored by the presence of aromatic groups at the 2-position of the oxazinic ring, since the imine formed will be stabilized by resonance [30]. Formation followed by evaporation of such low molecular weight by-product may represent one of the additional mass loss events observed in the Naft-H-Fu and Naft-Fu-Fu compounds. In Figure 10, a mechanism for the formation of this imine is proposed.

*3.4. Polymerization of Naphthoxazines Containing Furan Groups.* As previously discussed (Section 3.2.2), all the compounds undergo mass loss at temperatures lower than the onset polymerization temperatures ($T_{onset}$). In this sense, the use of catalysts was evaluated as a way of lowering the polymerization temperature of the compounds, making it possible to obtain the polymers without higher losses.

Several catalysts have been evaluated in this sense, such as phosphorus pentachloride ($PCl_5$) [33], lithium iodide (LiI) [34], and *p*-toluene-sulfonates [35]. However, these compounds are difficult to handle, expensive, and toxics, which make this catalyst not suitable for large-scale production. In the previous work by our group, Kotzebue and coworkers [23] have demonstrated that magnesium chloride ($MgCl_2$) can be used as an efficient, safe, and inexpensive catalyst in the polymerization of benzoxazines. Thus, this compound was evaluated as an alternative to catalyze the polymerization of the naphthoxazines studied in this work.

As seen in Figure 11, the use of $MgCl_2$ (1 mol%) reduced significantly the onset polymerization temperature to all naphthoxazines. Kotzebue and coworkers [23] explained that the catalyst could coordinate with oxygen atom of oxazinic ring, making the C-O bond more susceptible to be broken by a small amount of heat absorption.

Once the problem of mass loss during the polymerization process was overcome through the use of the catalyst, the polymerization of the naphthoxazines was performed. The temperature program used in this work (described in Section 2.4) was chosen in the sense to ensure the complete melting of the monomers before the polymerization (to avoid heterogeneity in the polymer) as well as to ensure the curing, once 210°C is higher than onset temperature of all catalyzed compounds (Figure 11).

The thermal stabilities of the poly(naphthoxazines) obtained were studied by TGA under nitrogen atmosphere. The thermograms (TGA and DTGA) of poly(naphthoxazines) are shown in Figure 12, and the obtained data are summarized in Table 3.

Among the polymers, poly(Naft-Fu-FU) has the highest thermal stability, while poly(Naft-Fu-H) and poly(Naft-H-Fu) have very similar $T_{5\%}$ and $T_{10\%}$ values. It was also observed that the poly(Naft-H-Fu) and poly(Naft-Fu-Fu) presented the highest char yield than poly(Naft-Fu-H). This fact can be attributed to the greater number of furan groups present in these compounds which are able to increase the cross-linked density in the polymers since their unsubstituted 5-position is electronically activated for electrophilic substitutions.

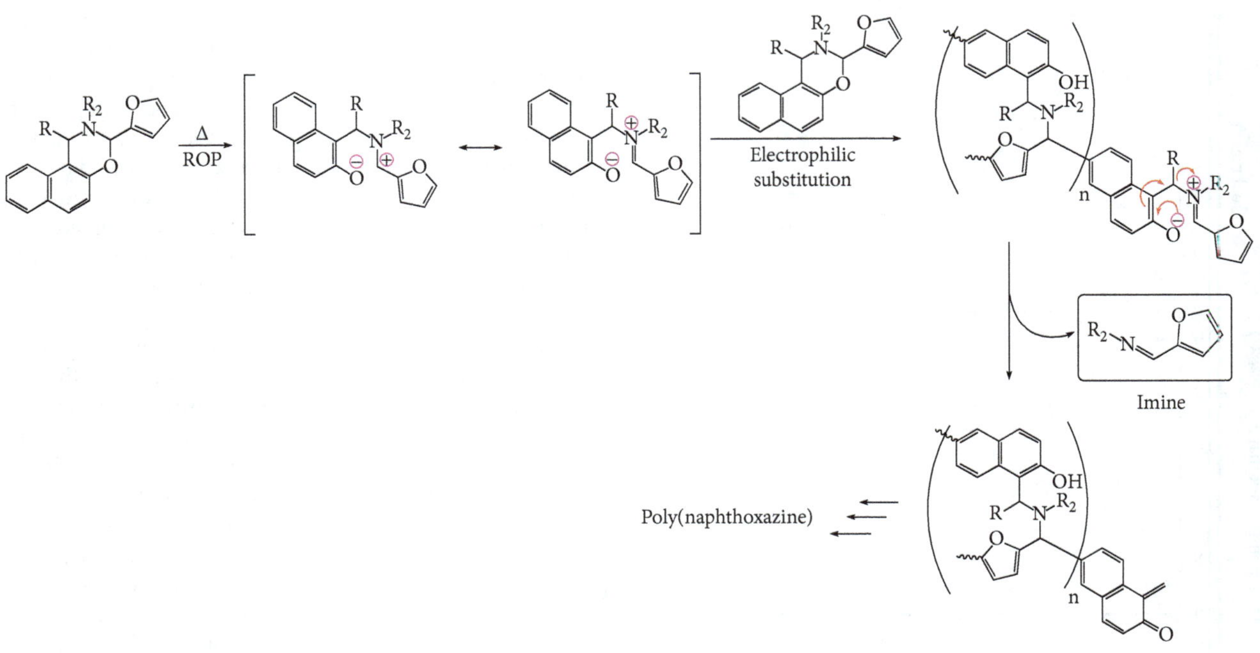


FIGURE 10: Proposed mechanism for the formation of an imine as a by-product of the polymerization process.

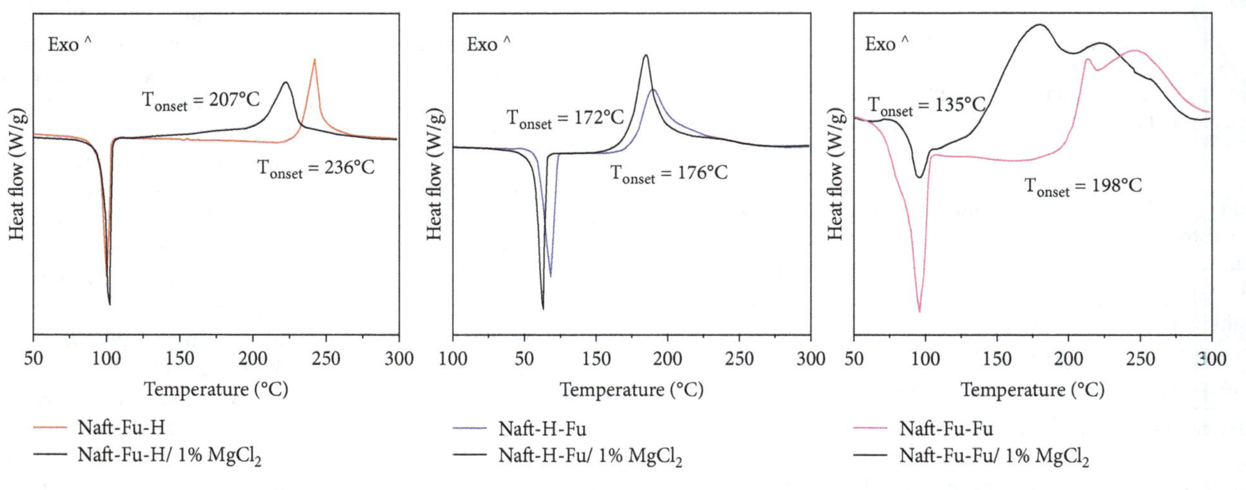


FIGURE 11: DSC curves of naphthoxazines with and without $MgCl_2$ (1 mol%).

The flame resistance of a polymer is a suitable property that can be evaluated by the determination of the limiting oxidation index (LOI), which is the minimum $O_2$ concentration to let the material ignite to cause flammability. Materials with LOI values $\leq 26$ are considered flammable, and an interesting way to estimate LOI values of halogen-free polymers is by using the char yields (CY) of TGA analyses using the van Krevelen equation [36] (1). The LOI determination was recently utilized to determine the flame resistance of polybenzoxazines [37], and considering the elevated values of char yield observed in Table 3, TGA results were also employed to evaluate the flame retardancy of the poly (naphthoxazines) in this work.

$$\mathrm{LOI} = 17.5 + 0.4(\mathrm{CY}). \tag{1}$$

The LOI values reported in Table 3 demonstrate that the poly(naphthoxazines) containing furan groups have outstanding flame resistance, since polymers having LOI values $> 28$ are considered self-extinguishing. Such high flame resistance and self-extinguishing properties are especially desirable in high-performance materials and become more attractive if considered in the use of biobased, halogen-, silane-, and phosphorus-free materials.

The polymerization of naphthoxazines containing furan groups was also confirmed by FTIR analysis (Figure 13). The main sign of polymerization of monomers is the dramatic decrease of characteristic bands around 930–940 $cm^{-1}$, attributed to the out-of-plane deformation mode of the $\alpha,\beta$-disubstituted naphthalene ring, which indicates the opening of the oxazinic ring. It was also noticed the appearance of the broad band at around 3320 $cm^{-1}$ (O-H stretching), which

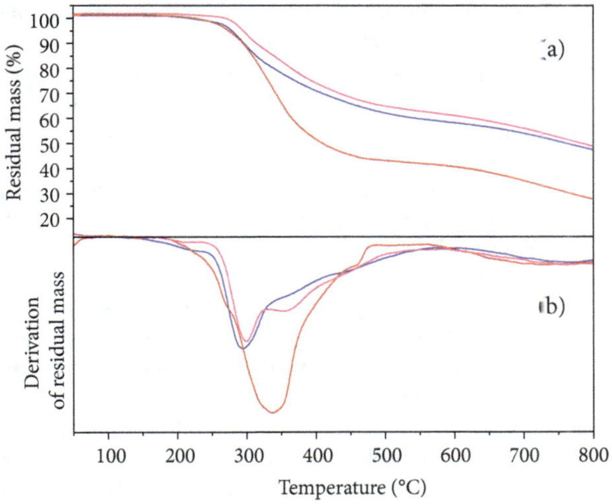


FIGURE 12: Thermograms of poly(naphthoxazines). (a) TGA, (b) DTGA.

TABLE 3: Summary of the values of TGA thermograms of the poly(naphthoxazines).

| Polymer | $T_{5\%}$ (°C) | $T_{10\%}$ (°C) | Char yield (%) | LOI* |
|---|---|---|---|---|
| Poly(Naft-Fu-H) | 273 | 294 | 27 | 28.3 |
| Poly(Naft-H-Fu) | 278 | 296 | 47 | 36.3 |
| Poly(Naft-Fu-Fu) | 295 | 313 | 48 | 36.7 |

*Calculated from the char yield at 800°C of TGA analysis.

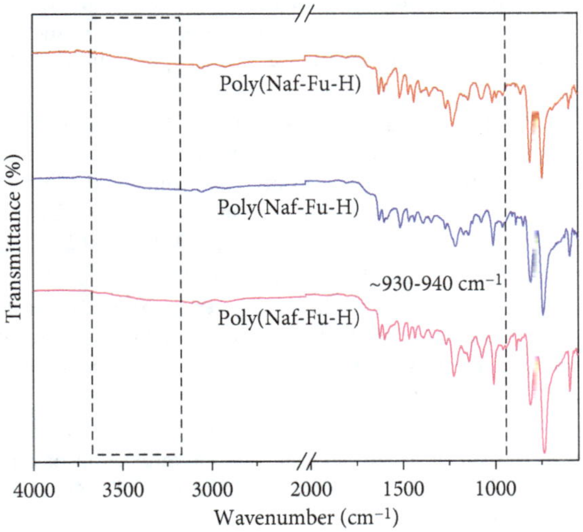


FIGURE 13: FTIR spectra of poly(naphthoxazines)

was completely absent in the monomer spectra, indicating that the polymer structure contains phenolic-type poly (naphthoxazines). The fact that was not observed an overall displacement of the bands associated with the stretching of the C=C bonds of the naphthalene ring in addition with the fact that the bands associated with furan groups becoming broad, consistent with the formation of disubstituted furan rings, indicates that the electrophilic substitutions occur mainly in the furan groups, since the OH groups also appear.

## 4. Conclusions

This work reported the synthesis of novel biobased poly (naphthoxazines) containing furan groups. The syntheses were carried out in a single step, without the use of solvents and catalysts, obtaining the products in satisfactory yields and high purity. The naphthoxazines had their chemical structures completely characterized by FTIR, $^{1}$H NMR, and $^{13}$C NMR techniques. The DSC analysis showed that all naphthoxazines exhibit exothermic typical polymerization events, making these compounds suitable for obtention of poly(naphthoxazine) resins. However, TGA analysis has shown that the compounds suffer significant mass losses at temperatures below the onset polymerization temperature. In this way, a catalyst (1 mol% $MgCl_2$) was used in order to allow the polymerization of the compounds before the mass loss events. The poly(naphthoxazines) obtained were studied by TGA analyses and exhibited high thermal stability, with high flame resistance and self-extinguishing properties (LOI $> 28$), which makes these compounds attractive in the development of high-performance materials.

## Conflicts of Interest

The authors declare that there is no conflict of interest regarding the publication of this paper.

## Acknowledgments

The authors acknowledge the National Council for Scientific and Technological Development (CNPq) (409814/2016-4), CAPES, and FUNCAP for the financial support and CENAUREMN (Centro Nordestino de Aplicação e Uso da Ressonância Magnética Nuclear at Fortaleza, Brazil,) for NMR analyses.

## References

[1] Y. L. Liu and C. I. Chou, "High performance benzoxazine monomers and polymers containing furan groups," *Journal of Polymer Science Part A: Polymer Chemistry*, vol. 43, no. 21, pp. 5267–5282, 2005.

[2] S. B. Shen and H. Ishida, "Synthesis and characterization of polyfunctional naphthoxazines and related polymers," *Journal of Applied Polymer Science*, vol. 61, no. 9, pp. 1595–1605, 1996.

[3] J. R. Oliveira, L. R. V. Kotzebue, F. W. M. Ribeiro et al., "Microwave-assisted solvent-free synthesis of novel benzoxazines: a faster and environmentally friendly route to the development of bio-based thermosetting resins," *Journal of Polymer Science Part A: Polymer Chemistry*, vol. 55, no. 21, pp. 3534–3544, 2017.

[4] H. Ishida and T. Agag, *Handbook of Benzoxazine Resins*, Elsevier, Amsterdam, 2011.

[5] Z. L. Tang, Z. W. Xia, S. H. Chang, and Z. X. Wang, "Synthesis and fungicidal activity of novel 2-aryl-3-(1,3,4-thiadiazolyl)-6(8)-methyl-1,3-benzoxazines," *Bioorganic & Medicinal Chemistry Letters*, vol. 25, no. 16, pp. 3378–3381, 2015.

[6] A. U. G. Gabbas, M. B. Ahmad, N. Zainuddin, and N. A. Ibrahim, "Synthesis and antimicrobial evaluation of new 3,4-dihydro-2H-benzo- and 1,3-oxazine derivatives," *Rasayan Journal of Chemistry*, vol. 9, pp. 3–9, 2016.

[7] Z. Tang, W. Chen, Z. Zhu, and H. Liu, "Synthesis of 2, 3-diaryl-3,4-dihydro-2H-1,3-benzoxazines and their fungicidal activities," *Journal of Heterocyclic Chemistry*, vol. 48, no. 2, pp. 255–260, 2011.

[8] Z. Tang, Z. Zhu, Z. Xia et al., "Synthesis and fungicidal activity of novel 2,3-disubstituted-1,3-benzoxazines," *Molecules*, vol. 17, no. 7, pp. 8174–8185, 2012.

[9] B. P. Mathew, A. Kumar, S. Sharma, P. K. Shukla, and M. Nath, "An eco-friendly synthesis and antimicrobial activities of dihydro-2H-benzo- and naphtho-1,3-oxazine derivatives," *European Journal of Medicinal Chemistry*, vol. 45, no. 4, pp. 1502–1507, 2010.

[10] M. J. Millan, M. Brocco, M. Papp et al., "S32504, a novel naphtoxazine agonist at dopamine D3/D2 receptors: III. Actions in models of potential antidepressive and anxiolytic activity in comparison with ropinirole," *The Journal of Pharmacology and Experimental Therapeutics*, vol. 309, no. 3, pp. 936–950, 2004.

[11] C. R. Arza, P. Froimowicz, L. Han, R. Graf, and H. Ishida, "Design, synthesis, characterization, and polymerization of fused-ring naphthoxazine resins," *Macromolecules*, vol. 50, no. 23, pp. 9249–9256, 2017.

[12] T. Agag, "Preparation and properties of some thermosets derived from allyl-functional naphthoxazines," *Journal of Applied Polymer Science*, vol. 100, no. 5, pp. 3769–3777, 2006.

[13] X. Ning and H. Ishida, "Phenolic materials via ring-opening polymerization: synthesis and characterization of bisphenol-A based benzoxazines and their polymers," *Journal of Polymer Science Part A: Polymer Chemistry*, vol. 32, no. 6, pp. 1121–1129, 1994.

[14] S. B. Shen and H. Ishida, "Dynamic mechanical and thermal characterization of high-performance polybenzoxazines," *Journal of Polymer Science Part A: Polymer Chemistry*, vol. 37, no. 23, pp. 3257–3268, 1999.

[15] T. Agag and T. Takeichi, "Synthesis and characterization of novel benzoxazine monomers containing allyl groups and their high performance thermosets," *Macromolecules*, vol. 36, no. 16, pp. 6010–6017, 2003.

[16] J. Wang and Q. Mao, "New thermosetting resin from benzoxazine and cyanate ester resin," *Advances in Polymer Technology*, vol. 32, pp. 474–485, 2012.

[17] T. Agag and T. Takeichi, "Novel benzoxazine monomers containing p-phenyl propargyl ether: polymerization of monomers and properties of polybenzoxazines," *Macromolecules*, vol. 34, no. 21, pp. 7257–7263, 2001.

[18] H. D. Kim and H. Ishida, "A study on hydrogen-bonded network structure of polybenzoxazines," *The Journal of Physical Chemistry. A*, vol. 106, no. 14, pp. 3271–3280, 2002.

[19] P. Froimowicz, K. Zhang, and H. Ishida, "Intramolecular hydrogen bonding in benzoxazines: when structural design becomes functional," *Chemistry - A European Journal*, vol. 22, no. 8, pp. 2691–2707, 2016.

[20] J. Liu, X. Lu, Z. Xin, and C. L. Zhou, "Surface properties and hydrogen bonds of mono-functional polybenzoxazines with different N-substituents," *Chinese Journal of Polymer Science*, vol. 34, no. 8, pp. 919–932, 2016.

[21] H. C. Erythropel, J. B. Zimmerman, T. M. de Winter et al., "The Green ChemisTREE: 20 years after taking root with the 12 principles," *Green Chemistry*, vol. 20, no. 9, pp. 1929–1961, 2018.

[22] L. Dumas, L. Bonnaud, M. Olivier, M. Poorteman, and P. Dubois, "High performance bio-based benzoxazine networks from resorcinol and hydroquinone," *European Polymer Journal*, vol. 75, pp. 486–494, 2016.

[23] L. R. V. Kotzebue, F. W. M. Ribeiro, V. G. Sombra et al., "Spectral and thermal studies on the synthesis and catalyzed oligomerization of novel cardanol-based benzoxazines," *Polymer*, vol. 92, pp. 189–200, 2016.

[24] X. Shen, J. Dai, Y. Liu, X. Liu, and J. Zhu, "Synthesis of high performance polybenzoxazine networks from bio-based furfurylamine: furan vs benzene ring," *Polymer*, vol. 122, pp. 258–269, 2017.

[25] S. A. Ozturkcan, K. Turhan, and Z. Turgut, "Synthesis and characterizations of new 1,3-oxazine derivatives," *Journal of The Chemical Society of Pakistan*, vol. 33, pp. 939–944, 2011.

[26] T. Uyar, J. Hacaloglu, and H. Ishida, "Synthesis, characterization, and thermal properties of alkyl-functional naphthoxazines," *Journal of Applied Polymer Science*, vol. 127, no. 4, pp. 3114–3123, 2013.

[27] R. Borah, A. K. Dutta, P. Sarma, C. Dutta, and B. Sarma, "Synthesis of anti-2,3-dihydro-1,2,3-trisubstituted-1H-naphth[1,2-e][1,3]oxazine derivatives via multicomponent approach," *RSC Advances*, vol. 4, no. 21, article 10912, 2014.

[28] S. Ohashi, V. Pandey, C. R. Arza, P. Froimowicz, and H. Ishida, "Simple and low energy consuming synthesis of cyanate ester functional naphthoxazines and their properties," *Polymer Chemistry*, vol. 7, no. 12, pp. 2245–2252, 2016.

[29] N. Sundaramoorthy Karthikeyan, G. Ramachandran, R. S. Rathore, and K. Iyer Sathiyanarayanan, "Selective synthesis and structural confirmation of 1,3-oxazines and Schiff bases with controllable substitution patterns," *Asian Journal of Organic Chemistry*, vol. 1, no. 2, pp. 173–179, 2012.

[30] S. Ohashi, F. Cassidy, S. Huang, K. Chiou, and H. Ishida, "Synthesis and ring-opening polymerization of 2-substituted 1,3-benzoxazine: the first observation of the polymerization of oxazine ring-substituted benzoxazines," *Polymer Chemistry*, vol. 7, no. 46, pp. 7177–7184, 2016.

[31] X. Li, X. Luo, M. Liu, Q. Ran, and Y. Gu, "The catalytic mechanism of benzoxazine to the polymerization of cyanate ester," *Materials Chemistry and Physics*, vol. 148, no. 1-2, pp. 328–334, 2014.

[32] R. Andreu, J. A. Reina, and J. C. Ronda, "Studies on the thermal polymerization of substituted benzoxazine monomers: electronic effects," *Journal of Polymer Science Part A: Polymer Chemistry*, vol. 46, no. 10, pp. 3353–3366, 2008.

[33] E. Calò, A. Maffezzoli, G. Mele et al., "Synthesis of a novel cardanol-based benzoxazine monomer and environmentally sustainable production of polymers and bio-composites," *Green Chemistry*, vol. 9, no. 7, pp. 754–759, 2007.

[34] C. Liu, D. Shen, R. M. Sebastián, J. Marquet, and R. Schönfeld, "Mechanistic studies on ring-opening polymerization of

benzoxazines: a mechanistically based catalyst design," *Macromolecules*, vol. 44, no. 12, pp. 4616–4622, 2011.

[35] Y. X. Wang and H. Ishida, "Cationic ring-opening polymerization of benzoxazines," *Polymer*, vol. 40, no. 16, pp. 4563–4570, 1999.

[36] D. W. van Krevelen, "Some basic aspects of flame resistance of polymeric materials," *Polymer*, vol. 16, no. 8, pp. 615–620, 1975.

[37] L. R. V. Kotzebue, J. R. De Oliveira, J. B. Da Silva, S. E. Mazzetto, H. Ishida, and D. Lomonaco, "Development of fully biobased high-performance bis-benzoxazine under environmentally friendly conditions," *ACS Sustainable Chemistry & Engineering*, vol. 6, no. 4, pp. 5485–5494, 2018.

# 14

# Preparation and Characterization of Rubber Blends for Industrial Tire Tread Fabrication

B. Mensah,[1,2] B. Agyei-Tuffour,[2] E. Nyankson,[2] Y. D. Bensah,[2] D. Dodoo-Arhin,[2] J. K. Bediako,[3] B. Onwona-Agyeman,[2] and A. Yaya[2]

[1]*932, Yun-Tech, ITECO 762, Deokpung-dong, Hanam-si, Gyeonggi-do, Republic of Korea*
[2]*Department of Materials Science and Engineering, CBAS, University of Ghana, Legon, Ghana*
[3]*Division of Semiconductor and Chemical Engineering, Chonbuk National University, Jeonju, Jeonbuk 561-756, Republic of Korea*

Correspondence should be addressed to A. Yaya; ayaya@ug.edu.gh

Academic Editor: Domenico Acierno

The physico-mechanical properties of variable rubber blends including epoxide natural rubber (ENR), polybutadiene rubber (BR), and solution polymerized styrene-butadiene rubber (SBR) filled with silanized silica and carbon black mixtures were explored. The tensile, hardness, resilience, abrasion, and fatigue behavior were investigated. An optimized composition involving 30 phr of ENR and 70 phr SBR filled with mixtures of carbon blacks and silanized silica was proposed to be a suitable composition for the future development of green passenger truck tires, with low rolling resistance (fuel saving ability), high wear resistance, and desired fatigue failure properties.

## 1. Introduction

Environmental threats caused by excessive fuel consumption have been a major problem in the automobile industry. It is well established that an automobile consumes almost 30% of the energy for a drive on a highway [1, 2]. Thus, the automobile industry is obliged to reduce its portion accordingly through multifaceted approaches including the fabrication of new tires with less rolling resistances. Aerodynamics, nature of roads, the type of vehicle, and their engine capacity as well as other factors may contribute to fuel loses in a typical vehicle; however, rolling resistance (RR), which is associated with the material design, is the main cause of fuel loss in motor vehicles (a condition related to high $CO_2$ emission) [2, 3]. The RR is a property that is due to the viscoelastic behavior of the elastomer(s) used in making the tire/tread [3]. As illustrated in Figure 1, the rotation of the tire in the direction d, under the weight of the vehicle W, creates repeated cycles of deformation (region sitting on the road, i.e., y) and recovery (i.e., region x and z region) resulting in energy dissipation called hysteresis loss. This loss can be overcome by external works which give rise to RR [3, 4]. Also, the interaction of the tire with the road is a major contributing factor for road accidents. Thus, the interaction should provide adequate force (traction) for acceleration and braking to prevent slipping [3, 4]. Hence, a green tire/tread with fuel-saving capacity (lower RR, excellent traction (T), and abrasion resistance (AR)) is of great priority in tire engineering [4–7].

The type of matrix and filler used, filler dispersions, and their interactions (Filler-polymer and filler-filler interactions) are the main controlling factors to attain a composition intended for green tire manufacturing [3, 8, 9]. Over the past decades, high strength and resilient natural rubber (NR) filled with reinforcement such as carbon blacks (CB) and/or silica vulcanizates was widely explored for vehicle tire fabrication, but the one-matrix (NR-CB or NR-silica) composites have shown unsatisfactory results, as they failed to meet the green tire requirements owing to the fact that NR itself is susceptible to ozone fracturing with undesirable wear resistance (WR), RR, and TR properties [8, 10, 11]. Recently, engineered functional elastomers and active reinforcements

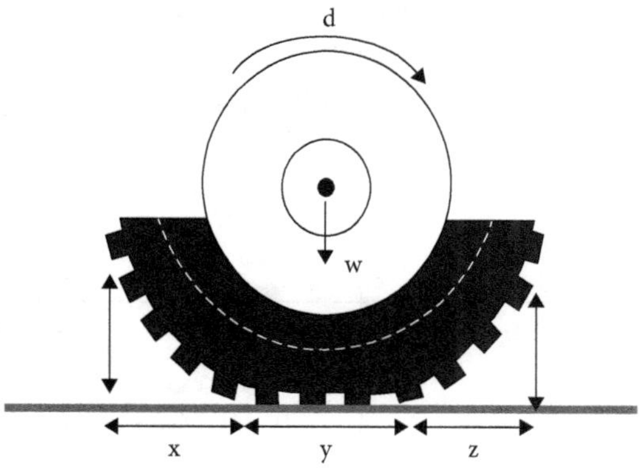


FIGURE 1: Illustration of the mechanism leading to hysteresis loss.

have been extensively investigated by various researchers aimed to attain efficient composition for the development of green tire/tread [7, 8, 10, 11]. In these attempts, blends of NR with at least one or more synthetic elastomers selected from the category of cis 1,4-polybutadiene (BR), NR/cis 1,4-polyisoprene rubber (IR), styrene-butadiene-based rubber (SBR), or their derivatives have been reinforced with mono or hybrid fillers such as CB and/or silanized silica as already demonstrated in previous studies [5, 7, 9]. The emergence of epoxidized natural rubber (ENR), which is reported to exhibit a higher polarity than NR, owing to the epoxide groups in its structure, has opened a new research area that seeks to examine its compatibility with silica for advanced applications including green tire fabrication [12, 13]. Sengloyluan et al. [14] have recently observed that increasing the mol% content of epoxide groups of ENR increases the tensile properties of ENR-silica-filled compounds than their counterpart NR silica-based compounds, although the curing times were prolonged. Other research [15] have also confirmed that ENR (epoxide group: 25% mol) had shown a glass transition temperature most suitable for tread application. They observed that ENR-silica tread compounds showed lower RR, hence reduced fuel consumption, with enhanced wet and ice traction compared with benchmark premium passenger or truck treads. Furthermore, Sarkawi et al. [12] had examined ENR-silica for the application of tire tread compound, in which ENR-silica was found to exhibit lower RR and better wet traction compared to NR-carbon black vulcanizates. However, there is currently little research work on ENR used as a component for green tire manufacturing. Nevertheless, a tire/tread generally constitutes a structure of various components with diverse properties and chemistry; hence, choosing each of these substances in their various amounts, the processing conditions and characterizations techniques adopted to arrive at conditions that satisfies the magic triangle of a green tire/tread (excellent WR and TR and low RR), has been a very challenging task. For this reason, further exploration in an attempt to obtain a composition for green tire/tread is inevitable. Presently, a detailed study on the blends of ENR, BR, and SSBR, with mixtures of carbon blacks and silanized silica, was investigated. The effect of maleic-anhydride (compatibilizer) on these blends was also examined. Wear resistance, flex fatigue life, hysteresis, tensile properties, and crosslinking densities were conducted to better understand the physico-mechanical properties of the vulcanizates. The current results present an optimized composition for green tire/tread preparation.

## 2. Experimental

*2.1. Chemicals and Compound Formulation.* In this study, ENR (epoxide group: 25–40% mol) was supplied by Doo Man Co. Ltd., South Korea; SSBR (SSBR-3323) was obtained from LG Chem, South Korea; and BR (BR-1208), carbon black N550 (with the specific surface of 95–105 $m^2/g$), and CORAX N330 (iodine adsorption 8.8 g/kg) were supplied by Infochems, South Korea. The silica (Z175MP, with the specific surface of 134.63 $m^2/g$ and a mean particle size of 440 nm) was supplied by IDONG TECH.; silane coupling agent (Kettlitz-Silanogran Si69/GR) was supplied by Kangshin Industrial Co. Ltd.; aromatic oil (TDAE, dark brown, KUKDONG OIL & CHEMICAL CO. Ltd.) and maleic anhydride (MA) crystals (purity 99%, YONGSAN CHEMICALS INC) were supplied by IDONG TECH. The antiozonant (6PPD) and antioxidant (TMQ) and the rest of the curatives: zinc oxide (ZnO), stearic acid (SA), sulfur (S), tetramethylthiuram disulfide (TMTD), and N-cyclohexyl 2-benzothiazole sulfonamide (CZ) were all obtained from Infochems Company Ltd. The compound formulation expressed as parts per hundred parts of rubber (phr) with their respective codes is listed in Table 1.

*2.2. Rubber Compounding Process.* The rubber compounding was done in an internal mixer (model: QPBV-300, QMESYST, South Korea) at 90°C and 30 rpm. First, the elastomers were masticated in the kneader for 1 min, except for sulfur; the other processing ingredients were simultaneously added and mixed for about 2 min. Later, the reinforcements were incorporated and mixed for an additional of 1 min. Finally, the sulfur was added and then mixed for less than 30 seconds. The composition was passed over a two-roll mill (QM300, QMESYSTEM) repeatedly for about 9 min and then sheeted out. A rectangular sheet of samples (15 cm × 15 cm × 2 mm) was molded by electrical hot press machine (model: TO-200, TESTONE, South Korea) at a pressure of 25 tons with the help of the optimum curing conditions obtained from the cure rheometer (MDR, model: PDR2030, TESTONE Ltd., South Korea) at 160°C. The samples were cut into standard shapes and subjected to characterization analysis.

## 3. Characterization

*3.1. Tensile Strength and Cyclic Loading.* The tensile strength (TS) measurement for uncured and cured specimens was carried out according to ASTM D412 standard by using QM100s machine (QMESYSTEM, South Korea) at a crosshead speed of 500 mm/min and 25°C temperatures. For the uncured compositions, the maximum tensile stress was taken as the green strength from the generated stress-strain curve, and at least three samples were tested for each composition and averaged.

Table 1: Compound formulation and their codes measured in (phr)[a].

| Chemicals | TP1 | TP2 | TP3 | TP4 | TP5 | TP6 | TP7 | TP8 |
|---|---|---|---|---|---|---|---|---|
| ENR | 15 | 70 | 15 | 30 | — | — | 30 | 30 |
| SBR | 15 | 15 | 70 | — | 30 | 30 | — | — |
| BR | 70 | 15 | 15 | 70 | 70 | 70 | 70 | 70 |
| ZnO | 5 | 5 | 5 | 5 | 5 | 5 | 5 | 5 |
| TMTD | 1 | 1 | 1 | 1 | 1 | 1 | 1 | 1 |
| CBS(CZ) | 0.5 | 0.5 | 0.5 | 0.5 | 0.5 | 0.5 | 0.5 | 0.5 |
| S/A | 1.4 | 1.4 | 1.4 | 1.4 | 1.4 | 1.4 | 1.4 | 1.4 |
| Sulfur | 2.1 | 2.1 | 2.1 | 2.1 | 2.1 | 2.5 | 2.5 | 2.5 |
| MA | 3 | 3 | 3 | 3 | 3 | — | 3 | 3 |
| N550 | 25 | 25 | 25 | 25 | 25 | 25 | 25 | 25 |
| N330 | 10 | 10 | 10 | 10 | 10 | 10 | — | 10 |
| TDAE | 10 | 10 | 10 | 10 | 10 | 10 | 10 | 10 |
| Silica | 10 | 10 | 10 | 10 | 10 | 10 | 10 | — |
| TESPT | 3 | 3 | 3 | 3 | 3 | 3 | 3 | — |
| TMQ | 1.5 | 1.5 | 1.5 | 1.5 | 1.5 | 1.5 | 1.5 | 1.5 |
| 6PPD | 1 | 1 | 1 | 1 | 1 | 1 | 1 | 1 |
| Retarder | 0.5 | 0.5 | 0.5 | 0.5 | 0.5 | 0.5 | 0.5 | 0.5 |
| Wax | 1 | 1 | 1 | 1 | 1 | 1 | 1 | 1 |

[a]Parts per hundred parts of rubber.

*3.2. Peel Test (Tack Strength).* The TS machine was used to carry out the peel test on the green samples. Sample preparation includes pressing the samples in between two mylar film for about 5 min. Prior to the test, the mylar film covering the faces of two rectangular samples strips (23 m × 65 mm × 2 mm) was removed and gently brought together. A pressure system of about 0.13 MPa was gently and uniformly applied on each side of the bonded strips at an equal rate. At least each press lasted for a contact time of about 10 seconds. The slight pressure given was to ensure interdiffusion and the removal of entrapped air bubbles at the interface of the bonded strips. Immediately after the pressing, the sample is subjected to the peel test. The peeling test was based on 180° orientation and at a speed of 50 mm/min. The tack strengths $G_a$ (N/m) of the samples were then estimated. The testing machine generates plots of the force (Newtons) required to separate the bonded strips against the distance of separation (mm). The average forces required for separating bonded strips were recorded and together with the samples width, the $G_a$ (N/m) was calculated using

$$G_a = \frac{2F}{d}, \tag{1}$$

where $F$ is the estimated average peel force and $d$ is the width (m) of the strips of the samples. For each composition, 3 samples were tested and averaged.

*3.3. Crosslinking Density.* Swelling experiments of cured composites were performed by equilibrating them in toluene at room temperature for 48 h. The swelling degree $Q_r$ was calculated using the relation

Table 2: Solubility parameters and calculated interaction parameters of solvent and polymers.

| Samples/parameters | $\delta$ (cal/cc)$^{1/2}$ | Estimated $\chi_1$ | Reference |
|---|---|---|---|
| *Toluene | 8.91 | — | [18] |
| *BR | 8.38 | 0.352 | [19] |
| *SSBR | 8.04 | 0.372 | [20] |
| *ENR | 8.10 | 0.368 | [21] |

*Data for solubility parameters ($\delta$) were taken from the corresponding literatures.

$$Q_r = \frac{W_s - W_i}{W_d}, \tag{2}$$

where $W_i$ is the weight of the rubber sample before immersion into the solvent; $W_s$ and $W_d$ are the weights of the sample in the swollen state and after drying in an oven at 80°C, respectively. The elastically active network chain density commonly referred to as the crosslink density $N$ was calculated from the Flory-Rehner equation given by equation (3) [16]

$$-\left[\text{In}(1 - v_r) + v_r + \chi_1 v_r^2\right] = v_s \left[\frac{v_r^{1/3} - v_r}{2}\right] N, \tag{3}$$

where $v_r(1/Qr)$ is the volume fraction of polymer in the swollen gel at equilibrium, $v_s$ is the molar volume of the solvent (106.3 mL/mol for toluene), and $\chi_1$ is the polymer-solvent parameter determined from Bristow-Watson equation (4) [17], where

$$\chi_1 = \beta_1 + \left(\frac{v_s}{RT}\right)\left[\delta_s - \delta_p\right]^2, \tag{4}$$

where $\beta$ is the lattice constant, usually taken as 0.34, $v_s$ is the molar volume of solvent, $R$ is the universal gas constant, $T$ is the absolute temperature, and $\delta$ is the solubility parameter, and the subscripts s and p refer to the swelling agent and polymer, respectively. The solubility parameters ($\delta$) with respective estimated interaction parameters ($\chi_1$) of BR, SBR, ENR, and solvent (toluene) are as shown in Table 2.

*3.4. Abrasion Resistance.* The abrasiveness of the various composites was examined by using the DIN abrasion instrument obtained from QMESYS Co. Ltd. A cylindrical-like rubber test specimen (~6 mm thickness) was made and abraded against an abrasive surface mounted on a rotating cylindrical drum. The amount of rubber reduced in the specimen due to abrasion is measured. At least, three samples were tested and averaged for one composition.

*3.5. Fatigue Behavior.* The flex-fatigue life of various compounds was measured using a De Mattia flexing machine (QM 650D) according to ASTM D430 [22]. The specimen shapes and dimensions are shown in Figure 2). The specimen thickness was measured close to the groove. The conditions used include conditioning the test specimens for at least 12 h at the test temperature of 23 ± 1°C and doing the test at a 300 ± 10 cycles/min. For each composite, three pieces were

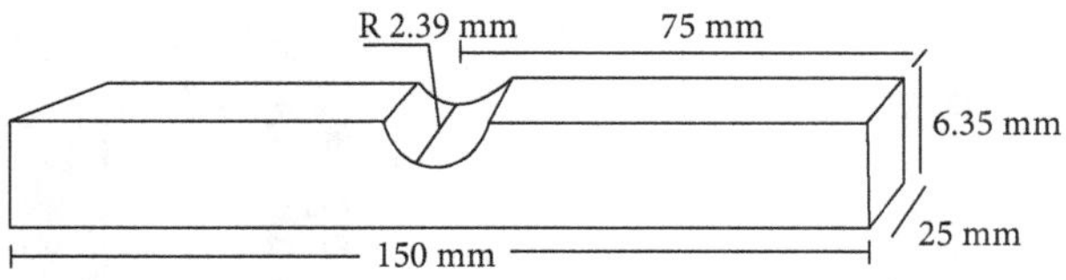


FIGURE 2: De Mattia flexing specimen with a semicircular groove that was used to measure the flex-fatigue life of the various samples [22].

tested, and the average flex-fatigue-life value was reported. The flex-fatigue life was defined by the cycles at which a visible crack (a grade-one crack) was observed.

*3.6. Resilience.* The resilience of the various compounds was tested by using DIN resilience tester (DIN 53512 and ISO 4662) obtained from QMESYS Company Ltd. The hammer of the instrument was allowed to strike the samples (~6 mm thickness) for about 5 times to eliminate internal defects; the values of the remaining hits were recorded. Three samples were tested for each composition; and their results were averaged.

*3.7. Hardness Test (Shore A).* A constant loader hardness tester (Shore A) with a load of 5 kg was used to test the hardness of the vulcanizates. At least 3 samples of about ~5 mm thickness were tested and averaged for one composition.

## 4. Results and Discussions

*4.1. Green and Tack Strength.* In the tire manufacturing technology, the various uncured components of the tire are brought together and gently assembled before vulcanization begins in the curing chamber; hence, self-adhesion (similar interface), adhesive tack (dissimilar interface), and green strength of the rubber compounds intended for industrial tire tread development are the crucial part of the tire/tread design. Here, we present the green strength (GS) and self-adhesion properties of the various tread-based compounds as shown in Figures 3(a) and 3(b). The samples labeled TP2 and TP4 showed inferior GS probably due to higher stiffness as a result of strong interactions (rubber-filler or filler-filler), especially with blends rich in ENR. The compounds with relatively higher stiffness such as sample TP2 have indicated inferior tack strength (see Figure 3(b)), which is related to the inability of the rubber molecules to flow or interdiffuse across the interface. Although sample TP7 contains 30 phr ENR, with only one carbon black (CB), N550 and silanized silica (Si) showed higher green strength compared to TP2 (70 phr ENR) with triple-filler mixtures and others. Also, the sample without silica (TP8) shows improved GS than the one with silanized silica inclusion (TP4). These observations could mean that the combination of a two-filler system (N550 and silanized silica) offers effective reinforcing action in low ENR compounds than the three-phase filler systems, probably due to extreme inhomogeneity. Also, the GS has improved for TP5 (with MA) than TP6 (without MA), indicating that MA induced a grafting process during the melt blending stage. When the GS and tack strength of all the compounds are compared, it can be observed that compositions enriched with high content of the low glass transition temperature ($T_g$) elastomers (BR and SBR) shows somewhat better GS and tack strength than those enrich in ENR. This is because the molecular movement of the low $T_g$ molecules is higher, although the filler inclusion restricts the mobility of their chains. However, the average depth of penetration of the interdiffused chains was still adequate to form entanglement on either side of the interface (Figure 3(c)) and thus offered greater resistance to peeling compared to compounds enriched with ENR which appears to have obtained higher reinforcement effect. Thus, the current compounds have exhibited higher GS than isobutylene-isoprene gum-montmorillonite loaded BIIR-MMT (0~16 phr) composites [23], probably due to high reinforcements. Additionally, BIIR-MMT compounds also show higher GS and tack strength than the current compounds, owing to lower $T_g$ and lower elasticity characteristics nature of BIIR gum [23, 24].

*4.2. Rheometry Properties.* The curing properties which include onset and optimum curing time ($t_{s2}$ and $T_{90}$), the minimum and maximum torque ($M_L$ and $M_H$), the differences in torque ($\Delta M = M_H - M_L$), and the cure rate index (CRI) ($100/(T_{90}-t_{s2})$), of the various compounds deduced from the rheo-curves at 160°C are plotted in Figures 4(a)–4(d). When the blend gets richer in SBR, there is an increase in the $T_{90}$ due to possible diffusion of curatives from SBR interface to ENR and/or BR interface as illustrated in Figures 5(a) and 5(b), suggesting higher crosslinking density accumulation at these phases [25]. Also, the presence of the MA was found to cause further delaying in $T_{90}$. For example, when MA was excluded from the blend (i.e., TP6), the $T_{90}$ dropped drastically with higher cure rate compared to its counterpart (TP5 with MA) and the rest of the composition containing MA, as seen in Figures 4(a) and 4(b). Although the MA is supposed to play a role of compatibilizer and grafting within the blends (example: ENR-MA-g-ENR, ENR-g-MA-BR, ENR-g-SBR, and BR-g-MA-SBR), the anhydrides groups on MA have acted as scavengers of free radicals and consequently retarded vulcanization of the compounds.

It is interesting to also observe that removal of the silanized silica from the composition (i.e., TP8) slightly minimized the $T_{90}$, suggesting that silanized silica took part in the curing reaction.

A similar trend was observed when NR was filled with silanized silica; the $t_{s2}$ and $T_{90}$ has shortened [26]. The $t_{s2}$ of TP2, TP5, and TP7 has slightly dropped compared to the other blends. The TP2 and TP7, respectively, contain 70 phr and 30 phr of ENR in the blends with epoxide groups that might have participated in the initial vulcanization reaction of the blends. Nelson and Kutty [27], after grafting MA to a whole tire, reclaim that both $t_{s2}$ and $T_{90}$ is delayed. The TP2 ENR (70 phr) reveal the highest $M_L$(viscosity index), $M_H$ (crosslinking density index), and consequently $\Delta M$ (strength index) compared to those with a lower content of ENR as well as other blends. This observation has confirmed the possible diffusion of curatives from SBR phase to the ENR phase with higher crosslinking density which is also observed by Goyanes et al. [25].

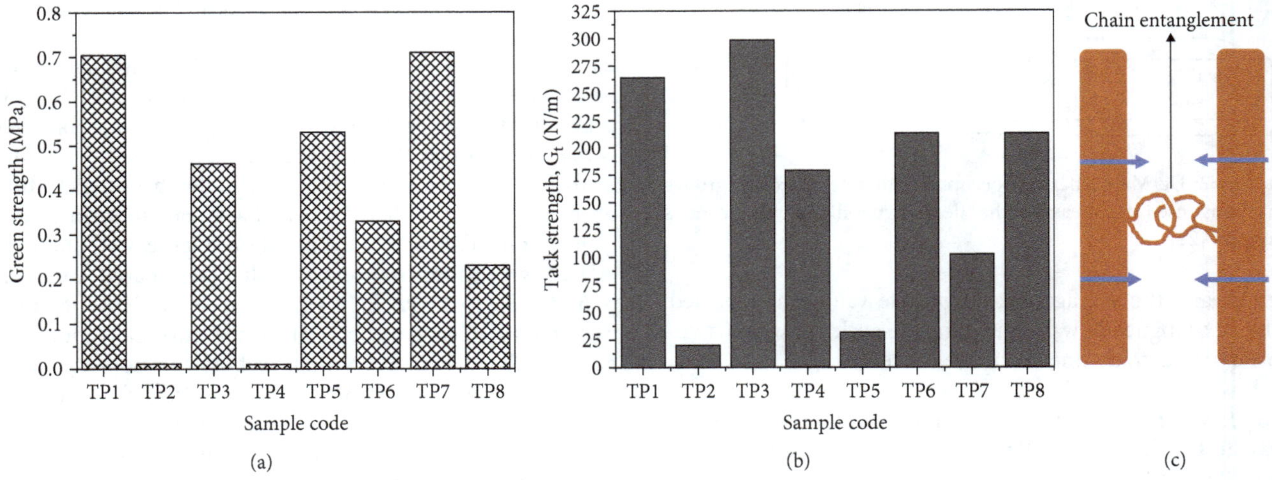


Figure 3: Green strength and autohesion behavior of uncured composites.

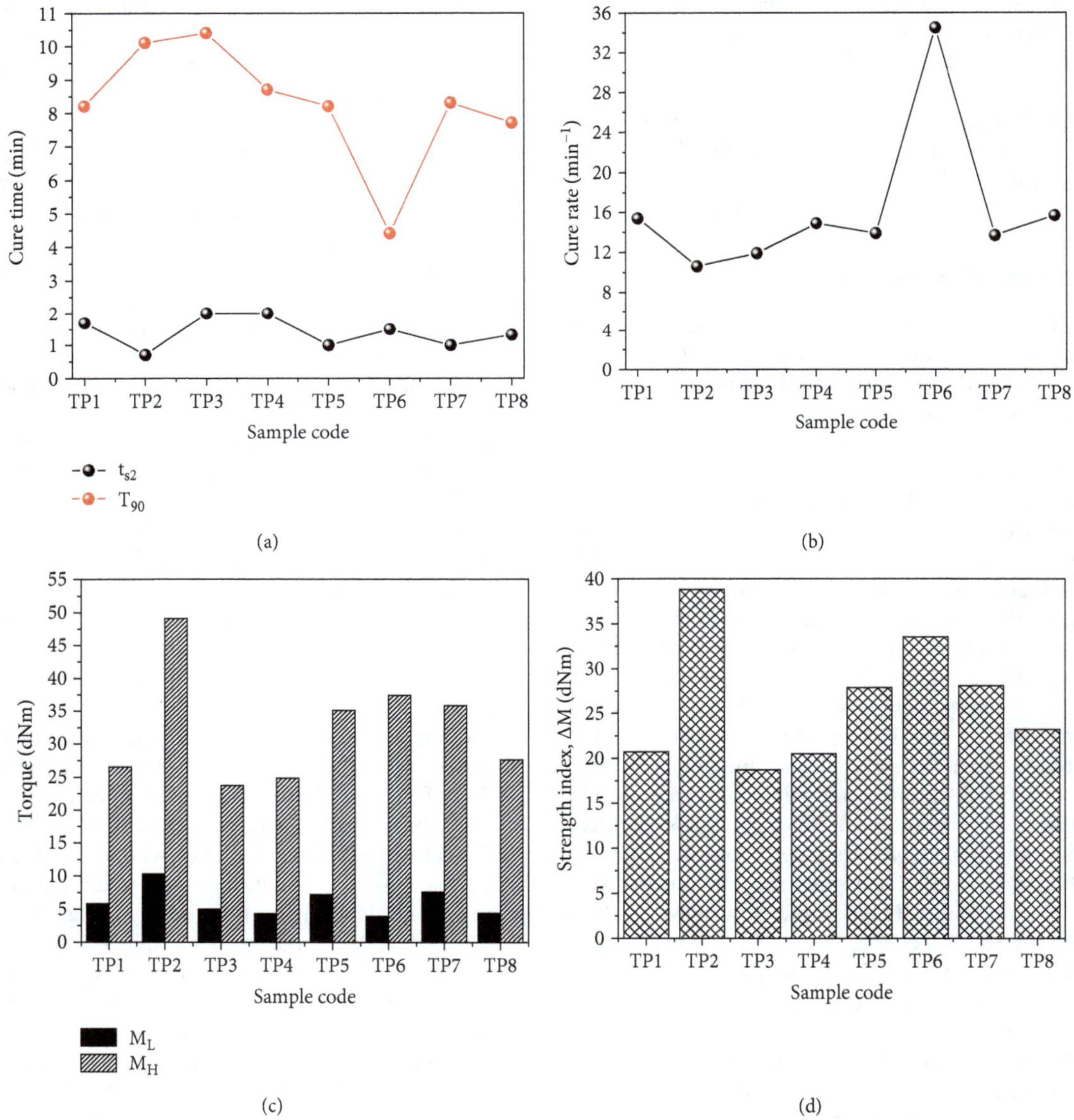


Figure 4: Rheometry properties of the various compounds cured at 160°C.

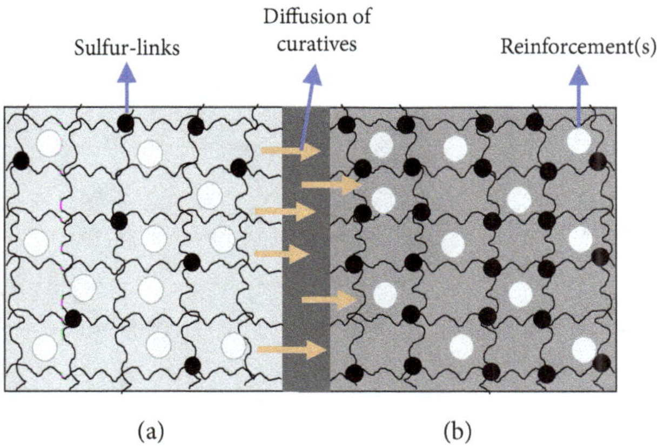


FIGURE 5: Illustration of interfacial diffusion of curatives from ENR-SBR or ENR-BR interface to ENR phase; (a) phase of SBR and (b) phase of ENR.

It is also interesting to know that the TP7 (only one CB) and TP8 (without silica) have improved torques ($M_H$ and $\Delta M$) than their counterpart blend TP4 (30 phr ENR, 70 phr BR with mixtures of CB and silanized silica). This can mean that a combination of the three fillers has not favored TP4 blend formation.

In general, the curing properties of the present tread-based compounds have exhibited better performances than those reported earlier for NR (75 phr)/BR (25 phr)-CB (50 phr)/silica (10 phr), NR (75 phr)/BR (25 phr)-CB (60 phr) [5] and NR-CB (53 phr), and NR (80 phr)/SBR (20 phr)-CB (57 phr) [6]. Also, the proposed passenger tire tread compounds by Atashi and Shiva [7] show severe delays in curing ($t_{s2}$ and $T_{90}$) compared to our current development. Again, when ENR was used as a compatibilizer in NR at varying amounts (5–20 phr), lengthening of curing times; $t_{s2}$: 4~11 min and $T_{90}$: 9~19 mins were recorded compared to our current work. It was assumed that the silica (Si) adsorbed the curing accelerator onto its surface as silanol which underwent hydrogen bonding with silanol groups [14]. However, it should be emphasized here that properties like $M_H$ and $\Delta M$ may not be subject to only chemical networks creation but also physical networks are possible; hence, further tests are presented to further examine physico-mechanical properties of the various vulcanizates.

*4.3. Crosslinking Density.* The degree of swelling of polymer composites in the desired solvent has been a useful technique to check the matrix-filler interaction of composites. For composites with one matrix system, only the calculated interaction parameter ($\chi_1$) of that matrix is used in the estimation of its crosslinking density ($N$) [28, 29]. Researchers have adopted the use of the mixtures rule [30] for blend systems or a method of averaging $\chi_1$ for each the polymer involved to compute the $N$ of composites with many matrix phases [18, 20, 31]. In this present study, we also adopted the latter method of averaging $\chi_1$ for the blend systems, for the computation of their respective $N$ values. The results of equilibrium swelling ratio ($Q_r$), the calculated solubility parameters ($\delta$), the interaction parameters as well as their respective crosslinking densities ($N$) for the blends are as presented in Table 3.

TABLE 3: Swelling properties and crosslinking densities of the various compounds.

| Samples/parameters | $Q_r$ | $\delta$ (cal/cc)$^{1/2}$ | $\chi_1$ | $N$ (mol/cm$^3$) |
|---|---|---|---|---|
| TP1 | 1.736 | 8.173 | 0.363 | 0.00412 |
| TP2 | 1.059 | 8.173 | 0.363 | 0.02997 |
| TP3 | 1.590 | 8.173 | 0.363 | 0.00378 |
| TP4 | 2.290 | 8.070 | 0.370 | 0.00116 |
| TP5 | 1.418 | 8.210 | 0.361 | 0.00588 |
| TP6 | 1.569 | 8.210 | 0.361 | 0.00399 |
| TP7 | 1.609 | 8.240 | 0.359 | 0.00366 |
| TP8 | 1.792 | 8.240 | 0.359 | 0.00253 |

The TP2, a triple matrix composites of SBR: 15 phr, BR: 15 phr, and ENR: 70 phr, has shown the least $Q_r$ with corresponding thrilling $N$ value, indicating a stronger interaction (filler-polymer and/or filler-filler) in comparison with the remaining samples. During the vulcanization process, the sulfur could form bridges between the organic tail of silanized silica and the backbone of ENR (see Figure 6i), whilst the carbon blacks and the ENR can also provide additional hydrogen bonding via polar-polar interactions (see Figure 6ii). Thus, the crosslinking reaction seems to favor the blends with richer ENR leading to an increase $N$. The next sample with the highest $N$ is TP5, and the order TP5 > TP6 and TP5 > TP8 suggests that MA and silanized silica play an effective role in the crosslinking reaction. The blend with mixtures of only N550 and silanized silica (TP7) attains comparable $N$ as TP5 (30 phr SBR/70 phr BR), but TP4 attains the poorest $N$ behavior although it contains 30 phr ENR with mixtures of carbon blacks and silanized silica.

This observation can be ascribed to the possible formation of agglomerates which dominates the filler-polymer interactions. Although the changes in the $\delta$ and their corresponding $\chi_1$ for a mono, dual, and triple matrix system could influence the $N$ values of the corresponding compounds; however, the pattern of variation of $N$ for the various blend systems in this present work has still remained unaffected. The influence of $N$ on the remaining physico-mechanical properties of the various compounds is presented in the proceeding sections.

*4.4. Tensile Properties and Hardness.* The tensile properties which include tensile strength (MPa), elongation at break (%), modulus at 200 and 300% (M200 and M300), and the Young's modulus (MPa) deduced at lower strain less than 5% of the various blends are presented in Figures 7(a)–7(d). The order of decreasing tensile strength is TP5 > TP1 > TP3 > TP6 > TP7 = TP4 > TP2 > TP8, with TP5 (30 phr SBR/70 phr BR) being the highest and TP8 (without silica) the least. It can be observed that the blends richer in BR and/or SBR seem to show higher tensile strength (see Figure 7(a)) and elongation at break (see Figure 7(b)) compared to the blends rich in ENR. This is because the high crosslink density in the ENR-based systems (refer to Figure 4 and Table 3)

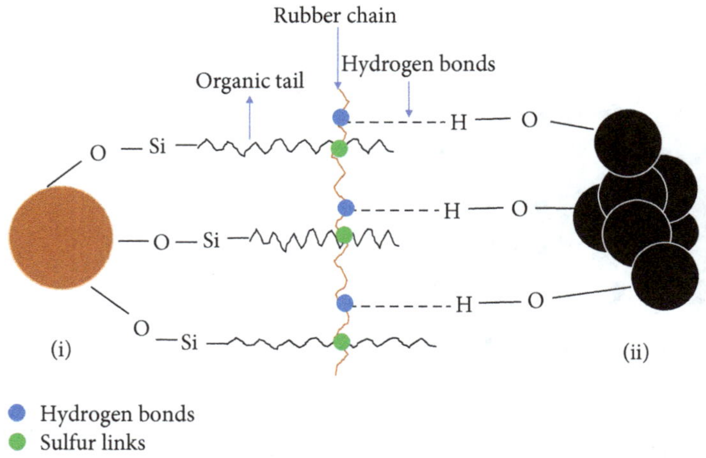


FIGURE 6: Illustration of possible bond formations between silanized silica, carbon blacks, and rubber molecules.

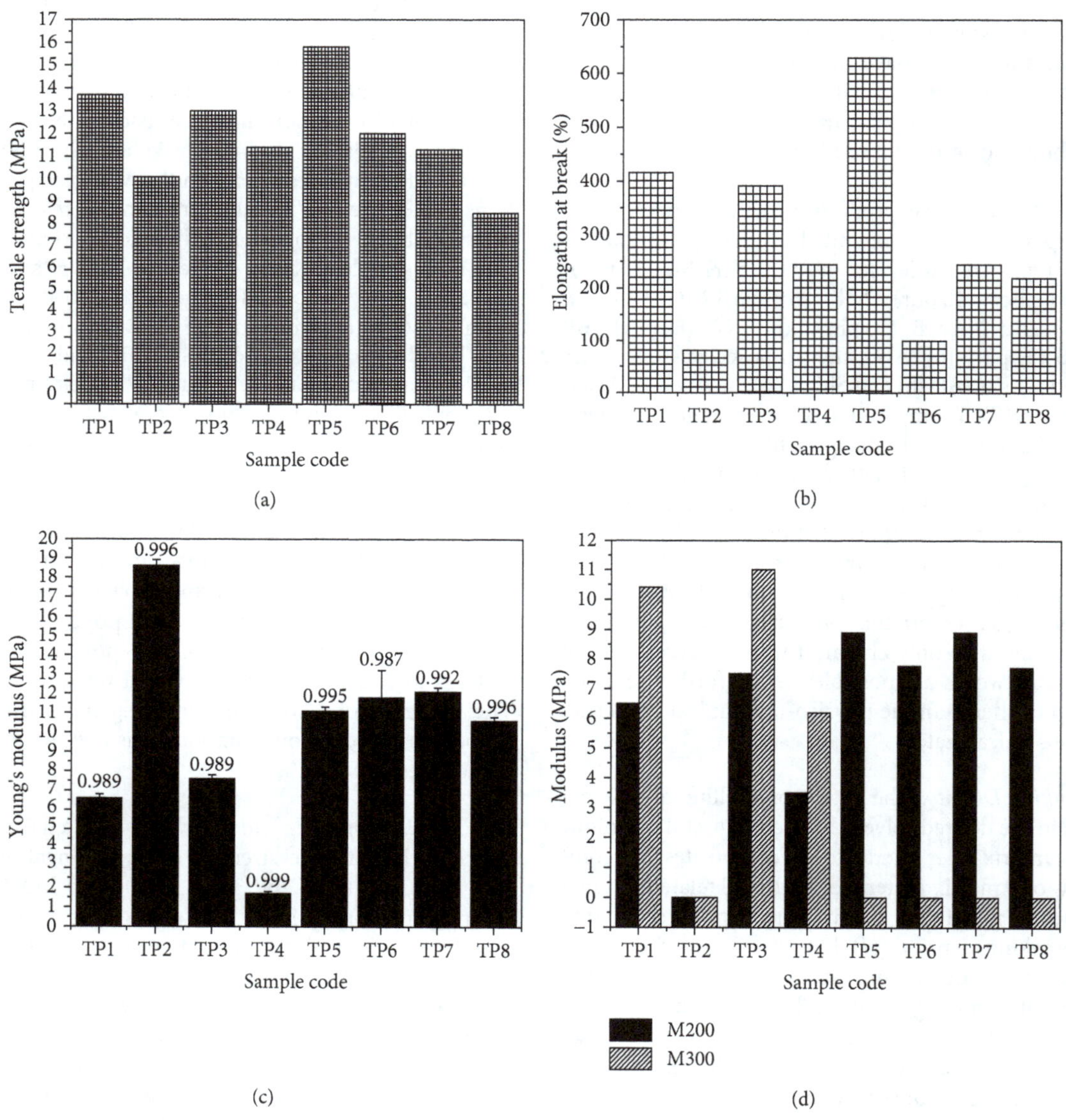


FIGURE 7: Tensile properties of vulcanizates: (a) tensile strengths, (b) modulus at 300 and 200%, (c) Young's modulus at strain less than 10%, and (d) elongation at break.

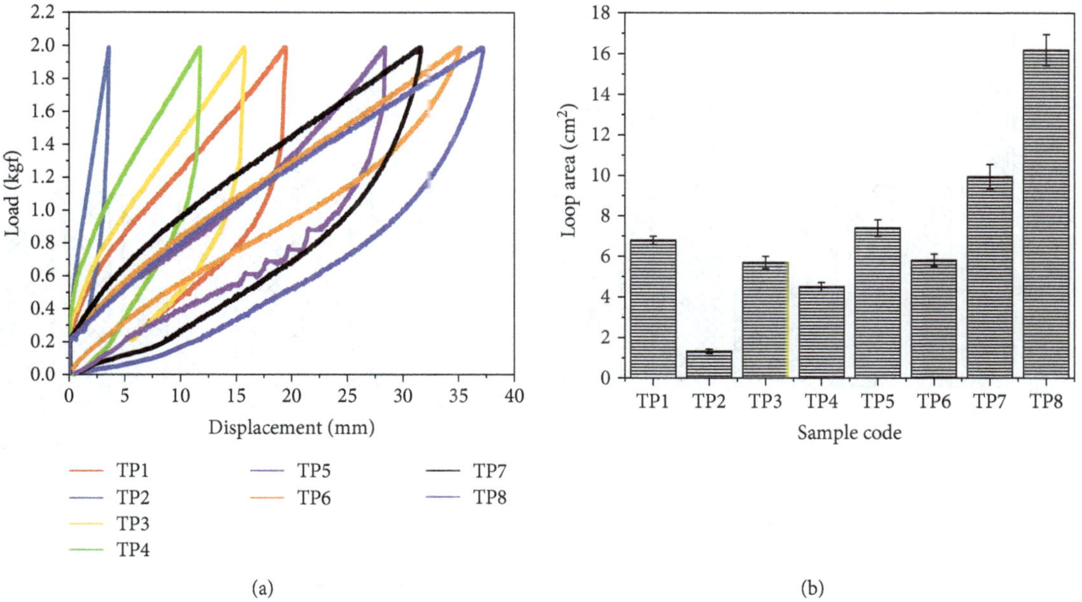


FIGURE 8: Cyclic testing of the vulcanizates; (a) the first loop after first cycles of vulcanizates and (b) area of the loop of the first cycle.

have made them much rigid and less stretchable (brittle-like) than the others (see Figures 7(c) and 7(d)).

Additionally, the $T_g$ of virgin ENR is generally higher due to the epoxide functionalities ($T_g$ dependent on epoxide content) [32, 33] and appears to increase further upon filler inclusion (becomes stiffer) compared to the low $T_g$ elastomers (BR and SBR) [34] of the same filler loading. One observation is that the inclusion of silanized silica to the blends promotes filler-polymer interaction resulting in improved tensile strength, suggesting the 35% increment in tensile strength in TP4 (30 phr ENR/70 phr BR) which contains 10 phr of silanized silica more than its counterpart TP8 (30 phr ENR/70 phr BR) without silica. It can also be observed that the inclusion of MA contributes to higher tensile strength, a reason why TP5 (30 phr SBR/70 phr ER) with MA has obtained more than 33% growth in tensile strength than TP6 (30 phr SBR/70 phr BR) of the same filler loading but without MA. Meanwhile, the tensile strength is not much affected when only one type of CB is used (in the case of TP7) compared to those with two mixtures of CBs (TP4). The tensile properties of this present compounds have generally improved compared to the truck-based tread compounds comprising of NR (75 phr)/BR (25 phr)-CB (50 phr)/Si (10 phr) and NR (75 phr)/BR (25 phr)-CB (60 phr) proposed earlier by Zafarmehrabian et al. [5].

*4.5. Hysteresis Loss.* The viscoelastic behavior of elastomer results in an energy loss during a cycle of extension-contraction; the energy loss is called hysteresis (or hysteretic loss). The hysteresis loss is related to the rolling resistance (RR) (fuel consumption) of tread tire; the higher the hysteresis loss, the higher the RR of a moving tire and the higher fuel is consumed [35]. The cyclic loading is indicated in stress-strain plots in Figure 8(a) whilst the area of loss is plotted in Figure 8(b). The decreasing order of the hysteresis loss is due to the estimated area of cyclic curves of the various blends and it decreases in the order: TP2 > TP4 > (TP3 = TP6) > TP1 > TP5 > TP7 > TP8. As shown, compounds rich in ENR content containing the three mixtures of fillers (N550, N330, and silanized silica) have indicated lower areas associated with improvement in hysteresis loss. Next is the compound with a high content of SBR and/or BR blend ratio containing all the mixture of fillers.

Blends with only one CB (25 phr of N550) with silanized silica (10 phr) do not contribute to the improvement in hysteresis compared to those with triple filler mixtures. For instance, TP7 (30 phr ENR/70 phr SBR) with 25 phr (N550) and silanized silica (10 phr) is 133% higher (poor RR effect) compared to TP4 (30 phr ENR/70 phr SBR) with mixtures of CBs (25 phr N550/10 phr N330) and silanized silica (10 phr). This means that complementary of the fillers enhances hysteresis loss, even though the formation and collapsing of aggregate (filler-filler accumulation) which leads to frictional energy is possible [35]. Nonetheless, it appears the TP4 (30 phr ENR/70 phr SBR) contributes more to the improvement in hysteresis loss compared to compounds without silica (TP8) which attains extremely high loop area (higher hysteresis loss) of about 277%. Therefore, it has been seen that lower hysteresis loss (lower RR) is consistent with the compounds with higher stiffness.

*4.6. Hardness and Wear Resistance.* Figures 9(a) and 9(b) represent the hardness and abrasion loss (mg) of the various blends. Again, the compound with higher crosslinking density ($N$) and Young's modulus with improved hysteresis loss (TP2) has attained higher hardness properties whilst TP4 showed the least. According to Zafarmehrabian et al. [5], abrasion behavior of rubbers is influenced by two crucial factors: modulus and $N$. That is, high modulus and $N$ lead to the enhancement of the abrasion strength of rubber compounds [5]. Contrary, the blend TP2 (with high ENR content) attained the inferior abrasion resistance (see Figure 9(b))

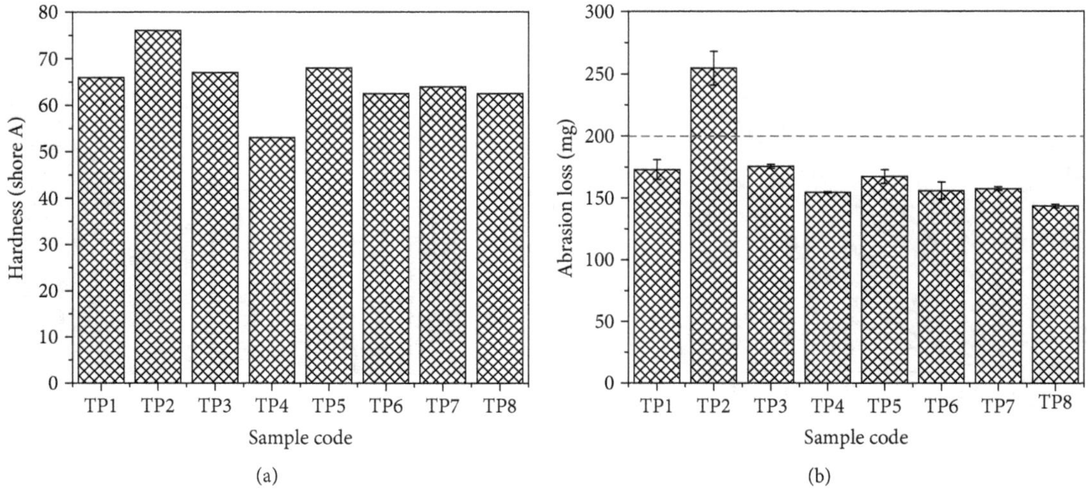


FIGURE 9: Hardness and wear resistance of the vulcanizates.

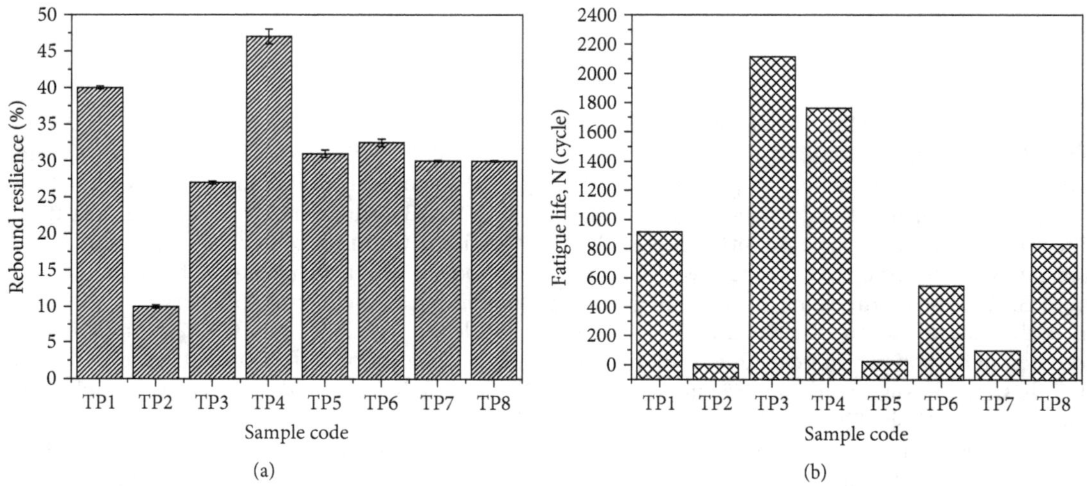


FIGURE 10: Rebound resilience and fatigue fracture properties of vulcanizates.

compared to the remaining samples which have shown relatively lower abrasion loss (improved wear resistance). The abrasion loss of these compounds is less than 200 mg (mass loss), a prerequisite for tire treads, shoe soles, and other abrasive surfaces. The inclusion of silanized silica was reported to improve the wear of rubber compounds due to strong filler-rubber interactions [7]. However, in this work, the compound without silica showed relatively improved wear resistance compared to those containing silica. Thus, it can be concluded that a critical level of modulus and $N$ are required to promote improved wear resistance. Therefore, an improvement in wear resistance must be explained in terms of the nature of network density such as filler-filler and/or filler-rubber and degree of $N$.

*4.7. Resilience and Fatigue Properties.* The resilience and fatigue properties of the various tread compounds are shown in Figures 10(a) and 10(b). The resilience seems to be dependent on the three factors: stiffness, hardness, and crosslinking density ($N$). The higher these properties, the lower the resilience; hence, TP2 (with extremely high $N$, hardness, and stiffness) shows the least resilience property. Another observation is that the resilience of the compounds seems to have been improved with compounds rich in BR content and less amount of ENR and/or SBR phases as in TP4 (30 phr ENR/70 phr BR) and TP1 (15 phr ENR/15 phr SBR/70 phr BR). This is because BR is well known for its higher resilience (the ability to recover size and shape after stress) compared to NR and SBR elastomers [36]. Generally, we have observed that the compounds TP1 and TP4 have shown higher resilience properties than the industrial tread compounds (80 phr NR/20 phr SBR/57 phr N375): 38.9%, (80 phr NR/20 phr SBR/57 phr N339): 38.4%, and (72 phr NR/28 phr SBR/55 phr N339): 34.5% reported earlier by Bijarimi et al. [6]. Also, apart from the rigid tread compound (TP2), the current compounds have still exhibited higher resilience properties than those reported for the blend (60 phr SBR/20 phr BR/20 phr NR/50 phr of N324), which obtained 20% as the sample with the best resilience properties by Atashi and Shiva et al. [7]. Furthermore, the fatigue

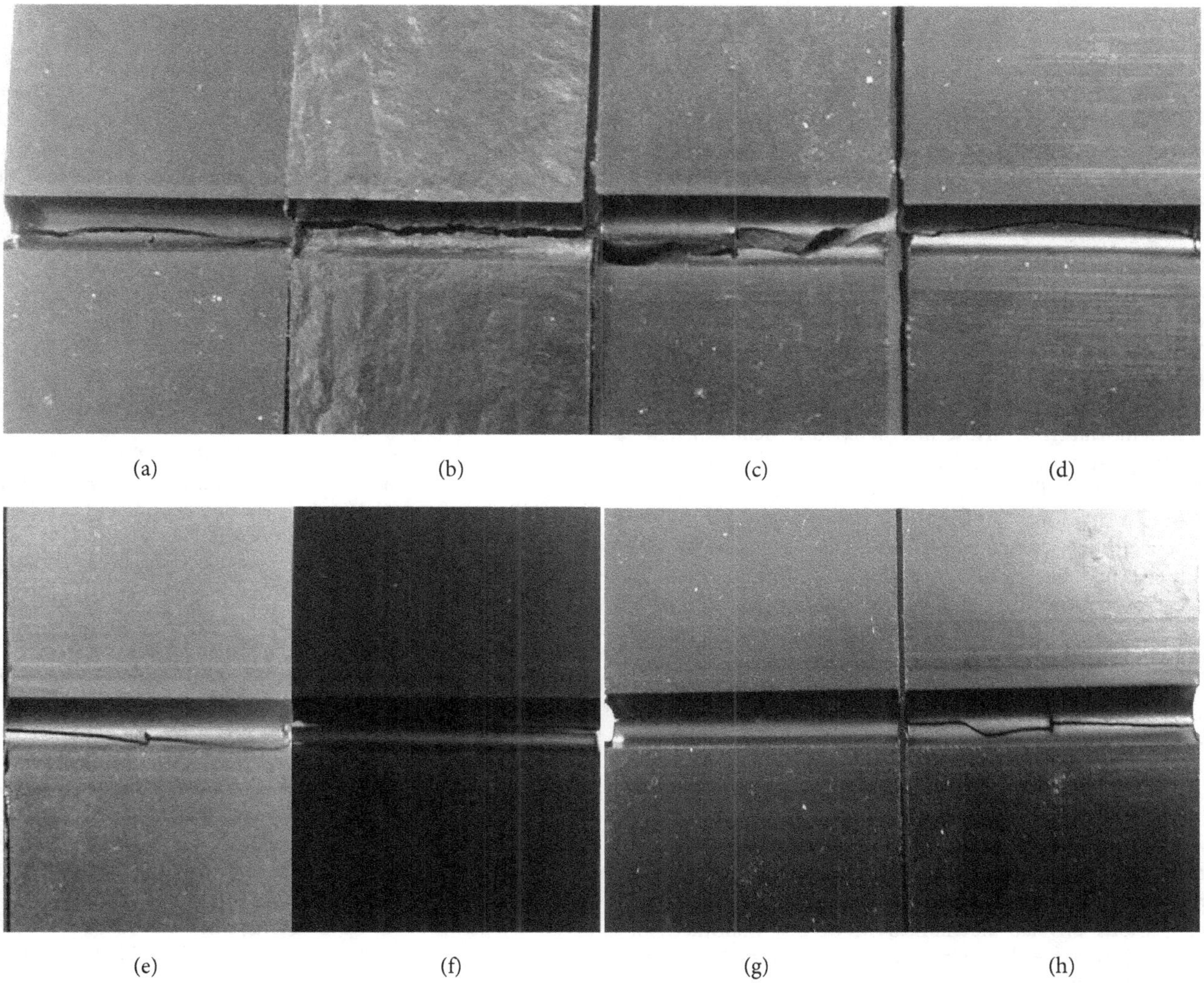

FIGURE 11: Morphology and mechanism of fatigue fracture of various vulcanizates; (a) TP1, (b) TP2, (c) TP3, (d) TP4, (e) TP5, (f) TP6, (g) TP7, and (h) TP8.

life which is generally related to the crack-growth characteristics of rubber materials has been presented in Figure 10(b).

The order of reducing in fatigue life is TP3 > TP4 > TP1 > TP8 > TP6 > TP7 > TP2 = TP5. The styrene content of the SBR sample has been observed to hinder the rotation of the backbone of SBR and also makes the backbone rigid [36]. Hence, high viscoelastic energy dissipation during the stretching and collapsing of the rigid structure of SBR may have introduced strain-induced crystallization that retards its failure. Again, it appears compounds with a high content of ENR and lower SBR phases show inferior fatigue properties due to extreme rigidity of bonds or higher networked density of the samples (TP2, TP5, and TP7). With the increased crosslinking density ($N$), the gel point of the blends is reached, and fatigue fracture (breaking of chemical bonds) becomes eminent. Thus, compounds with low molecular weight and substantial number of free-chain ends associated with large viscous flow portions such as TP3 (15 phr ENR/70 phr SBR/15 phr BR), TP4 (30 phr ENR/70 phr SSBR/25 phr N550/10 phr N330/10 phr silanized silica), and TP1 (15 phr ENR/15 phr SBR/70 phr BR/25 phr N550/10 phr N330/10 phr silanized silica) demonstrated the highest fatigue property compared to the compounds with restricted mobility of chains. When observed, TP4 (with silanized silica) is over 117% higher in fatigue resistance than its counterpart of the same composition except without silanized silica. This indicates an additional reinforcement offered by the silanized silica. Also, next sample which attained good fatigue property is TP8 (without silica). When the compound TP7 is compared with TP4 and TP8 of the same compositions except for only one carbon black filler and silanized silica, it is obvious that the high surface area carbon black does not enhance fatigue properties at this loading level as low surface area carbon black will do, because it is reported that as the surface area of carbon black increases, the maximum fatigue resistance shifts towards lower filler loading [36, 37]. Another observation is that when the MA was eliminated from the compounds (in the case of TP6), it has attained over 2000% increment in fatigue resistance compared to the compound of the same composition without MA, confirming that MA contributed to the formation of tightly crosslinks structures (filler-rubber links). The mode of fatigue of the various compounds has been discussed in the next sections for better understanding.

*4.7.1. Mechanism of Fatigue Failure of Composites.* The fracture mechanism and process can be viewed and analyzed physically by using microscopic techniques such as SEM, TEM, and AFM to analyze the morphology of suspicious vulcanizates. Here, we present a photo of the cracked surface of the various blends as presented in Figures 11(a)–11(h). In general, some samples (with higher crosslinking densities) such as TP2, TP5, and TP7 show a kind of stable brittle-like crack path associated with the low fatigue resistance (order of decreasing inferiority: TP2 > TP5 > TP7). Also, TP2 with the superior crosslinking density demonstrates a rocky-like surface, indicating a brittle-like behavior. Meanwhile, the remaining samples with relatively low crosslinking densities (TP1, TP3, TP4, TP6, and TP8) exhibit a kind of unstable brittle-like fracture paths, with somewhat high failure resistance when compared. In particular, the sample TP3 (Figure 11(c)) has exhibited a kind of nonunidirectional crack propagation path popularly called knotty crack growth, a mechanism linked to redistribution of loads by either crystalline sites in the composites or well aligned reinforcements, thereby reducing the sudden failure of the bulk matrix [35, 38]. This could explain why TP3 exhibited high fatigue strength in relation to its counterparts.

## 5. Conclusions

Various rubber blends filled with carbon black mixtures and silanized silica were prepared by melt mixing process and their various physico-mechanical properties examined for the suitability of tread applications. It was observed that high content of ENR (~70 phr) greatly improves the crosslinking density, Young's modulus, and hysteresis loss (lower rolling resistance), and at the same time, poor abrasion resistance and fatigue failure properties are obtained. Although it is challenging to obtain all the properties for green tire (WT, RR, and WR) at some instance due to compromising effects of properties, however, adopting blends with small amount of ENR content such as TP4 (30 phr ENR/70 phr SBR) with mixtures of CBs (25 phr N550/10 phr N330) and silanized silica (10 phr) present a suitable composition for developing tread with fuel-saving capabilities can be achieved.

## Conflicts of Interest

The authors declare that they have no conflicts of interest.

## Acknowledgments

A. Yaya acknowledges the financial support from the University of Ghana Building a New Generation of Academics in Africa (BANGA-Africa) Project, with funding from the Carnegie Corporation of New York. The assistance offered by YUTECH Company (South Korea) for granting us access to use their facilities for this work is also acknowledged.

## References

[1] S. H. Cadle and R. L. Williams, "Gas and particle emissions from automobile tires in laboratory and field studies," *Journal of the Air Pollution Control Association*, vol. 28, no. 5, pp. 502–507, 1978.

[2] D. J. Schuring, "Effect of tire rolling loss on vehicle fuel consumption," *Tire Science and Technology*, vol. 22, no. 3, pp. 148–161, 1994.

[3] A. Y. C. Lou, "Relationship of tire rolling resistance to the viscoelastic properties of the tread rubber," *Tire Science and Technology*, vol. 6, no. 3, pp. 176–188, 1978.

[4] T. Freudenmann, H.-J. Unrau, and M. El-Haji, "Experimental determination of the effect of the surface curvature on rolling resistance measurements," *Tire Science and Technology*, vol. 37, no. 4, pp. 254–278, 2009.

[5] R. Zafarmehrabian, S. T. Gangali, M. H. R. Ghoreishy, and M. Davallu, "The effects of silica/carbon black ratio on the dynamic properties of the tread compounds in truck tires," *Journal of Chemistry*, vol. 9, no. 3, p. 1112, 2012.

[6] M. Bijarimi, H. Zulkafli, and M. D. H. Beg, "Mechanical properties of industrial tyre rubber compounds," *Journal of Applied Sciences*, vol. 10, no. 13, pp. 1345–1348, 2010.

[7] H. Atashi and M. Shiva, "Formulation for passenger tread tire compound based on styrene butadiene rubber/cis-butadiene rubber/natural rubber blend and semi-dispersible silica/carbon black filler system," *Asian Journal of Chemistry*, vol. 22, pp. 7519–7530, 2010.

[8] G. Kraus and H. R. Hunt, "Carbon black for low-hysteresis rubber compositions," 1978, Google Patents.

[9] W. K. Dierkes, L. A. Reuvekamp, A. J. Ten Brinke, and J. W. Noordermeer, *Silanes and Other Coupling Agents*, vol. 3, p. 89, 2004.

[10] G. J. Lake, "Ozone cracking and protection of rubber," *Rubber Chemistry and Technology*, vol. 43, no. 5, pp. 1230–1254, 1970.

[11] Y. Okado, M. Kawamura, and T. Tomioka, "Rubber composition comprising furnace carbon black," 1982, Google Patents.

[12] S. S. Sarkawi, A. K. C. Aziz, R. A. Rahim, R. A. Ghani, and A. N. Kamaruddin, "Properties of epoxidized natural rubber tread compound: the hybrid reinforcing effect of silica and silane system," *Polymers & Polymer Composites*, vol. 24, p. 775, 2016.

[13] W. Kaewsakul, K. Sahakaro, W. K. Dierkes, and J. W. Noordermeer, "Verification of interactions between silica and epoxidised squalene as a model for epoxidised natural rubber," *Journal of Rubber Research*, vol. 17, p. 129, 2014.

[14] K. Sengloyluan, K. Sahakaro, and J. W. Noordermeer, "Use of epoxidized natural rubber as a compatibilizer in silica-filled natural rubber compounds," in *28th International Conference of the Polymer Processing Society 2012*, Nuremberg, Germany, 2012.

[15] P. J. Martin, P. Brown, A. V. Chapman, and S. Cook, "Silica-reinforced epoxidized natural rubber tire treads — performance and durability," *Rubber Chemistry and Technology*, vol. 88, no. 3, pp. 390–411, 2015.

[16] P. J. Flory and J. Rehner Jr, "Statistical mechanics of cross-linked polymer networks I. rubberlike elasticity," *The Journal of Chemical Physics*, vol. 11, no. 11, pp. 512–520, 1943.

[17] G. M. Bristow and W. F. Watson, "Cohesive energy densities of polymers. Part 1.—Cohesive energy densities of rubbers by swelling measurements," *Transactions of the Faraday Society*,

vol. 54, pp. 1731–1741, 1958.

[18] B. Moon, J. Lee, S. Park, and C.-S. Seok, "Study on the aging behavior of natural rubber/butadiene rubber (NR/BR) blends using a parallel spring model," *Polymer*, vol. 10, no. 6, p. 658, 2018.

[19] B. A. Miller-Chou and J. L. Koenig, "A review of polymer dissolution," *Progress in Polymer Science*, vol. 28, no. 8, pp. 1223–1270, 2003.

[20] S. El-Sabbagh and A. Yehia, "Detection of crosslink density by different methods for natural rubber blended with SBR and NBR," *Egyptian Journal of Solids*, vol. 30, p. 157, 2007.

[21] J. B. Nagode and C. M. Roland, "Miscible mixtures of polychloroprene and epoxidized polyisoprene," *Polymer*, vol. 32, no. 3, pp. 505–509, 1991.

[22] Y. P. Wu, W. Zhao, and L. Q. Zhang, "Improvement of flex-fatigue life of carbon-black-filled styrene-butadiene rubber by addition of nanodispersed clay," *Macromolecular Materials and Engineering*, vol. 291, no. 8, pp. 944–949, 2006.

[23] B. Mensah, S. Kim, D. H. Lee, H. G. Kim, J. G. Oh, and C. Nah, "Autohesion behavior of brominated-isobutylene-isoprene gum nanocomposites with layered clay," *Elastomers and Composites*, vol. 49, no. 1, pp. 43–52, 2014.

[24] K. Chino, M. Ashiura, J. Natori, M. Ikawa, and T. Kawazura, "Thermoreversible crosslinking rubber using supramolecular hydrogen bonding networks," *Rubber Chemistry and Technology*, vol. 75, no. 4, pp. 713–723, 2002.

[25] S. Goyanes, C. C. Lopez, G. H. Rubiolo, F. Quasso, and A. J. Marzocca, "Thermal properties in cured natural rubber/styrene butadiene rubber blends," *European Polymer Journal*, vol. 44, no. 5, pp. 1525–1534, 2008.

[26] S. S. Idrus, H. Ismail, and S. Palaniandy, "The effects of silanized ultrafine silica on the curing characteristics, tensile properties, and morphological study of natural rubber compounds," *Polymer-Plastics Technology and Engineering*, vol. 50, no. 1, pp. 1–7, 2011.

[27] P. Nelson and S. Kutty, "Studies on the utilisation of rubber reclaim in elastomers," *Progress in Rubber Plastics and Recycling Technology*, vol. 19, p. 171, 2003.

[28] B. Mensah, S. Kim, S. Arepalli, and C. Nah, "A study of graphene oxide-reinforced rubber nanocomposite," *Journal of Applied Polymer Science*, vol. 131, no. 16, 2014.

[29] C. Nah, S. G. Kim, G. S. Shibulal et al., "Effects of curing systems on the mechanical and chemical ageing resistance properties of gasket compounds based on ethylene-propylene-diene-termonomer rubber in a simulated fuel cell environment," *International Journal of Hydrogen Energy*, vol. 40, no. 33, pp. 10627–10635, 2015.

[30] M. A. Mansilla, L. Silva, W. Salgueiro, A. J. Marzocca, and A. Somoza, "A study about the structure of vulcanized natural rubber/styrene butadiene rubber blends and the glass transition behavior," *Journal of Applied Polymer Science*, vol. 125, no. 2, pp. 992–999, 2012.

[31] H. S. Ahmad, H. Ismail, and A. Abd. Rashid, "ENR-50 compatibilized natural rubber/recycled acrylonitrile-butadiene rubber blends," *Sains Malaysiana*, vol. 44, no. 6, pp. 835–842, 2015.

[32] V. Tanrattanakul, B. Wattanathai, A. Tiangjunya, and P. Muhamud, "*In situ* epoxidized natural rubber: improved oil resistance of natural rubber," *Journal of Applied Polymer Science*, vol. 90, no. 1, pp. 261–269, 2003.

[33] N. A. Kinasih, M. I. Fathurrohman, and D. A. Winarto, "Swelling behaviour in n-pentane and mechanical properties of epoxidized natural rubber with different epoxide content," *IOP Conference Series: Materials Science and Engineering*, vol. 223, article 012002, 2017.

[34] W. M. Groenewoud, *Characterisation of Polymers by Thermal Analysis*, Elsevier Science, 2001.

[35] J. de Castro, R. Zargar, M. Habibi et al., "Nonmonotonic fracture behavior of polymer nanocomposites," *Applied Physics Letters*, vol. 106, article 221904, no. 22, 2015.

[36] J. Zhao and G. N. Ghebremeskel, "A review of some of the factors affecting fracture and fatigue in sbr and br vulcanizates," *Rubber Chemistry and Technology*, vol. 74, no. 3, pp. 409–427, 2001.

[37] E. E. Auer, K. W. Doak, and I. J. Schaffner, "Factors affecting laboratory cut-growth resistance of cold sbr tread stocks," *Rubber Chemistry and Technology*, vol. 31, no. 1, pp. 185–201, 1958.

[38] K. Cho, W. J. Jang, D. Lee, H. Chun, and Y.-W. Chang, "Fatigue crack growth of elastomers in the swollen state," *Polymer*, vol. 41, no. 1, pp. 179–183, 2000.

# 15

# Analytical Methods to Characterize and Purify Polymeric Nanoparticles

**Cristina Fornaguera**[1] **and Conxita Solans**[2]

[1]Grup d'Enginyeria de Materials, Institut Químic de Sarrià (IQS), Universitat Ramon Llull and Sagetis-Biotech, Via Augusta 390, 08017 Barcelona, Spain
[2]Institute of Advanced Chemistry of Catalonia (IQAC-CSIC) and Networking Centre on Bioengineering, Biomaterials and Nanomedicine (CIBER-BBN), C/Jordi Girona 18-26, 08034 Barcelona, Spain

Correspondence should be addressed to Cristina Fornaguera; cfornaguera@gmail.com

Academic Editor: Zhonghua Peng

Advances in polymeric nanoparticles as novel nanomedicines have opened a new class of diagnostic and therapeutic tools for many diseases. However, although the benchtop research studies in the nanoworld are numerous, their translation to currently marketed products is still limited. This lack of transference can be attributed, among other factors, to problems with nanomedicine characterization. Characterization techniques at the nanoscale could be divided in three categories: characterization of physicochemical properties (e.g., size and surface charge), characterization of nanomaterials interactions with biological components (e.g., proteins from the blood), and analytical characterization and purification methods. Currently available literature of this last group only describes methodologies applied for a specific type of nanomaterial or even methods used for bulk materials, which are not completely applicable to nanomaterials. For this reason, the current review aims to become a scholastic guide for those scientists starting in the nanoworld, giving them a description of analytical characterization techniques aimed to analyze polymers forming nanoparticles and possible forms to purify them before being used in preclinical and clinical applications.

## 1. Introduction

The field of nanotechnology and, more specifically, nanomedicine emerged about 20 years ago and since then, it experienced an exponential progress both in the fundamental study of nanosystems and in their multiple applications. Specifically, studies on polymeric nanoparticles have gained attention due to the multiple advantages that are attributed to this kind of nanosystems in terms of safety, versatility, and robustness [1–3]. Although the number of benchtop research studies developing novel nanosystems intended for biomedical applications is enormous, their use as clinically effective products is still limited. One of the main concerns of pharmaceutical industries for the production of novel formulations based on polymeric nanoparticles is the complexity of a deep characterization, which would enable their safe production and use in humans [4]. Therefore, a wide characterization of nanomedicines is a must before their testing in preclinical and clinical stages. However, the current technology is challenging in the sense that many characterization techniques have been applied directly from those methods used for bulk materials (not at the nanoscale) or using conditions that do not simulate biological environment [1, 2]. Therefore, many efforts must be devoted to the enhancement of the performance of current techniques. Nanomedicine characterization can be divided in three steps: first, an analytical characterization, useful for characterizing the materials they are composed of as well as to find out the impurities present and develop purification processes; second, a physicochemical characterization of the main parameters that will define the performance of nanomaterials *in vivo*, such as size, surface charge, and stability in biological conditions; and third, the study of their interaction with biological components (Figure 1). Although many reviews

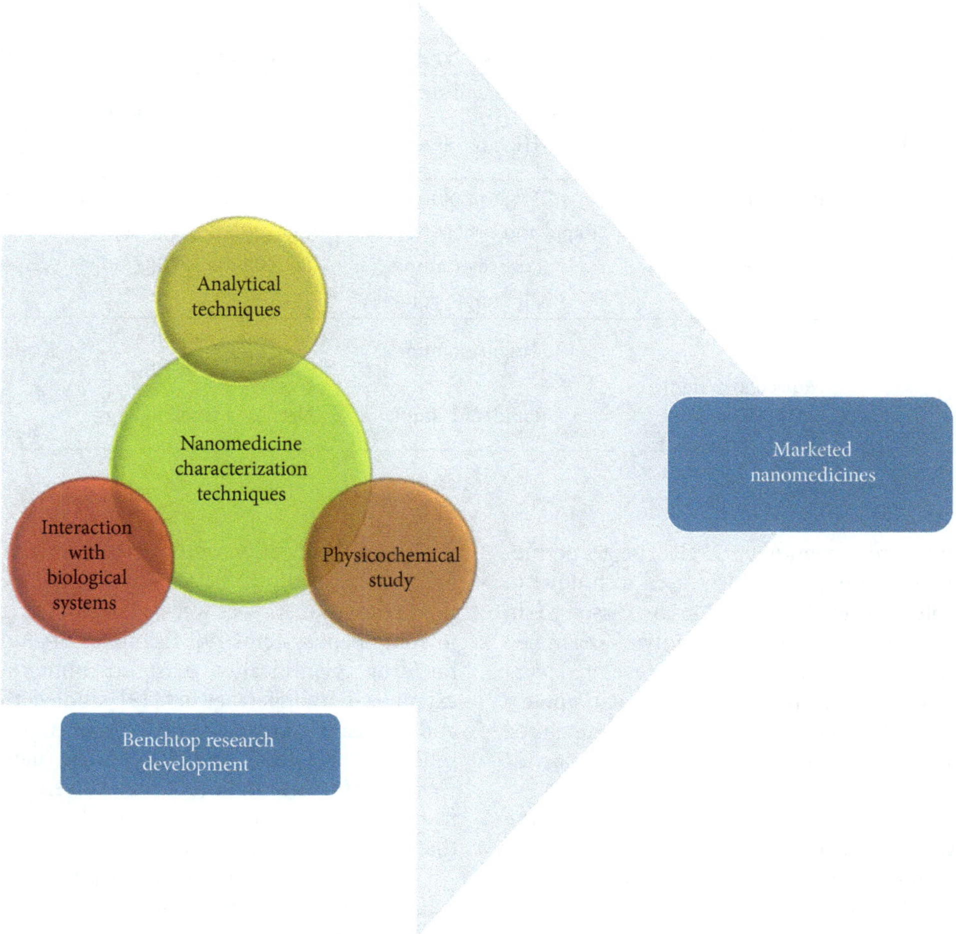


FIGURE 1: Schematic representation of the process of nanomedicine characterization before translating to pharmaceutical production.

for the characterization of polymeric nanoparticles designed as nanomedicines exist, most of them give a particular point of view, signaling only some techniques [2, 5, 6]. Therefore, scientists working on the development of novel nanoformulations find themselves lost in the huge but dispersed existent bibliography. This is the reason that motivated the authors to write a series of three reviews with a scholastic character, to enable those scientists starting in the nanoworld to have guidelines for the correct characterization of polymeric nanoparticles. The first review was devoted to the characterization of nanomaterial interactions with biological systems [2]; the second one describes the physicochemical characterization techniques at the nanoscale, to assess size, surface charge, and stability of polymeric nanoparticles [7]; and the present one (third one) devoted to describe analytical characterization and purification techniques useful for nanomedicine study of polymeric nanoparticles (see a schematic representation of these techniques in the SI). Therefore, the purpose of the present review is to be a first practical guideline for those scientists initiating their studies in the nanoworld. It should be noted that it was not the objective of the authors to get deep into description of each individual technique but rather describe briefly each methodology to help the readers to select the most appropriate technique for their study and look for more specific information in the numerous references given for each technique.

These methods have been classified not as a function of what is characterized but as a function of the technique: chromatographic, spectroscopic, calorimetric, and purification techniques. Authors will guide the reader through them with the objective to help in the selection of one or other technique depending on the parameter to study. Physicochemical techniques, mainly used to characterize size (e.g., light scattering or microscopy), are out of the scope of the present review [7].

## 2. Chromatographic Techniques

Chromatographic techniques, in general, are a group of techniques devoted to the separation of various compounds [8] as a step prior to their characterization. Their advantages are related with their high power of separation of substances, and the easy and simple manipulation, although they have as drawback the nonspecific interactions and the difficulty in method optimization [8, 9]. In this work, various types of chromatographic techniques are briefly described (Table 1). All of them could be also classified as purification techniques (see Section 5), since they are also able to separate compounds.

Table 1: Analytical chromatographic techniques.

| Technique | Characteristics that analyses | Advantages | Disadvantages | References |
|---|---|---|---|---|
| Gel permeation chromatography (GPC) | Molecular weight | Rapid and simple<br>High resolution | Interaction sample with column filling | Williams [10]<br>Neverova and Van Eyk [8]<br>Cho et al. [1] |
| High-performance liquid chromatography (HPLC) | Quantification of actives<br>Purification | High resolution<br>Rapid and easy performance<br>Low cost/sample<br>Small sample volumes | Interaction sample with column filling | Neverova and Van Eyk [8]<br>Sapsford et al. [9] |
| Size exclusion chromatography (SEC) | Molecular weight | High resolution<br>Rapid and simple | Interaction sample with column filling<br>Need of a labelling tag | Kostanski et al. [25]<br>Sapsford et al. [9]<br>Rebolj et al. [26] |

*2.1. Gel Permeation Chromatography (GPC).* The gel permeation chromatography (GPC) is a widely used technique to determine the molecular weight of materials dissolved in organic solvents as well as the physical stability of assembled nanomaterials [1, 10]. The nanomaterials are eluted as a function of their molecular weight: the bigger the nanomaterials, the faster the elution. The quantification of the eluted samples is performed by means of UV-Vis absorption or changes in the refractive index.

This technique could be useful, for example, to study the stability of a polymer dissolved in an organic solvent. After various periods of time, the dissolved polymer should be analyzed through the GPC and its molecular weight assessed by using a calibration curve. Other examples of the use of GPC in the nanoscale could be the assessment of the polymer molecular weight, the degree of polymerization of a synthesized polymer, or even the number of monomer subunits that a polymer contains [11–14]. In some cases, GPC has also been used for the purification of quantum dots or carbon nanotubes, for example [15, 16].

GPC is advantageous in terms of short-time experiments. However, an important drawback of this technique is the possible interaction between the nanomaterials and the column filling, which could interfere the size assessment [1, 8].

*2.2. High-Performance Liquid Chromatography (HPLC).* High-performance liquid chromatography (HPLC) is the most used type of chromatography not only for colloidal nanosystem studies but also for other type of materials (e.g., proteins). In the vast majority of studies, it is used for the fine quantification and separation (purification) of actives, such as drugs [5, 8, 9]. Briefly, it consists in the injection of the liquid sample using a pump that introduces it to a flow (mobile phase) that passes through a separative column (stationary phase), which entraps the molecules depending on their nature. The more interactions the molecules have with the column filling, the later they will be eluted. Further, molecules are eluted in a characteristic pattern for each compound. It results a chromatogram with the peaks of each compound (Figure 2).

The quantification of the actives is required in any study of the encapsulation efficiency of drugs in the nanosystems or their release kinetics, as well as the percentage of conjugation to some nanosystems [9]. Examples of studies using HPLC for drug quantification exist are numerous [17–19]. For example, Fornaguera et al. [19] studied the encapsulation and release kinetics of dexamethasone (an anti-inflammatory drug) from polymeric nanoparticles. They were able to determine very low concentrations of the drug in a release study receptor solution, due to the high sensibility that offers the HPLC technique.

The resolution of the HPLC depends on the filling of the column (on the stationary phase properties), which is commonly composed of silica with attached alkyl chains, being the reversed phase $C_{18}$-type columns the most widely used, since it enables a differential retentionship depending on the polarity of the compounds [8, 9].

The advantages of HPLC are the high resolution, the low volumes required, and an easy, rapid, and economic manipulation [8, 9]. However, some drawbacks derived from the interaction of the samples with the stationary phase (column filing) could take place [9].

In 2004, it appeared the ultra-HPLC (UHPLC) technique, with many advantages among the traditional HPLC. It uses a column filling of particles of sub-2 micron size, while conventional HPLC uses particles between 2.5 and 5 microns. This reduction on the filling particle size enables a finer separation of similar compounds. In addition, the working pressure of UHPLC equipment is markedly higher than that supported by conventional HPLC, which enables more rapid flow rates, resulting in shorter elution times and decrease on the solvent amount used [20, 21].

*2.3. Size Exclusion Chromatography (SEC).* Size exclusion chromatography (SEC), together with ion-exchange and affinity chromatography are classified as low-pressure liquid chromatography [9]. It is useful to characterize and separate (purify) nanomaterials with different properties, dispersed in an aqueous solution. For example, it has been useful to separate antibody-conjugated nanoparticles from the free

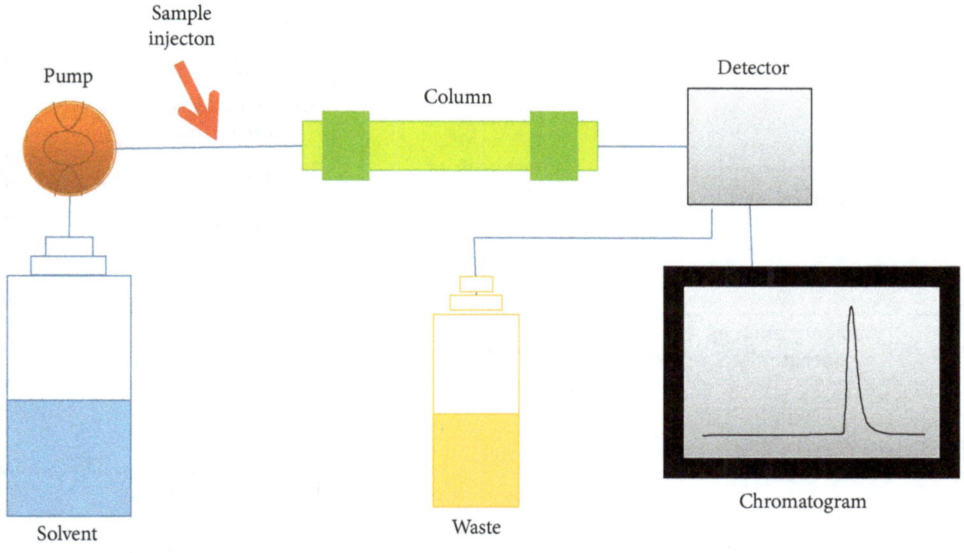


FIGURE 2: Schematic representation of a HPLC system.

antibody and nanoparticles [22, 23]. Another example of its use is the characterization of the molecular weight of nanoobjects, such as proteins or polymers [24]. The separation of the compounds depends not only on their molecular weight but also on their 3D dimensions, due to differences in pore permeation [25, 26]. It has the same advantages than other chromatographic techniques.

## 3. Spectroscopic Techniques

Spectroscopic techniques are those that give information on the interaction of an electromagnetic radiation with a sample, thus resulting in an absorption that depends on the excitation wavelength. Therefore, a wavelength spectrum with absorption/emission peaks that depend on the material is produced [9].

These techniques, summarized in Table 2, in general, are useful to characterize materials, which includes also nanomaterials, since spectra are characteristic of each material. A specific use of them, for example, could be the determination of the formation of bioconjugates between nanomaterials and conjugated moieties [9].

*3.1. Infrared Spectroscopy.* Fourier transformed infrared spectroscopy (FTIR) is a spectroscopic technique based on the measurement of vibrational transitions between different excitation states of molecules [27]. IR radiation is absorbed by molecules with dipoles that oscillate with the same frequency of the incident IR light. The IR absorption is an energy transfer of molecules and, if a change on their vibration occurs (by stretching, bending, or twisting of the bonds), the IR absorption changes; and this change can be characterized (e.g., covalent attachment of a carboxylic group to an amine group that results in an amide group) [5, 9].

It is a widely used technique, not only in the nanomedicine field but also in a variety of scientific fields. It has many advantages (Table 2), such as the fast and inexpensive performance and the obtaining of characteristic IR fingerprints of each compound, since they are a combination of the vibrational state of each atom [5]. In addition, it can be a quantitative tool in some specific conditions [27]. However, as drawbacks, there is the sample preparation (dried samples required, although some apparatus are prepared for liquid samples, but in this last case, a high concentration of the compounds to analyze is required), a requirement of technique optimization in most cases and interference with water molecules, which strongly absorb and need of a background signal [9]. Concerning nanomedicines experimentation, Sapsford et al. [9] and Lin et al., [5] for example, used FTIR to confirm the attachment of biomolecules onto nanomaterials surface, since FTIR results in a band pattern as a function of the chemical groups.

*3.2. UV-Visible Spectroscopy.* UV-Visible spectroscopy is a spectroscopy type that emits radiation of wavelength between 190 and 800 nm, widely used for the quantification of compounds concentration, and, in some cases, even size and shape, since each material absorbs at a determined wavelength and changes in the spectra could be related with changes in the aggregation of nanomaterials [5, 9]. It has been used, for example, to determine the conjugation and ratio of conjugation of biomolecules to nanomedicines [2, 5, 6, 8, 9].

It is a simple, fast, and cost-effective technique that can be applied to a variety of nanomaterials (Table 2). However, since most of the materials produce absorption at some wavelength, interferences between absorption of different compounds have to be taken into account [9].

*3.3. Fluorimetry.* Fluorimetry is a very sensitive spectroscopic technique that consists in the quantitative measurement of fluorescent signals, usually produced by aromatic molecules,

TABLE 2: Summary of the main advantages and disadvantages of spectroscopic techniques.

| Technique | Use | Advantages | Disadvantages | References |
|---|---|---|---|---|
| Fourier transformed infrared (FTIR) | Chemical composition | Fast and inexpensive<br>Characteristic of each material<br>Possible quantification | Complicated sample preparation<br>Interference of water<br>Relatively low sensitivity<br>Requires dried samples<br>Destructive | Alben and Fiamingo [27]<br>Pretsch 2003<br>Sapsford et al. [9]<br>Lin et al. [5] |
| UV-Vis | Quantification of concentration<br>Size and shape determination | Cost effective<br>Simple and fast<br>Useful for a variety of compounds | Interference between materials | Sapsford et al. [9]<br>Lin et al. 2014 [5] |
| Fluorimetry | Quantitative fluorescence determination | High sensitivity<br>Compound specificity | Limited to fluorescent compounds<br>Limited fluorescence lifetime | Lakowicz [28] |

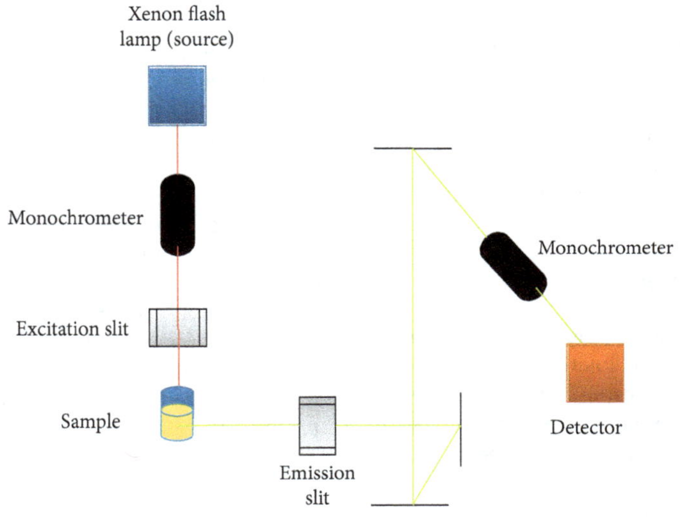


FIGURE 3: Schematic representation of a fluorescence spectrophotometer model.

for the detection and characterization of organic and inorganic compounds thanks to the application of a fluorescent laser to a sample (Figure 3) [28]. The fluorescent signal is composed of

(i) Excitation: it is produced by the absorption of the electromagnetic radiation by the sample to study.

(ii) Losses of vibrational energy: they are produced after the absorption of energy and before the emission of fluorescence, due to the internal collision between the molecules of the sample.

(iii) Emission: it consists of the energy produced by a molecule of the sample when it drops to a lower vibrational level (corresponding to lower energy and longer wavelengths), thus emitting energy in the form of fluorescence. The fluorescence is obtained as a spectrum and not a single band because not all the molecules of the sample drop to the same vibrational level.

This technique has been used for various applications, all of them taking advantage of the capacity of the studied materials to be excited under a fluorescent laser and emit fluorescence of another wavelength. For example, Lin et al. [5] used fluorescence to study conformational changes of biomaterials and their conjugation with nanomaterials. Fornaguera et al. [29], in contrast, took advantage of the fluorescent properties of galantamine drug to quantify its encapsulation efficiency in polymeric nanoparticles and its release. Chen et al. [30] used fluorimetry to confirm covalent bond formation onto polymeric nanoparticle surface. In addition, since the excitation/emission are wavelength

TABLE 3: Summary of the main advantages and disadvantages of calorimetric techniques.

| Technique | Use | Advantages | Disadvantages | References |
|---|---|---|---|---|
| Differential scanning calorimetry (DSC) | Glass transition<br>Melting temperature | Low amounts of sample<br>High precision and sensitivity | Requires sample preparation<br>Requires an appropriate reference | Hunt [36]<br>Haq [35]<br>Lu and Li [37]<br>Sapsford et al. [9] |
| Thermogravimetry (TGA) | Weight loss | Low amounts of sample<br>High precision and sensitivity | Requires sample preparation<br>Requires an appropriate reference | Berger et al. [38]<br>Sapsford et al. [9]<br>Lin et al. [5] |

specific, it can be also used to quantify two colocalized fluorescent dyes, due to the simultaneous detection of the different wavelengths.

Although the varied uses of fluorescence, it has two main drawbacks (Table 2): only some materials have fluorescence and the fluorescence lifetime is limited, with a wavelength dependent on the specific compound [28].

Recently, advanced techniques combining the advantages of fluorescent signal have appeared. Of special importance is the Förster resonance energy transfer or FRET technology. It consists in the combination of two fluorescent probes labeling pairs of two compounds, the first one called the donor and the second one called the acceptor. Fluorescence of excitation wavelength of the donor is directed to the compound, which, under the specific fluorescence will emit fluorescence in another wavelength. The acceptor is excited specifically by the emission wavelength of the donor (energy transference) and emits fluorescence in another wavelength. Therefore, if both compounds are very close (<10 nm), when exciting the donor, only fluorescence signal of the acceptor emission wavelength will be detected. In contrast, if both molecules are not close enough, when exciting the donor, fluorescence emission of this donor wavelength will be detected. Therefore, this technique is very useful to study two compound aggregation, and it is starting to play an important role in nanosystem studies [31–33]. Lai et al. [31], for example, studied the drug release from porous silica nanoparticles using this technique. Liu et al. [32], in contrast, used FRET technology as a nanodiagnostic system to detect the presence of chrome in urine, which produces the dissociation of FRET pairs, specifically designed for this purpose.

## 4. Calorimetric Techniques

Calorimetric techniques are a type of techniques that apply a temperature change to the samples to study physical phenomena, such as the crystalline transition, fusion, vaporization, sublimation, absorption, adsorption, and desorption and chemical phenomena, such as chemisorptions, desolvation, decomposition, oxidative degradation, solid-state, and solid-gas reactions [34]. Two main calorimetric techniques will be described in this review (Table 3).

*4.1. Differential Scanning Calorimetry (DSC).* Differential scanning calorimetry (DSC) is a technique that continuously measures the apparent specific heat of a system as a function of the temperature [35]. It is useful for the measurement of the glass transition and melting temperatures of materials including nanosystems. The glass transition temperature (Tg) is the temperature at which a material in a solid state changes its conformation. The melting temperature is always higher than the Tg [36, 37]. Therefore, in some cases, the determination of the melting temperature is not performed because it requires too high temperatures.

It is a useful technique to determine the structure and stability of nanomaterials, as well as their conformation, since material transitions will change depending on the nanomaterial composition [5, 35].

*4.2. Thermogravimetric Assays (TGA).* Thermogravimetric assays (TGA) are another type of calorimetric techniques which measure the weight loss of the samples [34, 38].

It is a useful technique to determine the amount of nanoconjugation, since the change on the nanomaterial composition produces changes in the temperature weight loss [5, 9]. For example, Fornaguera et al. [39] used TGA analysis to confirm the covalent attachment of a dendritic carbosilane wedge to polymeric nanoparticles.

Since both calorimetric techniques have a similar performance, their advantages and disadvantages have been summarized together. Calorimetric techniques are advantageous in terms of small amount of sample required, high precision and sensitivity. However, an appropriate reference as well as the preparation of the sample is required [9, 35, 38].

## 5. Purification Techniques

The composition of colloidal nanomaterials is a key point, since it affects not only its transport, delivery, and biodistribution *in vivo*, but, most importantly, it can contribute to toxicity-related problems [4–6, 9]. For this reason, to ensure a safe formulation, free of contaminants, a purification step is strongly recommended, followed by a physicochemical characterization, before starting preclinical and clinical analysis.

Although various methods exist for the colloidal nanomaterial purification, such as the magnetic separation for magnetic nanoparticles, for example [40, 41], in most cases, when nanosystems do not possess any specific inherent property that facilitates purification, this step may represent a difficult and tedious process, being sometimes impossible to confirm the presence of a purified compound [9]. In the following, common purification techniques, useful for a variety of nanosystems, are summarized (Table 4). Specific

Table 4: Summary of the main advantages and disadvantages of the described purification techniques.

| Technique | Use | Advantages | Disadvantages | References |
|---|---|---|---|---|
| Filtration | Purification<br>Sterilization | Useful for thermolabile compounds | Time consuming | Scopes [44] |
| | Concentration | Rapid and simple | Bigger sizes determined by the cut-off size | Sapsford et al. [9] |
| | Dispersant changement | Commercially available devices | Single-use devices | Roy et al. [43] |
| | Reduce size polydispersity | Cost effective | | Dobrovolskaia and McNeil [42] |
| Centrifugation | Purification | High efficiency | Special equipment required for large volumes | Scopes [44] |
| | Concentration | Rapid, facile, and economic | Difficult to resuspend soft matter | Sapsford et al. [9] |
| | Dispersant changement | Low amounts of sample | | |
| | | Appropriate for different kinds of nanomaterials | | |
| Dialysis | Purification | Rapid, facile, and economic | Limited to the membranes MWCO | Vauthier et al. [46] |
| | Concentration | Commercially available devices | High receptor solution volumes | Sapsford et al. [9] |
| | Dispersant changement | No sample pretreatment | | |
| Electrophoresis | Purification | Simple and economic | Postelectrophoresis purification steps | Sapsford et al. [9] |
| | | High resolution and sensitivity | | Jason [51] |
| | | | Need of charged compounds | Scopes [45] |

techniques, useful only for a determined material have not been described in this review.

It is worth remarking the contamination by endotoxins. Although methodology to purify from endotoxins is out of the scope of the present review, authors would like to remark the importance of producing nanomedicines clean from this type of contamination, which could be produced from many lipopolysaccharides, of the omnipresent Gram-negative bacteria. It is important to clean all material used before starting the production (e.g., by cleaning with sodium hydroxide or heating treatments) of nanomedicines and to confirm that resulting nanomedicines are clean from endotoxins [6].

*5.1. Filtration.* Filtration is a purification method, specifically used to sterilize nanomaterial colloidal dispersions. This method is advantageous as a sterilization technique for thermolabile compounds [42]. In addition, it represents a rapid, commercially available, cost-effective, and simple technique. Although it can be performed under atmospheric pressure, usually, it is performed taking advantage of special devices with filters, which are centrifuged, thus increasing the speed and efficiency of the process [9, 43]. An aspect that could be considered as a drawback is that filtration of large volumes could produce clogging of the filters, reason for which filters are single-use devices [44].

In addition, it can also be useful for other purposes, such as for the concentration of colloid nanomedicines or the reduction of their polydispersity. For example, Roy et al. [43] produced Au and CdS nanoparticles and passed them through a multiwall carbon nanotube, which was used as a filter to control the maximum nanoparticle size, thus eliminating the bigger ones and reducing nanoparticle size polydispersity [43, 45–50].

*5.2. Centrifugation.* Centrifugation is another technique useful for the purification of nanomaterials. It consists of the application of a centrifugal force to enhance the precipitation of nanomaterials due to the increased gravitational field [44]. Different kinds of centrifugation exist, such as the conventional centrifugation, ultracentrifugation, and gradient centrifugation, whose use depends on the objective of the study and on the nanomaterials type, specifically on their size [9].

Centrifugation is more efficient than filtration. It is a rapid, facile, and economic technique, able to be used for different kinds of nanomedicines. In addition, low amounts of sample are required. However, the centrifugation of large volumes requires special equipment and in some cases, difficulties on resuspending sediment nanomaterials appear, specifically when working with soft matter, which are not always possible to recover their dispersion liquid state [9, 44].

Apart from the use of centrifugation to purify nanomedicines, it has been also used to concentrate nanomedicines, to change their dispersant, and to separate conjugated nanomaterials from those nonconjugated [9]. For example, Fornaguera et al. [19] centrifuged PLGA nanoparticles to purify them from the surfactant traces and to concentrate them.

*5.3. Dialysis.* Dialysis consist on changing the nanomaterial dispersant by means of submerging a semipermeable dialysis bag (or a dialysis dispositive) filled with the nanomaterial dispersion, in a receptor solution. Therefore, it is not only useful for the nanomaterial purification but also for nanomedicine concentration and to change the dispersant to achieve the desired properties (e.g., dialysis with PBS to achieve the physiological pH and osmolality). The liquid diffuses through the membrane from the more concentrated solution (sample or receptor solution) to the less concentrated one, to achieve an osmotic equilibrium. Therefore, osmotic conditions have to be completely controlled to avoid volume changes; except in the case of nanosystem concentration, where a decrease of the sample volume is required and it can be achieved using a hypotonic receptor solution [9]. It is worth noting that not only the liquid but also molecules with smaller molecular weight than the molecular weight cut-off of the membrane can also diffuse to achieve the osmotic equilibrium. For example, Vauthier et al. [46] used and demonstrated inverse dialysis as an appropriate methodology to concentrate polymeric nanoparticles using a receptor solution with a high osmotic pressure.

This technique is advantageous in terms of minimal sample manipulation, without the need of any pretreatment, but it is limited to the existent dialysis membranes. If the sample to dialyse is expected to interact with the membrane or if the molecular weight cut-off is not appropriate, this technique cannot be used. In addition, high volumes of the receptor solution are usually required [9, 46].

*5.4. Electrophoresis.* Electrophoretic techniques consist of the application of an electric field to a polymeric gel (composed mainly of agarose or polyacrylamide) submerged in a liquid buffer; through which charged molecules run depending on their charge and/or on their molecular weight. Although electrophoresis is used with many objective, one of the uses of this technique is the separation and purification of determined nanomaterials, which has been widely reported [9]. General examples of the purification use of electrophoresis are the following. Meyer et al. [47] reported the use of PAGE (polyacrylamide gel) electrophoresis for the purification of RNAs extracted from enzymatic synthesis or from cells, for example. They also demonstrated both native and denaturing electrophoretic conditions for the selection of RNAs of the required size.

The main advantages of this technique are its economic and simple performance together with the high resolution and sensitivity. However, only charged compounds can be purified through electrophoresis, and once the compound is separated in the gel, further purification steps are required to extract it from the gel [9, 45, 51].

Although it is not specifically a technique for the purification of nanosystems, it was considered appropriate to remark an electrophoretic type widely used to characterize the formation of nanocomplexes by the electrostatic interactions between their components. This technique is called electrophoretic mobility shift assay (EMSA). EMSA consists in a native (nondenaturing) polyacrylamide electrophoresis where samples migrate depending on their charge, under an electric field [48]. It is widely used in the nanomedicine field for the determination of the complexation ratio between two compounds, for example, polymeric nanoparticles and antisense oligonucleotides (Figure 4) [39]. The complexation is detected due to a retardation of the oligonucleotide band migration when complexes are formed, since complexes migrate slower than free nucleic acids [48].

Its rapid performance, simplicity, robustness, and high sensitivity make this methodology the choice for the study of electrostatic complexes formation of a wide range of compounds. Although EMSA is usually applied as a qualitative technique, under the appropriate conditions, it can be also useful to quantify the stoichiometry of complexation. It is also remarkable that multiple EMSA variants exist for various purposes, such as the time-course EMSA to measure the dissociation kinetics or the circular permutation to measure the DNA bending [48–50]. However, EMSA has also some disadvantages, such as the performance of the assay in non-chemical equilibrium [48].

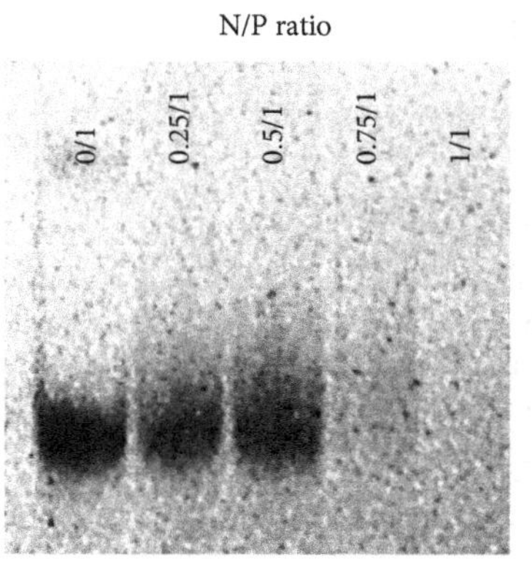


Figure 4: Example of an EMSA of polymeric nanoparticles conjugated with antisense oligonucleotides at different nanoparticle/oligonucleotide (N/P) charge ratios. When the complexation is achieved (zero surface charge), the band is diffused (at 0.75/1 in this case).

## 6. Conclusions

A complete characterization of nanomaterials intended for biomedical purposes (diagnostic, treatment, or theragnostic) is a must before the translation to preclinical and clinical studies. Classical chemistry analytical techniques can be applied for the characterization of some aspects of nanomaterials, since nanomaterial properties usually change from those of their components. Chromatographic techniques can be useful for the determination of the molecular weight of nanosystems, as well as for the purification of the produced nanoobject from raw materials and impurities. Spectroscopic techniques, in parallel, can be very useful for the confirmation of the formation of the nanosystems, assessing the materials of which it is composed as well as its aggregation/attachment state. Calorimetric techniques can be useful for the study of the nanomaterial behavior when submitting them to temperature changes. Finally, it is always required a purification step to ensure the obtaining of a safe nanosystem, free of impurities and raw materials. Therefore, as a general conclusion of this review, it is strongly recommended to take a look on the variety of existent techniques to look for all the aspects that must be known before the translation of a novel nanomaterial to a human diagnostic or therapeutic formulation.

## Conflicts of Interest

The authors declare that they have no conflicts of interest.

## Acknowledgments

This work was funded by Spanish Ministry of Economy and Competitivity, MINECO (grant CTQ2014-52687) and Generalitat de Catalunya (grant 2014-SGR-1655). CIBER-BBN is an initiative funded by the VI National R&D&I Plan 2008–2011, Iniciativa Ingenio 2010, Consolider Program, CIBER Actions and financed by the Instituto de Salud Carlos III with assistance from the European Regional Development Fund. Cristina Fornaguera is grateful to MINECO for their PTQ2015 grant.

## References

[1] E. J. Cho, H. Holback, K. C. Liu, S. A. Abouelmagd, J. Park, and Y. Yeo, "Nanoparticle characterization: state of the art, challenges, and emerging technologies," *Molecular Pharmaceutics*, vol. 10, no. 6, pp. 2093–2110, 2013.

[2] C. Fornaguera and C. Solans, "Methods for the in vitro characterization of nanomedicines—biological component interaction," *Journal of Personalized Medicine*, vol. 7, no. 1, p. 2, 2017.

[3] N. Segovia, P. Dosta, A. Cascante, V. Ramos, and S. Borrós, "Oligopeptide-terminated poly ($\beta$-amino ester) s for highly efficient gene delivery and intracellular localization," *Acta Biomaterialia*, vol. 10, no. 5, pp. 2147–2158, 2014.

[4] M. A. Dobrovolskaia, M. Shurin, and A. A. Shvedova, "Current understanding of interactions between nanoparticles and the immune system," *Toxicology and Applied Pharmacology*, vol. 299, pp. 78–89, 2016.

[5] P. C. Lin, S. Lin, P. C. Wang, and R. Sridhar, "Techniques for physicochemical characterization of nanomaterials," *Biotechnology Advances*, vol. 32, no. 4, pp. 711–726, 2014.

[6] R. M. Crist, J. H. Grossman, A. K. Patri et al., "Common pitfalls in nanotechnology: lessons learned from NCI's Nanotechnology Characterization Laboratory," *Integrative Biology*, vol. 5, no. 1, pp. 66–73, 2012.

[7] C. Fornaguera and C. Solans, "Characterization of polymeric nanoparticle dispersions for biomedical applications: size, surface charge and stability," *Pharmaceutical Nanotechnology*, vol. 6, 2018.

[8] I. Neverova and J. E. Van Eyk, "Role of chromatographic techniques in proteomic analysis," *Journal of Chromatography, B: Analytical Technologies in the Biomedical and Life Sciences*, vol. 815, no. 1-2, pp. 51–63, 2005.

[9] K. E. Sapsford, K. M. Tyner, B. J. Dair, J. R. Deschamps, and I. L. Medintz, "Analyzing nanomaterial bioconjugates: a review of current and emerging purification and characterization techniques," *Analytical Chemistry*, vol. 83, no. 12, pp. 4453–4488, 2011.

[10] T. Williams, "Gel permeation chromatography: a review," *Journal of Materials Science. Materials in Medicine*, vol. 5, no. 9, pp. 811–820, 1970.

[11] H. Determann, *Chromatography Gel Filtration, Gel Permea*, Springer Science and Business Media, 2012.

[12] J. C. Moore, "Gel permeation chromatography. I. A new method for molecular weight distribution of high polymers," *Journal of Polymer Sciences*, vol. 2, pp. 835–843, 1964.

[13] E. Temyanko, P. S. Russo, and H. Ricks, "Study of rodlike homopolypeptides by gel permeation chromatography with light scattering detection: validity of universal calibration and stiffness assessment," *Macromolecules*, vol. 34, no. 3, pp. 582–586, 2001.

[14] A. Rahimi, A. Ulbrich, J. J. Coon, and S. S. Stahl, "Formic-acid-induced depolymerization of oxidized lignin to aromatics," *Nature*, vol. 515, no. 7526, pp. 249–252, 2014.

[15] Y. Shen, M. Y. Gee, R. Tan, P. J. Pellechia, and A. B. Greytak, "Purification of quantum dots by gel permeation chromatography and the effect of excess ligands on shell growth and ligand exchange," *Chemistry of Materials*, vol. 25, no. 14, pp. 2838–2848, 2013.

[16] B. S. Flavel, K. E. Moore, M. Pfohl, M. M. Kappes, and F. Hennrich, "Separation of single-walled carbon nanotubes with a gel permeation chromatography system," *ACS Nano*, vol. 8, no. 2, pp. 1817–1826, 2014.

[17] T. Radhakrishna, D. S. Rao, and G. O. Reddy, "Determination of pioglitazone hydrochloride in bulk and pharmaceutical formulations by HPLC and MEKC methods," *Journal of Pharmaceutical and Biomedical Analysis*, vol. 29, no. 4, pp. 593–607, 2002.

[18] A. E. Gulyaev, S. E. Gelperina, I. N. Skidan, A. S. Antropov, G. Y. Kivman, and J. Kreuter, "Significant transport of doxorubicin into the brain with polysorbate 80-coated nanoparticles," *Pharmaceutical Research*, vol. 16, no. 10, pp. 1564–1569, 1999.

[19] C. Fornaguera, M. Llinàs, C. Solans, and G. Calderó, "Design and *in vitro* evaluation of biocompatible dexamethasone-loaded nanoparticle dispersions, obtained from nano-emulsions, for inhalatory therapy," *Colloids and Surfaces B: Biointerfaces*, vol. 125, pp. 58–64, 2015.

[20] J. Cielecka-Piontek, P. Zalewski, A. Jelińska, and P. Garbacki, "UHPLC: the greening face of liquid chromatography," *Chromatographia*, vol. 76, no. 21-22, pp. 1429–1437, 2013.

[21] S. Fekete, J. Schappler, J.-L. Veuthey, and D. Guillarme, "Current and future trends in UHPLC," *TrAC Trends in Analytical Chemistry*, vol. 63, pp. 2–13, 2014.

[22] K. Ulbrich, T. Hekmatara, E. Herbert, and J. Kreuter, "Transferrin- and transferrin-receptor-antibody-modified nanoparticles enable drug delivery across the blood-brain barrier (BBB)," *European Journal of Pharmaceutics and Biopharmaceutics*, vol. 71, no. 2, pp. 251–256, 2009.

[23] S. Wagner, F. Rothweiler, M. G. Anhorn et al., "Enhanced drug targeting by attachment of an anti αv integrin antibody to doxorubicin loaded human serum albumin nanoparticles," *Biomaterials*, vol. 31, no. 8, pp. 2388–2398, 2010.

[24] M. Sadao and H. G. Barth, *Size Exclusion Chromatography*, Springer Science and Business Media, 2013.

[25] L. K. Kostanski, D. M. Keller, and A. E. Hamielec, "Size-exclusion chromatography - a review of calibration methodologies," *Journal of Biochemical and Biophysical Methods*, vol. 58, no. 2, pp. 159–186, 2004.

[26] K. Rebolj, D. Pahovnik, and E. Žagar, "Characterization of a protein conjugate using an asymmetrical-flow field-flow fractionation and a size-exclusion chromatography with multi-detection system," *Analytical Chemistry*, vol. 84, no. 17, pp. 7374–7383, 2012.

[27] J. Alben and F. Fiamingo, "Fourier transformed infrared spectroscopy," in *Optical Techniques in Biological Research*, D. Rousseau, Ed., pp. 133–178, Bell telephone laboratories Inc., 1984.

[28] J. Lakowicz, *Principles of Fluorescence Spectroscopy*, Springer, 2010.

[29] C. Fornaguera, N. Feiner-Gracia, G. Calderó, M. J. Garcia-Celma, and C. Solans, "Galantamine-loaded PLGA nanoparticles, from nano-emulsion templating, as novel advanced drug delivery systems to treat neurodegenerative diseases," *Nanoscale*, vol. 7, no. 28, pp. 12076–12084, 2015.

[30] H. Chen, J. Gao, Y. Lu et al., "Preparation and characterization of PE38KDEL-loaded anti-HER2 nanoparticles for targeted cancer therapy," *Journal of Controlled Release*, vol. 128, no. 3, pp. 209–216, 2008.

[31] J. Lai, B. P. Shah, E. Garfunkel, and K. Lee, "Energy transfer-based mesoporous silica nanoparticles for real-time monitoring of drug release," *ACS Nano*, vol. 7, no. 3, pp. 2741–2750, 2013.

[32] B. Liu, H. Tan, and Y. Chen, "Upconversion nanoparticle-based fluorescence resonance energy transfer assay for Cr (III) ions in urine," *Analytica Chimica Acta*, vol. 761, pp. 178–185, 2013.

[33] J. Shi, F. Tian, J. Lyu, and M. Yang, "Nanoparticle based fluorescence resonance energy transfer (FRET) for biosensing applications," *Journal of Materials Chemistry B*, vol. 3, no. 35, pp. 6989–7005, 2015.

[34] A. Coats and J. Redfern, "Thermogravimetric analysis," *The Analyst*, vol. 88, no. 1053, pp. 906–924, 1963.

[35] I. Haq, D. Chowdhry, and T. Jenkins, "Calorimetric techniques in the study of high-order DNA-drug interactions," *Methods in Enzymology*, vol. 340, pp. 109–149, 2001.

[36] A. Hunt, "A simple connection between the melting temperature and the glass temperature in a kinetic theory of the glass transition," *Journal of Physics. Condensed Matter*, vol. 4, no. 32, pp. L429–L431, 1992.

[37] Z. Lu and J. Li, "Correlation between average melting temperature and glass transition temperature in metallic glasses," *Applied Physics Letters*, vol. 94, no. 6, article 061913, 2009.

[38] R. Berger, H. P. Lang, C. Gerber et al., "Micromechanical thermogravimetry," *Chemical Physics Letters*, vol. 294, no. 4-5, pp. 363–369, 1998.

[39] C. Fornaguera, S. Grijalvo, M. Galán et al., "Novel non-viral gene delivery systems composed of carbosilane dendron functionalized nanoparticles prepared from nano-emulsions as non-viral carriers for antisense oligonucleotides," *International Journal of Pharmaceutics*, vol. 478, no. 1, pp. 113–123, 2015.

[40] M. Colombo, S. Carregal-Romero, M. F. Casula et al., "Biological applications of magnetic nanoparticles," *Chemical Society Reviews*, vol. 41, no. 11, pp. 4306–4334, 2012.

[41] L. H. Reddy, J. L. Arias, J. Nicolas, and P. Couvreur, "Magnetic nanoparticles: design and characterization, toxicity and biocompatibility, pharmaceutical and biomedical applications," *Chemical Reviews*, vol. 112, no. 11, pp. 5818–5878, 2012.

[42] M. A. Dobrovolskaia and S. E. McNeil, "Handbook of immunological properties of engineered nanomaterials," in *Frontiers in Nanobiomedical Research*, SAIC-Frederick Inc., 1st edition, 2013.

[43] S. Roy, V. Jain, R. Bajpai et al., "Formation of carbon nanotube bucky paper and feasibility study for filtration at the nano and molecular scale," *Journal of Physical Chemistry C*, vol. 116, no. 35, pp. 19025–19031, 2012.

[44] R. Scopes, *Protein Purification: Principles and Practice*, Springer, New York, NY, USA, 1994.

[45] R. K. Scopes, *Protein Purification: Principles and Practice*, Springer Science and Business Media, 2013.

[46] C. Vauthier, B. Cabane, and D. Labarre, "How to concentrate nanoparticles and avoid aggregation?," *European Journal of Pharmaceutics and Biopharmaceutics*, vol. 69, no. 2, pp. 466–475, 2008.

[47] M. Meyer and B. Masquida, "Polyacrylamide gel electrophoresis for purification of large amounts of RNA," in *Nucleic Acid Crystallography*, E. Ennifar, Ed., vol. 1320 of Methods in Molecular Biology, Humana Press, New York, NY, USA, 2016.

[48] L. M. Hellman and M. G. Fried, "Electrophoretic mobility shift assay (EMSA) for detecting protein-nucleic acid interactions," *Nature Protocols*, vol. 2, no. 8, pp. 1849–1861, 2007.

[49] J. Gerstle and M. G. Fried, "Measurement of binding kinetics using the gel electrophoresis mobility shift assay," *Electrophoresis*, vol. 14, no. 1, pp. 725–731, 1993.

[50] D. Senear and M. Brenowitz, "DNA bending in protein-DNA complexes," *Methods in Enzymology*, vol. 208, pp. 118–146, 1991.

[51] J. C. Jason, *Protein Purification: Principles, High Resolution Methods, and Applications*, John Wiley and Sons, 2012.

16

# Study on Microstructure Effect of Carbon Black Particles in Filled Rubber Composites

**Li Hong Huang**,[1,2] **Xiaoxiang Yang**,[1,3] **and Jianhong Gao**[1,3]

[1]*College of Chemical Engineering, Fuzhou University, Fuzhou, Fujian 350108, China*
[2]*Fuzhou University Zhicheng College, Fuzhou, Fujian 350002, China*
[3]*Quanzhou Normal University, Quanzhou, Fujian 362000, China*

Correspondence should be addressed to Xiaoxiang Yang; yangxx@fzu.edu.cn

Academic Editor: György Bánhegyi

The cross sections of blended natural/styrene-butadiene (NSBR) composites filled with different volume fractions of carbon particles were observed using a Quanta 250 scanning electron microscope. In addition, the sizes and distributions of the carbon particles were analyzed using Nano Measurer. A two-dimensional representative volume element model (RVE) for a rubber composite reinforced with circular carbon particles was established, and the uniaxial tensile behaviors of polymer nanocomposites with different particle size distribution patterns were simulated using the ABAQUS software. The results showed the following. (1) For the random models, if the difference of particle size was larger and particle distance was closer, stress distribution would be denser as well as the stress concentration would become greater. However, if the difference of particle size was small, for the case of same particle volume fraction, the particle size has little influence on the macromechanical properties whether the average size is large or small. (2) The correlation between the volume fraction and distribution of the carbon particles revealed that when the volume fraction of carbon black particles was larger than 12%, clusters between carbon particles in the polymer nanocomposites could not be avoided and the modulus of the composites increased with an increase in the cluster number.

## 1. Introduction

As a nanoparticle enhancer, carbon black has been widely used to improve the strength and toughness of polymer nanocomposites [1, 2]. The reinforcing effect mainly depends on the size, shape, and dispersion degree of the particles in the matrix, as well as the interface adhesion between the particle and matrix. The mixing process for carbon black-filled rubber includes four stages: mixing, dispersing, blending, and plasticizing. After the mixing process, the carbon black particles are fully mixed and then dispersed to form new polycarbon particles with a size of 0.2 $\mu$m. The size of the carbon black particles is changed by the mixing process. The bonding and adsorption degree between the carbon black particles and matrix depend on the new size of the carbon particles [3]. Additionally, because of the limited technological conditions and interaction of the particles' Van der Waals forces, carbon black particles will form a large number of aggregates with the nonuniform structure filler. The shear modulus and elastic modulus of the rubber will be greatly changed according to the differences in the concentration and number of carbon black particles. When the distribution of the carbon black in the composition is bad, the aggregates will increase and the rubber modulus will decrease, which is the Payne effect [4].

Finite element modeling can be used to explain the mechanical properties of the polymer composites. These models are used to reveal the concrete features of the nanostructure, enhancer-matrix interactions, and debonding on the macro- and microstructure. Because of the complexity of the calculations, finite element modeling provides a geometry-based periodic unit cell model called a representative volume element (RVE) to analyze relatively small volumes and deformations, instead of large deformations and complex boundary conditions. The single particle model [5–7] and multiparticle model [8, 9] have been widely used to

TABLE 1: Ingredients of carbon black-filled NSBR composites.

| Rubber | NR | SBR | N220 | ZnO | Octadecanoic acid | Age resister 4020 | Brim stone | Accelerant |
|---|---|---|---|---|---|---|---|---|
| NSBR1 | 80 | 20 | 0 | 5 | 2 | 0.7 | 2.6 | 1.3 |
| NSBR3 | 80 | 20 | 20 | 5 | 2 | 0.7 | 2.6 | 1.3 |
| NSBR4 | 80 | 20 | 30 | 5 | 2 | 0.7 | 2.6 | 1.3 |
| NSBR6 | 80 | 20 | 50 | 5 | 2 | 0.7 | 2.6 | 1.3 |

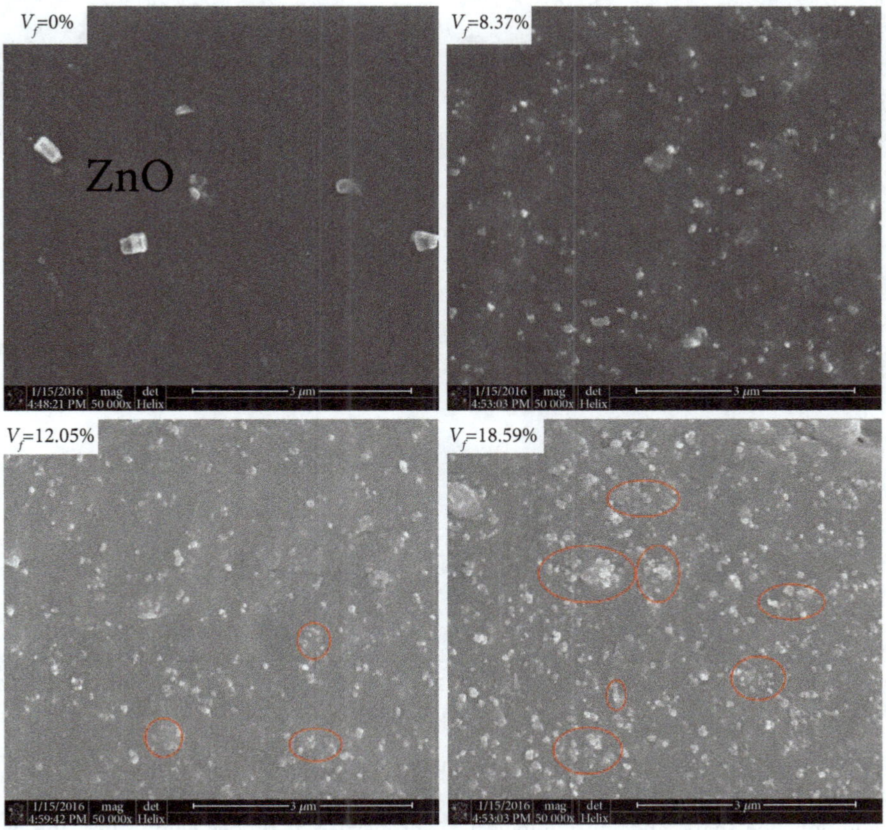


FIGURE 1: SEM photos of rubber composites with different carbon black volume fractions.

analyze the mechanical behaviors of polymer composites. The particle random model [10] and random sequential adsorption model [11] are often used for particle-filled composites. They can effectively embody the macroscopic mechanical properties but ignore the effect of the real mesostructure particle distribution. The new size of the carbon black particles and the number and size of the aggregates in polymer nanocomposites have a great impact on the mechanical properties of the composites.

In the present work, scanning electron microscope (SEM) observations of natural/styrene-butadiene blend rubber (NSBR) composites filled with different volume fractions of carbon particles were performed, as reported in the next section. The distribution of the carbon black particles and whether they are clustered in the polymer nanocomposites were investigated using Nano Measurer, as discussed in Section 3. Structural Modeling presents the results of a numerical simulation using the ABAQUS software, where the results for different carbon structure RVE models are discussed and compared. At the same time, the mechanical properties of the blends were studied. The final section presents the conclusions.

## 2. SEM Analysis

*2.1. Rubber Ingredient.* In order to investigate the mixing of rubber composites with different carbon black volume fractions, four kinds of specimens were prepared according to the GB528-2009 standard. The ingredients listed in Table 1 show the use of 0, 20, 30, and 50 phr of carbon black in the composites, corresponding to filler volume fractions of 0%, 8.37%, 12.05%, and 18.59%, respectively.

*2.2. SEM Analysis.* A morphological study was carried out using the Quanta 250 SEM. The test specimens came from the same batch of rubber. First, they were cooled down to the glass transition temperature by liquid nitrogen. They were then cut into two parts with smooth surfaces in the frozen state, and the fracture surface morphological structures were observed after the application of special spray

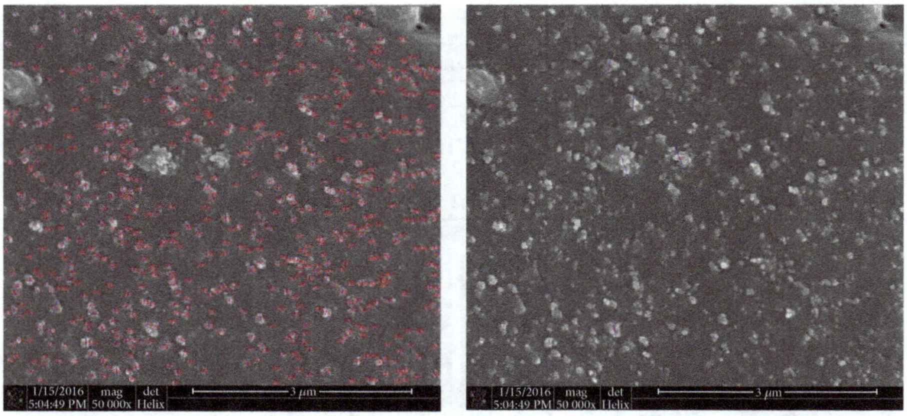


(a) Particles except clusters (b) Clusters

FIGURE 2: Particle size and number analysis of NSBR6 rubber.

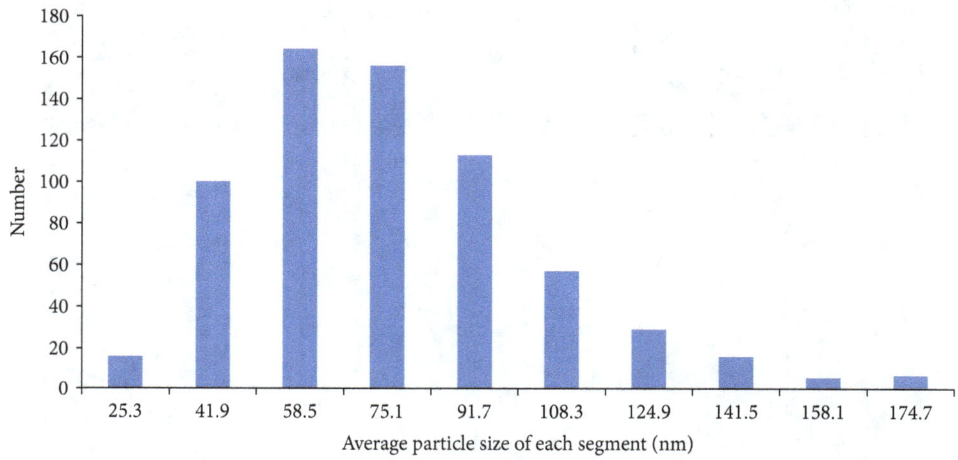


FIGURE 3: Particle size and diameter statistics of NSBR6 rubber.

gold craft paint. In addition, the shape and morphological distribution of the carbon black and whether defects such as holes or cracks existed were observed. The results of the SEM observations of the four kinds of rubber composites with different carbon black volume fractions are shown in Figure 1.

As seen in the SEM image in Figure 1, the two kinds of rubber matrixes, the natural rubber and styrene-butadiene rubber with no carbon black particles filled, are well mixed. The white particles in Figure 1 ($V_f = 0\%$) are ZnO, which had an average size of 200 nm. When the filler volume fraction is approximately 8.37%, the added carbon black (N220) is uniformly dispersed and the particle size is consistent, except for the ZnO. When the filler volume fraction is approximately 12.05%, the carbon black particles are uniformly dispersed, while a small amount of carbon black particles is aggregated. When the filler volume fraction is approximately 18.59%, aggregation appears, with most of the aggregates are small clusters or multiparticle agglomerations. This is a result of the increase in the carbon black volume fraction. In order to discuss the effects of the carbon black morphological distribution and shape on the macroscopic mechanical properties of rubber composites, the particle size distribution will be investigated statistically.

## 3. Particle Size and Distribution Statistics

Figure 2 shows the particle size and distribution of the carbon black particles of NSBR6 found using Nano Measurer. The statistical results for all the particles except clusters are given in Figure 2(a), with those for the clusters are provided in Figure 2(b). As shown in Figure 3 and Table 2, the average particle size of all the carbon black particles in the SEM picture is 75.88 nm and it is composed of 662 different particles. To investigate the influence of different diameters on the structure, different diameters were counted in ten segments with various sizes and numbers in the range of 17.45–182.15 nm. The average particle sizes of the clusters are 250–460 nm.

## 4. Structural Modeling

*4.1. RVE Size Determination and Boundary Conditions.* The particle-filled composite was approximately regarded as a periodic set of RVEs, and the mesomechanical properties of

TABLE 2: Statistics results for particle size.

| Diameter range (nm) | 17–33.6 | 33.6–50.2 | 50.2–66.8 | 66.8–83.4 | 83.4–100 | 100–116.6 | 116.6–133.2 | 133.2–149.8 | 149.8–166.4 | 166.4–183 |
|---|---|---|---|---|---|---|---|---|---|---|
| Average particle size (SEM) (nm) | 25.3 | 41.9 | 58.5 | 75.1 | 91.7 | 108.3 | 124.9 | 141.5 | 158.1 | 174.7 |
| Particle number (SEM) | 16 | 100 | 164 | 156 | 113 | 57 | 29 | 16 | 5 | 6 |
| Percentage | 2.4% | 15.1% | 24.8% | 23.6% | 17.1% | 8.6% | 4.4% | 2.4% | 0.8% | 0.9% |
| Total number | | | | | | 662 | | | | |
| Total average particle size (nm) | | | | | | 75.88 | | | | |

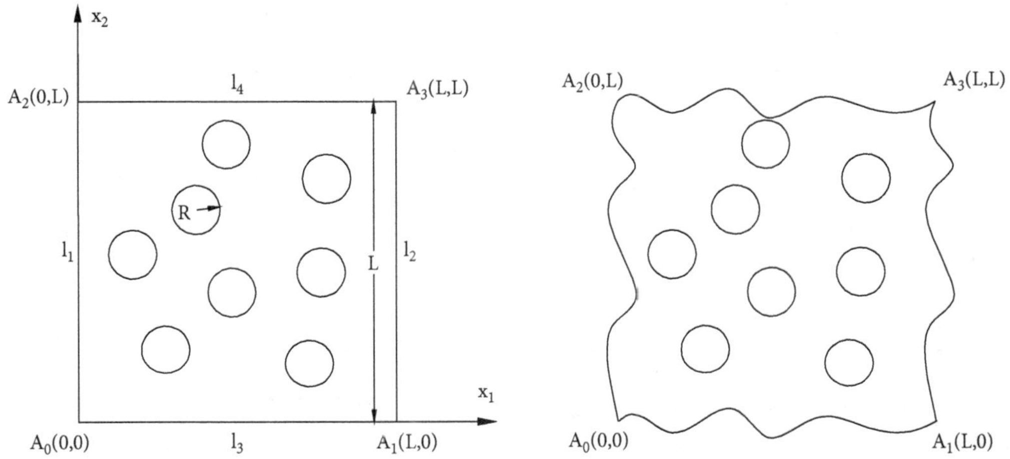


FIGURE 4: Schematic of the 2D RVE model.

the composite were studied by applying periodic boundary conditions to the RVE models. A large number of studies have been effectively performed on the macroscopic properties of composite materials [12–14]. A 3D model is the best choice for numerical simulations. However, because of computational limitations, the 3D multiparticle RVE model could only converge in a very small strain. Therefore, a 2D RVE model was adopted. Because the area fraction used in the model represented the volume fraction of the composite, the results calculated with the 2D RVE model in ABAQUS deviated from the experimental results. However, the simulation results could be used for a qualitative discussion when the particle size distribution was the same. The size of the RVE elements and the choice of boundary conditions have previously been discussed [9, 15].

During the simulation process, the physical properties of the material, elastic modulus, and Poisson's ratio were related to the ratio $L/R$, in which $L$ was the side length of the RVE and $R$ was the radius of a particle. In other words, the particle and RVE sizes determined the reliability of the simulation. When $L/R$ was very small, the numerical solution greatly deviated from the test value. In contrast, when $L/R$ was large enough, the solution was more stable. The two-dimensional RVE model is shown in Figure 4. In order to verify the stability of the model, RVE models with different $L/R$ values were established, with a volume fraction $V_f$ of carbon black particles in NSBR of 10%. The equivalent elastic modulus values

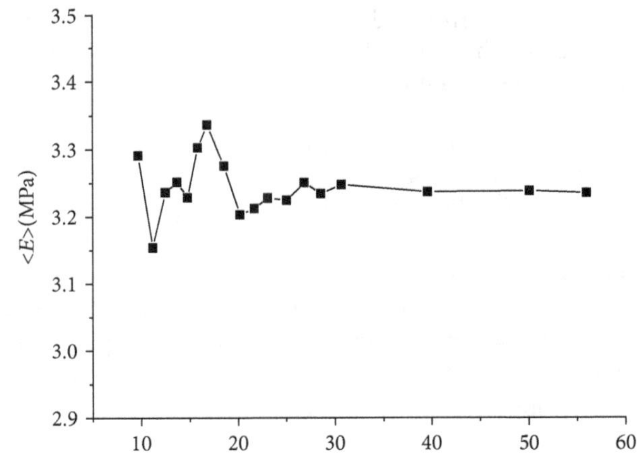


FIGURE 5: Equivalent elastic modulus <E> values of NSBR rubber composites ($V_f$ = 10%) with different $L/R$ ratios.

of the composite materials <E> [5] in the RVE with different $L/R$ values were calculated.

Figure 5 shows the equivalent elastic modulus <E> values of the NSBR rubber composites with different $L/R$ ratios at $V_f$ = 10%. It was found that when the value of $L/R$ was small, the equivalent elastic modulus had a range of 3.15–3.35. When $L/R$ reached to 25, the equivalent elastic modulus tended to be stable. According to the statistical results, RVE

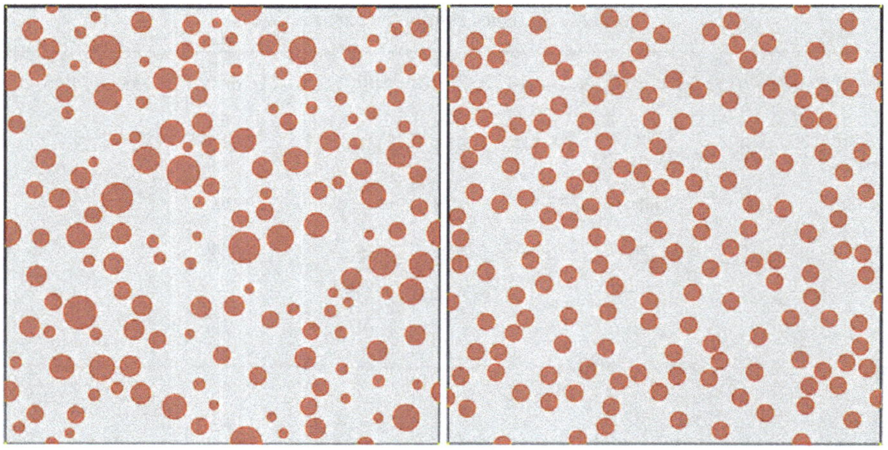

FIGURE 6: Two kinds of RVE random adsorption sequence models.

TABLE 3: Results for particle size and diameter in RVE ($V_f = 18.59\%$).

| Diameter range (mm) | 33.6–50.2 | 50.2–66.8 | 66.8–83.4 | 83.4–100 | 100–116.6 | 116.6–133.2 | 133.2–149.8 | 149.8–166.4 |
|---|---|---|---|---|---|---|---|---|
| Average particle size (SEM) (nm) | 41.9 | 58.5 | 75.1 | 91.7 | 108.3 | 124.9 | 141.5 | 158.1 |
| Particle number (SEM) | 100 | 164 | 156 | 113 | 57 | 29 | 16 | 5 |
| Average particle size (RVE) (nm) | 40 | 50 | 75 | 90 | 110 | 120 | 140 | 150 |
| Particle number (RVE) | 23 | 40 | 36 | 28 | 13 | 7 | 4 | 2 |

side length $L$ was on the micrometer scale and $L/R$ had a value greater than 25.

The periodic boundary conditions must be applied to the RVE model [16]. The nodes on the borders repeat those on the opposite side borders, and the displacements and deformations of the corresponding nodes on two opposite boundaries remain the same. For this purpose, the boundary displacement constraint is considered as follows:

$$\begin{aligned} u(x_1, L) &= u(x_1, 0) + \left(u_{A_2} - u_{A_0}\right), \\ u(L, x_2) &= u(0, x_2) + \left(u_{A_1} - u_{A_0}\right), \end{aligned} \quad (1)$$

where $u_{A_i}$ are the displacement vectors of $A_i$, $i = 0,1,2$. Respectively, $l_1$, $l_2$, $l_3$, and $l_4$ are the borders of $x_1 = 0$, $x_1 = L$, $x_2 = 0$, and $x_2 = L$, as shown in Figure 4. The displacement vector of the points in $l_i$ are $u_{l_i}$, which satisfy the following conditions:

$$\begin{aligned} u_{l_4} &= u_{l_3} + \left(u_{A_2} - u_{A_0}\right), \\ u_{l_2} &= u_{l_1} + \left(u_{A_1} - u_{A_0}\right). \end{aligned} \quad (2)$$

### 4.2. Structural Modeling

*4.2.1. Modeling.* The geometric model of the 2D RVE of the composite material was established according to the statistical results for the carbon black in the NSBR6 rubber presented in Figure 3 and Table 2. The side length of the RVE $L = 2\ \mu$m, because the sizes of RVE unit and SEM unit are different; in order to ensure the number of particles is integer, the average diameter of particles is slightly different from that of SEM but make sure that the area fraction is the same and the numerical results are not affected. For the same volume fraction $V_f = 18.59\%$, the numbers of particles with different diameters in the RVE are calculated and listed in Table 2. To improve the convergence and accuracy, the maximum and minimum particle diameters are deleted. The average particle diameter of each segment is set for the different diameters in the RVE, and the ratio $L/R \geq 26.7$ for RVE side length $L = 2\ \mu$m, maximum particle size $D = 150$ nm, and $R = 75$ nm. The solution is stable. Two kinds of RVE random adsorption sequence models are shown in Figure 6. For the random adsorption sequence model of the carbon particle diameter segment, the average particle diameter of each segment is listed in Table 3. The average diameter of all particles is 75.88 nm, while the number of particles is 168 when the side length of the RVE is $L = 2\ \mu$m and the volume fraction is 18.59%. The number of particles and mesh generation of the two kinds of models is shown in Table 4.

TABLE 4: Comparison of two kinds of models.

| | Number of particles | Number of grids |
|---|---|---|
| Same particle size model | 168 | 52683 |
| Different particle size model | 153 | 52995 |

The main mechanical parameters used in the model are as follows. Carbon black was supposed to be a linear and elastic material. Its elastic modulus $E_f = 500$ MPa, and Poisson's ratio $\mu_f = 0.3$ [5]. The parameters of the experimental data are fitted based on the multistep relaxation stretching cyclic

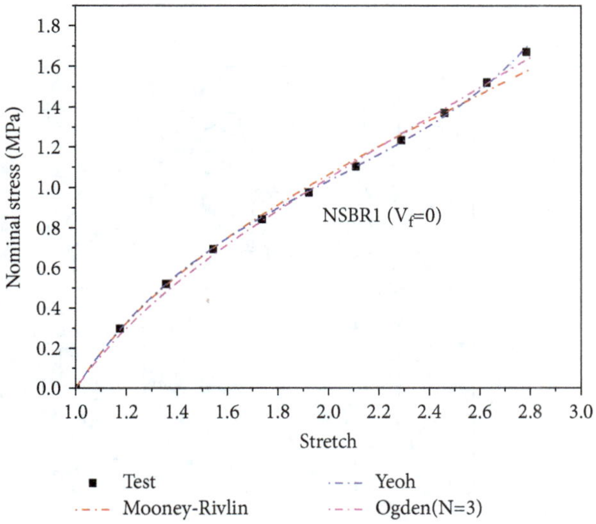


FIGURE 7: Fitting results of some constitutive models for tensile experimental data of NSBR1.

TABLE 5: Material parameters of Yeoh models (NSBR1).

| Rubber | $C_{10}$ (MPa) | $C_{20}$ (MPa) | $C_{30}$ (MPa) |
|---|---|---|---|
| NSBR1 | 0.652 | −0.0119 | 0.001367 |

loading and unloading test of the NSBR blends at room temperature using the automatic control testing machine of SHIMADZU AG-plus 50 kN, with the results shown in Figure 7. The Yeoh constitutive model is adopted for the matrix material, and the material parameters are summarized in Table 5. The Yeoh strain energy function under uniaxial extension is given by (3). The adaptive grid type is the quadratic complete integral of the plane strain eight-node quadrilateral hybrid unit CPE8H for the matrix and CPE8 for the carbon. Simulated uniaxial tension tests are performed using the periodic boundary conditions.

$$W_{\text{Yeoh}} = C_{10}(I_1 - 3) + C_{20}(I_1 - 3)^2 + C_{30}(I_1 - 3)^3. \quad (3)$$

In order to investigate these different random causes, the same analysis of the different random models was performed for a volume fraction of 18.59%. All of the stress-strain curves of the composites (Figures 8 and 9) show that these results are almost coincident and the random error is small.

*4.2.2. Results.* Figures 10(a) and 10(b) show the distribution of the Mises stress isolines with the different particle size random model and same particle size random model in the case of the uniaxial tensile test with displacement $x = 2\ \mu$m; models are created with four RVEs in the $x$ and $y$ directions in an RVE array. As seen in Figure 9, the stress isoline at the joint boundary of two adjacent RVEs is continuous and closed in the loading process. In other words, the stress isoline of the boundary in the RVE is continuous, which satisfies the uniform stress in the boundary of the corresponding node of the RVE and conforms to the stress continuity of the periodical structure.

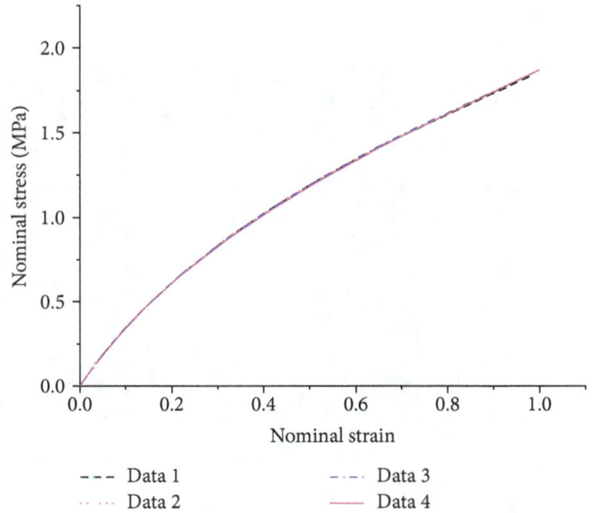


FIGURE 8: Stress-strain curves of different stochastic models with same particle size for NSBR6 in case of uniaxial tensile.

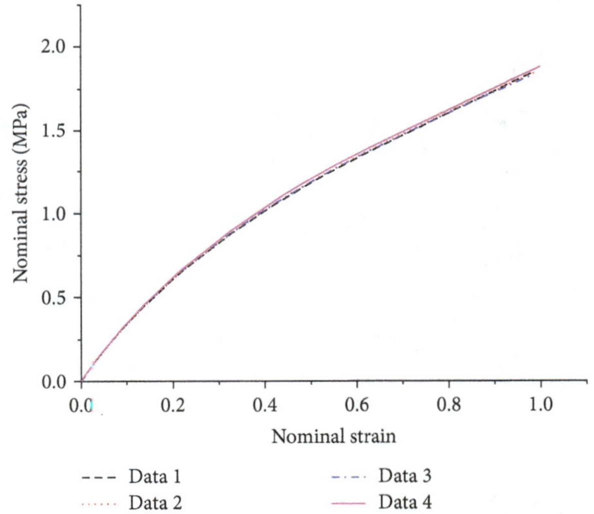


FIGURE 9: Stress-strain curves of different stochastic models with different particle sizes for NSBR6 in case of uniaxial tensile.

Figures 11(a) and 11(b) show the deformations and Mises stress isolines with the different particle size random model and same particle size random model in uniaxial tensile with displacement $x = 2\ \mu$m. By comparison with Figure 6, it can be found that the Mises stress distributions of the two kinds of models are not uniform. In the dense particle distribution area, the stress concentration is more obvious and the stress isolines are denser when the particle positions are closer. Because of the different particle sizes, the stress concentration is very serious in the dense region of the large particles. Smaller particles show more diversity in the particle size difference and a smaller stress gradient. In the same particle size random model, the stress isolines are sparse, which indicates that the stress concentration of the different particle size stochastic models is more obvious than that of the same particle size random models.

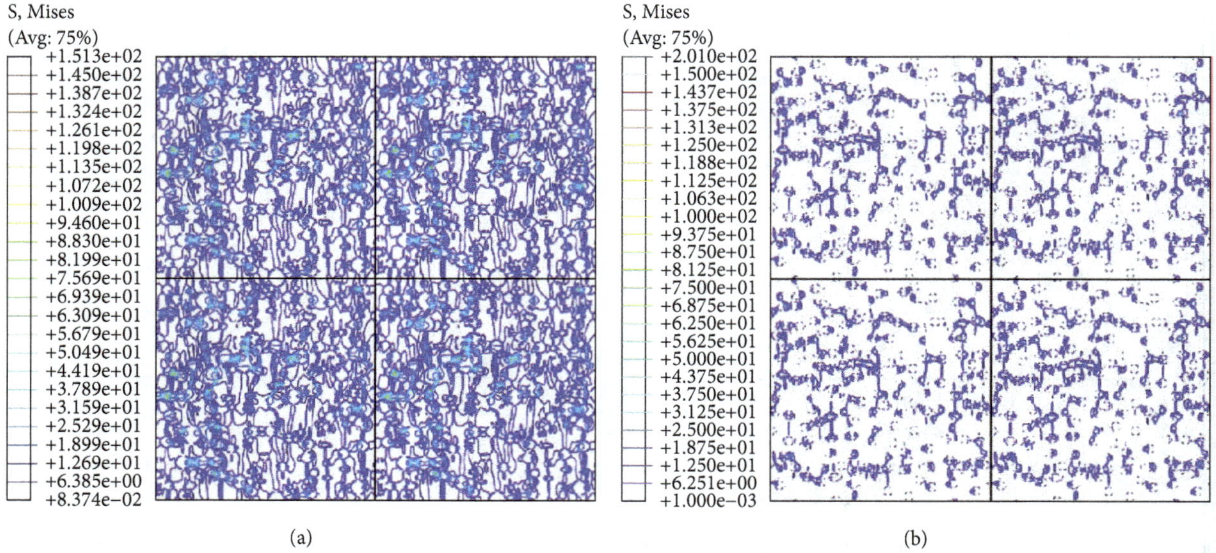


FIGURE 10: Distribution of Mises stress in two kinds of RVE models: (a) different particle size random model and (b) same particle size random model.

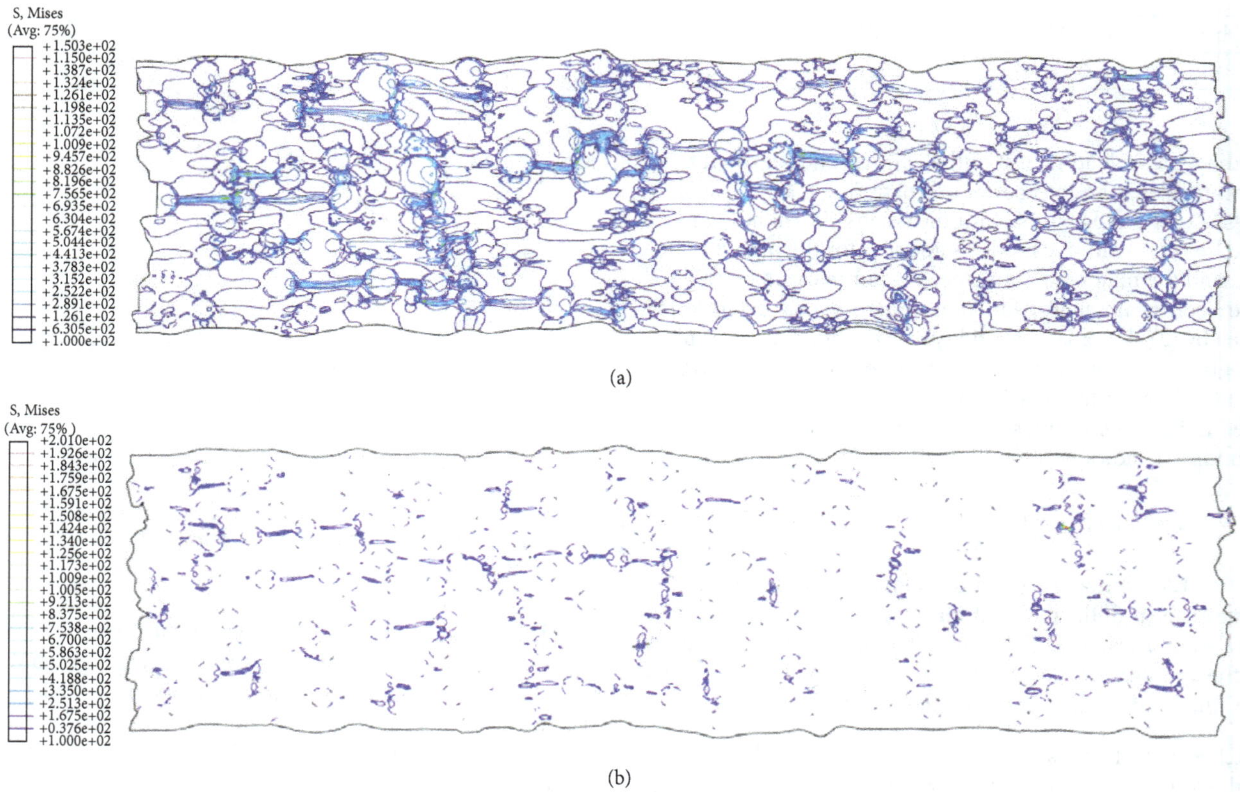


FIGURE 11: Deformation and distribution of Mises stress in two kinds of RVE models: ($\varepsilon = 1$) (a) different particle size random model and (b) same particle size random model.

Figure 11(a) plots the deformation in the uniaxial tensile state. The stress-deformation relationship between the matrix and particles is complex because of the influence of the distance between the particles and the particle size. It is not difficult to find that a series of shear bands [17] is formed when the $x$ direction is in uniaxial tensile. In the continuous shear zone, the stress isolines in the particles and matrix are dense and the maximum stress lies in it. At the same time, when the dense shear zone appears, the stress concentration will also occur here. When the differences in particle size are greater or a close particle has a larger particle size, the stress shear zone will appear, the stress gradient becomes

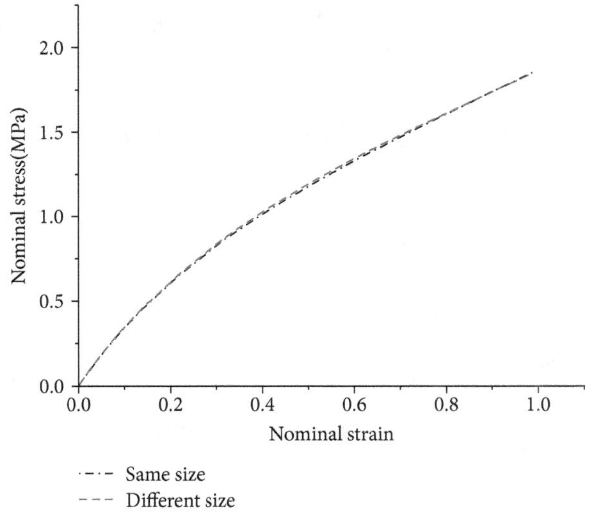


FIGURE 12: Nominal stress-strain curves for two kinds of RVE models.

larger, and the stress concentration is obvious. This indicates that the stress concentration is mainly determined by the differences in the positions and sizes of the particles in the tensile direction and also influenced by the distribution of the particles in the vertical direction.

The deformation is displayed in Figure 11(b). A similar shear band between the particles is formed, but it is not obvious. This is because there is no difference in particle size. In addition, the maximum stress is 151.3 MPa in the different particle size random model, which is located between two similar large particles in the shear band. However, the maximum stress is 201 MPa in the same particle size random model, which is located between the two nearest particles with $x$ direction distance $x = 0.0108$ $\mu$m. Therefore, it can be concluded that the complex stress-deformation relationship between the matrix and particles mainly depends on the distance between the particles in the stretching direction, the relative position, and the particle size. When the particle distance in the tensile direction is closer, it is easier to form the shear zone in the tensile direction and the larger difference in the particle size will lead to an obvious stress concentration.

The nominal stress-strain curves of the two kinds of models are shown in Figure 12. The macroscopic stress-strain curves of the different random models are basically coincident in the case with the same numbers of meshes and particles. The results show that the particle size difference for the rubber specimen prepared in this study will not have a great influence on the macroscopic tensile properties, but interfacial debonding occurs between the matrix and carbon particles during the tensile process and the stress concentration has a significant effect on the stress-strain after this interfacial debonding.

### 4.3. Effects of Different Particle Clusters

*4.3.1. Modeling.* Based on the above analysis, the distance between the particles has a great influence on the stress concentration of the composites. Here, the random sequence adsorption (RSA) algorithm [18] is adopted to discuss the distribution pattern of the particles instead of the ordinary random model in the mesomodel because the RSA ensures that the shortest distance between the particles is $\geq 0.07R$ and the convergence of the calculation is faster. Taking into account the spacing of the particles when they aggregate, the spacing between any two particles in the clusters is set to be the shortest distance $0.07R$. The average particle size of the clusters is 250–460 nm (Figure 2), and 19 particles form a cluster. Figure 13 presents the random models of different numbers of clusters. The numbers of clusters and grids of the different cluster models are listed in Table 6. The tensile behaviors of the filled rubber with different numbers of clusters will be discussed.

*4.3.2. Result.* Figure 14 shows the deformation and von Mises stress isoline (same particle size) of the NSBR6 for a nominal strain $\varepsilon = 0.4$. The von Mises stress contour in Figure 14 shows that the stress concentrations in composites with clusters are more dramatic than that with no clustered particles in uniaxial tensile. The maximum von Mises stress values are different for different numbers of clusters. Specifically, the maximum stress is 47.82 MPa in the RVE model with no clustered particles, and the maximum stress values are 52.58 MPa, 53.94 MPa, 55.97 MPa, 54.61 MPa, 58.56 MPa, and 101.7 MPa in the RVE models with 1, 2, 3, 4, 5, and 6 clusters, respectively. The maximum stress increases with the number of clusters. The contour density of the other particles is much smaller than that of the clusters. At the same time, the stress distribution of particles near the clusters is also affected. The contour distribution of the other particles is narrower when the number of clusters is greater. The stress in the tensile direction of the thin layer rubber matrix between the particles in the clusters is relatively larger, and it is possible for interfacial debonding or cracking to appear.

Figure 15 shows the nominal stress-strain curves of NSBR6 with different numbers of clusters. The curves show that the elastic modulus of the filled rubber slightly increases when the number of clusters is greater. Therefore, for the same strain, the nominal stress increases with the number of clusters. For example, at nominal strain $\varepsilon = 0.4$, the nominal stress is 1.026 MPa in the RVE model with no clustered particles and the nominal stress values are 1.035 MPa, 1.048 MPa, 1.093 MPa, 1.128 MPa, 1.133 MPa, and 1.230 MPa in the RVE models with 1, 2, 3, 4, 5, and 6 clusters, respectively. Comparing Figures 13 and 14, it can be concluded that the macroscopic properties of the composites have a meaningful relationship to whether the particles cluster in the material and to the number of clusters. A significant stress concentration in the composites will cause the material to generate microscopic and macroscopic defects, and the convergence will be reduced. Because the results calculated with the 2D RVE model in ABAQUS were deviated from the experimental results as we discussed before, the simulation results were used for a qualitative discussion. The modulus of the calculated curve increased as the number of clusters increased, and experimental results were more close to those from the

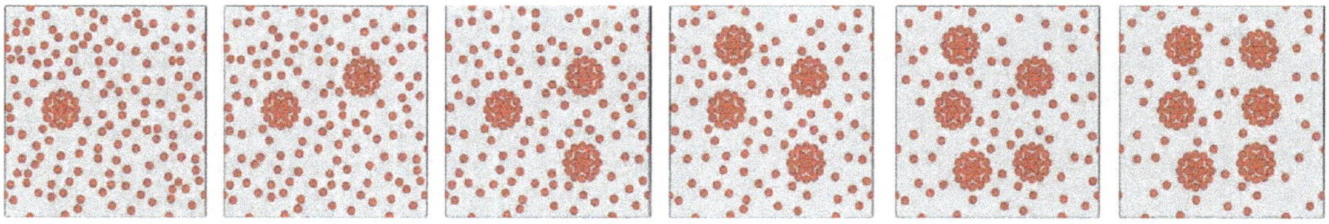

FIGURE 13: Different carbon cluster numbers in same particle size models.

TABLE 6: Grid numbers of different cluster models.

| Size | Same particle size | | | | | | |
|---|---|---|---|---|---|---|---|
| Cluster number | 0 | 1 | 2 | 3 | 4 | 5 | 6 |
| Grid number | 52683 | 52818 | 52714 | 52948 | 52930 | 53182 | 53366 |

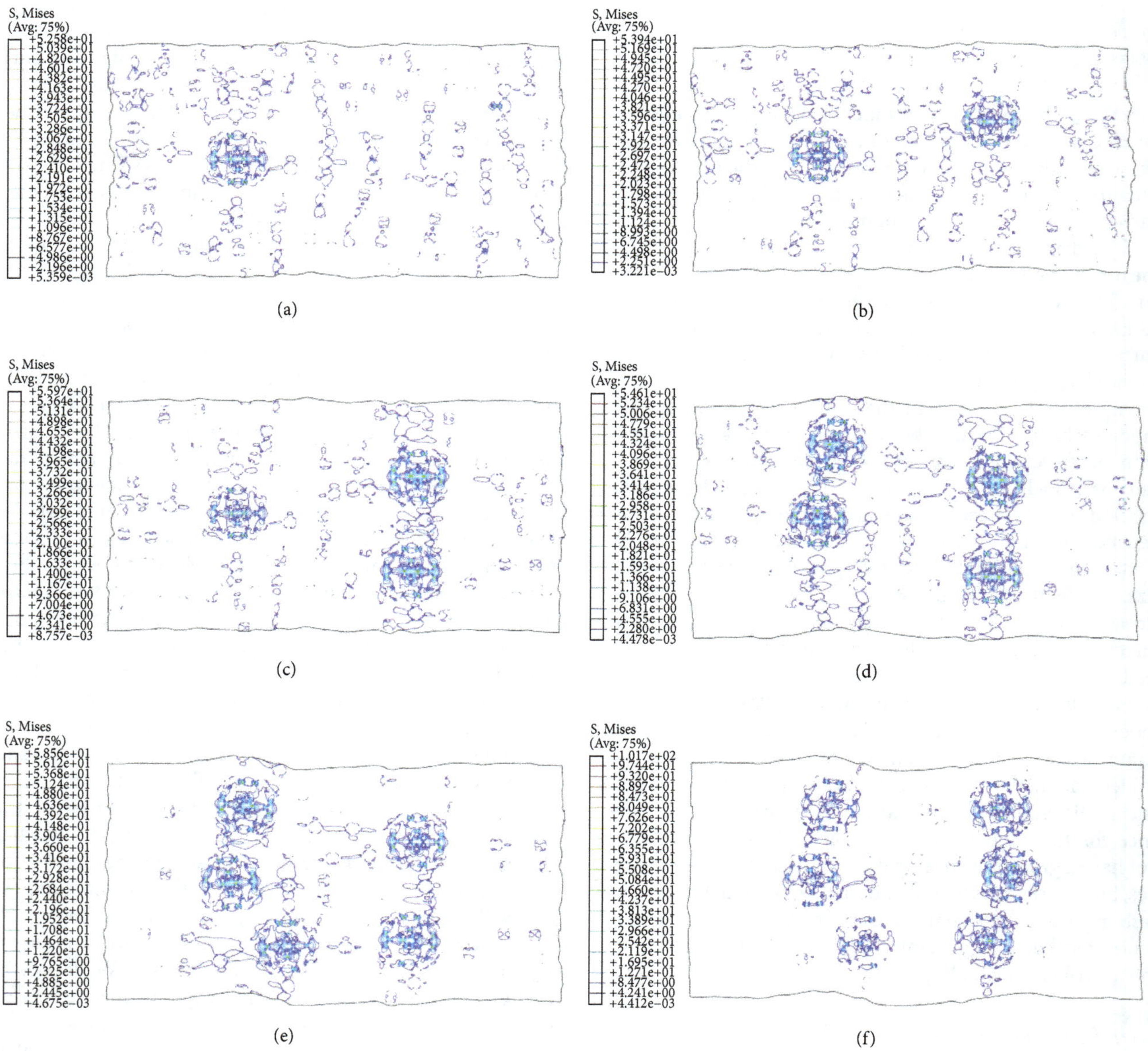


FIGURE 14: Deformation and distribution of Mises stress in the RVE model with different numbers of clusters ($\varepsilon = 0.4$) (same size): (a) one cluster, (b) two clusters, (c) three clusters, (d) four clusters, (e) five clusters, and (f) six clusters.

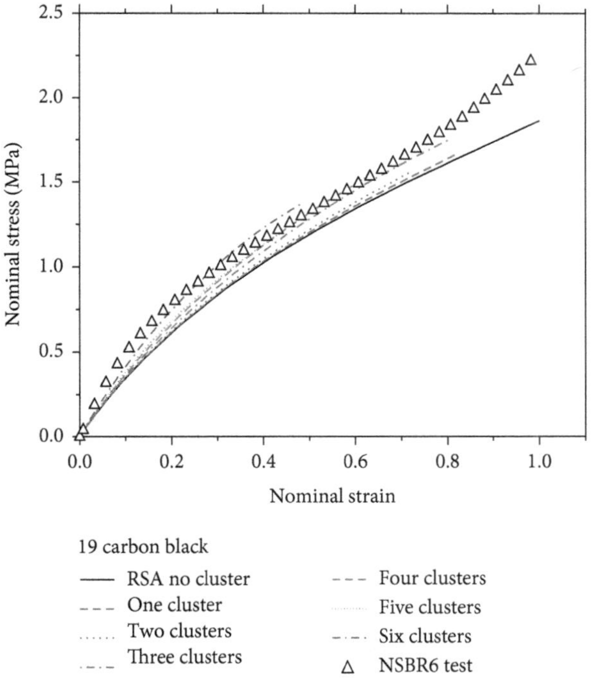


FIGURE 15: Stress-strain curves of NSBR6 with different numbers of clusters (same particle size).

simulations, which took into account clusters with a smaller deviation.

## 5. Conclusions

The SEM results for NSBR composites showed that the carbon black is dispersed more and more unevenly in the mixing process with an increase in the carbon black content. Therefore, the agglomeration phenomenon cannot be neglected when the carbon black N220 particle volume fraction reaches 12%.

For this reason, a combined statistical-numerical approach was adopted to determine the RVE size. When $L/R$ reaches a certain value, the equivalent elastic modulus tends to be stable. In addition, the stress distribution and stress concentration are related to the size, spacing, and relative positions of particles on the mesoscale. However, there is no difference in the macroscopic stress-strain curve.

This is caused by the fact that the material with localized cluster particles loses homogeneity. The simulated results indicated that the elastic modulus of a composite increased slightly when the clusters were taken into account. The macroscopic properties of the composites have a meaningful relationship to the size and amount of the particle clusters. A significant stress concentration would lead to inner defects.

## Conflicts of Interest

The authors declare that they have no conflicts of interest.

## Acknowledgments

The support from NSFC (Grant no. 11372074) and Education Science and Technology Project JAT170850 of Fujian Province is appreciated.

## References

[1] B. C. Kim, S. W. Park, and D. G. Lee, "Fracture toughness of the nano-particle reinforced epoxy composite," *Composite Structures*, vol. 86, no. 1-3, pp. 69–77, 2008.

[2] N. Zheng, Y. Huang, H.-Y. Liu, J. Gao, and Y.-W. Mai, "Improvement of interlaminar fracture toughness in carbon fiber/epoxy composites with carbon nanotubes/polysulfone interleaves," *Composites Science and Technology*, vol. 140, pp. 8–15, 2017.

[3] B. F. Guan, G. F. Ma, R. M. Wei, and B. Y. Chen, "Rubber fatigue failure model based on the dispersion degree of carbon black in filler rubber," *World Rubber Industry*, vol. 38, no. 6, pp. 20–23, 2011.

[4] G. Heinrich and T. A. Vilgis, "A statistical mechanical approach to the Payne effect in filled rubbers," *Express Polymer Letters*, vol. 9, no. 3, pp. 291–299, 2015.

[5] J. S. Bergstrom and M. C. Boyce, "Mechanical behavior of particle filled elastomers," *Rubber Chemistry and Technology*, vol. 72, no. 4, pp. 633–656, 1999.

[6] Y. Fukahori and W. Seki, "Stress analysis of elastomeric materials at large extensions using the finite element method. Part II. Stress and strain distribution around rigid spherical particles," *Journal of Materials Science*, vol. 28, no. 16, pp. 4471–4482, 1993.

[7] V. V. Moshev and L. L. Kozhevnikova, "Structural cell of particulate elastomeric composites under extension and compression," *International Journal of Solids and Structures*, vol. 39, no. 2, pp. 449–465, 2002.

[8] M. D. Rintoul and S. Torquato, "Reconstruction of the structure of dispersions," *Journal of Colloid and Interface Science*, vol. 186, no. 2, pp. 467–476, 1997.

[9] R. J. M. Smit, W. A. M. Brekelmans, and H. E. H. Meijer, "Prediction of the mechanical behavior of nonlinear heterogeneous systems by multi-level finite element modeling," *Computer Methods in Applied Mechanics and Engineering*, vol. 155, no. 1-2, pp. 181–192, 1998.

[10] C. Andraud, A. Beghdadi, E. Haslund, R. Hilfer, J. Lafait, and B. Virgin, "Local entropy characterization of correlated random microstructures," *Physica A: Statistical Mechanics and its Applications*, vol. 235, no. 3-4, pp. 307–318, 1997.

[11] J. LLorca and J. Segurado, "Three-dimensional multiparticle cell simulations of deformation and damage in sphere-reinforced composites," *Materials Science and Engineering: A*, vol. 365, no. 1-2, pp. 267–274, 2004.

[12] M. Kulkarni, *Finite Element Analysis of 2-D Representative Volume Element*, Michigan Technological University, 2012.

[13] Y. Cao, *Representative Volume Element (RVE) FiniteElement Analysis (FEA) of Al Metal-Matrix Composites*, University of Wisconsin-Milwaukee, 2016.

[14] J. C. Yun and S. J. Park, "Development of representative volume element homogenization model for predicting transversely isotropic elasticity of lithium-ion batteries," *Journal of Engineering Materials and Technology*, vol. 139, no. 4, article 041008, 2017.

[15] I. M. Gitman, H. Askes, and L. J. Sluys, "Representative volume: existence and size determination," *Engineering Fracture Mechanics*, vol. 74, no. 16, pp. 2518–2534, 2007.

[16] R. J. M. Smit, W. A. M. Brekelmans, and H. E. H. Meijer, "Prediction of the large-strain mechanical response of heterogeneous polymer systems: local and global deformation behaviour of a representative volume element of voided polycarbonate," *Journal of the Mechanics and Physics of Solids*, vol. 47, no. 2, pp. 201–221, 1999.

[17] K. Qiu, Y. P. Jiang, X. P. Shi, L. G. Sun, and Q. Q. Wu, "Microstructure effect of metallic glass matrix composites," *Acta Materiae Compositae Sinica*, vol. 34, no. 6, pp. 1316–1324, 2017.

[18] Q. Li and X. Yang, "Three dimensional numerical simulation for mechanical behavior of carbon black filled rubber composites," *Chinese Journal of Solid Mechanics*, vol. 34, no. 6, pp. 541–549, 2013.

17

# Thermal and Rheological Formulation and Evaluation of Synthetic Bitumen from Reprocessed Polypropylene and Oil

**Ricardo S. Souza**,[1] **Leila L. Y. Visconte**,[1,2] **Ana L. N. da Silva**,[1] **and Valéria G. Costa**[3]

[1]*Instituto de Macromoléculas Professora Eloisa Mano (IMA), Programa em Ciência e Tecnologia de Polímeros, Universidade Federal do Rio de Janeiro (UFRJ), Avenida Horácio Macedo 2.030, Centro de Tecnologia-Bloco J, Ilha do Fundão, 21941-598 Rio de Janeiro, Brazil*

[2]*Escola Politécnica/Programa de Engenharia Ambiental, Universidade Federal do Rio de Janeiro (UFRJ), Av. Athos da Silveira Ramos 149, Centro de Tecnologia-Bloco A, Ilha do Fundão, 21941-909 Rio de Janeiro, RJ, Brazil*

[3]*Departamento de Processamento e Caracterização de Materiais (DPCM), Laboratório de Materiais Poliméricos (LAMAP), Instituto Nacional de Tecnologia (INT), Av. Venezuela 82, Praça Mauá, 20081-312 Rio de Janeiro, RJ, Brazil*

Correspondence should be addressed to Ricardo S. Souza; ricardossouza21@gmail.com

Academic Editor: Bernabé L. Rivas

Bitumen from the catalytic cracking of petroleum is widely used in the construction industry as a binder for mineral aggregates paving asphalt cement (PAC). However, this material has serious limitations for use at different temperatures. Modifications with polymers have been widely studied to improve the properties of this material and increase its useful life. As a result of the increased environmental risk, the volume of solid waste (MSW) can be used as an alternative to the best use of this material. One of these is the formulation of synthetic bitumen (binders) from mixtures of lubricant oil/polymer that can show better when compared to traditional and modified bitumen. In this work, a mixture of lubricating oil and recycled polypropylene (PP) was made for the formulation of synthetic bitumens. The rheological analyses carried out on the material indicate that at low temperatures the formulated material shows high loading modulus values, indicating high mechanical strength, and the material also presented high resistance to permanent deformation; the viscosity profile indicates that the material is highly pseudoplastic presenting high viscosity at zero shear rate and low visibility at high shear rates, highlighting the samples containing 10 and 20% of polymer material in the formation as the best for use as one of the components of the pavement cement.

## 1. Introduction

Bitumen is a construction material used as binder in flexible pavements. The increase in road transport flow, environmental and external factors, and the failures incurred during the construction phase can reduce the useful life of the pavements to the greater occurrence of permanent faults in the asphalt pavement. Thus, solutions such as the modification of bitumen with polymeric materials have been used to improve the bitumen properties of the petrochemical cracking, with respect to the adhesiveness, elasticity, and thermomechanical resistance, thus the behavior of pavements [1–6].

Bitumens originating from petroleum may be actively or passively modified. Passive modifiers are those that are added to the base material but do not react with it, such as polypropylene (PP), polyethylene (high-density HDPE and low-density LDPE), ethylene-vinyl acetate (EVA), styrene butadiene rubber (SBR) and styrene butadiene styrene (SBS), acrylonitrile-butadiene-styrene (ABS), and crumb rubber. Active modifiers are known to react chemically with the base polymers; however, the main concern with respect to bitumen/polymer blends is the lack of stability during prolonged storage at elevated temperatures. The physicochemical properties of polymer-modified bitumen (PMB) are strongly dominated by the chemical characteristic of the base bitumen [7–12].

An excellent bitumen should behave as an elastic solid at low temperatures or during rapid loading and as a viscous liquid at high temperatures or slow loading. In other words,

Table 1: PP reprocessing conditions in double screw extruder.

| Sample | Temperature (°C) Zone | | | | | | | Speed (rpm) | |
|---|---|---|---|---|---|---|---|---|---|
| | 1 | 2 | 3 | 4 | 5 | 6 | 7 | Extruder | Granulator |
| PP extrusion | 80 | 150 | 170 | 180 | 190 | 200 | 80 | 100 | 40 |

at low temperatures, around 0°C, the bitumen needs to be able to withstand repeated loads and discharges without failure or fatigue leading to cracks, at moderate temperatures, around 50°C, bitumens must withstand loading to avoid permanent deformations, storing the deformation in the form of energy, and at high temperatures (around 135°C), the material must be easy workability, easy transportation, compactness, applicability, and storage. Another behavior that should be avoided is the loss of adhesion between the bitumen and the mineral aggregates that make up the paving cement [13–17].

An alternative in the substitution of petroleum bitumen has been discussed, being denominated synthetic bitumen. They are thermorheologically complex materials since the mixture between the oil and the polymer can form complex three-dimensional structures. This material is sensitive to temperature changes that modify the viscosity. Techniques such as rheometry can be used to study properties and characteristics with respect to dynamic viscosity, viscoelasticity, complexity modulus, and activation energy to synthetic mixtures [18, 19].

The present work is aimed at formulating synthetic bitumens from reprocessed and reused materials that can be used as one of the components of asphalt paving cement and performing the rheological characterization to determine its potential as one of the asphalt components.

## 2. Materials and Methods

The oil used to prepare the synthetic bitumen compositions was the postconsumed AC Delco-5w30 API SL (group III mineral type), removed from a vehicle after it was changed. The polymer used to carry out the work was PP H301 from Braskem company. The characteristics of PP can be observed in Table 1.

The polymer was reprocessed in the Extrusa Brasil-DRC 22 double screw type extruder; the aim of which was to simulate the formulation of synthetic bitumen using post-consumed polymeric material. The material was extruded 5 times under the same conditions, presented in 1, and the flow rate used was 5 kg/h.

The binary mixtures between the automotive paraffinic oil and the PP were performed on the IKA C-MAG HS7 agitator plate with the IKA ETS-D7 heating controller. The temperature used in the process was maintained at 165°C. A mechanical stirrer with a propeller-type stirring rod (four blades) with a rotation of 450 rpm was used. The mixtures were stirred for 2 hours. The mixing proportions were 5, 10, 15, and 20% (*w*/*v*) of PP in paraffinic oil.

*2.1. Rheological Tests.* Rheological tests including dynamic oscillatory, creep-recovery, and steady shear tests were performed on a HAAKE MARS II, Thermo Electron Corporation rheometer, using a parallel-plate geometry set of 35 mm and gap adjusted to 2 mm.

Dynamic frequency sweep tests were conducted in the frequency range of 0.01 to 100 rad·s$^{-1}$ with a constant applied stress of 1 Pa, which is within the linear viscoelastic region. The analyses were carried out in the range of 25, 50, 75, 100, 125, and 150°C.

In creep-recovery experiments, the constant shear stress was set to 1 Pa. The creep time and recovery time were set at as 400 s each. The analysis temperatures were 15 and 50°C.

The recovery measurements were started immediately after the end of the creep phase.

The steady shear measurements of synthetic bitumen samples were evaluated by a controlled shear rate method. The shear stress and shear viscosity values were determined at 160°C over the shear rate range of 0.1 to 300 s$^{-1}$.

The dynamic melt rheological properties of the synthetic bitumens were measured in order to gain a fundamental understanding of elastic and viscous material behavior at different temperatures and thus to determine the better composition to be used as an asphaltic binder. In oscillatory analysis, $G'(\omega)$ and $G''(\omega)$ values represent the elastic and viscous behavior of samples, respectively. $G'(\omega)$ is called the "dynamic storage modulus," and $G''(\omega)$ is called the "dynamic loss modulus." The relationships between these properties with complex viscosity are showed below:

$$\eta^* = \frac{\sigma^*}{\dot{\gamma}^*} = \frac{\eta_0}{1 + i\lambda\omega} = \frac{\eta_0}{1 + \lambda^2\omega^2} - i\frac{\eta_0\lambda\omega}{1 + \lambda^2\omega^2}. \quad (1)$$

The two components of $\eta^*$, $\eta'$ (dynamic viscosity) and $\eta''$ (imaginary viscosity), can be expressed by the following equations:

$$\begin{aligned} \eta'(\omega) &= \frac{\eta_0}{1 + \lambda^2\omega^2}, \\ \eta''(\omega) &= \frac{\eta_0\lambda\omega}{1 + \lambda^2\omega^2}, \\ G'(\omega) &= \frac{\eta_0\lambda\omega}{1 + \lambda^2\omega^2}, \\ G''(\omega) &= \frac{\eta_0}{1 + \lambda^2\omega^2}, \end{aligned} \quad (2)$$

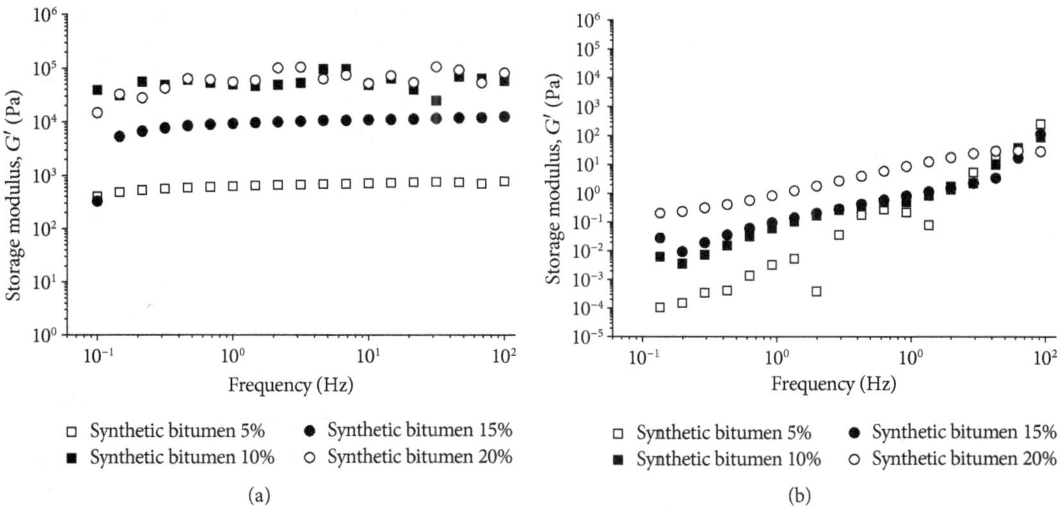


FIGURE 1: Storage modulus, $G'(\omega)$, (a) for synthetic bitumen compositions at 25°C and (b) for synthetic bitumen compositions at 150°C.

where $G'(\omega)$ and $G''(\omega)$ are the real and imaginary components, respectively, of a "complex modulus" defined as follows:

$$G^*(\omega) \equiv G'(\omega) + iG''(\omega). \quad (3)$$

## 3. Results and Discussion

*3.1. Dynamic Oscillatory Tests.* The results of dynamic frequency sweep tests are shown in Figure 1, which presents the storage modulus, $G'(\omega)$, for synthetic bitumen compositions at 25 and 150°C.

At low temperature, the analyzed synthetic bitumen compositions displayed a Newtonian behavior. It can be also observed that both 10% and 20% samples showed high storage modulus, $G'(\omega)$, at low temperature. At high temperature, 150°C, and lower frequency range, the sample with 10% presented the lower $G'(\omega)$ values, indicating its higher viscous behavior in relation to 15% and 20% samples. This characteristic is desirable because at high temperature, the material must not withstand to applied deformation. The sample with 5% showed lower $G'(\omega)$ values, both at low and high temperatures, indicating that this composition did not present a synergistic effect on the viscoelastic property. The results obtained at temperatures of 50, 75, and 100°C showed no significant change compared to 25°C.

Figure 2 represents $G^*$ versus temperature (at $10\,\text{rad}\cdot\text{s}^{-1}$) for synthetic bitumens; $G^*$ can reveal aspects of the material structure indicating strength of the material [20].

The figure shows that samples with 10% and 20% present the highest $G^*$ values, indicating that these compositions have a positive synergistic effect of improving viscoelastic properties. Xia et al. [21] carried out a frequency sweep test in pure bitumen at a temperature of 60°C; the material exhibits a value of $G^*$ around $10^3$ Pa at a frequency of $10\,\text{rad}\cdot\text{s}^{-1}$, 2 decades smaller than synthetic earnings with 10 and 20% of PP presented in this work.

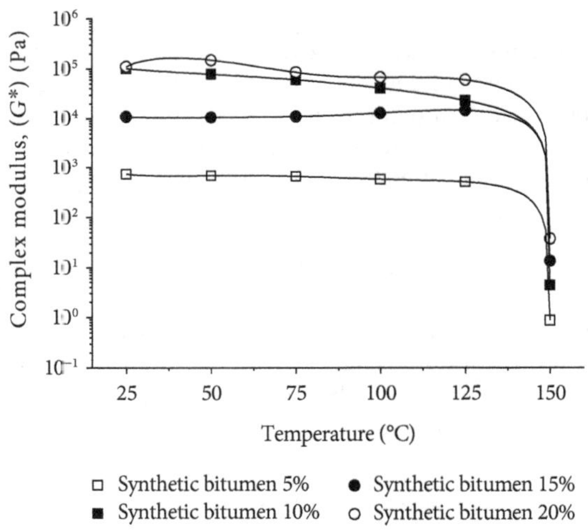


FIGURE 2: $G^*$ versus temperature for synthetic bitumens.

Figure 3 shows the results of $\eta^*(1/T)$ obtained from temperature ramp tests for synthetic bitumens. It can be observed that only for 10% and 20% samples $\eta^*$ increases. According to Parvez et al., high $G^*$ values are expected at low temperature. As $G^* = \eta^* \cdot \omega$, high values of $\eta^*$ indicate high rutting resistance at high temperature. The relationship between viscosity and temperature for synthetic bitumens is expressed by the Arrhenius equation as follows:

$$\frac{G^*}{\omega} = \eta^* = \text{A}e^{E_a/RT}, \quad (4)$$

where $E_a$ is the flow activation energy, $A$ is the preexponential parameter, and $R$ is the universal gas constant. It is known that $E_a$ is an important factor that strongly influences the viscosity.

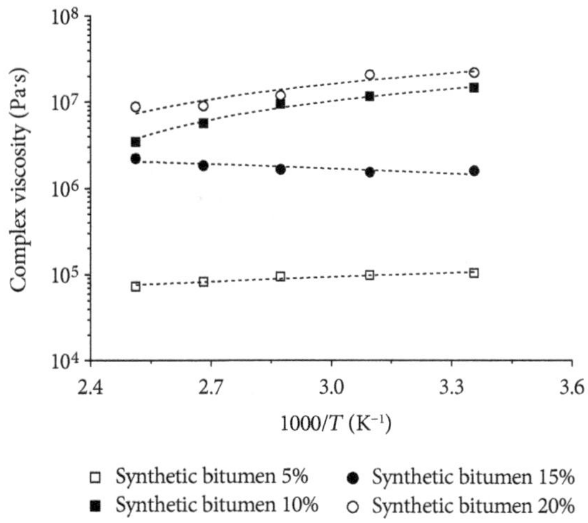


FIGURE 3: Complex viscosity, $\eta^*(1/T)$, for sample compositions.

Table 2 presents the $E_a$ values, calculated from (4). The data given in Figure 3 show good fit to the Arrhenius model for 10% and 20% binder compositions.

It is observed that the activation energy of 10% composition is lower than 20%. This suggests that the change in the viscosity with temperature is slow for 10% when compared with 20%. Hence, the high temperature resistance of 10% composition is better than 20%. Activation energy for pure asphalt is 112.53 kJ/mol [13].

*3.2. Creep-Recovery Experiments.* In a creep-recovery test, a constant stress ($\sigma_0$) is applied at initial time ($t = 0$) for $t$ seconds such that steady state is not reached and then the recovery is monitored. During a creep test, slow development of deformations ($\varepsilon_0$) is observed. The shear strain ($\varepsilon(t)$) is then monitored as a function of time as follows:

$$\varepsilon(t) = \varepsilon_0(t, \sigma_0) + \Psi(t, \sigma_0) + \frac{t}{\eta(\sigma_0)}\sigma_0, \qquad (5)$$

where $\varepsilon_0$ is an instantaneous deformation, $\Psi(t, \sigma_0)$ is a function describing delayed development of deformations, $\eta(\sigma_0)$ is viscosity, which can depend on stress, and $t$ is the current time. The creep-recovery experiments were conducted in order to measure the viscoelastic properties of the synthetic bitumen, that is, the efficiency that binder composition has to withstand deformations induced by a constant applied stress. In case of occurring deformation, it also evaluated the material ability to recover from this deformation. Figure 4 shows the results of creep tests for all sample compositions analyzed at 25 and 50°C.

Figure 4(a) shows that the 5% synthetic bitumen composition presents the lower stress strength in relation to other compositions. This behavior is in accordance with dynamic oscillatory results. For this composition, when the stress is removed, it can be observed that part of deformation is recovered. For 20% synthetic bitumen composition, it was not possible to obtain deformation data, indicating that the stress value used in the experiment was not sufficient to provoke the material deformation. Figure 4(b) presents the creep-recovery behavior for all sample compositions analyzed at 50°C. It is possible to highlight the deformation increase observed for 5% synthetic bitumen—around 0.5% to 1.1% at high temperature. To other compositions, the creep-recovery behavior remained constant. The addition of higher contents of polymeric material in the synthetic asphalt improves the deformation strength behavior, reduces the effect of asphalt flow at successive stresses, and increases the useful life of the asphalt.

TABLE 2: Parameters for the Arrhenius model.

| Samples | $E_a$ (kJ/mol) | $R^2$ |
|---|---|---|
| Synthetic bitumen 5% | — | — |
| Synthetic bitumen 10% | 83 | 0.90 |
| Synthetic bitumen 15% | — | — |
| Synthetic bitumen 20% | 166 | 0.98 |

*3.3. Steady Shear Tests.* The steady shear tests were done to measure the flow behavior of the synthetic bitumen compositions. The samples were analyzed at 160°C to ensure its complete melt. Figure 5 shows the curves of stress versus the shear rate for all compositions.

The results show that as PP content increases, the viscosity values increase. Figure 5 also shows that compositions present a pseudoplastic behavior, which is well described by an Ostwald-de-Waele model, which can be written as follows:

$$\sigma = K \cdot \dot{\gamma}^n, \qquad (6)$$

where $K$ and $n$ are empirical parameters.

The Ostwald-de-Waele model parameters calculated from linear regression of the rheological data for different materials are presented in Table 3.

Analyzing the parameters determined from the Ostwald-de-Waele model, the data was applicable to all samples analyzed ($R^2 > 0.9$). The addition of PP increased the elastic behavior of the final composition, as seen earlier. The decrease in the $n$ parameter indicated a more pronounced shear thinning behavior at lower shear rates. These results corroborate the data discussed previously in this paper. Figure 6 shows the viscosity versus shear rate curves for all samples analyzed.

The results from Figure 6 show that at lower shear rates, the synthetic asphalts tend to present higher viscosity values, except the 5% composition, which shows a Newtonian behavior, presenting very low viscosity values in the whole shear rate range evaluated. This behavior is not desirable because the viscosity values at lower frequency ($\eta$ at $\dot{\gamma} \to 0 = \eta_0$) must be high, if not, the mineral aggregates of the asphalt would be easily sedimented due to the low flow resistance offered by the material. As polymeric material content increases, the viscosity values also increase at lower shear rate conditions, improving the flow resistance, which can keep the aggregates suspended. As shear rates increase, the viscosity values tend to decrease. This behavior suggests easy manipulation of the material during use

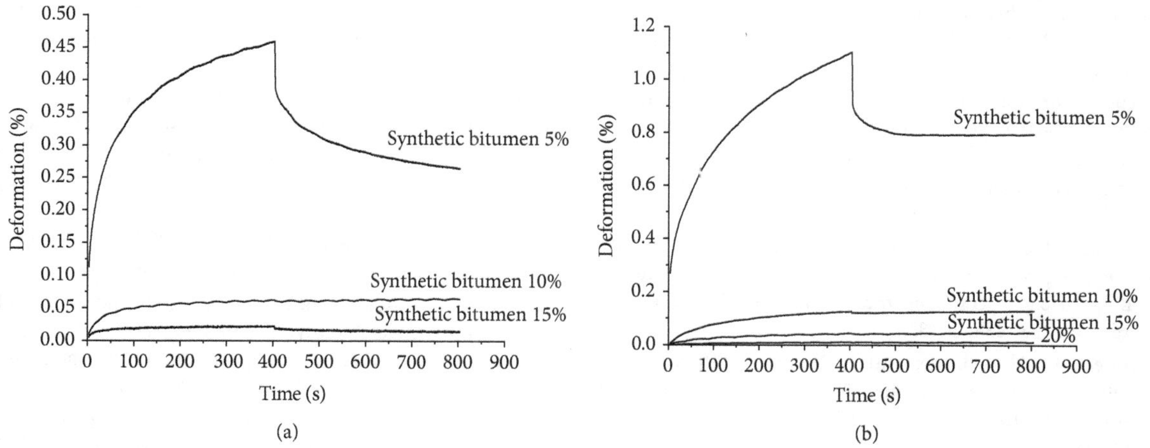


Figure 4: Creep-recovery tests for synthetic bitumen composition at (a) 25°C and (b) 50°C.

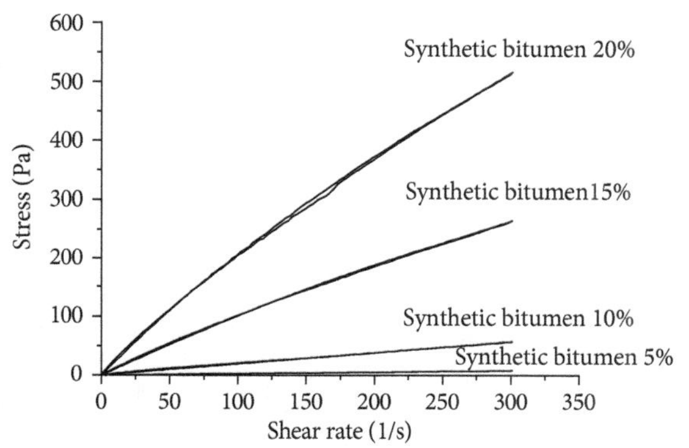


Figure 5: Stress versus shear rate for synthetic bitumens.

Table 3: Linear regression parameters for the Ostwald-de-Waele model.

| Binder composition | $K$ (Pa·s) | $n$ | $R^2$ |
|---|---|---|---|
| Synthetic bitumen 5% | 0.03 | 0.97 | 0.99 |
| Synthetic bitumen 10% | 0.27 | 0.94 | 0.99 |
| Synthetic bitumen 15% | 1.76 | 0.88 | 0.99 |
| Synthetic bitumen 20% | 4.20 | 0.84 | 0.99 |

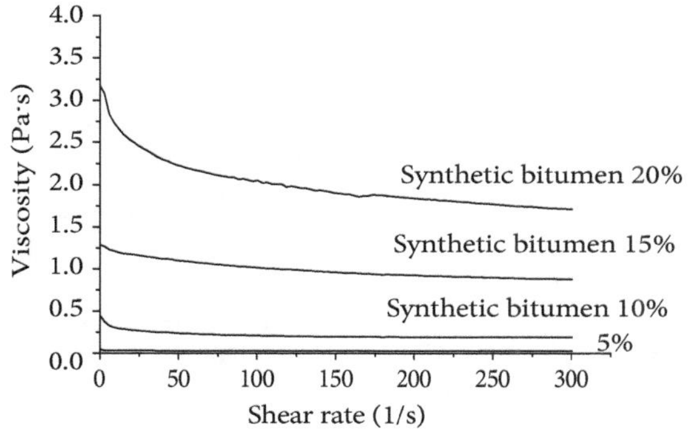


Figure 6: Viscosity versus shear rate for synthetic bitumens.

since it is submitted to high rates in its transport and during spreading.

## 4. Conclusions

The formulation of synthetic asphalt represents a great alternative for the improvement of pavement cement, as well as a man-made solid waste disposal. The analyses showed that the formulated materials presented high mechanical resistance, giving high values of $|G^*|$ at low temperature. These materials also have high resistance to deformations, with some samples reaching values of 0.02% of total deformation. At high temperatures (170°C), the material presents a distinct flow behavior, presenting low viscosities and ease of flow, an aspect that can contribute in a positive way, facilitating transportation and application.

## Conflicts of Interest

The authors, Ricardo S. Souza, Leila L.Y. Visconte, Ana L.N. da Silva, and Valéria G. Costa, declare that there is no conflict of interest regarding the publication of this paper.

## Acknowledgments

This work was funded by the Conselho Nacional de Desenvolvimento Científico e Tecnológico (CNPq) in a postgraduate scholarship format for author Ricardo S. Souza.

## References

[1] G. D. Airey, "Rheological evaluation of ethylene vinyl acetate polymer modified bitumens," *Construction and Building Materials*, vol. 16, no. 8, pp. 473–487, 2002.

[2] A. Pérez-Lepe, F. J. Martínez-boza, and C. Gallegos, "High temperature stability of different polymer-modified bitumens:

a rheological evaluation," *Journal of Applied Polymer Science*, vol. 103, no. 2, pp. 1166–1174, 2007.

[3] M. R. S. Fernandes, M. M. C. Forte, and L. F. M. Leite, "Rheological evaluation of polymer-modified asphalt binders," *Materials Research*, vol. 11, no. 3, pp. 381–386, 2008.

[4] M. García-Morales, P. Partal, F. J. Navarro et al., "Viscous properties and microstructure of recycled eva modified bitumen," *Fuel*, vol. 83, no. 1, pp. 31–38, 2004.

[5] M. Çalısıcı, M. Gürü, M. K. Çubuk, and Ö. F. Cansız, "Improving the aging resistance of bitumen by addition of polymer based diethylene glycol," *Construction and Building Materials*, vol. 169, pp. 283–288, 2018.

[6] S. Köfteci, P. Ahmedzade, and B. Kultayev, "Performance evaluation of bitumen modified by various types of waste plastics," *Construction and Building Materials*, vol. 73, pp. 592–602, 2014.

[7] O. González, M. E. Muñoz, A. Santamaría, M. García-Morales, F. J. Navarro, and P. Partal, "Rheology and stability of bitumen/EVA blends," *European Polymer Journal*, vol. 40, no. 10, pp. 2365–2372, 2004.

[8] M. Shivokhin, M. García-Morales, P. Partal, A. A. Cuadri, and C. Gallegos, "Rheological behaviour of polymer-modified bituminous mastics: a comparative analysis between physical and chemical modification," *Construction and Building Materials*, vol. 27, no. 1, pp. 234–240, 2012.

[9] L. M. B. Costa, H. M. R. D. Silva, J. R. M. Oliveiran, and S. R. M. Fernandes, "Incorporation of waste plastic in asphalt binders to improve their performance in the pavement," *International Journal of Pavement Research and Technology*, vol. 6, no. 4, pp. 457–464, 2013.

[10] P. Ayar, "Effects of additives on the mechanical performance in recycled mixtures with bitumen emulsion: an overview," *Construction and Building Materials*, vol. 178, pp. 551–561, 2018.

[11] S. Fernandes, L. Costa, H. Silva, and J. Oliveira, "Effect of incorporating different waste materials in bitumen," *Ciência & Tecnologia dos Materiais*, vol. 29, no. 1, pp. e204–e209, 2017.

[12] C. Oliviero Rossi, A. Spadafora, B. Teltayev, G. Izmailova, Y. Amerbayev, and V. Bortolotti, "Polymer modified bitumen: rheological properties and structural characterization," *Colloids and Surfaces A: Physicochemical and Engineering Aspects*, vol. 480, pp. 390–397, 2015.

[13] M. A. Parvez, H. I. Al-Abdul Wahhab, I. A. Hussein, and M. Al-Mehthel, "Thermorheology of polyethylene wax modified sulfur asphalt," *International Polymer Processing*, vol. 30, no. 2, pp. 202–209, 2015.

[14] V. González, F. J. Martínez-Boza, F. J. Navarro, C. Gallegos, A. Pérez-Lepe, and A. Páez, "Thermomechanical properties of bitumen modified with crumb tire rubber and polymeric additives," *Fuel Processing Technology*, vol. 91, no. 9, pp. 1033–1039, 2010.

[15] S. Desseaux, S. dos Santos, T. Geiger et al., "Improved mechanical properties of bitumen modified with acetylated cellulose fibers," *Composites Part B: Engineering*, vol. 140, pp. 139–144, 2018.

[16] M. Ranieri and C. Celauro, "Improvement of high modulus asphalt mixtures with average quality aggregate and bitumen by application of polymeric additives," *Construction and Building Materials*, vol. 178, pp. 183–194, 2018.

[17] D. Grossegger, B. Gomez-Meijide, S. Vansteenkiste, and A. Garcia, "Influence of rheological and physical bitumen properties on heat-induced self-healing of asphalt mastic beams," *Construction and Building Materials*, vol. 182, pp. 298–308, 2018.

[18] C. Fuentes-Audén, F. J. Martínez-Boza, F. J. Navarro, P. Partal, and C. Gallegos, "Formulation of new synthetic binders: thermo-mechanical properties of recycled polymer/oil blends," *Polymer Testing*, vol. 26, no. 3, pp. 323–332, 2007.

[19] J. W. Goodwin and R. W. Hughes, *Rheology for Chemists. An Introduction*, RSCPublishing, 2ª edition, 2008.

[20] T. Xia, L. Zhou, J. Xu, Y. Qin, W. Chen, and J. Dai, "Rheology and thermal stability of polymer modified bitumen with coexistence of amorphous phase and crystalline phase," *Construction and Building Materials*, vol. 178, pp. 272–279, 2018.

[21] T. Xia, J. Xu, T. Huang et al., "Viscoelastic phase behavior in SBS modified bitumen studied by morphology evolution and viscoelasticity change," *Construction and Building Materials*, vol. 105, pp. 589–594, 2016.

# 18

# Analysis and Design of Short FRP-Confined Concrete-Encased Arbitrarily Shaped Steel Columns under Biaxial Loading

**Wei He,[1,2] Bing Fu,[3,4] and Feng-Chen An[5]**

[1]*School of Water Conservancy and Environment, Zhengzhou University, Zhengzhou, Henan Province, China*
[2]*Zhengzhou Metro Co. Ltd., Zhengzhou, Henan Province, China*
[3]*School of Civil and Transportation Engineering, Guangdong University of Technology, Guangzhou, Guangdong Province, China*
[4]*Department of Civil and Environmental Engineering, The Hong Kong Polytechnic University, Hung Hom, Hong Kong*
[5]*College of Petroleum Engineering, China University of Petroleum, Beijing, China*

Correspondence should be addressed to Bing Fu; cefubing@gdut.edu.cn

Academic Editor: Mehdi Salami-Kalajahi

The FRP-confined concrete-encased steel column is a new form of hybrid column, which integrates advantages of all the constituent materials. Its structural performance, including load carrying capacity, ductility, and corrosion resistance, has been demonstrated to be excellent by limited experimental investigation. Currently, no systematic procedure, particularly for that with reinforced structural steel of arbitrary shapes, has been proposed for the sectional analysis and design for such novel hybrid columns under biaxial loading. The present paper aims at filling this research gap by proposing an approach for the rapid section analysis and providing rationale basis for FRP-confined concrete-encased arbitrarily shaped steel columns. A robust iterative scheme has been used with a traditional so-called fiber element method. The presented numerical examples demonstrated the validity and accuracy of the proposed approach.

## 1. Introduction

Lateral confinement leads to the compressed concrete under multiaxial compression and results in enhancements in both ductility and strength of the compressed concrete [1]. Such a feature is highly preferable for design and constructing columns in a region of high seismic risk, where adequate ductility of columns is necessary to ensure high moment redistribution capacity of structures and avoid collapse of structures due to the shaking from large earthquakes [2]. In conventional reinforced concrete structures, the lateral confinement to the compressed concrete is mainly provided by the transverse steel reinforcement in the form of either spirals or hoops. Concrete-filled steel tubular columns, which have been widely used in high-rise buildings, bridges, etc., are also utilizing the increased strength and deformability of confined concrete to achieve a high structural performance [3, 4]. In such a composite column, an outer steel tube is used to replace longitudinal and transverse steel reinforcements in conventional reinforced concrete columns and to provide continuous confinement to concrete infilled. However, such a concrete-filled steel tubular column, especially in a harsh environment, is susceptible to severe corrosion problem due to the direct exposure of the outer steel tube to ambient environment.

Fiber reinforced polymer (FRP) composites in the form of wraps or jackets have been widely used to serve as a confining device for seismic retrofit of existing RC columns [5, 6]. Its wide application is mainly attributed to the superior properties of FRP composites, such as high strength-to-weight ratio and excellent corrosion resistance. Recently, combining FRP composites with traditional construction materials (e.g., steel and concrete) has gained increasing research attention to form a hybrid column to achieve high structural performance by integrating advantages of the constituent materials. A successful example is the hybrid FRP-concrete double-skin tubular columns (DSTCs), which was proposed by Teng et al. [7] and then received a great deal of follow-on research (e.g., [8–13]). A DSTC consists of an FRP outer tube, a steel inner tube, and a layer of concrete sandwiched between them [7]. The FRP tube offers mechanical resistance primarily in the hoop direction to confine the concrete and to enhance the shear resistance of the member; the steel tube provides the main longitudinal reinforcement and prevents the concrete from inward spalling and the steel tube from outward local buckling deformations. Optimal use of FRP, steel, and concrete in the manner of DSTC therefore makes it an economical and corrosion-resistant column form.

Another successful example that integrates advantages of all the constituent materials into a hybrid column is the FRP-confined concrete-encased steel composite columns (FCSCs), which was first proposed by Liu et al. [14] for retrofit of existing steel columns. In their study, five notched steel columns to simulate the corroded section were encased using FRP-confined concrete to enhance its load carrying capacity, and tests results demonstrated the feasibility of such a retrofit technique. Karimi and his coauthors [4, 15, 16] then introduced the FCSCs for the new construction of a column and conducted a systematic experimental study on the compressive behavior of both short and slender FCSCs. Recently, Yu et al. [17] presented a combined experimental and theoretical study on the behavior of FCSCs under concentric and eccentric compression and revealed that two flanges of H-shaped steel could provide additional confinement to infilled concrete thus enhancing the ductility and load carrying of the composite columns. In order to further enhance the confinement efficiency for FRP-confined concrete-encased steel composite square columns and inspired by Yu et al. [17], Huang and his coauthors [18] innovatively proposed a new form of FCSCs, in which a cross-shaped steel section was used to replace H-shaped (or I-shaped) steel. Their test results demonstrated that despite concrete infill in square section, it was effectively confined by both outer tube and cross-shaped steel and justified the rationales of the proposed new form of composite columns. In some cases, such as corner columns of buildings, irregular cross section or regular cross section placed asymmetrically is often encountered [19]. Despite the fact that the structural performance of FCSCs has been demonstrated by a number of concentric of eccentric compression tests, no approach has been proposed for a rapid section analysis and design of FCSCs with arbitrarily shaped steel and under biaxial loading.

Against the above background, the present paper proposes an approach for the rapid section analysis and provides rationale basis for FRP-confined concrete-encased arbitrarily shaped steel columns. The robust iterative scheme proposed by Chen et al. [19] has been adopted with a traditional so-called fiber element method, where all the constitutive materials were treated as fiber elements avoiding the integration for the stress block of the FRP-confined concrete. The presented numerical examples demonstrated the validity and accuracy of the proposed approach.

## 2. Methodology

A proper approach is necessary for robust convergence of the exact location of the neutral axis. The iterative procedure proposed by Chen et al. [19] has been adopted herein, in which the iterative quasi-Newton procedure is employed within the Regula-Falsi numerical scheme for the solution of equilibrium equations. In addition, the use of the plastic centroidal axes of the cross section as the reference axes of loading guarantees the convergence of solution in the iterative process. All the constituent materials, including FRP-confined concrete under the combination of axial compression and bending, are treated as fiber elements to avoid the definition of stress block for the FRP-confined concrete, which is difficult to analytically determine. Appropriate constitutive models were used to simulate the mechanical properties of the constituent materials. For instance, strain gradient along the section has been taken into account in the constitutive model of the confined concrete, which is simulated by using a so-called variable confinement stress-strain model proposed in the Chinese code (GB50608-2010).

*2.1. Basic Assumptions.* The sectional analyses and design in the present paper were conducted based on the following basic assumptions:

(1) Section plane remains plane after loading; this assumption ensures that the strain at any point of the cross section is proportional to its distance from the neutral axis

(2) Failure limit state is only defined by the attainment of the strain $\varepsilon_{cu}$ of the extreme compression fiber; this means no failure mode of steel/steel bar rupture has been considered

(3) Tensile strength of confined concrete is neglected

(4) No contribution of FRP tubes to compressive strength of composite columns has been taken directly into account in the section analysis

*2.2. Definition of the Reference-Loading Axes.* If the usual definition of reference-loading axes (i.e., geometric centroid as its origin) is used, it is possible that the origin of the loading may fall outside the $M_x - M_y$ interaction curve, especially when the axial load is close to the axial load capacity of columns with irregular structural steel under concentric compression [19]. To overcome such divergence difficulty, Chen et al. [19] accepted the plastic centroid of the cross section of columns as the origin of the reference-loading axes. Such definition of the origin of the reference-loading axes ensures $\alpha_m$, which is the inclination of resultant bending moment resistance and is equal to arctan $(M_y/M_x)$ and increases monotonically from 0 to $2\pi$ with increase of $\theta_n$, orientation of neutral axis from 0 to $2\pi$. In this way, the existence and uniqueness of $\theta_n$ and convergence are guaranteed.

For any arbitrary cross section, the definition of the plastic centroid (i.e., the origin of loading-reference axes in this study) was given in (1) and (2):

$$X_{\text{pc}} = \frac{X_c A_c f_c/\gamma_c + X_s A_s f_s/\gamma_s + X_r A_r f_r/\gamma_r}{X_c A_c f_c + X_s A_s f_s + X_r A_r f_r}, \quad (1)$$

$$Y_{\text{pc}} = \frac{Y_c A_c f_c/\gamma_c + Y_s A_s f/\gamma_s + Y_r A_r f_r/Y_r A_r f_r}{Y_c A_c f_c + Y_s A_s f_s + Y_r A_r f_r}, \quad (2)$$

in which $A_c$, $A_s$, and $A_r$ are areas of concrete, shaped steel, and steel rebars, respectively; $\gamma_c$, $\gamma_s$, and $\gamma_r$ are the safety of confined concrete, shaped steel, and steel rebars, respectively; and $f_c$, $f_s$, and $f_r$ are the respective characteristic strength of confined concrete, shaped steel, and steel rebars and set to be the ultimate strength of confined concrete, yielding strength of shaped steel and steel rebars, respectively.

*2.3. Fiber Element Method Adopted.* In the present paper, all the structural components in the composite columns, including the confined concrete, structural steel, and steel bars, have been treated as fiber elements to calculate the stress resultants. This approach avoids the difficulty of the determination of the stress block for confined concrete under bending with/without axial compression. Cross section of any shape has been first meshed to determine the stress resultants of each component. The total resultants of each component can be obtained by a summation over all the fiber elements given by (3), (4), and (5):

$$N_z = \sum_{i=1}^{m_c} \sigma_{\text{ci}} A_{\text{ci}} + \sum_{j=1}^{m_s} \sigma_{\text{sj}} A_{\text{sj}} + \sum_{k=1}^{m_r} \sigma_{\text{rk}} A_{\text{rk}}, \quad (3)$$

$$M_x = -\sum_{i=1}^{m_c} \sigma_{\text{ci}} A_{\text{ci}} y_{\text{ci}} - \sum_{j=1}^{m_s} \sigma_{\text{sj}} A_{\text{sj}} y_{\text{sj}} - \sum_{k=1}^{m_r} \sigma_{\text{rk}} A_{\text{rk}} y_{\text{rk}}, \quad (4)$$

$$M_y = -\sum_{i=1}^{m_c} \sigma_{\text{ci}} A_{\text{ci}} x_{\text{ci}} - \sum_{j=1}^{m_s} \sigma_{\text{sj}} A_{\text{sj}} x_{\text{sj}} - \sum_{k=1}^{m_r} \sigma_{\text{rk}} A_{\text{rk}} x_{\text{rk}}, \quad (5)$$

where $N_z$, $M_x$, and $M_y$ are the stress resultants of axial compression, moment over $x$-axis, and moment over $y$-axis, respectively; $m_c$, $m_s$, and $m_r$ are the numbers of concrete fibers, structural steel fibers, and reinforcing bar fibers, respectively; $\sigma_{ci}$, $\sigma_{sj}$, and $\sigma_{rk}$ are the stresses of each concrete fiber, structural steel fiber, and reinforcing bar fiber, respectively; and $A_{ci}$, $A_{sj}$, and $A_{rk}$ are areas of respective concrete fiber, structural steel fiber, and reinforcing bar fiber, respectively.

*2.4. Modelling of Confined Concrete.* The behavior of concrete confined by FRP jacketing has been extensively investigated, and a number of strength models have been developed (e.g., [5, 20–22]). All the strength models, which give the stress-strain relationship explicitly, can be used in the proposed approach; however, a simple and generally accurate model is more preferable for such a design purpose. A stress-strain model of confined concrete in the Chinese code (GB50608-2010), which reflects the effect of strain gradient along the section, has therefore been adopted in the present section analysis. In this model, the slope of the second portion of the stress-strain curve (i.e., $E_2$) is defined as a function of the load eccentricity and equal to be that of the confined model in Lam and Teng [5] for the concrete under the concentric compression (i.e., a zero load eccentricity) and to be zero for the concrete under pure bending (i.e., an infinite load eccentricity).

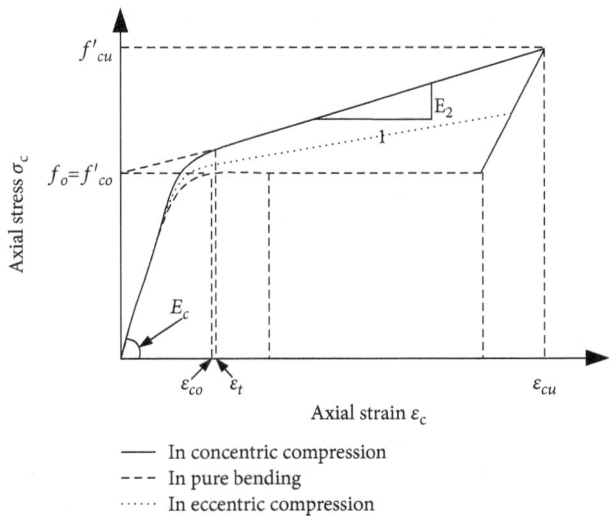


FIGURE 1: Stress-strain models for concrete in FRP-jacketed composite columns.

As shown in Figure 1, the strain-stress relationship of confined subjected to pure bending is formulated by (6) and (7), while that under eccentric compression is given in (8) and (9).

$$\sigma_{\text{cc}} = E_1 \varepsilon_{\text{cc}} - \frac{E_1^2}{4 f_c} \varepsilon_{\text{cc}}^2 \quad \text{for } 0 \le \varepsilon_{\text{cc}} \le \frac{2 f_c}{E_1}, \quad (6)$$

$$\sigma_{\text{cc}} = f_c \quad \text{for } \frac{2 f_c}{E_1} \le \varepsilon_{\text{cc}} \le \varepsilon_{\text{cc,ub}}, \quad (7)$$

in which $\sigma_{\text{cc}}$ and $\varepsilon_{\text{cc}}$ are the axial stress and axial strain, respectively; $\varepsilon_{\text{cc,ub}}$ is the design ultimate compressive strain of the concrete subjected to pure bending; $E_1$ is the elastic modulus of unconfined concrete; and $f_c$ is the design compressive strength of unconfined concrete.

$$\sigma_{\text{cc}} = E_1 \varepsilon_{\text{cc}} - \frac{(E_1 - E_{2,\text{ec}})^2}{4 f_c} \varepsilon_{\text{cc}}^2 \quad \text{for } 0 \le \varepsilon_{\text{cc}} \le \varepsilon_t, \quad (8)$$

$$\sigma_{\text{cc}} = f_c + E_{2,\text{ec}} \varepsilon_{\text{cc}} \quad \text{for } \varepsilon_t \le \varepsilon_{\text{cc}} \le \varepsilon_{\text{cc,uec}}, \quad (9)$$

in which $\varepsilon_{\text{cc,uec}}$ is the design ultimate compressive strain of the concrete in sections subjected to eccentric compression and $E_{2,ec}$ is the slope of the second linear portion of the stress-strain curve. Related parameters above are determined by using the following equations:

$$\varepsilon_t = \frac{2f_c}{(E_1 - E_{2,\text{ec}})},$$
$$E_{2,\text{ec}} = E_2 \frac{d}{d + e_i},$$
$$E_2 = \frac{f'_{\text{cc}} - f_c}{\varepsilon_{\text{cu}}}, \tag{10}$$
$$\varepsilon_{\text{cc,uec}} = (\varepsilon_{\text{cc},u} - \varepsilon_{\text{cc,ub}}) \frac{d}{d + e_i} + \varepsilon_{\text{cc,ub}},$$

in which $E_2$ is the slope of the second portion of the stress-strain curves for the FRP-confined concrete under axial compression; $f'_{cc}$ is the ultimate strength of the confined concrete under concentric compression; $d$ is the diameter of the concrete core; $e_i$ is the load eccentricity; and $\varepsilon_{\text{cc,ub}}$ is the design ultimate compressive strain of the concrete subjected to pure bending. Such a parameter should be determined from tests and is defined as the smaller value of the design ultimate strains obtained from the axial compression tests on concrete-filled FRP tubes and hollow FRP tubes, respectively. Due to lack of test data, the present manuscript assumes $\varepsilon_{\text{cc,ub}}$ be equal to the corresponding ultimate strain of FRP-confined concrete under concentric compression. Such an assumption cannot lead to a significant loss in accuracy and will not adversely affect the main purpose of the present manuscript, which is to validate the proposed approach for the rapid section analysis of FRP-confined concrete-encased arbitrarily shaped steel columns. $\varepsilon_{\text{cu}}$, the ultimate strain of FRP-confined concrete under concentric compression, can be determined following [5] as follows:

$$\frac{\varepsilon_{\text{cu}}}{\varepsilon_{\text{co}}} = 1.75 + 12\left(\frac{f_{l,a}}{f_c}\right)\left(\frac{\varepsilon_{h,\text{rup}}}{\varepsilon_{\text{co}}}\right)^{0.45}, \tag{11}$$

in which $\varepsilon_{\text{cu}}$ and $\varepsilon_{co}$ are the ultimate strains of the confined concrete under concentric compression and unconfined concrete, respectively; $\varepsilon_{h,\text{rup}}$ is the ultimate hoop strain of the FRP reinforcement; and $f_{l,a}$ is the actual maximum confining pressure given in the following:

$$f_{l,a} = \frac{2E_{\text{FRP}} t \varepsilon_{h,\text{rup}}}{d}, \tag{12}$$

where $E_{\text{FRP}}$ is elastic modulus of the FRP composite.

As revealed by Lam and Teng [23], the ultimate hoop strains of FRP measured in tests on FRP-confined concrete cylinders are substantially below those from flat coupon tensile tests. As a result, a reduction factor for the determination of hoop strains of FRP based on the ultimate rupture strain of FRP from coupon tensile tests should be included, e.g., 0.586 for CFRP and 0.624 for GFRP in the present paper.

*2.5. Modelling of Shaped Steel and FRP Tubes.* In the sectional analysis, the structural steel and/or steel bars were also treated as fiber elements with perfectly plastic property and the plane section assumption. The plane section assumption is generally valid, because compression force resisted by the column leads to the expansion of concrete and therefore significant interaction between different constituents. The size of elements should match that of concrete elements. The effect of FRP tubes on structural performance has been taken into account by modifying the constitutive of the confined concrete with the amount of FRP tubes. This means that no axial loading was carried by FRP tubes and no elements in analysis represent the FRP tubes. This makes sense as the fiber of tubes is hoop or predominantly hoop, mainly providing confinement to the concrete core. Neglecting the contribution of FRP tubes to axial load resistance is therefore reasonable and leads to a slightly conservative prediction, which is preferable for a design purpose.

*2.6. Iterative Procedure for the Sectional Analysis.* For a given composite column, a $M_x - M_y$ interaction curve under a given axial load can be determined by the sectional analysis. The series of such $M_x - M_y$ interaction curves under different axial loads can be utilized to judge whether one composite column satisfies the load carrying requirement. For example, a point representing a design load set ($N_{zd}$, $M_{xd}$, $M_{yd}$) that falls outside the interaction curves means the design column does not satisfy the load carrying requirement; otherwise it is appropriate for use. Detailed iterative procedure for the sectional analysis of a FRP-confined column with structural steel of any shape is given in Figure 2(a). In this iterative procedure, the iterative solution method for parameter $d_n$ is adopted that was presented in Chen et al. [19], and similar approach is used for the eccentricity $e$.

*2.7. Design of Required Dimensions of Hybrid Columns.* It is a common task for civil engineers to design the dimensions for a composite column with a given cross section under known loads $N_{zd}$, $M_{xd}$, and $M_{yd}$. In this task, all structural parameters, except the size of structural steel, are specified, and its purpose is to determine the size of structural steel for such a column. The iterative procedure presented in Figure 2(b) is used for this task.

## 3. Effect of Constitutive Model of Confined Concrete

As mentioned in previous sections, different constitutive models have been proposed by researchers and adopted for the sectional analysis for the composite columns under the combination of axial compression and bending. Good agreement between analytical results and test results on hybrid FRP double-skin composite columns was obtained by [24], where the constitutive models for concrete in pure bending were adopted for analysis. However, such simplification may result in too conservative results, especially for the columns confined by FRPs of a large amount and under eccentric compression with a small eccentricity. In this section, sectional analyses with different constitutive models

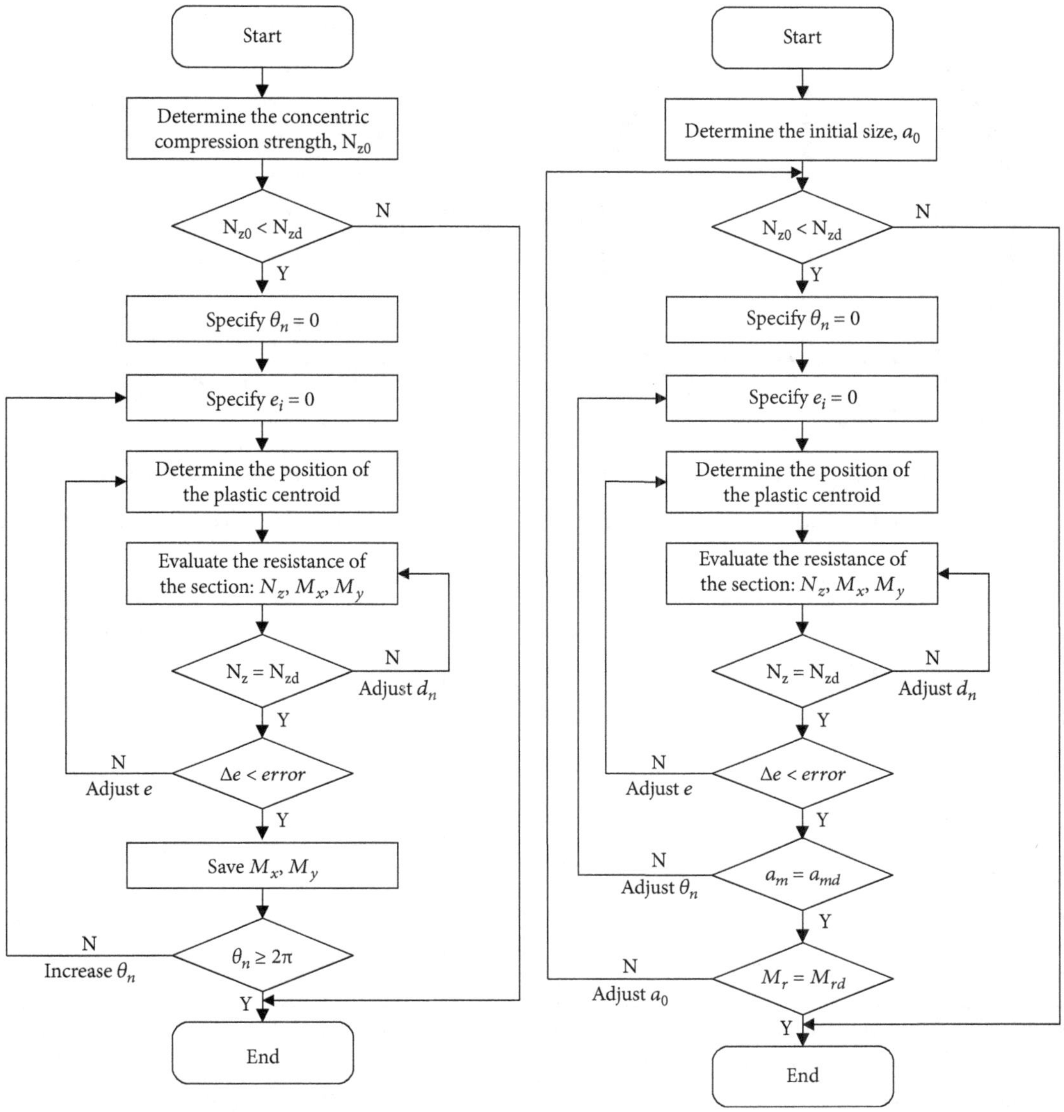


(a) For section analysis

(b) For section design

FIGURE 2: Flowcharts for section analysis and design of composite columns under biaxial loading.

for the column have been conducted to investigate the discrepancies induced by different uses of constitutive models for the confined concrete under eccentric compression. Three representative models for confined concrete in Figure 1 were adopted herein for the composite columns under eccentric compression with different eccentricities. It is believed that the sectional analysis with the constitutive models for the confined concrete under pure bending provides the lower limit prediction (denoted as lower limit in following figures), while that from the constitutive model for the confined concrete under axial compression leads to the lower limit prediction (denoted as the lower limit in following figures).

A composite column, which consists of a T-shaped steel, an outer FRP tube, and concrete infilled between them (Figure 3(a)), was analyzed with the mesh given in Figure 3(b) using the proposed approach. Three different constitutive models illustrated in Figure 1 were used for confined concrete infilled in the section analysis, leading to three curves for each case, respectively, under the axial compression loads of 1500 kN and 2100 kN (Figure 4). The results from the section analysis indicate that bending moments calculated with use of the lower limit model are significantly lower than those from other two models by about 33%, and such differences tend to be more significant in the case with the axial compression load of 2100 kN. However, differences between bending moments calculated using the upper model and middle mode are minor and can be neglected.

Some conclusions drawn from the sectional analysis above are as follows: (1) the discrepancies between the sectional analysis with the upper model and that with the middle model are small and can be neglected if the eccentricity is

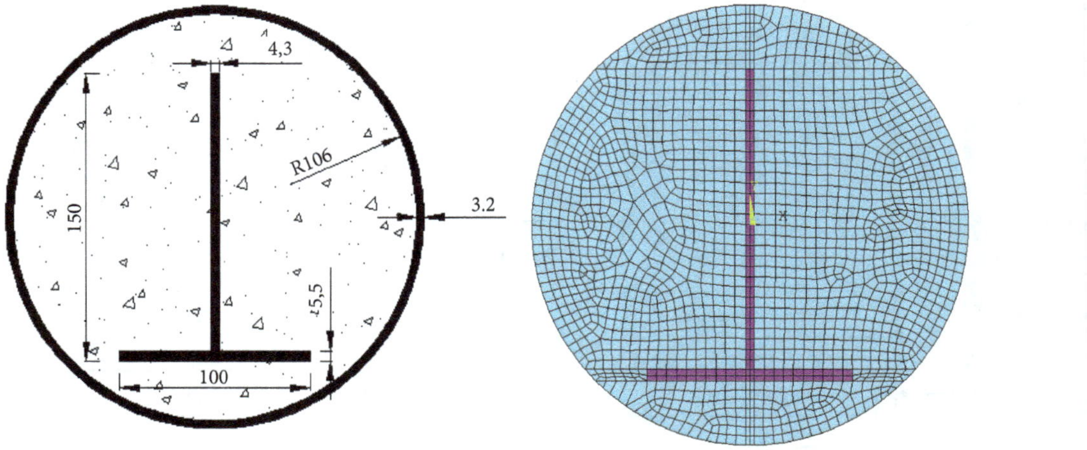


(a) Dimensions of the composite column

(b) Mesh of the composite column

FIGURE 3: Dimensions and mesh of the predefined column; effect of stress-strain models of confined concrete.

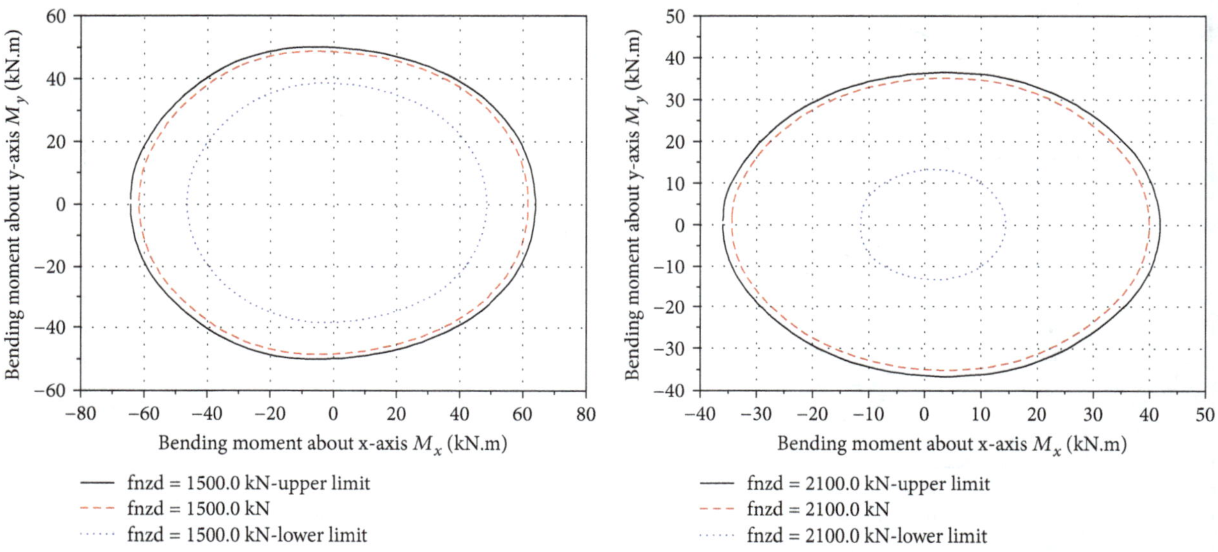


(a) Under the axial load of 1500 kN

(b) Under the axial load of 2100 kN

FIGURE 4: $M_x - M_y$ interaction curves of the predefined column; effect of stress-strain models of confined concrete.

small, and (2) the lower model may be too conservative for the sectional analysis of the columns under eccentric compression with small eccentricity and high compression load.

## 4. Effect of Mesh Schemes

As in all analysis with the fiber element approach, element size plays a significant role in the accuracy and efficiency of the analysis. Coarse mesh for the fiber element may result in inaccuracy in the analysis, while high computational cost may be induced by the refined mesh of the fiber elements. To find an optimal mesh for subsequent calculation, the effect of mesh on the sectional analysis was conducted.

To achieve this aim, four different mesh schemes in Figure 5 were adopted herein (i.e., element sizes of 2 mm, 5 mm, 20 mm, and 40 mm, respectively).

The sectional analysis for the column with four different mesh schemes in Figure 5 was given in Figure 6. The results from the sectional analysis with mesh scheme-2 are almost the same as those from the sectional analysis with mesh scheme-1, and thought to be pretty precise. The results from the mesh scheme-3 are also good with minor discrepancies compared with those from mesh scheme-3. As shown in Figure 6, large discrepancies can be found in the results from the mesh scheme-4 due to the use of coarse fiber elements. It also should be mentioned here that the computational cost of that with mesh scheme-1

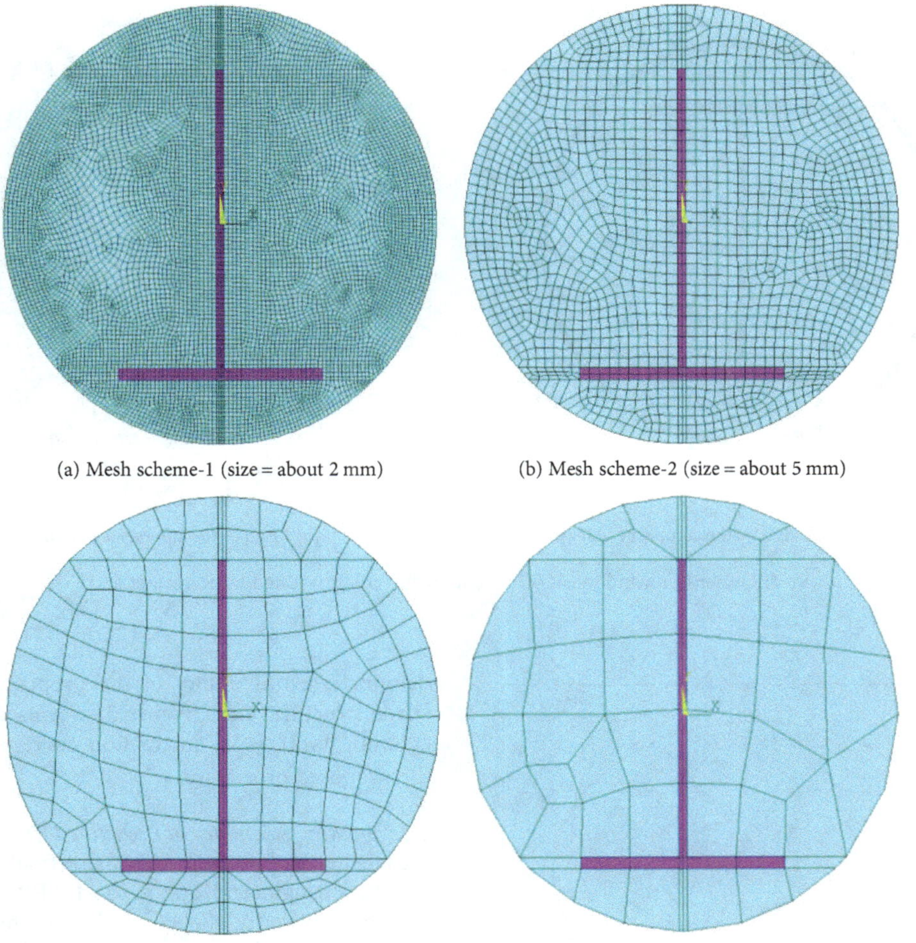

(a) Mesh scheme-1 (size = about 2 mm)

(b) Mesh scheme-2 (size = about 5 mm)

(c) Mesh scheme-3 (size = about 20 mm)

(d) Mesh scheme-4 (size = about 40 mm)

FIGURE 5: Mesh schemes for the predefined column.

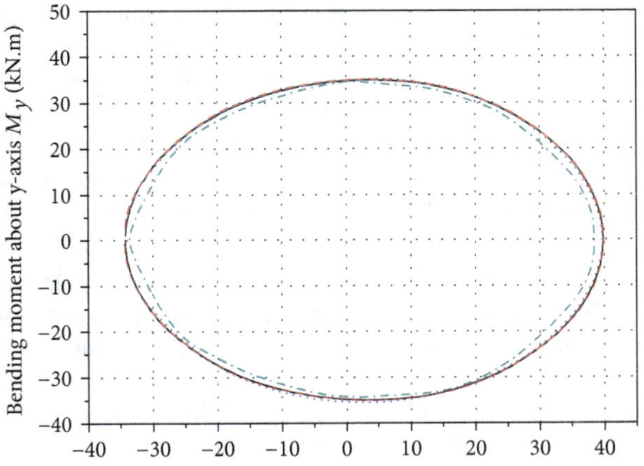


FIGURE 6: $M_x - M_y$ interaction curve with different mesh schemes.

is high. The mesh scheme-2 will be used for the subsequent sectional analysis, which combines the advantage of accuracy and computational cost.

## 5. Numerical Examples

Several numerical examples, including sectional analysis for a given composite short column, check of the adequacy of predefined cross section, and design of required structural steel, are presented herein to demonstrate the validity and efficiency of the proposed approach.

*5.1. Analysis of Designed Cross Section.* A FRP-confined concrete-encased I-shaped steel column tested by Karimi et al. [15] is first analyzed in this section. The GFRP tube with fiber orienting in the circumferential direction and its thickness is 3.2 mm. The structural steel is of I-shape and 500 mm long W150 × 14 section. For details of dimension, the readers can refer to Figure 7(a). Mesh of fiber elements was given in Figure 7(b). The $M_x - M_y$ interaction curve (i.e., the isoload contour) of the design cross section (Figure 8) under different axial loads can be easily obtained by the proposed approach. As shown in Figure 8, the resistance along the $x$-axis is stronger than that along the $y$-axis, and the $x$-axis can be taken as the strong axis in design.

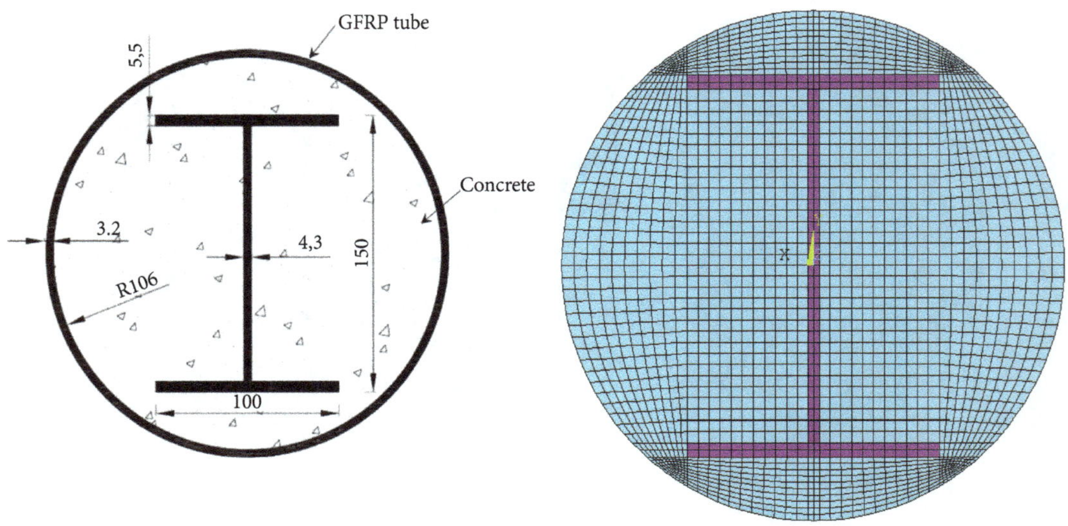


(a) Dimensions of the composite column (b) Mesh of the composite column

FIGURE 7: Dimensions and mesh of the composite column tested by Karimi et al. [15].

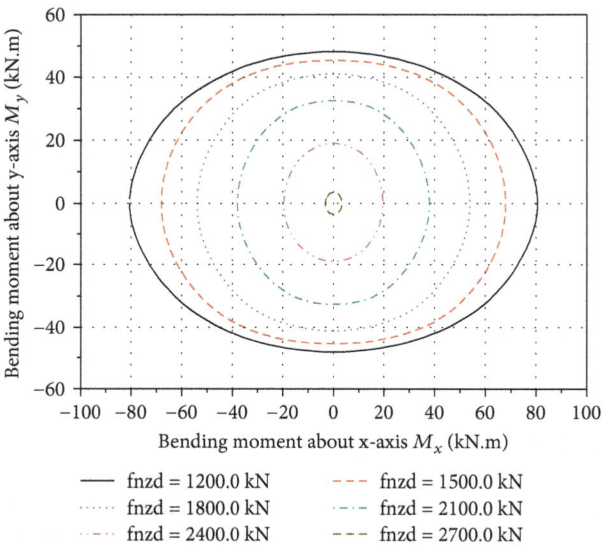


FIGURE 8: $M_x - M_y$ interaction curves of the composite column tested by Karimi et al. [15] under different axial load levels.

To demonstrate the robust ability of the proposed approach for different cross sections, one composite column with more symmetric structural steel was analyzed using this proposed procedure. As shown in Figure 9, a cross-shaped steel was placed in the composite column. As shown in Figure 10, the more symmetric placement of structural steel results in a more symmetric $M_x - M_y$ interaction curve (i.e., the isoload contour); the ultimate resistance along both two axes was almost equal. If no strong axis is required, such placement of structural steel may be preferred for designers.

As mentioned above, the proposed iterative procedure can also be conducted for the columns with structural steel of arbitrary shapes. The following example was used to demonstrate this ability of the proposed approach. A T-shape structural steel (Figure 3) was used in the composite column; its $M_x - M_y$ interaction curve was obtained using the proposed procedure.

*5.2. Checking Adequacy of Predefined Cross Section.* Checking the adequacy of predefined cross section is a routine task for structural engineers. For a short FRP-confined concrete-encased arbitrarily shaped steel columns, such a task can be accomplished by comparing the design load with the $M_x - M_y$ interaction curve produced from the proposed section analysis approach. If the load point falls outside the interaction curve, the composite column cannot carry the design load, otherwise indicating the adequacy of the predefined column. For example, the design task is to check whether a composite column with the predefined section in Figure 11 can carry the design load (2200.0 kN, 40.0 kN.m, and 40.0 kN.m). The checking procedure includes (1) spotting the bending moment pair (i.e., 40.0 kN.m, 40 kN.m) in Figure 11, (2) drawing the part of the $M_x - M_y$ interaction curve of 2200.0 kN in the axial load near the design bending moment pair, and (3) checking whether the design bending moment pair falls within the interaction curve or not. It is found herein that this bending moment pair falls outside the interaction curve, indicating the inadequacy of the section and a stronger cross section in need.

*5.3. Design of Required Steel Size.* The design task is to find a required structural steel size for a composite column to accommodate a specific load set $(N_{zd}, M_{xd}, M_{yd})$. The iterative procedure in Figure 2(b) can be followed to accomplish this task. For example, dimensions of the composite column, including the type of structural shape, were given in Figure 7. Determine the size of structural steel which can accommodate the load set (2400.0 kN, 71.0 kN.m,

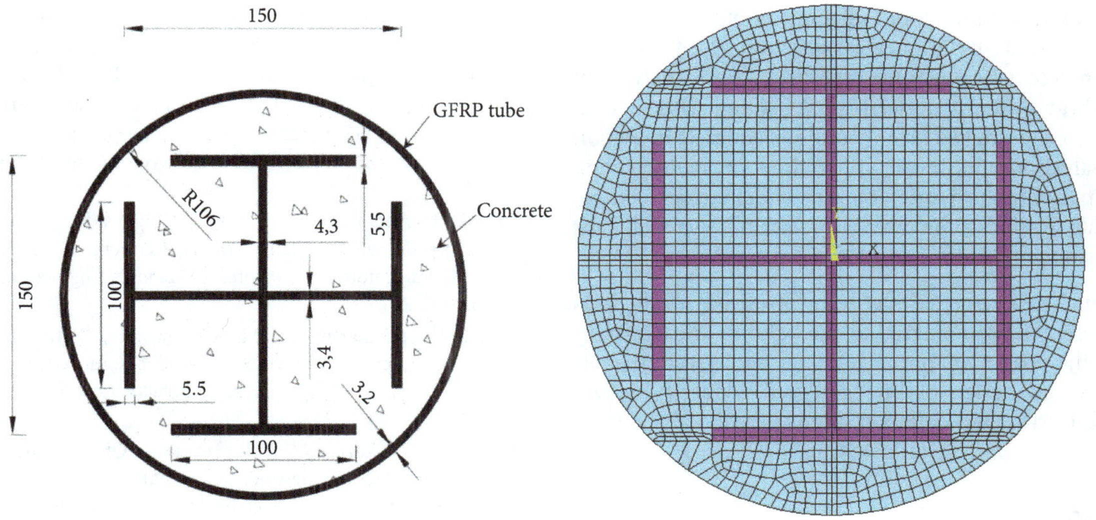


(a) Dimensions of the composite column (b) Mesh of the composite column

FIGURE 9: Dimensions and mesh of the composite column with a cross-shaped steel.

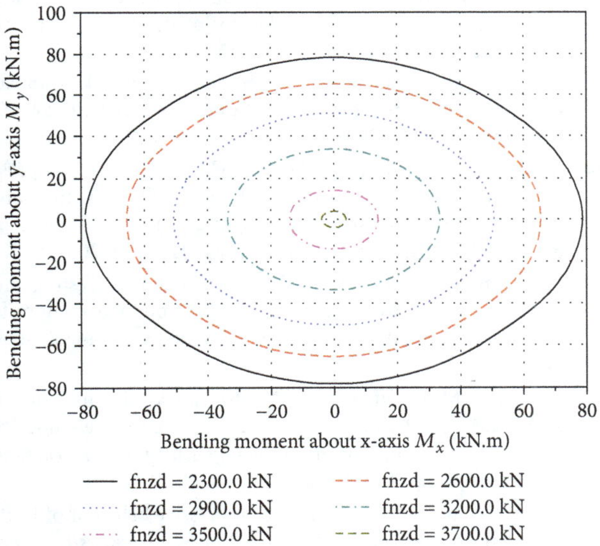


FIGURE 10: $M_x - M_y$ interaction curves of the composite column with a cross-shaped steel under different axial load levels.

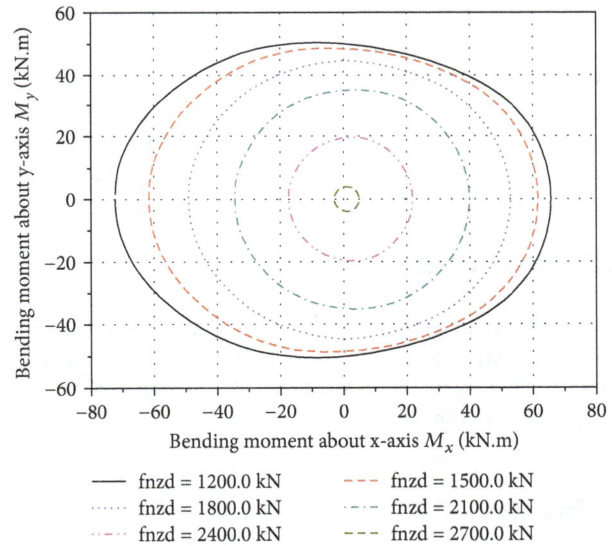


FIGURE 11: $M_x - M_y$ interaction curves of the composite column with a T-shaped steel under different axial load levels.

and −47.9 kN.m). Following the iterative procedure in Figure 2(b), the required size of steel should be 3.316 times that used in Figure 11. In addition, specific dimensions for shaped steel are given in different national codes, so dimensions should also satisfy such requirement.

## 6. Conclusions

This paper presents an approach for the rapid sectional analysis and design of short FRP-confined concrete-encased arbitrarily shaped steel columns under biaxial loading. A robust iterative scheme has been adopted with a traditional so-called fiber element method, where all the constituent materials were treated as fiber elements avoiding the integration for the stress block of the FRP-confined concrete. Several numerical examples were presented leading to the conclusions summarized as below:

(1) Use of the plastic centroid as the origin of reference-loading axes guarantees the convergence in the iterative process. Such a definition of reference-loading axes avoids the possibility that the origin of the reference-loading axes falls outside the $M_x - M_y$ interaction curve, even for cases with irregular structural steel under eccentric compression

(2) Strain gradient along section should be properly considered in the strain-stress model of confined concrete when conducting the section analysis of hybrid columns under biaxial loading. Use of Lam and Teng's model [5] for confined concrete in section analysis results in an upper limit of load carrying capacity of the hybrid columns under biaxial loading, thus leading to an unsafe prediction, especially for cases with a large eccentricity.

(3) Quick convergence has been observed in all cases presented, demonstrating the validity and accuracy of the proposed approach for the rapid section analysis and design of FRP-confined concrete-encased arbitrarily shaped steel columns.

## Disclosure

Publication of the present paper has been approved by Professor J.G. Teng at the Hong Kong Polytechnic University, China, who supervised the second author for part of the work in the present paper.

## Conflicts of Interest

The authors declare that there are no conflicts of interest regarding the possible publication of the present paper.

## Acknowledgments

The authors are grateful for the financial support received from the Nature Science Foundation of China (Project no. 51608130) and China Postdoctoral Science Foundation (Project no. 2016M602441).

## References

[1] H. Kupfer, H. K. Hilsdorf, and H. Rusch, "Behavior of concrete under biaxial stresses," *ACI Journal Proceedings*, vol. 66, no. 8, pp. 656–666, 1969.

[2] J. B. Mander, M. J. N. Priestley, and R. Park, "Theoretical stress-strain model for confined concrete," *Journal of Structural Engineering, ASCE*, vol. 114, no. 8, pp. 1804–1826, 1988.

[3] L. H. Han, W. Li, and R. Bjorhovde, "Developments and advanced applications of concrete-filled steel tubular (CFST) structures: members," *Journal of Constructional Steel Research*, vol. 100, pp. 211–228, 2014.

[4] K. Karimi, W. W. El-Dakhakhni, and M. J. Tait, "Performance enhancement of steel columns using concrete-filled composite jackets," *Journal of Performance of Constructed Facilities*, vol. 25, no. 3, pp. 189-201, 2011.

[5] L. Lam and J. G. Teng, "Design-oriented stress–strain model for FRP-confined concrete," *Construction and Building Materials*, vol. 17, no. 6-7, pp. 471–489, 2003.

[6] M. Samaan, A. Mirmiran, and M. Shahawy, "Model of concrete confined by fiber composites," *Journal of Structural Engineering*, vol. 124, no. 9, pp. 1025–1031, 1998.

[7] J. G. Teng, T. Yu, Y. L. Wong, and S. L. Dong, "Hybrid FRP–concrete–steel tubular columns: concept and behavior," *Construction and Building Materials*, vol. 21, no. 4, pp. 846–854, 2007.

[8] O. I. Abdelkarim and M. A. Elgawady, "Analytical and finite-element modeling of FRP-concrete-steel double-skin tubular columns," *Journal of Bridge Engineering*, vol. 20, no. 8, 2015.

[9] O. I. Abdelkarim and M. A. Elgawady, "Behavior of hollow FRP–concrete–steel columns under static cyclic axial compressive loading," *Engineering Structures*, vol. 123, pp. 77–88, 2016.

[10] O. I. Abdelkarim, M. A. Elgawady, A. Gheni, S. Anumolu, and M. Abdulazeez, "Seismic performance of innovative hollow-core FRP–concrete–steel bridge columns," *Journal of Bridge Engineering*, vol. 22, no. 2, 2017.

[11] T. Ozbakkaloglu and Y. Idris, "Seismic behavior of FRP-high-strength concrete–steel double-skin tubular columns," *Journal of Structural Engineering*, vol. 140, no. 6, 2014.

[12] S. B. Talaeitaba, M. Halabian, and M. Ebrahim Torki, "Nonlinear behavior of FRP-reinforced concrete-filled double-skin tubular columns using finite element analysis," *Thin-Walled Structures*, vol. 95, pp. 389–407, 2015.

[13] Y. F. Yang and L. H. Han, "Concrete-filled double-skin tubular columns under fire," *Magazine of Concrete Research*, vol. 60, no. 3, pp. 211–222, 2008.

[14] X. Liu, A. Nanni, and P. F. Silva, "Rehabilitation of compression steel members using FRP pipes filled with non-expansive and expansive light-weight concrete," *Advances in Structural Engineering*, vol. 8, no. 2, pp. 129–142, 2005.

[15] K. Karimi, M. J. Tait, and W. W. El-Dakhakhni, "Testing and modeling of a novel frp-encased steel–concrete composite column," *Composite Structures*, vol. 93, no. 5, pp. 1463–1473, 2011.

[16] K. Karimi, M. J. Tait, and W. W. El-Dakhakhni, "Influence of slenderness on the behavior of a FRP-encased steel-concrete composite column," *Journal of Composites for Construction*, vol. 16, no. 1, pp. 100–109, 2012.

[17] T. Yu, G. Lin, and S. S. Zhang, "Compressive behavior of FRP-confined concrete-encased steel columns," *Composite Structures*, vol. 154, pp. 493–506, 2016.

[18] L. Huang, T. Yu, S. S. Zhang, and Z. Y. Wang, "FRP-confined concrete-encased cross-shaped steel columns: concept and behaviour," *Engineering Structures*, vol. 152, pp. 348–358, 2017.

[19] S. F. Chen, J. G. Teng, and S. L. Chan, "Design of biaxially loaded short composite columns of arbitrary section," *Journal of Structural Engineering*, vol. 127, no. 6, pp. 678–685, 2001.

[20] T. Ozbakkaloglu and J. C. Lim, "Axial compressive behavior of FRP-confined concrete: experimental test database and a new design-oriented model," *Composites Part B: Engineering*, vol. 55, no. 12, pp. 607–634, 2013.

[21] Y. F. Wu and Y. G. Cao, "Effect of load path on behavior of FRP-confined concrete," *Journal of Composites for Construction*, vol. 21, no. 4, 2017.

[22] Y. F. Wu and Y. Wei, "General stress-strain model for steel- and FRP-confined concrete," *Journal of Composites for Construction*, vol. 19, no. 4, 2015.

# 19

# Tailor-Made, Linear, and "Comb-Like" Polyester-Based Copolymers: Synthesis, Characterization, and Thermal Behavior of Potential 3D-Printing/Electrospinning Candidates

**Franco L. Redondo,[1,2] Mario D. Ninago,[3,4] Augusto G. O. de Freitas,[5] Cristiano Giacomelli,[6] Andrés E. Ciolino,[1,2] and Marcelo A. Villar[1,2]**

[1]*Planta Piloto de Ingeniería Química, PLAPIQUI (UNS-CONICET), CP: 8000, Camino "La Carrindanga" Km 7, Bahía Blanca, Buenos Aires, Argentina*
[2]*Departamento de Ingeniería Química, Universidad Nacional del Sur (UNS), CP: 8000, Av. Alem 1253, Bahía Blanca, Buenos Aires, Argentina*
[3]*Departamento de Ingeniería Química, Facultad de Ciencias Aplicadas a la Industria (FCAI), Universidad Nacional de Cuyo (UNCuyo), CP: 5600, Bernardo de Irigoyen 375, San Rafael, Mendoza, Argentina*
[4]*Consejo Nacional de Investigaciones Científicas y Técnicas (CONICET), Godoy Cruz, 2290 Ciudad Autónoma de Buenos Aires, Argentina*
[5]*Universidade Federal do Pampa, Unipampa (97650-000), Itaqui, RS, Brazil*
[6]*Departamento de Química, Universidade Federal de Santa María, 97105-900 Santa María, RS, Brazil*

Correspondence should be addressed to Mario D. Ninago; mninago@plapiqui.edu.ar

Academic Editor: Cornelia Vasile

Tailor-made, linear, and "comb-like" poly(ε-caprolactone)-based copolymers were synthesized by employing a combination of controlled polymerization techniques. Poly(dimethylsiloxane-*block*-ε-caprolactone) copolymers (SCL#) were synthesized by a combination of anionic and ring-opening polymerization (ROP), whereas "comb-like" poly(hydroxyethylmethacrylate-*co*-(hydroxyethylmethacrylate-*graft*-ε-caprolactone)-*block*-ε-caprolactone) (HEMACL#) were synthesized through simultaneous ROP and reversible addition fragmentation chain transfer (RAFT) polymerization. Copolymers were characterized by hydrogen nuclear magnetic resonance ($^1$H-NMR), size exclusion chromatography (SEC), and Fourier transform infrared (FTIR) spectroscopy. All polymers exhibited narrow molar masses distributions ($M_w/M_n < 1.54$), and their thermal properties were analyzed by isothermal crystallization kinetics (Avrami's theory, by using differential scanning calorimetry (DSC)) and by employing modulated thermogravimetric analysis (MTGA). The macromolecular structure exerts a noticeable effect on the PCL block behavior when compared to the PCL homopolymer, at least for the temperature range studied (16–24°C): less differences in thermal properties were observed for linear block copolymers, whereas for "comb-like" graft copolymers their final crystallization capacity strongly depends on the presence of branches. For both sets of copolymers, the decrease in the resulting melting temperatures and the increase in the half-life crystallization time values might be useful processing parameters, particularly if these copolymers are planned for using as an alternative source of 3D printing or electrospinning materials.

## 1. Introduction

During the last decades, the development of new polymerization techniques (which involve less sophisticated procedures from a practical point of view) has noticeably increased the potential of polymer materials to expand their uses [1]. Complex macromolecular structures are now easily achieved, and a broad range of applications for new polymeric materials can be proposed at levels scarcely envisioned a few decades ago [2, 3]. Reversible deactivation radical polymerization

(RDRP) techniques, such as atom transfer polymerization (ATRP), nitroxide-mediated polymerization (NMP), and reversible addition-fragmentation chain-transfer (RAFT) polymerization, are now well-established, and their uses in the synthesis of macromolecules are widely employed in synthetic laboratories worldwide.

For a long time, graft and block copolymers have received the attention of polymer scientists due to their attractive macromolecular characteristics and properties. This interest lies on the combination of thermodynamically incompatible macromolecular species linked together, which imposes restrictions regarding their properties and final applications. Moreover, the processability of these materials is a crucial point that will justify their further uses. In such a sense, many processing parameters such as viscosity-temperature dependence, glass-transition temperature, thermal stability, and melting temperature of semicrystalline blocks or segments are critical to decide whether or not using them [4].

Recent developments regarding biocompatible materials have enabled 3D bioprinting of functional living tissues, mainly applied to regenerative medicine to address the need for organ transplantation [5–7]. In this aspect, 3D bioprinting materials are limited to biocompatible synthetic and natural polymers, such as targeted synthetic diblock copolymers, sodium alginate, chitosan, and acrylate-based polymers, among others [6]. In fact, many ink formulations or 3D-printable materials for biomedical purposes are made from copolymers or from the combination of complex macromolecular structures, which can be obtained by adjusting well-known polymerization pathways [8, 9].

Another area of interest in which biocompatible materials deserve special interest is electrospinning. By using this technique, the preparation of nanofibers from different materials (synthetic or natural polymers, composites, ceramics, and even inorganic oxides) can be strategically targeted in order to obtain specific properties [10]. Medical studies have used this technique to design nanofiber for drug delivery systems as well as scaffolds for tissue engineering [11]. The processability of the polymeric materials, their solubility behavior, and their concentration are some of the parameters of interest to produce targeted nanofiber mats or meshes.

Poly(ε-caprolactone) (PCL) is an aliphatic and hydrophobic polyester with good biodegradable and mechanical properties. This semicrystalline polymer has a relative low melting temperature ($T_m$~60°C) and an appropriate glass-transition temperature for technical applications ($T_g$~ – 60°C). In addition, the aliphatic nature of the main chain explains its compatibility with a large variety of polymers [12], whereas its lack of toxicity is a strong reason to use it as an excellent candidate in a wide range of biomedical applications such as long-term implantable devices, scaffolds for tissue growth, and drug delivery systems, among others [13–15]. PCL can be easily synthesized by the ring-opening polymerization (ROP) of ε-caprolactone (ε-CL) monomer promoted by suitable catalysts, such as tin(II) octanoate and a hydroxyl initiator [12]. Regarding its use as 3D-printing material, it has been reported that PCL nanoparticles have been used as ink that can be printed on a bioadhesive film for the local treatment of cervical cancer, corroborating that the printing process does not disrupt the structure of nanoparticles nor inclusion complexes [16].

Poly(2-hydryoxyethyl methacrylate) (PHEMA) is another biocompatible polymer, which is widely employed to produce hydrogel materials for ocular devices and soft contact lens [17]. These applications are possible due to the hydrophilicity of the hydroxyl moiety, which imposes to PHEMA-based hydrogel physical properties that resemble those of living tissue: high water content, soft and rubbery consistency, and low interfacial tension [18]. HEMA monomer can be copolymerized with other biocompatible monomers (such as ε-CL), by employing conventional free-radical polymerization mechanisms or RDRP techniques. Particularly, PCL chains grafted onto the PHEMA backbone can be obtained by RAFT polymerization [19–22], since this method allows tuning the molar masses distribution and the polydispersity indexes of the resulting complex macromolecular architectures like stars, blocks, or graft copolymers without needing sophisticated experimental procedures [23].

Poly(siloxanes) belong to the family of organic/inorganic materials, such as poly(phosphazenes), and they can be obtained from cyclic siloxane monomers. Particularly, poly(-dimethylsiloxane) (PDMS) is the homopolymer made from siloxane cycles whose chemical structure is $[-(CH_3)_2SiO-]_n$, in which $n$ denotes 3 or 4 repeating units. PDMS has unique properties such as high thermal and electric stability, and low glass-transition temperature ($T_g$~ – 123°C). Besides, PDMS has no chemical reactivity and displays a marked hydrophobicity and excellent physiological compatibility [24, 25]. The conventional synthesis of PDMS is carried out by hydrolysis and condensation reactions of dimethyldichlorosilanes in alkaline or acid media, by producing cyclic and linear polymers without a good control of the homogeneity [26]. A better control of PDMS molecular parameters can be achieved by using anionic ROP of cyclic siloxanes [27–29]. Although PDMS and its derivatives are widely employed in technical applications, their use as 3D-printing materials has been proved challenging due to poor mechanical properties and need for supporting during the process [30, 31]. Consequently, the searching of likely alternatives for these materials constitutes an area of broad interest.

In previous papers, we have reported the synthesis of PCL-based copolymers by employing controlled polymerization techniques [27, 28, 32]. In an effort to expand the understanding of their physicochemical characteristics, the aim of this work was focused on the thermal behavior (at moderate temperatures) of two block copolymers series based on ε-CL monomer. In this sense, poly(dimethylsiloxane-*block*-ε-caprolactone), P (DMS-*b*-ε-CL), poly(hydroxyethylmethacrylate-*co*-(hydroxyethylmethacrylate-*graft*-ε-caprolactone)-*block*-ε-caprolactone), and P (HEMA-*co*-(HEMA-*g*-ε-CL)-*b*-ε-CL) copolymers were synthesized by combining anionic, ROP, and RAFT polymerizations. Besides synthesis, the work is focused on the crystallization and thermal behavior of these polyester-based copolymers, both containing biocompatible blocks.

PCL has been employed as 3D bioprinting material, and it is expected that biocompatible block copolymers based on PCL might expand its applications in the field. As it is known, in material extrusion techniques (such as 3D-printing or electrospinning devices), polymers are directly pushed out through a nozzle when a constant pressure is applied. The extruded material is deposited at a constant speed and fully solidifies on the substrate after coming out from the nozzle [6]. In this sense, the understanding of the crystallization behavior of branched and linear copolymers based on PCL at temperatures lower than that corresponding to the linear homopolymer could lead to changes in processing parameters, especially if they are targeted for specific purposes. In such a sense, the thermal stabilities of the studied polymers were analyzed, and Avrami's isothermal crystallization model was employed to calculate kinetic parameters from the crystallization process [32–36]. It is inferred that data obtained from this analysis will provide useful information for further applications, particularly if these materials or combinations of them are planned for using as potential inks in 3D-printing or electrospinning devices [5–7, 16].

## 2. Materials and Methods

*2.1. Materials.* Hexamethyl(cyclotrisiloxane) monomer ($D_3$, Sigma-Aldrich) was purified according to conventional routines for high-vacuum anionic polymerization techniques. The initiator employed for anionic $D_3$ polymerization was *sec*-butyllithium (*sec*-$Bu^-Li^+$), freshly prepared in vacuum from *sec*-butyl chloride (Fluka) and lithium metal (Fluka) by following conventional synthetic procedures already reported [37]. Tetrahydrofuran (THF, Ciccarelli) was used as polymerization promoter, whereas cyclohexane (Dorwill) and degassed methanol (Química Industrial) were used as solvent and protonating agent, respectively [38–41].

ε-Caprolactone monomer (ε-CL, Aldrich) was employed as received. Its polymerization was performed by using tin(II) octanoate ($Sn(Oct)_2$, Aldrich) as catalyst, following the procedure previously reported by Satti et al. [42]. 2-Hydroxyethyl(methacrylate) (HEMA, Aldrich), diphenyl phosphate (DPP), 1,1′-azobis(cyclohexanecarbonitrile) (Vazo catalyst 88, Aldrich, 98%), toluene (Aldrich), methanol (Química Industrial), dimethylformamide (DMF), chloroform (Aldrich), deuterated chloroform ($CDCl_3$), and petroleum ether were used without any further purification. 2-(Benzylsulfanylthiocarbonylsulfanyl)ethanol (BSTSE, chain transfer agent for RAFT polymerization and hydroxyl initiator for ROP) was prepared following a one-pot procedure, purified by column chromatography on silica using petroleum ether as eluent and recrystallized under cold conditions, as it was previously described [20, 43, 44].

*2.2. Synthesis of P (DMS-b-ε-CL) Block Copolymers (SCL#).* Polymerizations were carried out in two steps. First, a model poly(dimethylsiloxane) hydroxyl-ended homopolymer (PDMS-OH) was synthesized by employing anionic polymerization and hand-made polymerization apparatuses (glass-blowing techniques), by a vacuum manifold (VM), and by following the methodology reported in our previous works [3, 39, 40]. Briefly, the reactor was attached to the VM and checked for pinholes. Then, the stopcock was closed and approximately 5 mL of commercial *n*-butyllithium solution in hexane (*n*-$Bu^-Li^+$ 2 M, Aldrich) was injected into the purge section flask. After that, the stopcock of the VM was opened and the apparatus was pumped during 30 minutes in order to remove traces of hexane and air inside. Then, the reaction solvent was distilled from a collection flask already attached to the VM. After two freezing-thawing cycles, the apparatus was removed from the VM by heat-sealing, and it was washed with the diluted *n*-$Bu^-Li^+$ solution inside. After washing, the solvent was distilled from purge section to the reactor, and once distillation was completed, the purge section was removed by heat-sealing procedures, leaving the clean reactor ready for polymerization. Then, the $D_3$ monomer was poured into the reactor flask, followed by the addition of *sec*-$Bu^-Li^+$. The reaction was left to proceed at room temperature during ~20 hours [39]. Finally, THF was added and $D_3$ polymerization was left to proceed, at room temperature, during 20 hours. The reaction was finished by the addition of well-degassed methanol. The resulting PDMS was precipitated in cold methanol, traces of solvents were removed by employing a roto-evaporator (~60°C), and the final product was kept in vacuum until constant weight.

Copolymerization of PDMS-OH with ε-CL monomer was carried out in a specially designed Schlenk-type reactor by following the methodology already published [42–44]. In this sense, copolymers with a chemical composition ranging from low to intermediate PCL content were obtained by ROP of cyclic monomer. Copolymerization was performed under nitrogen atmosphere, employing degassed toluene as reaction solvent and a catalyst/PDMS-OH ratio of 0.5. Reactions were carried out at ~100–110°C during 24 h. The resulting copolymers were precipitated in cold methanol, filtered, and vacuum-dried, at 20°C, until constant weight. Since in this synthetic procedure the amount of ε-CL monomer was adjusted in each reaction, P (DMS-*b*-ε-CL) copolymers were labelled as SCL#, in which # refers to PCL content from low (1) to high (4) wt%.

*2.3. Synthesis of P (HEMA-co-(HEMA-g-ε-CL)-b-ε-CL) Copolymers (HEMACL#).* Copolymerizations were carried out in a one-pot procedure through a combination of simultaneous RAFT and ROP polymerizations, by following a methodology already published [28]. BSTSE, HEMA and ε-CL monomers, Vazo-88 radical initiator, dry toluene, and DMF were placed in a dry Schlenk flask equipped with a stirrer. The tube was closed, subjected to three freeze–pump–thaw cycles in the VM, and subsequently backfilled with nitrogen gas. The ROP catalyst DPP was then added under gentle nitrogen flow. Subsequently, the flask was closed and immediately immersed in an oil bath settled at 100°C to promote the simultaneous copolymerization reaction. The reaction was left to proceed during 24, 2, or 1 h, and the final product was precipitated in cold methanol and dried in vacuum until constant weight [27, 28]. Since in this synthetic procedure the amount of grafted chains depends on the reaction time, P (HEMA-*co*-(HEMA-*g*-ε-CL)-*b*-ε-CL) copolymers were labelled as HEMACL#, in which # denotes low (1) to high (3) reaction times.

*2.4. Characterization*

*2.4.1. Fourier Transform Infrared Spectroscopy (FTIR).* Fourier transform infrared (FTIR) spectroscopy spectra of selected samples were obtained on a Nicolet® FTIR 520 spectrometer. Cast films from diluted solutions of the samples (1 wt% in THF) were obtained onto NaCl windows. FTIR spectra were recorded at 4 cm$^{-1}$ resolution, over the 4000–400 cm$^{-1}$ range, 64 scans, and dried air as background.

*2.4.2. Nuclear Magnetic Resonance ($^1$H-NMR).* $^1$H-NMR spectra of SCL# copolymers were recorded on a Bruker® 300 MHz instrument, by using deuterated chloroform ($CDCl_3$, Aldrich) as solvent. About 5–10 mg of sample was diluted directly in the NMR tubes, at room temperature. The chemical composition of synthesized copolymers was obtained by using the values of integrated areas from characteristic $^1$H signals of each block. For HEMACL# copolymers, $^1$H-NMR spectra were acquired using an Avance DPX-400 spectrometer (400 MHz for H, 100 MHz for C). The spectra were obtained by dissolving a small quantity of the polymer sample (~10 mg), at room temperature, by using $CDCl_3$ as solvent.

*2.4.3. Size Exclusion Chromatography (SEC).* SCL# samples were characterized by SEC on a system built with a Waters® 515 HPLC pump and a Waters® model 410 differential refractometer, equipped with three mixed-bed PLgel linear columns and a precolumn with 5 $\mu$m bead size (PLgel). Elution solvent was toluene (Aldrich), flowing at a rate of 1 mL min$^{-1}$ at room temperature. Injection volume was 200 $\mu$L, and PS standards were used for calibration. Mark-Houwink calibration constants used were $K_{PS} = 0.012$ mL g$^{-1}$, $\alpha_{PS} = 0.71$ and $K_{PDMS} = 0.0136$ mL g$^{-1}$, $\alpha_{PDMS} = 0.69$ for PDMS [32, 45]. SCL# were also analyzed by employing the same equipment by using THF (Ciccarelli) as solvent in order to distinguish PCL blocks in the resulting copolymers. In this case, the Mark-Houwink calibration constants used for PCL were $K_{PCL} = 1.395 \times 10^{-4}$ mL g$^{-1}$ and $\alpha_{PCL} = 0.786$ [46]. For HEMACL# samples, toluene was used as solvent employing the aforementioned equipment. Mark-Houwink constants employed for PCL were $K_{PCL} = 0.01298$ mL g$^{-1}$, $\alpha_{PCL} = 0.828$ [47].

*2.4.4. Differential Scanning Calorimetry (DSC).* Thermal analysis was performed by employing DSC Pyris 1 Perkin-Elmer® equipment. Samples were measured under He atmosphere, by using approximately 10 mg of each copolymer. For SCL# copolymers, samples were heated from −90 to 100°C. Then, they were kept at 100°C during 5 minutes in order to avoid the influence of the previous thermal history. After cooling at 10°C min$^{-1}$, they were heated again from −90 to 100°C at a heating rate of 10°C min$^{-1}$. From the second heating, glass-transition temperature ($T_g$) of PCL block and melting temperature ($T_m$) of PDMS and PCL blocks were determined. For HEMACL# copolymers, samples were heated from 15 to 100°C. Then, they were kept at 100°C during 5 minutes in order to avoid the influence of the previous thermal history. After cooling at 10°C min$^{-1}$, they were heated again from 15 to 100°C at a heating rate of 10°C min$^{-1}$. From the second heating, $T_m$ of PCL blocks and $T_g$ of PHEMA blocks were determined.

*2.4.5. Isothermal Crystallization Tests.* Isothermal crystallization tests were carried out by using the above-mentioned equipment and by employing sequential steps described as follows: first, samples were heated from 15 to 90°C at 10°C min$^{-1}$. Then, they were kept at 90°C during 5 min and rapidly cooled (60°C min$^{-1}$) to selected crystallization temperatures ($T_c$) according to the methodology reported in previous works [32, 34, 48, 49]. Isothermal crystallization tests were carried out for temperatures between 16 and 24°C, by using 2°C steps. Finally, after isothermal crystallization, samples were heated up to 90°C (10°C min$^{-1}$) in order to determine the heat of fusion ($\Delta H_f$). In all cases, the heat of crystallization ($\Delta H_c$) as a function of time ($t$) was recorded at different isothermal $T_c$.

*2.4.6. Modulated Thermogravimetric Analysis (MTGA).* Thermal stability was analyzed by using Discovery TA Instruments TGA 5500 equipment. Tests were carried out under nitrogen atmosphere, with a flow rate of 25 mL min$^{-1}$ and 2°C min$^{-1}$ heating rate, in the 40–700°C range. Weight loss curves as a function of temperature were registered. Besides, modulated thermogravimetry assays were performed to obtain continuous kinetic information regarding decomposition reactions. MTGA tests were made by using an oscillatory temperature program. Samples were analyzed by triplicate, and the resulting data (three values average) are informed with their associated standard deviation. The results obtained were compared with those for linear PCL in order to analyze the influence of the neighbor block and random branches on the thermal stability of the resulting copolymers.

## 3. Results and Discussion

SCL# copolymers were synthesized by combining classical anionic polymerization followed by ROP. PDMS-OH homopolymer was employed as macroinitiator for ROP of ε-CL monomer. Using this synthetic pathway, four different copolymers from low to intermediate PCL content were obtained. Scheme 1 shows the reaction pathways employed for this purpose.

Regarding HEMACL#, these copolymers were synthesized through a combination of RAFT and ROP polymerizations in a one-pot reaction, as it was reported in our previous work [28]. Scheme 2 shows the reaction pathway employed for this purpose, in which ROP of ε-CL simultaneously takes place from the hydroxyl moieties from HEMA monomer and BSTSE RAFT agent, respectively.

The reaction proposed in Scheme 2 produces graft PCL branches in the resulting HEMACL# copolymers, whose distribution will strongly depend on reaction time (2, 1, and 24 hours for HEMACL1, HEMACL2, and HEMACL3, respectively). The macromolecular architecture obtained can be thought as a "densed and/or un-densed comb-like copolymers," in which the main skeleton is a block

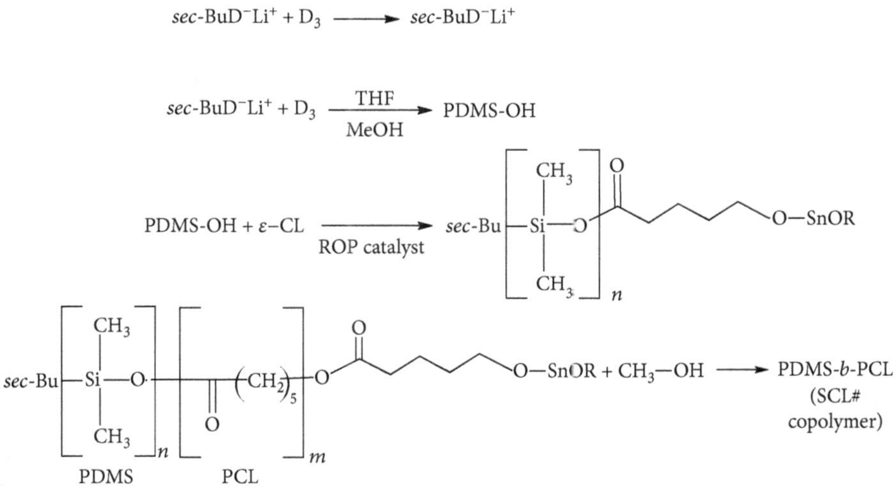

SCHEME 1: Synthesis of SCL# copolymers.

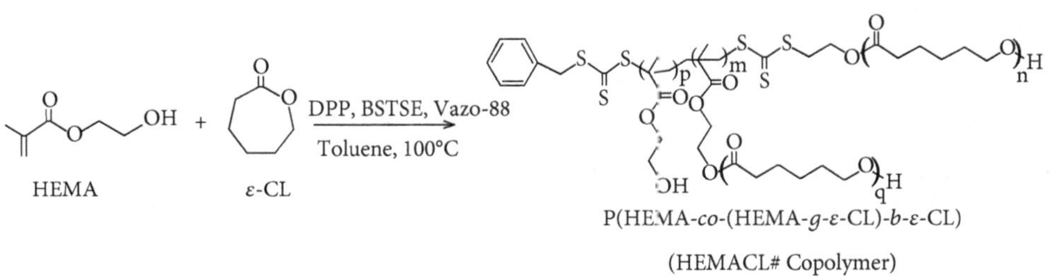

SCHEME 2: Synthesis of HEMACL# copolymers.

copolymer based on HEMA and ε-CL, and teeth are random PCL blocks pending from HEMA monomer units. Consequently, HEMACL# copolymers can be thought as "pseudo" branched PCLs.

According to chemical characterization, Figure 1 shows selected FTIR spectra from PDMS-OH and PCL homopolymers, as well as SCL1 and HEMACL3 copolymers. PDMS-OH and PCL spectra were shifted in the transmittance axis in order to display the common absorption bands that appear in both copolymers. The broad absorption band at 3900–3200 cm$^{-1}$ is associated with the stretching of the -OH group in PDMS-OH homopolymer. The spectrum of this polymer also shows an absorption band at 2963 cm$^{-1}$ associated to the vibration bonds of C-H attached to Si atoms [50]. At 1261 cm$^{-1}$, the out-of-phase vibrations of Si-$(CH_3)_2$ and O-Si-OR are also detected [1]. Besides, at 1093 cm$^{-1}$ and 1024 cm$^{-1}$, absorption bands associated to the vibration of Si-O-Si and C-Si-C bonds, respectively, can be observed [3, 32]. The PCL spectrum shows an absorption band at 1724 cm$^{-1}$ associated to the stretching vibrations from carbonyl >C=O groups, as well as at 1240 cm$^{-1}$ (vibrations of O-C=O) [33, 42]. On the other hand, FTIR spectra from SCL1 and HEMACL3 copolymers exhibit the typical absorption bands from their constitutive blocks, as it was reported by several authors [32, 42, 51]. In this sense, PCL absorption signals were detected at 1724 cm$^{-1}$ and 1240 cm$^{-1}$. In addition, for HEMACL# copolymer a strong stretching band at

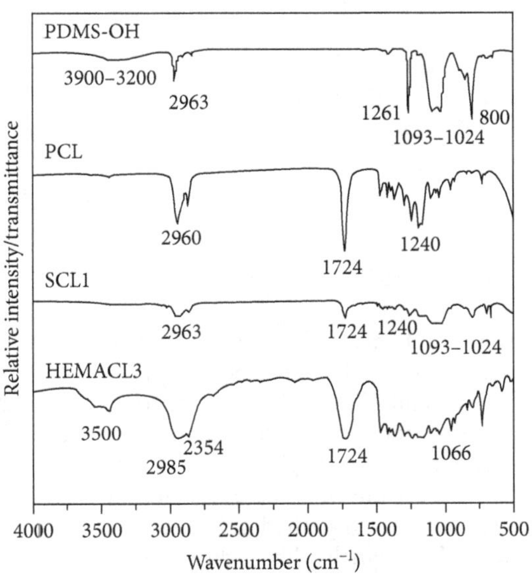


FIGURE 1: FTIR spectra of PDMS-OH, linear PCL, P (DMS-*b*-ε-CL) SCL1, and P (HEMA-*co*-(HEMA-*g*-ε-CL)-*b*-ε-CL) HEMACL3.

3500–3100 cm$^{-1}$ is observed, which should be assigned to -OH vibration as it was already reported by Sahiner and Demirci [51]. Besides, aliphatic C-H stretching bands at 2985 cm$^{-1}$ and 2854 cm$^{-1}$ are also detected [51], together

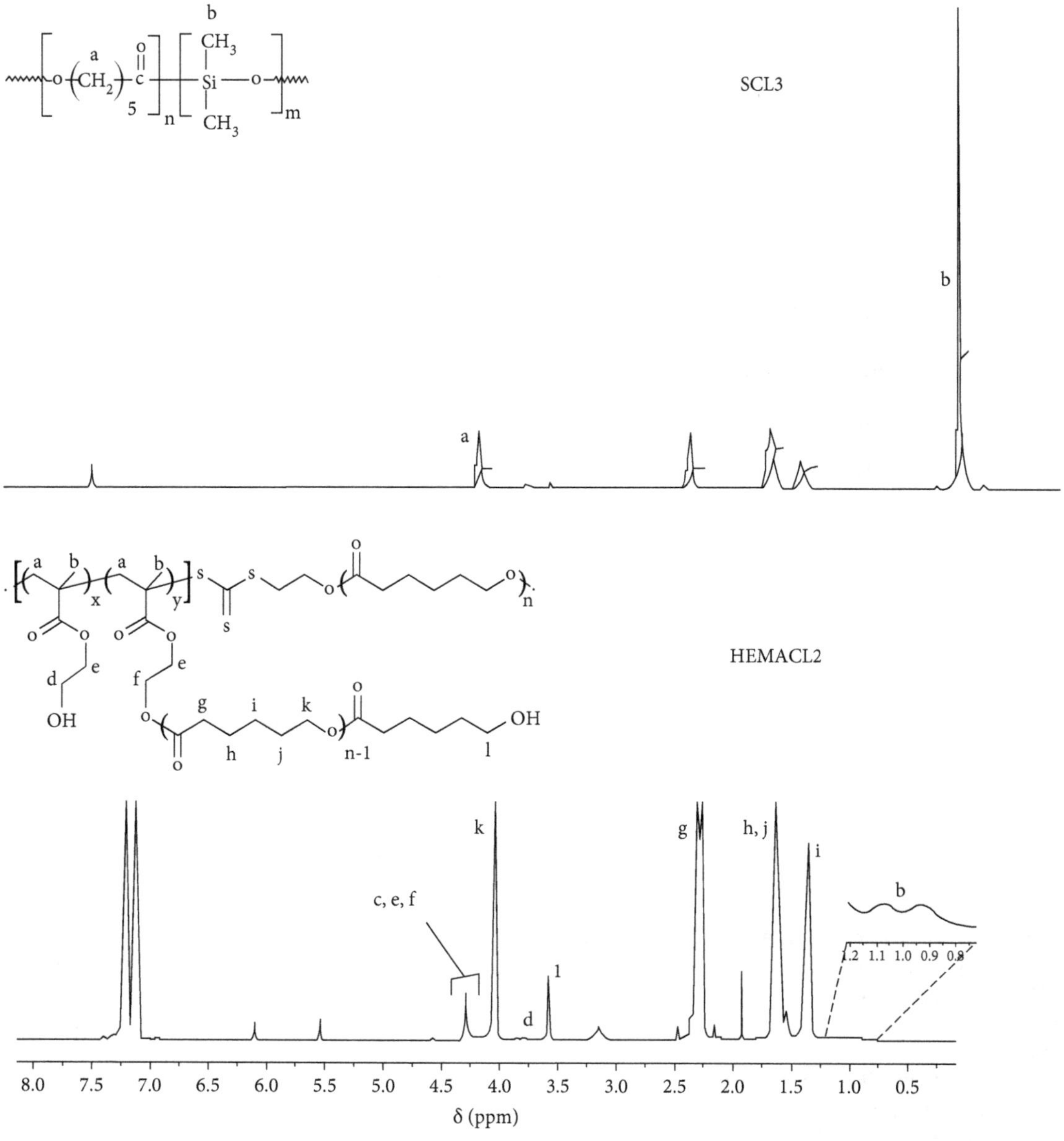


FIGURE 2: [1]H-NMR spectra of P (DMS-*b*-ε-CL) SCL3 and P (HEMA-*co*-(HEMA-*g*-ε-CL)-*b*-ε-CL) HEMACL2 copolymers.

with absorption bands assigned to the transfer agent BSTSE at 1066 $cm^{-1}$ [27, 28]. Figure 2 shows typical [1]H-NMR spectra from SCL3 and HEMACL2 copolymers. The characteristic signals from constitutive blocks and the corresponding $\delta$ (ppm) assignments are displayed in the spectra. According to the NMR spectrum provided in Figure 2 for HEMACL2, the difference between grafted and ungrafted HEMA units is the *d* and *f* signals that belong to methylene units attached to free –OH groups (d) or reacted –OH groups by esterification (f).

For SEC analysis, Figure 3 displays SEC chromatograms for PDMS-OH macroinitiator and SCL4 copolymer (toluene as solvent). The PDMS homopolymer was a model polymer, with a low polydispersity index ($M_n = 13{,}000\,g\,mol^{-1}$, $M_w/M_n = 1.07$, Table 1). On the other hand, the SCL4 copolymer shows an increase in $M_n$ value ($M_n = 65{,}000\,g\,mol^{-1}$), with $M_w/M_n = 1.54$ (Table 1). For the other copolymers, $M_n$ values show an increasing trend as soon as $w_{PCL}$ increases, and $M_w/M_n$ indexes display values between 1.36 (SCL1) and 1.45 (SCL3). Although $M_w/M_n$ values for SCL# seem to be far from model samples, they are quite good by taking into account that ε-CL monomer was employed as received in order to simplify synthetic procedures. As it can be seen from Table 1 and Figure 3, SCL4 displays a broad molar masses distribution (when compared to PDMS-OH) and a small shoulder at higher molar masses values. These facts might be assigned to the presence of water or hydroxyl impurities from commercial ε-CL monomer as well as inaccuracies in the refractive index detector for PCL blocks for the solvent employed (toluene). The presence of water or hydroxyl impurities might lead to uncontrolled ROP due to the formation of stable tin alkoxides [52, 53], whereas the use of toluene as solvent results in a less defined chromatogram when compared to that from the initial

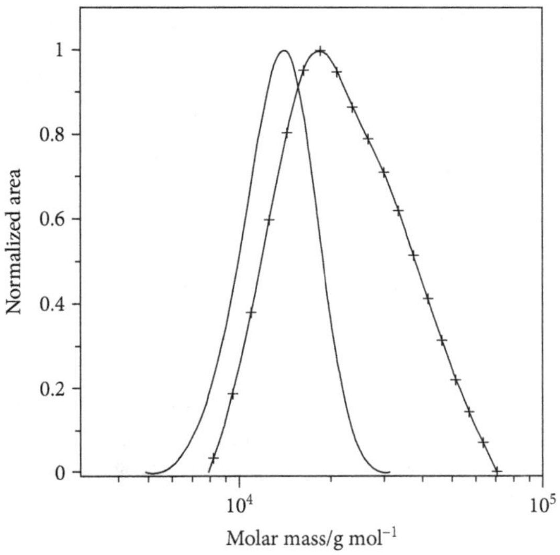


FIGURE 3: SEC chromatogram of PDMS homopolymer and P(DMS-$b$-ε-CL) SCL4 block copolymer. Symbols: (–) PDMS and (+) SCL4.

PDMS-OH macroinitiator. It is inferred that, with an additional purification step of commercial ε-CL monomer (for example, a previous treatment with calcium hydride), the presence of water or hydroxyl impurities will be diminished or suppressed. HEMACL# copolymers exhibited $M_n$ values between 32,300 g mol$^{-1}$ and 46,000 g mol$^{-1}$, and $M_w/M_n$ lower than 1.2, indicating a good control of the process over the chemical characteristics of the resulting macromolecules (Table 1). This fact proves the efficiency of the one-pot procedure to produce macromolecular architectures by employing RAFT and ROP simultaneously. As an additional fact, the expected molar masses can be achieved at reasonable reaction times (at least, 1 h) which is another feature that supports the efficiency of the proposed procedure. At this point, it is worth mentioning that HEMACL# copolymers deserve an additional purification process. In such a sense, Figure 4 displays SEC chromatograms from the crude reaction product and the resulting HEMACL3 copolymer after the purification process. As it can be seen, the crude reaction product contains residual PCL homopolymer (from BSTSE as well as from HEMA monomer) that disappears by successive precipitation employing controlled temperature and nonsolvent (methanol) fractionation.

DSC measurements were performed in order to evaluate thermal transitions from linear and "graft-like" PCL-based copolymers when compared to linear PCL (at this point, it should be mentioned that the temperature range under study was selected by taking into account the $T_m$ of the PCL block from HEMACL# copolymers, which is lower than that corresponding to the PCL homopolymer employed for comparison purposes; for more details, please refer to Table 1). For SCL# samples, three transitions are detected (Figures 5(a) and 5(b)): $T_g$ from the PCL block ($T_{gPCL}$~ – 60°C), $T_m$ of the semicrystalline PDMS block ($T_{mPDMS}$~ – 45°C), and $T_m$ of the PCL block ($T_{mPCL}$~54°C). On the other hand, for linear PCL, $T_{mPCL}$ is detected at ~56°C (Figure 5(c)), whereas for HEMACL# copolymers, $T_{mPCL}$ from PCL grafts is observed at ~40°C (Figure 5(d)). This reduction on the $T_m$ of PCL could be explained by taking into account the presence of random grafted PCL chains along the PHEMA backbone [27, 28, 54]. It is reasonable to suppose that longer reaction times allow higher density of PCL grafts (HEMACL3), whereas lower PCL grafts are expected for shorter reaction times (HEMACL2). This is quite reasonable since BSTSE acts as both a RAFT and ROP initiator. For short reaction times, few PCL grafts are attached from the PHEMA block because reaction time is not enough to achieve a good RAFT conversion. On the contrary, at higher reaction times, the probability to have randomly distributed grafts along PHEMA blocks increases. In addition, for these copolymers, a final thermal event around 68°C (associated to $T_{gPHEMA}$ from PHEMA block) was also observed (Figure 5(d) and Table 1) [55].

Isothermal crystallization curves of linear PCL, SCL#, and HEMACL# copolymers are displayed in Figure 6, in which the evolution of endothermic heat flow was recorded as a function of time. In Figure 6(a), the isothermal crystallization curves of linear PCL are plotted at different $T_c$ values. As it can be seen, crystallization time is increased at higher $T_c$ values, indicating that nucleation is the limiting factor for crystallization in this temperature range [56]. A similar behavior was observed for SCL# and HEMACL# copolymers. As examples, Figure 6(b) shows the crystallization evolution of linear PCL, SCL2, and HEMACL2 at $T_c = 20$°C. It is important to note that the completed crystallization processes of SCL2 and HEMACL2 copolymers occur at higher times when compared to linear PCL. For SCL2, this rise can be associated with the noncrystallizable PDMS block (Table 2). In this sense, Lovinger et al. [57] reported that the PDMS block avoids the regular crystallization of PCL chains because of the higher mobility of the viscoelastic, amorphous PDMS block throughout the crystallization process. On the other hand, HEMACL# copolymers show a wider difference, because the presence of random, grafted branches hinders the formation of PCL crystallites by increasing crystallization time values (Table 2) [12].

As it is reported in the literature [35, 58], isothermal crystallization kinetics can be better visualized by considering the relative degree of crystalline conversion as a function of time at different crystallization temperatures (Figure 7). Then, relative crystallinity, related to the crystalline volume ($V_c$), vs. different crystallization times ($t$), can be calculated from [1, 32, 35, 49]

$$V_c = \frac{\Delta H_t}{\Delta H_\infty} = \frac{\int_{t_0}^{t} (dH_c/dt)dt}{\int_{t_0}^{t_\infty} (dH_c/dt)dt}, \quad (1)$$

where $\Delta H_t$ and $\Delta H_\infty$ are the heat generated at time $t$ and at long time periods, respectively, and $dHc/dt$ is heat evolution rate. $V_c$ is obtained from the area of the exothermic peak on isothermal crystallization and is plotted vs. time (Figure 7). The slope of the graphs appears as an S-like shape. The S-like shape in the graphs with higher slopes reveals higher

TABLE 1: Chemical and thermal characterization of synthesized linear PCL and PDMS-OH homopolymers, and PCL-based copolymers.

| Polymer | $M_n$ [(a)]/g mol$^{-1}$ | $M_w/M_n$ [(a)] | $w_{PCL}$ [(b)] | $T_{gPCL}$ [(c)]/°C | $T_{gPHEMA}$ [(c)]/°C | $T_{mPDMS}$ [(c)]/°C | $T_{mPCL}$ [(c)]/°C | $T_{0.05}$ [(d)]/°C | $E_{ad}$ [(d)]/kJ mol$^{-1}$ |
|---|---|---|---|---|---|---|---|---|---|
| PCL | 11,000 | 1.14 | 1 | −66.0 | — | — | 55.9 | 341.5 | 217.6 |
| PDMS-OH | 13,000 | 1.07 | — | — | — | −44.9 | — | 340.0 | 102.1 |
| SCL1 | 20,000 | 1.36 | 0.35 | −59.1 | — | −42.4 | 50.4 | 205.1 | 147.4 |
| SCL2 | 21,300 | 1.33 | 0.39 | −59.6 | — | −43.4 | 51.1 | 247.3 | 142.3 |
| SCL3 | 25,000 | 1.45 | 0.48 | −59.9 | — | −43.9 | 47.3 | 241.9 | 152.6 |
| SCL4 | 65,000 | 1.54 | 0.80 | −62.1 | — | −45.2 | 54.8 | 195.7 | 172.7 |
| HEMACL1 | 32,300 | 1.19 | 0.89 | −63.3 | 68.7 | — | 41.1 | 231.6 | 132.5 |
| HEMACL2 | 38,600 | 1.13 | 0.93 | −63.1 | 67.9 | — | 38.2 | 227.4 | 134.7 |
| HEMACL3 | 46,000 | 1.11 | 0.97 | −63.4 | 68.2 | — | 42.9 | 225.3 | 131.4 |

[(a)]Number average molar mass ($M_n$) and polydispersity ($M_w/M_n$) determined by SEC and $^1$H-NMR; [(b)]weight fraction of PCL in the block copolymer determined by $^1$H-NMR; [(c)]glass-transition temperature ($T_g$) and melting temperature ($T_m$) determined by DSC; [(d)]5% thermal degradation temperature ($T_{0.05}$) and activation energy for degradation ($E_{ad}$) determined by MTGA.

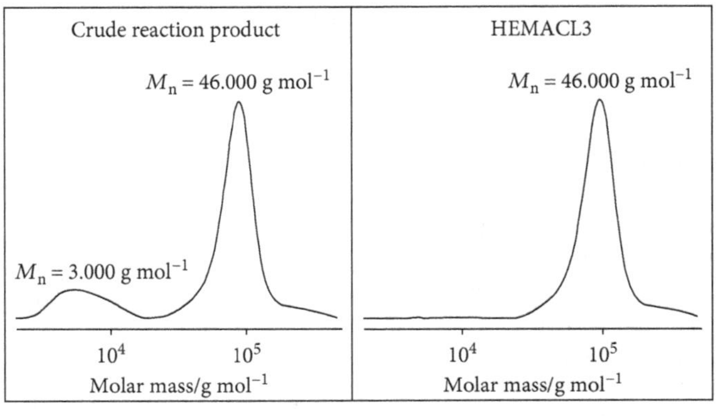


FIGURE 4: SEC chromatograms from the crude reaction product and the resulting HEMACL3 copolymer after the purification process.

rates for relative crystallization vs. time, indicating a faster crystallization rate at lower $T_c$ values (Figures 7(a)–7(c)). As examples, at a given $T_c$ value, SCL# and HEMACL# copolymers crystallize at a lower rate when compared to linear PCL as it can be deduced by exploring time values at a given relative crystallinity % (Figure 7(d)).

The above-described results allow employing Avrami's theory to obtain kinetics parameters of interest such as the half-life time for crystallization ($t_{1/2}$), the bulk crystallization constant ($k$), and Avrami's exponent ($n$), by following [2, 15, 49]

$$V_c = 1 - \exp(-kt^n). \quad (2)$$

This equation can be rewritten in a double logarithmic form as follows:

$$\log[-\ln(1 - V_c)] = \log k + n \log(t - t_0), \quad (3)$$

where $V_c$ is related to the relative crystallinity from polymers at different temperatures or times, and $t - t_0$ is the crystallization period involved in the process. The induction time (which includes the stabilization time) has been named $t_0$, and it was calculated by drawing a horizontal line from a point after the finishing crystallization process to a point at the beginning of the crystallization curve, by following the methodology reported by Lorenzo et al. [34]. In addition, $n$ represents crystallization types regarding dimensions (one-, two-, or three-dimensional) and $k$ is a growth rate constant that accounts for both nucleation and growth rate parameters.

Plots of $\log[-\ln(1 - V_c)]$ vs. $\log(t - t_0)$ for each crystallization temperature ($T_c$) ((3)) are shown in Figure 8. Avrami's exponent $n$ and crystallization rate $k$ are obtained from the slope and intercept of the curves, respectively. From these curves, $t_{1/2}$ is defined as the time required for developing 50% of final crystallinity. Usually, lower $t_{1/2}$ values are expected at higher crystallization rates (Table 2).

As example, for HEMACL3, all lines were almost straight and parallel to each other, shifting to a higher time when the crystallization temperature increases (Figure 8(a)). In addition, at $T_c$ = 20°C, the corrected Avrami's fit was evidenced for both linear PCL and copolymers (Figure 8(b)). Moreover, for SCL# copolymers, $n$ values were similar to those obtained for linear PCL, with calculated values between 2.4 and 3.4 (see Table 2). It is important to highlight that these values refer to a three-dimensional heterogeneous nucleation and they are in good agreement with values already reported [33, 36, 56]. On the other hand, HEMACL# copolymers show values of $n$ exponent from 1.8 to 2.6,

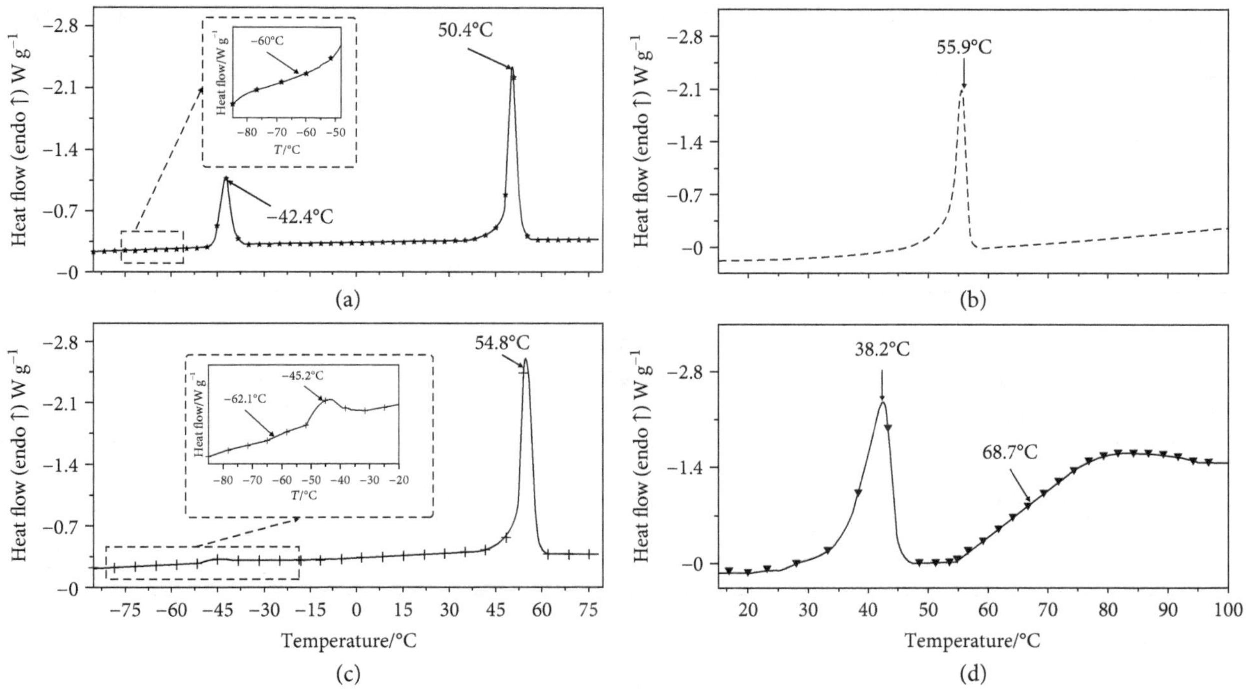


FIGURE 5: DSC curves of (a) P (DMS-*b*-ε-CL) SCL1, (b) P (DMS-*b*-ε-CL) SCL4, (c) linear PCL homopolymer, and (d) P (HEMA-*co*-(HEMA-*g*-ε-CL)-*b*-ε-CL) HEMACL3 copolymers.

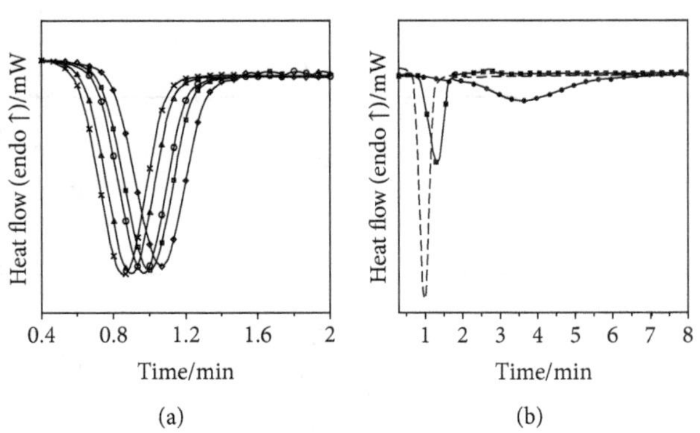


FIGURE 6: Crystallization endothermic curves for (a) linear poly(ε-caprolactone) homopolymer (PCL), symbols: (×) $T_c = 16°C$, (Δ) $T_c = 18°C$, (O) $T_c = 20°C$, (□) $T_c = 22°C$ and (◊) $T_c = 24°C$; and (b) linear PCL, SCL2 and HEMACL2 copolymers, crystallized at $T_c = 20°C$, symbols: (--) linear PCL, (●) SCL2 and (■) HEMACL2.

revealing that these "comb-like" copolymers adopt a two-dimensional crystallization growth with heterogeneous nucleation [32, 35, 36, 59]. The effect is strongly dependent on the number of branches, as it can be easily deduced by comparing the resulting curves for HEMACL1 and HEMACL3. It is reasonable to suppose that the presence of branches imposes dimensional restrictions over the resulting crystallites, by changing from three (linear) to two (comb-like) dimensions.

In Avrami's equation, *k* values are related with nucleation rate and growth processes. For SCL# and HEMACL# copolymers, a decrease in this parameter with an increase in $T_c$ was observed. The presence of the PDMS block in SCL# or grafted, random PCL branches in HEMACL# lead to a reduction in *k* values in all the studied samples (see Table 2). This reduction could be explained by taking into account the spatial configuration of PDMS block or grafted PCL branches. During the folding and growth processes of crystals, the viscoelastic silicon block as well as the random PCL branches hinder the normal crystallization event, leading to lower *k* values. The effect can be easily appreciated in Figure 9, in which the evolution of $t_{1/2}$ at each $T_c$ value is shown.

For SCL# copolymers, $t_{1/2}$ values exhibited an increase of 1.6 times when compared to linear PCL (Figure 9(a)). On the other hand, for HEMACL# copolymers, a difference in $t_{1/2}$ values appears at crystallization temperatures closer to $T_{mPCL}$. In this sense, HEMACL2 and HEMACL3 show $t_{1/2}$ values of 18 and 42 times higher when compared to linear PCL, respectively (as it can be appreciated for temperatures between 20 and 24°C, Figure 9(b)). From Figure 9, it is clear that "flat" curves can be obtained for SCL# copolymers whereas marked changes were observed for HEMACL# copolymers when compared to linear PCL. The "comb-like" structure of these copolymers noticeably affects the crystallization behavior of the semicrystalline PCL block since random grafts require more time for crystallization.

Finally, activation energy for isothermal crystallization was evaluated by using the Arrhenius equation [35, 56, 59]:

$$\left(\frac{1}{n}\right) \ln K = \ln K_0 - \frac{\Delta E_a}{RT_c}, \quad (4)$$

TABLE 2: Values of $t_{1/2}$ (min), $n$, $k$ (min$^{-1}$), and $\Delta E_a$ at different crystallization temperatures ($T_c$) for linear PCL, SCL#, and HEMACL# copolymers.

| | $T_c$/°C 16 | 18 | 20 | 22 | 24 | $\Delta E_a$/kJ mol$^{-1}$ |
|---|---|---|---|---|---|---|
| PCL | | | | | | |
| $t_{1/2}$ | 0.3 | 0.3 | 0.3 | 0.3 | 0.3 | −8.3 |
| $n$ | 2.8 | 2.7 | 2.7 | 2.7 | 2.7 | |
| $k$ | 21.1 | 19.2 | 18.5 | 18.5 | 17.8 | |
| SCL1 | | | | | | |
| $t_{1/2}$ | 0.4 | 0.5 | 0.5 | 0.5 | 0.5 | −1.9 |
| $n$ | 3.1 | 2.6 | 2.8 | 2.5 | 2.4 | |
| $k$ | 10.3 | 7.9 | 7.5 | 6.1 | 5.7 | |
| SCL2 | | | | | | |
| $t_{1/2}$ | 0.4 | 0.4 | 0.5 | 0.5 | 0.5 | −6.9 |
| $n$ | 2.6 | 2.6 | 2.6 | 2.5 | 2.3 | |
| $k$ | 5.5 | 5.4 | 4.4 | 4.4 | 3.8 | |
| SCL3 | | | | | | |
| $t_{1/2}$ | 0.4 | 0.4 | 0.5 | 0.4 | 0.4 | −8.1 |
| $n$ | 3.1 | 2.9 | 2.7 | 2.4 | 2.5 | |
| $k$ | 8.4 | 7.4 | 5.6 | 5.1 | 4.4 | |
| SCL4 | | | | | | |
| $t_{1/2}$ | 0.4 | 0.4 | 0.4 | 0.4 | 0.5 | −34.3 |
| $n$ | 3.4 | 3.2 | 2.9 | 2.7 | 2.8 | |
| $k$ | 15.9 | 13.3 | 11.8 | 8.2 | 5.6 | |
| HEMACL1 | | | | | | |
| $t_{1/2}$ | 0.6 | 0.6 | 0.6 | 0.6 | 0.7 | −35.2 |
| $n$ | 2.5 | 2.5 | 2.6 | 2.3 | 2.3 | |
| $k$ | 2.6 | 2.8 | 2.9 | 1.8 | 1.3 | |
| HEMACL2 | | | | | | |
| $t_{1/2}$ | 1.7 | 2.0 | 2.5 | 3.2 | 5.4 | −81.3 |
| $n$ | 2.5 | 2.6 | 2.3 | 2.0 | 1.9 | |
| $k$ | 1.4 $10^{-1}$ | 1.1 $10^{-1}$ | 8.5 $10^{-2}$ | 6.5 $10^{-2}$ | 2.9 $10^{-2}$ | |
| HEMACL3 | | | | | | |
| $t_{1/2}$ | 1.5 | 2. 3 | 4.6 | 7.9 | 12.6 | −200.7 |
| $n$ | 1.9 | 2.0 | 1.8 | 1.8 | 1.8 | |
| $k$ | 3.2 $10^{-1}$ | 1.4 $10^{-1}$ | 5.3 $10^{-2}$ | 2 $10^{-2}$ | 9.5 $10^{-3}$ | |

where $K_0$ is a temperature-independent preexponential factor, $\Delta E_a$ is the total activation energy, and $R$ is the universal gas constant.

Arrhenius plots of $n^{-1}$ ln $K$ versus $T_c^{-1}$ for linear PCL, SCL#, and HEMACL# copolymers were linear, and $\Delta E_a$ was determined from the slope of the curves (Figure 10). Values are summarized in Table 2. For all samples, the calculated $\Delta E_a$ values were negatives. In this sense, Dhanvijay et al. [36] refer this behavior to the released energy during melt to crystalline state transformation. By comparing linear PCL with SCL# and HEMACL# copolymers, the $\Delta E_a$ values of HEMACL# copolymers are the highest. $\Delta E_a$ values are strongly dependent on the co-monomer presence, whose chemical nature and macromolecular structure affect PCL chains. Therefore, collisions and entanglements of PDMS and PCL blocks are important when the amount of PCL increases in SCL# copolymers. On the other hand, the random grafted PCL chains hinder movement and rearrangement of the semicrystalline PCL block by reducing their crystallization ability. These results

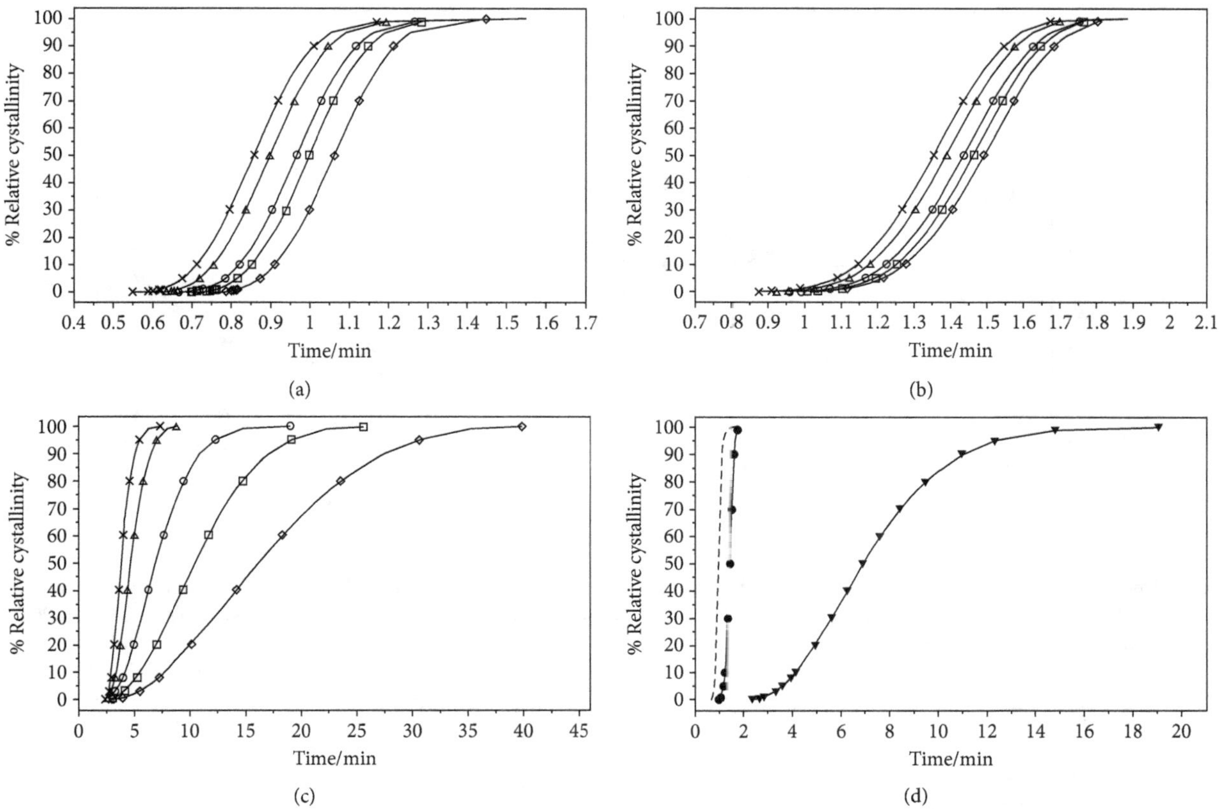


FIGURE 7: Relative Crystallinity (%) as a function of time (min) for: (a) linear poly (ε-caprolactone) PCL, (b) SCL2, (c) HEMACL3. Symbols: (×) $T_c = 16°C$, (Δ) $T_c = 18°C$, (○) $T_c = 20°C$, (□) $T_c = 22°C$, and (◊) $T_c = 24°C$. (d) Linear PCL, SCL2, and HEMACL3 copolymers, crystallized at $T_c = 20°C$, symbols: (–) linear PCL, (●) SCL2, and (▼) HEMACL3.

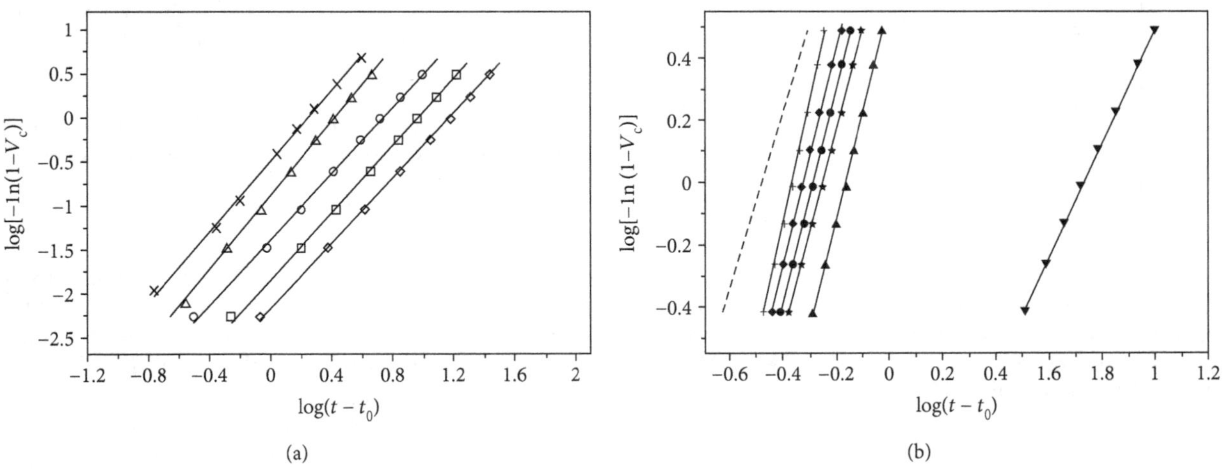


FIGURE 8: Experimental data obtained from the isothermal crystallization of (a) HEMACL3 at different crystallization temperature ($T_c$), symbols: (×) $T_c = 16°C$, (Δ) $T_c = 18°C$, (○) $T_c = 20°C$, (□) $T_c = 22°C$, and (◊) $T_c = 24°C$ and (b) linear poly(ε-caprolactone) homopolymer PCL and SCL# and HEMACL# copolymers crystallized at $T_c = 20°C$, symbols: (–) linear PCL, (+) SCL4, (◆) SCL3, (●) SCL2, (★) SCL1, (▲) HEMACL1, and (▼) HEMACL3.

are in accordance with the behavior reported by Cai et al. [59] when isothermal crystallization of PCL composites was analyzed.

According to the thermogravimetric analysis, weight loss events associated to linear and "comb-like" copolymers as function of temperature were registered (Figure 11). In addition, values of temperature at 5% of thermal degradation ($T_{0.05}$) were included in Table 1 for each sample. It is important to note that $T_{0.05}$ values denote the beginning of the thermal degradation process, which is a characteristic

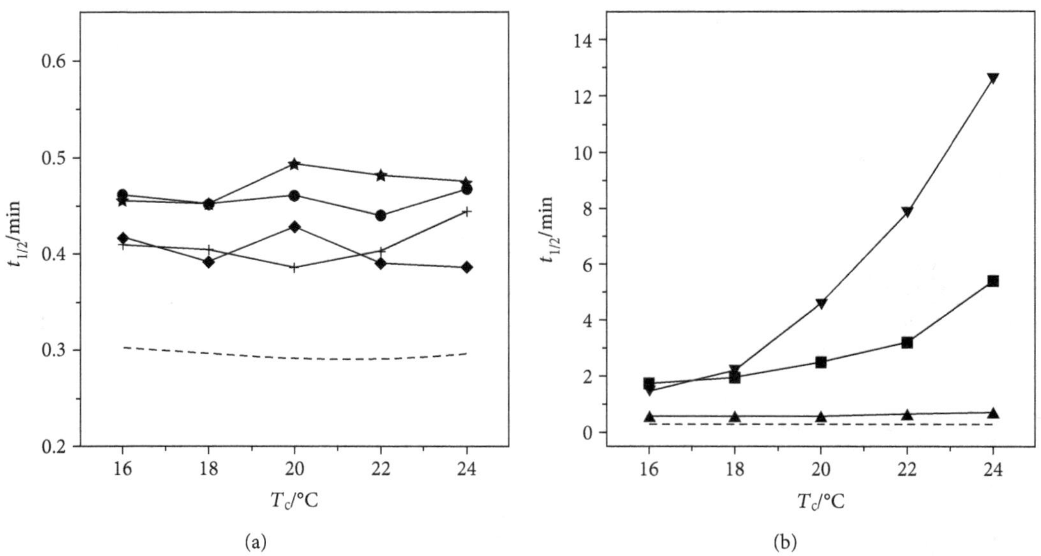


Figure 9: Dependence of $t_{1/2}$ with the isothermal crystallization temperature ($T_c$) for (a) linear PCL and SCL# copolymers, symbols: (–) linear PCL, (+) SCL4, (♦) SCL3, (●) SCL2, and (★) SCL1 and (b) linear PCL and HEMACL# copolymers, symbols: (–) linear PCL, (▲) HEMACL1, (■) HEMACL2, and (▼) HEMACL3.

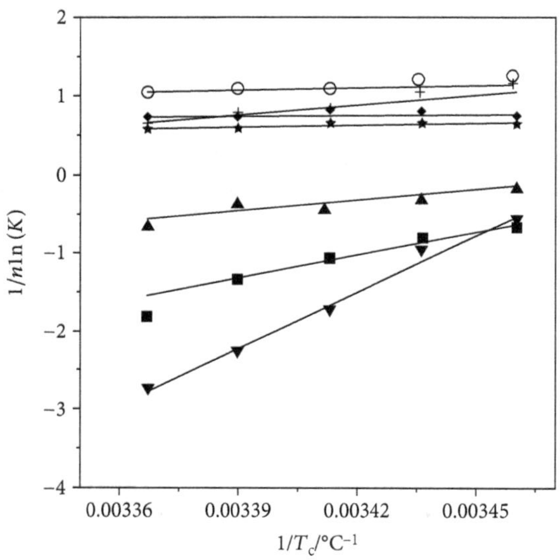


Figure 10: Arrhenius plots $1/n$ ln ($K$) vs. $1/T_c$ for linear PCL homopolymer and SCL# and HEMACL# copolymers. Symbols: (○) linear PCL, (+) SCL4, (♦) SCL3, (★) SCL1, (▲) HEMACL1, (■) HEMACL2, and (▼) HEMACL3.

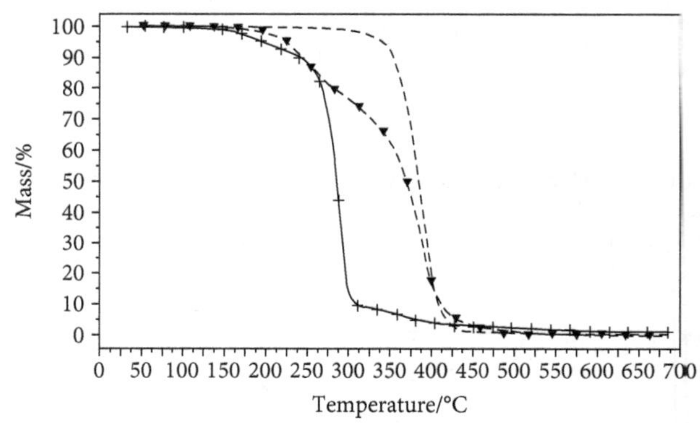


Figure 11: MTGA curves of linear PCL homopolymer and SCL# and HEMACL# copolymers. Symbols: (–) linear PCL, (+) SCL4, and (▼) HEMACL3.

value for each polymer that depends on many macromolecular features.

Figure 11 shows MTGA curves corresponding to linear PCL homopolymer and SCL4 and HEMACL3 copolymers, respectively. As it can be seen, linear PCL presents a typical weight loss step which is associated to PCL decomposition at 341°C [59, 60]. By analyzing the first derivative curve from the MTGA thermogram (not shown), for the SCL4 copolymer three thermal degradation events were detected. These events were observed at 193°C (rupture of polyester chains via ester pyrolysis reaction producing $H_2O$, $CO_2$, and 5-hexenoic acid); at 289°C (PCL block decomposition), and at 366°C (PDMS block degradation) [32, 61]. On the other hand, HEMACL# copolymers showed the thermal decomposition associated to PHEMA and PCL homopolymers already reported [28, 62].

All the studied materials exhibited a reduction in the maximum degradation temperature when compared to linear PCL. In this sense, thermal degradation was faster for SCL# than HEMACL# copolymers, denoting that the PCL block is less thermally stable in the 150–450°C range. A plausible explanation might be due to the hydroxyl-end group (–OH). For block copolymers, it is reasonable to assume one –OH group per molecule (PCL block), whereas the amount increases for comb-like copolymers (–OH groups from PCL grafts and PCL block). Since the chemical nature of end-groups strongly influences the thermal behavior of PCL [63], the differences observed can be reasonable explained. The calculated activation energy values ($E_{ad}$) for

SCL# and HEMACL# copolymers were lower than the corresponding value for linear PCL (Table 1). These results agree well with previous works regarding the thermal degradation of PCL-based copolymers reported by Ninago et al. [32].

## 4. Conclusions

Linear and block copolymers based on ε-CL were successfully synthesized through a combination of anionic, RAFT, and ROP methods. Samples showed $M_w/M_n < 1.54$, which reveal a good control over macromolecular architectures. Isothermal DSC experiments (combined with Avrami's theory) were performed in order to obtain kinetics parameters regarding crystallization behavior. For the temperature range studied (from $T_c = 16°C$ to $T_c = 24°C$), $k$ values were reduced whereas $t_{1/2}$ values were increased when compared to the linear PCL homopolymer. The effect was more notorious for HEMACL# copolymers, in which branch density might affect crystallization processes due to less chain mobility and spatial hindrance. Regarding thermal stabilities, the presence of –OH groups from PCL branches reduces the energy barrier of the thermal degradation process for the corresponding HEMACL# copolymers.

## Conflicts of Interest

The authors declare that there is no conflict of interest regarding the publication of this paper.

## Acknowledgments

We express our gratitude to the Consejo Nacional de Investigaciones Científicas y Técnicas (CONICET, Argentina, grant PIP 0428), the Fondo para la Investigación Científica y Tecnológica (FONCyT, Argentina, grant PICT-2016-0181) and the Universidad Nacional de Cuyo (UNCuyo, Argentina) for their financial support.

## References

[1] N. A. Agudelo and L. D. Perez, "Synthesis and characterization of polydimethylsiloxane end-modified polystyrene from poly (styrene – co –vinyltriethoxysilane) copolymers," *Materials Research*, vol. 19, no. 2, pp. 459–465, 2016.

[2] M. B. Runge, S. Dutta, and N. B. Bowden, "Synthesis of comb block copolymers by ROMP, ATRP, and ROP and their assembly in the solid state," *Macromolecules*, vol. 39, no. 2, pp. 498–508, 2006.

[3] M. D. Ninago, V. Hanazumi, M. G. Passaretti, D. A. Vega, A. E. Ciolino, and M. A. Villar, "Enhancement of mechanical and optical performance of commercial polystyrenes by blending with siloxane-based copolymers," *Journal of Applied Polymer Science*, vol. 134, no. 30, 2017.

[4] N. Hadjichristidis, S. Pispas, and G. Floudas, *Block Copolymers: Synthetic Strategies, Physical Properties, and Applications*, John Wiley & Sons, Inc., Hoboken, NJ, USA, 2003.

[5] S. V. Murphy and A. Atala, "3D bioprinting of tissues and organs," *Nature Biotechnology*, vol. 32, no. 8, pp. 773–785, 2014.

[6] J. Y. Lee, J. An, and C. K. Chua, "Fundamentals and applications of 3D printing for novel materials," *Applied Materials Today*, vol. 7, pp. 120–133, 2017.

[7] X. Qu, P. Xia, J. He, and D. Li, "Microscale electrohydrodynamic printing of biomimetic PCL/nHA composite scaffolds for bone tissue engineering," *Materials Letters*, vol. 185, pp. 554–557, 2016.

[8] Y. He, C. J. Tuck, E. Prina et al., "A new photocrosslinkable polycaprolactone-based ink for three-dimensional inkjet printing," *Journal of Biomedical Materials Research Part B: Applied Biomaterials*, vol. 105, no. 6, pp. 1645–1657, 2017.

[9] H. N. Chia and B. M. Wu, "Recent advances in 3D printing of biomaterials," *Journal of Biological Engineering*, vol. 9, no. 1, p. 4, 2015.

[10] S. Agarwal, A. Greiner, and J. H. Wendorff, "Functional materials by electrospinning of polymers," *Progress in Polymer Science*, vol. 38, no. 6, pp. 963–991, 2013.

[11] T. J. Sill and H. A. von Recum, "Electrospinning: applications in drug delivery and tissue engineering," *Biomaterials*, vol. 29, no. 13, pp. 1989–2006, 2008.

[12] J. L. Wietor, D. J. M. Van Beek, G. W. Peters, E. Mendes, and R. P. Sijbesma, "Effects of branching and crystallization on rheology of polycaprolactone supramolecular polymers with ureidopyrimidinone end groups," *Macromolecules*, vol. 44, no. 5, pp. 1211–1219, 2011.

[13] M. Yazdimamaghani, M. Razavi, D. Vashaee, V. R. Pothineni, J. Rajadas, and L. Tayebi, "Significant degradability enhancement in multilayer coating of polycaprolactone-bioactive glass/gelatin-bioactive glass on magnesium scaffold for tissue engineering applications," *Applied Surface Science*, vol. 338, pp. 137–145, 2015.

[14] J. Z Liang, L. Zhou, C. Y. Tang, and C. P. Tsui, "Crystallization properties of polycaprolactone composites filled with nanometer calcium carbonate," *Journal of Applied Polymer Science*, vol. 128, no. 5, pp. 2940–2944, 2013.

[15] E. C. Chen and T. M. Wu, "Isothermal crystallization kinetics and thermal behavior of poly (ε-caprolactone)/multi-walled carbon nanotube composites," *Polymer Degradation and Stability*, vol. 92, no. 6, pp. 1009–1015, 2007.

[16] C. Varan, H. Wickström, N. Sandler, Y. Aktaş, and E. Bilensoy, "Inkjet printing of antiviral PCL nanoparticles and anticancer cyclodextrin inclusion complexes on bioadhesive film for cervical administration," *International Journal of Pharmaceutics*, vol. 531, no. 2, pp. 701–713, 2017.

[17] T. V. Chirila, I. J. Constable, G. J. Crawford et al., "Poly (2-hydroxyethyl methacrylate) sponges as implant materials: in vivo and in vitro evaluation of cellular invasion," *Biomaterials*, vol. 14, no. 1, pp. 26–38, 1993.

[18] N. A. Peppas, H. J. Moynihan, and L. M. Lucht, "The structure of highly crosslinked poly (2-hydroxyethyl methacrylate) hydrogels," *Journal of Biomedical Materials Research Part A*, vol. 19, no. 4, pp. 397–411, 1985.

[19] J. L. Zhu, X. Z. Zhang, H. Cheng, Y. Y. Li, S. X. Cheng, and R. X. Zhuo, "Synthesis and characterization of well-defined, amphiphilic poly (N-isopropylacrylamide)-b-[2-hydroxyethyl

methacrylate-poly (ε-caprolactone)]n graft copolymers by RAFT polymerization and macromonomer method," *Journal of Polymer Science Part A: Polymer Chemistry*, vol. 45, no. 22, pp. 5354–5364, 2007.

[20] A. G. O. de Freitas, S. G. Trindade, P. I. R. Muraro et al., "Controlled one-pot synthesis of polystyrene-block-polycaprolactone copolymers by simultaneous RAFT and ROP," *Macromolecular Chemistry and Physics*, vol. 214, no. 20, pp. 2336–2344, 2013.

[21] M. Le Hellaye, C. Lefay, T. P. Davis, M. H. Stenzel, and C. Barner-Kowollik, "Simultaneous reversible addition fragmentation chain transfer and ring-opening polymerization," *Journal of Polymer Science Part A: Polymer Chemistry*, vol. 46, no. 9, pp. 3058–3067, 2008.

[22] X. Qin, Y. Li, F. Zhou, L. Ren, Y. Zhao, and X. Yuan, "Polydimethylsiloxane-polymethacrylate block copolymers tethering quaternary ammonium salt groups for antimicrobial coating," *Applied Surface Science*, vol. 328, pp. 183–192, 2015.

[23] G. Moad, Y. K. Chong, A. Postma, E. Rizzardo, and S. H. Thang, "Advances in RAFT polymerization: the synthesis of polymers with defined end-groups," *Polymer*, vol. 46, no. 19, pp. 8458–8468, 2005.

[24] F. Azemar, F. Faÿ, K. Réhel, and I. Linossier, "Control of hydration and degradation properties of triblock copolymers polycaprolactone-*b*-polydimethylsiloxane-*b*-polycaprolactone," *Journal of Applied Polymer Science*, vol. 12, p. 131, 2014.

[25] G. Gu, Z. Wu, Z. Zhang, and F. Qing, "Tribological properties of fluorine-containing additives of silicone oil," *Tribology International*, vol. 42, no. 3, pp. 397–402, 2009.

[26] D. R. Anderson and A. L. Smith, *Analysis of Silicones*, Wiley-Interscience, New York, NY, USA, 1974.

[27] M. D. Ninago, A. G. O. de Freitas, V. Hanazumi et al., "Synthesis of grafted block copolymers based on ε-caprolactone: influence of branches on their thermal behavior," *Macromolecular Chemistry and Physics*, vol. 216, no. 24, pp. 2331–2343, 2015.

[28] M. D. Ninago, A. E. Ciolino, M. A. Villar, A. G. O. de Freitas, P. I. R. Muraro, and C. Giacomelli, "Thermal characterization of "comb-like" block copolymers based on PCL obtained by combining ROP and RAFT polymerizations," *Macromolecular Symposia*, vol. 368, no. 1, pp. 84–92, 2016.

[29] M. D. Ninago, M. L. S. Marani, V. A. González et al., "Lithium naphthalenides in non-polar or in low-polarity media: some insights regarding their use as initiators in anionic polymerizations," *Polymer Bulletin*, vol. 74, no. 2, pp. 307–323, 2017.

[30] T. J. Hinton, A. Hudson, K. Pusch, A. Lee, and A. W. Feinberg, "3D printing PDMS elastomer in a hydrophilic support bath via freeform reversible embedding," *ACS Biomaterials Science & Engineering*, vol. 2, no. 10, pp. 1781–1786, 2016.

[31] V. Ozbolat, M. Dey, B. Ayan, A. Povilianskas, M. C. Demirel, and I. T. Ozbolat, "3D printing of PDMS improves its mechanical and cell adhesion properties," *ACS Biomaterials Science & Engineering*, vol. 4, no. 2, pp. 682–693, 2018.

[32] M. D. Ninago, A. J. Satti, A. E. Ciolino, and M. A. Villar, "Influence of amorphous block on the thermal behavior of well-defined block copolymers based on ε-caprolactone," *Journal of Thermal Analysis and Calorimetry*, vol. 112, no. 3, pp. 1277–1287, 2013.

[33] M. Lanfranconi, V. A. Alvarez, and L. N. Ludueña, "Isothermal crystallization of polycaprolactone/modified clay biodegradable nanocomposites," *Journal of Thermal Analysis and Calorimetry*, vol. 126, no. 3, pp. 1273–1280, 2016.

[34] A. T. Lorenzo, M. L. Arnal, J. Albuerne, and A. J. Müller, "DSC isothermal polymer crystallization kinetics measurements and the use of the Avrami equation to fit the data: guidelines to avoid common problems," *Polymer Testing*, vol. 26, no. 2, pp. 222–231, 2007.

[35] H. Liu, Y. Huang, L. Yuan et al., "Isothermal crystallization kinetics of modified bamboo cellulose/PCL composites," *Carbohydrate Polymers*, vol. 79, no. 3, pp. 513–519, 2010.

[36] P. U. Dhanvijay, V. V. Shertukde, and A. K. Kalkar, "Isothermal and nonisothermal crystallization kinetics of poly (ε-caprolactone)," *Journal of Applied Polymer Science*, vol. 124, no. 2, pp. 1333–1343, 2012.

[37] N. Hadjichristidis, H. Iatrou, S. Pispas, and M. Pitsikalis, "Anionic polymerization: high vacuum techniques," *Journal of Polymer Science Part A: Polymer Chemistry*, vol. 38, no. 18, pp. 3211–3234, 2000.

[38] A. E. Ciolino, M. A. Villar, E. M. Vallés, and N. Hadjichristidis, "Synthesis and characterization of model polybutadiene-1, 4-b-polydimethylsiloxane-b-polybutadiene-1, 4 copolymers," *Journal of Polymer Science Part A: Polymer Chemistry*, vol. 45, no. 13, pp. 2726–2733, 2007.

[39] M. D. Ninago, A. J. Satti, J. A. Ressia et al., "Self-assembly of well-defined PS-b-PDMS copolymers in bulk and in selective solvent," *Chemical Engineering Transactions*, vol. 17, pp. 1807–1811, 2009.

[40] M. D. Ninago, A. J. Satti, J. A. Ressia, A. E. Ciolino, M. A. Villar, and E. M. Vallés, "Controlled synthesis of poly (dimethylsiloxane) homopolymers using high-vacuum anionic polymerization techniques," *Journal of Polymer Science Part A: Polymer Chemistry*, vol. 47, no. 18, pp. 4774–4783, 2009.

[41] M. D. Ninago, A. J. Satti, A. E. Ciolino et al., "Synthesis and morphology of model PS-b-PDMS copolymers," *Journal of Polymer Science Part A: Polymer Chemistry*, vol. 48, no. 14, pp. 3119–3127, 2010.

[42] A. J. Satti, A. G. O. de Freitas, M. L. S. Marani et al., "Anionic ring opening polymerization of ε-caprolactone initiated by lithium silanolates," *Australian Journal of Chemistry*, vol. 70, no. 1, pp. 106–112, 2017.

[43] W. Yuan, J. Yuan, F. Zhang, X. Xie, and C. Pan, "Synthesis, characterization, crystalline morphologies, and hydrophilicity of brush copolymers with double crystallizable side chains," *Macromolecules*, vol. 40, no. 25, pp. 9094–9102, 2007.

[44] J. Skey and R. K. O'Reilly, "Facile one pot synthesis of a range of reversible addition-fragmentation chain transfer (RAFT) agents," *Chemical Communications*, no. 35, pp. 4183–4185, 2008.

[45] J. Brandrup, E. H. Immergut, and E. A. Grulke, *Polymer Handbook*, Wiley, New York, NY, USA, 4th edition, 1999.

[46] T. Hamaide, M. Pantiru, H. Fessi, and P. Boullanger, "Ring-opening polymerisation of ε-caprolactone with monosaccharides as transfer agents. A novel route to functionalised nanoparticles," *Macromolecular Rapid Communications*, vol. 22, no. 9, pp. 659–663, 2001.

[47] H. Sun, L. Mei, C. Song, X. Cui, and P. Wang, "The in vivo degradation, absorption and excretion of PCL-based implant," *Biomaterials*, vol. 27, no. 9, pp. 1735–1740, 2006.

[48] S. G. Nanaki, G. Z. Papageorgiou, and D. N. Bikiaris, "Crystallization of novel poly (ε-caprolactone)-*block*-poly (propylene adipate) copolymers," *Journal of Thermal Analysis and Calorimetry*, vol. 108, no. 2, pp. 633–645, 2012.

[49] W. Phetwarotai and D. Aht-Ong, "Isothermal crystallization behaviors and kinetics of nucleated polylactide/poly (butylene adipate-co-terephthalate) blend films with talc," *Journal of Thermal Analysis and Calorimetry*, vol. 126, no. 3 pp. 1797–1808, 2016.

[50] N. Wu, L. Huang, and A. Zheng, "Synthesis and properties of polystyrene/polydimethylsiloxane graft copolymers," *Frontiers of Chemistry in China*, vol. 1, no. 3, pp. 350–356, 2006.

[51] N. Sahiner and S. Demirci, "The use of graphene oxide-embedded superporous poly (2-hydroxyethylmethacrylate) cryogels for p(aniline) conductive polymer synthesis and their use in sensor applications," *Materials & Design*, vol. 120, pp. 47–55, 2017.

[52] S. H. Hyon, K. Jamshidi, and Y. Ikada, "Synthesis of polylactides with different molecular weights," *Biomaterials*, vol. 18, no. 22, pp. 1503–1508, 1997.

[53] R. F. Storey and J. W. Sherman, "Kinetics and mechanism of the stannous octoate-catalyzed bulk polymerization of ε-caprolactone," *Macromolecules*, vol. 35, no. 5, pp. 1504–1512, 2002.

[54] J. Choi, S. W. Chun, and S. Y. Kwak, "Non-isothermal crystallization of hyperbranched poly(ε-caprolactone)s and their linear counterpart," *Macromolecular Chemistry and Physics*, vol. 207, no. 13, pp. 1166–1173, 2006.

[55] N. A. Peppas, "Poly (2-hydroxyethyl methacrylate)," *Polymer Data Handbook*, pp. 598-599, 1999.

[56] G. Siqueira, C. Fraschini, J. Bras, A. Dufresne, R. Prud'homme, and M. P. Laborie, "Impact of the nature and shape of cellulosic nanoparticles on the isothermal crystallization kinetics of poly (ε-caprolactone)," *European Polymer Journal*, vol. 47, no. 12, pp. 2216–2227, 2011.

[57] A. J. Lovinger, B. J. Han, F. J. Padden, and P. A. Mirau, "Morphology and properties of polycaprolactone-poly (dimethyl siloxane)-polycaprolactone triblock copolymers," *Journal of Polymer Science Part B: Polymer Physics*, vol. 31, no. 2, pp. 115–123, 1993.

[58] J. Cai, M. Liu, L. Wang, K. Yao, S. Li, and H. Xiong, "Isothermal crystallization kinetics of thermoplastic starch/poly (lactic acid) composites," *Carbohydrate Polymers*, vol. 86, no. 2, pp. 941–947, 2011.

[59] J. Cai, Z. Xiong, M. Zhou et al., "Thermal properties and crystallization behavior of thermoplastic starch/poly (ε-caprolactone) composites," *Carbohydrate Polymers*, vol. 102, pp. 746–754, 2014.

[60] M. D. Ninago, O. V. López, M. M. S. Lencina et al., "Enhancement of thermoplastic starch final properties by blending with poly (ε-caprolactone)," *Carbohydrate Polymers*, vol. 134, pp. 205–212, 2015.

[61] O. Persenaire, M. Alexandre, P. Degée, and P. Dubois, "Mechanisms and kinetics of thermal degradation of poly (ε-caprolactone)," *Biomacromolecules*, vol. 2, no. 1, pp. 288–294, 2001.

[62] A. G. P. R. Figueiredo, A. R. P. Figueiredo, A. Alonso-Varona et al., "Biocompatible bacterial cellulose-poly (2-hydroxyethyl methacrylate) nanocomposite films," *BioMed Research International*, vol. 2013, Article ID 698141, 14 pages, 2013.

[63] G. Gorrasi, L. Vertuccio, L. Annunziata, C. Pellecchia, and D. Pappalardo, "Correlations between microstructural characterization and thermal properties of well defined poly(ε-caprolactone) samples by ring opening polymerization with neutral and cationic bis(2,4,6-triisopropylphenyl)tin(IV) compounds," *Reactive and Functional Polymers*, vol. 70, no. 3, pp. 151–158, 2010.

# Synthesis of Controllable Carboxylated Polystyrene Microspheres by Two-Step Dispersion Polymerization with Hydrocarbon Alcohols

**Bin Zhang,[1,2] Jiazhen Lu,[1] Xin Liu,[1] Haibo Jin,[1,2] Guangxiang He,[1,2] and Xiaoyan Guo[1,2]**

[1]*College of Chemical Engineering, Beijing Institute of Petrochemical Technology, Beijing 102617, China*
[2]*Beijing Key Laboratory of Fuels Cleaning and Advanced Catalytic Emission Reduction Technology, Beijing 102617, China*

Correspondence should be addressed to Guangxiang He; hgx@bipt.edu.cn and Xiaoyan Guo; guoxiaoyan@bipt.edu.cn

Academic Editor: Bernabé L. Rivas

Carboxylated polystyrene (PS) microspheres were synthesized by the two-step dispersion polymerization of styrene in hydrocarbon alcohols ($C_nH_{2n+1}OH$, $n = 1 - 5$) in the presence of acrylic acid (AA) as a functional comonomer, 2,2′-azobis-(2-methylbutyronitrile) (AMBN) as the initiator and polyvinylpyrrolidone (PVP55) as the dispersant. The effects of solvent type, AA concentration, and first reaction time on the carboxyl content on the microsphere surfaces were investigated. The PS microspheres were characterized by scanning electron microscopy, infrared spectroscopy, nuclear magnetic resonance, and gel permeation chromatography. The results showed that the first reaction time of the two-stage dispersion polymerization had a great influence on the nucleation process, and appropriately prolonging the first-step reaction time had a great influence on the surface carboxyl content. The effect of the solvent on the surface carboxyl content of PS microspheres was significant. With n-butanol as the solvent, the carboxyl group content on the surface of the microspheres reached 57.05 mg/g.

## 1. Introduction

In recent years, as the emphasis on health has increased, the technical requirements for the detection of various indicators in serum have become increasingly high. Dry chemical in vitro diagnostic reagents have the advantages of convenience, portability, accuracy, and no pollution [1, 2] and have attracted the attention of many researchers. A dry chemical in vitro diagnostic reagent usually consists of a diffusion layer, reagent layer, indicator layer, and substrate layer. After the tested sample diffuses through the diffusion layer and then reacts with the reagent layer, it is detected and converted into a signal value after the indicator layer is colored. Therefore, the diffusion effect of the diffusion layer determines the accuracy of the detection. Good diffusion results require fast and uniform diffusion. At present, dry chemical in vitro diagnostic reagents independently developed at home and abroad mostly utilize diffusion papers such as filter papers [3] and polysulfonic acid [4]. The diffusion layers using these types of materials cause inaccurate test results because of their poor diffusion and reactivity. In addition, the diffusion layer often has the function of immobilizing solid reagents and biological ligands. Therefore, the selection of materials with high surface reactivity and uniform diffusivity is necessary for the preparation of dry chemical in vitro diagnostic reagents.

Carboxylated polystyrene (PS) microspheres have high reactivity with various ligands, particularly biomolecules, and have widespread applications. The surface modification with carboxyl groups can improve the colloidal stability of the microsphere dispersion and prevent particle aggregation [5]. Carboxylated microspheres are often used in biomedical and diagnostic fields such as drug delivery, sensors, and biological detection. For example, polypyrrole-coated PS particles are used to adsorb proteins as a new diagnostic material [6–10].

There are many polymerization methods for preparing PS microspheres, such as precipitation polymerization, seed polymerization, microemulsion, and dispersion polymerization.

Precipitation polymerization is a heterogeneous polymerization method, and the prepared particles have the advantages of low viscosity and clean surfaces but low solid contents and irregular geometries [11, 12]. Seed polymerization is a typical preparation method for core-shell emulsions. The particle size distribution obtained by this method is narrow, and the stability is good. However, the preparation process is cumbersome and complex, the period is long, and functional monomers easily become embedded in the particles [13]. Microemulsion is a method of forming an emulsion by using an emulsifier and forming particles in microbubbles; it can produce particles with uniform particle size and good stability, but the smaller particle size is still several nanometers [14]. In recent years, dispersion polymerization has gradually become the main method for synthesizing PS microspheres. Compared to other synthetic methods, dispersion polymerization has the advantages of easy functionalization, simple operation, uniform particle size, and an appropriate reaction period [15]. Song et al. used dispersion polymerization to discuss the effect of the amount of added acid on the carboxyl content on the surface of the microsphere [16] and summarized the effect of the amount of acid added on the appearance and surface carboxyl groups of microspheres, but they only discussed the solvent of ethanol. Lee and Ha et al. [17] used dispersion polymerization to prepare PS microspheres using fluorinated alcohol as the solvent; the synthesized microspheres had high carboxyl contents but small particle sizes. Bai et al. [18] prepared PS microspheres in an ionic liquid by dispersion polymerization, which expanded the types of solvents, but the sphericity was irregular and the dispersibility was poor. Many problems remain in dispersion polymerization, restricting the applications of carboxyl microspheres in the biomedical and diagnostic fields; these problems include low surface carboxyl content, nonuniform particle sizes, and irregular sphericity. The preparation of microspheres with regular spherical shapes and controllable carboxyl contents on the surface remains challenging.

In this study, carboxylated PS microspheres were prepared for use as the diffusion layers of in vitro diagnostic reagents to immobilize some curing reagents, such as phosphocholine. A two-step dispersion polymerization method was used to prepare carboxylated PS microspheres with controllable surface carboxyl groups by adjusting the solvent, the concentration of acrylic acid (AA) added, and the reaction time of the first step. The influence of the above preparation conditions on the surface carboxyl content of the microspheres was analyzed.

## 2. Experiment

*2.1. Materials and Instrumentations.* Styrene (St) was purchased from Sinopharm. Methanol, ethanol, propanol, butanol, and pentanol were purchased from Beijing Chemical Corporation. Polyvinylpyrrolidone (PVP, $M_w$ ~55000), Triton (TX-100), acrylic acid (AA), and 2,2′-azobis-(2-methylbutyronitrile) (AMBN) were purchased from J&K Scientific Ltd. All of the reagents were used without further purification.

A Tensor-27 Fourier transform infrared spectroscopy (FTIR) made in Bruker (Germany) was utilized to measure the carboxyl group content on carboxylated polystyrene surface. A Zeiss-Supra 55 field emission scanning electron microscope (SEM) made in Carl Zeiss (Germany) was utilized at operation voltages from 2 to 5 kV to observe particle morphology. The molecular weight was measured by the Waters-e2695 gel permeation chromatography (GPC) made in Waters (America) and the carboxyl group was measured by the Waters nuclear magnetic resonance (NMR) made in Waters (America).

*2.2. Synthesis of Poly(St-co-AA) Microspheres.* The microspheres were prepared by two-step dispersion polymerization. In the first step, the dispersant (PVP), codispersant (TX100), initiator (AMBN), monomer (St), and the solvent were added to a 250 mL four-necked reaction flask equipped with a nitrogen inlet, a thermometer, and a condenser. After a homogeneous solution was formed at room temperature, the solution was deoxygenated by bubbling nitrogen at room temperature for 10 min. The flask was then placed in an oil bath at 70°C and subjected to mechanical stirring for h (h = 1–5) hour(s) at a rate of 180 rpm. In the second step, the AA was dissolved in a mixed solution of the equivalent St and the equivalent solvent at 70°C under nitrogen atmosphere. After the first reaction time, the acrylic solution was added to a four-necked flask. The reaction was stopped after (24 h) hours of continuous reaction. After the reaction was complete, the sample was ultrasonically dispersed in ethanol and centrifuged; the supernatant was discarded. After washing three times, the sample was stored in a centrifuge tube and sealed with ethanol to prepare for later use.

*2.3. Characterization of Carboxylated PS Microspheres.* The surface morphology and dispersion of microspheres were characterized by scanning electron microscopy (SEM, ZEISS-SUPRA55). The molecular weight of the microspheres was measured by high-performance gel permeation chromatography (GPC).

The surface carboxyl content of the microspheres was determined by the conductivity method. The mass of $m$ microspheres was weighted, ultrasonically dispersed in a sodium hydroxide solution of 25 mmol/L in concentration, and then put aside for 2 hours to allow full reaction of the carboxyl and sodium hydroxide. Then, the sample was titrated with 9 mmol/L hydrochloric acid solution to determine the change in conductivity. The conductivity decreased with the addition of hydrochloric acid solution, then tended to plateau, before increasing finally.

The conductivity declined with excessive sodium hydroxide. In this period, the hydrochloric acid and sodium hydroxide neutralization reaction occurred. The volume of consumed hydrochloric acid volume was denoted as $V_1$.

In the constant-conductivity phase of the hydrochloric acid and sodium carboxylate reaction, the volume of hydrochloric acid consumed is denoted as $V_2$.

The rising-conductivity phase is excessive in hydrochloric acid solution.

Carboxyl content on surface (%/g) = $M_{COOH} \times (V_2 - V_1) \times C_{HCl}/m$.

## 3. Results and Discussion

*3.1. Carboxylation of Polystyrene Microspheres.* Figure 1 is a NMR characterization diagram. It can be seen that the peak of 11 ppm which is a characteristic peak of the hydrogen atom on the carboxyl group from the Figure 1 indicating that the carboxyl group is successfully modified to the surface of polystyrene. Figure 2 is a FTIR characterization diagram. The peak of 1700 cm-1 which is a characteristic peak of the C=O bond on the carboxyl group can also be seen. It indicates that the carboxyl group is successfully modified to the surface of polystyrene.

*3.2. The Effect of First-Step Reaction Time on the Carboxyl Content of the Surface.* Table 1 presents a table with all reaction conditions in this part for the synthesis experiments conducted.

Figure 3 shows that the carboxyl content on the surface of the microspheres first increases and then decreases with increases in the first-step reaction time, reaching a maximum at 3 h. This is related to the nucleation process of the microspheres. When the initiator radicals and monomers are polymerized, the probability of polymerization of free radicals and monomers with high solubility is greater [19]. In the first step, the reaction time is short. There are many unreacted St monomers, and the formed microspheres are smaller in volume and quantity. In the second-step reaction, although the St monomer concentration is relatively high compared to that of AA and the probability of polymerization is greater, the acrylic monomer can still participate in the polymerization. At this time, the volume of the microspheres grows faster, and most of the carboxyl groups on the long chain of the polymer are embedded in the microspheres. As the reaction progresses, the diameter of the microspheres tends to increase gradually. At this time, the carboxyl groups on the long chain of the polymer are exposed on the surfaces of the microspheres. With the prolongation of the first-step reaction time, the St monomer added in the first step consumes more, the formed microspheres are larger in diameter and more in number, and the surface area is larger. In the second-step reaction, the newly generated long-chain carboxyl-containing polymer is more exposed to the surface of microspheres, and thus the surface carboxyl content increases. However, when the first-step reaction time is too long, the initiation efficiency of the initiator decreases, the active center decreases, the polymerization rate decreases, and the surface carboxyl content decreases. Therefore, as the first reaction time increases, the surface carboxyl content increases first and then decreases.

Figure 4 shows the Fourier transform infrared (FTIR) spectra of the PS microspheres prepared with different first-step reaction times. The respective absorption peaks are described in Table 2. The peak at 1700 $cm^{-1}$ is a stretching vibration peak of the C=O bond on the carboxyl group, characteristic of the carboxyl group. Compared to the characteristic peaks of PS adjacent to it (the stretching vibration peaks at 1452 $cm^{-1}$, 1601 $cm^{-1}$, and 1493 $cm^{-1}$ on the benzene ring -C=C-), the intensity of the characteristic peak of the carboxyl group first increases and then decreases with increasing of the reaction time of the first step, reaching a maximum at 3 h, which is consistent with the results obtained by the conductivity method.

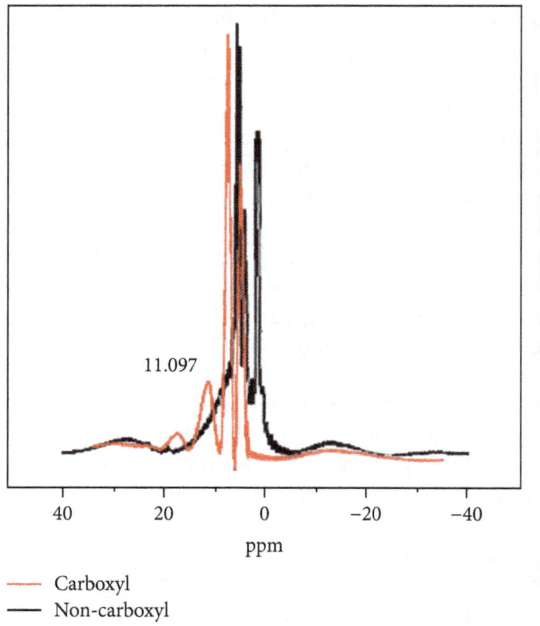


FIGURE 1: NMR spectra of carboxyl and noncarboxyl PS microspheres prepared.

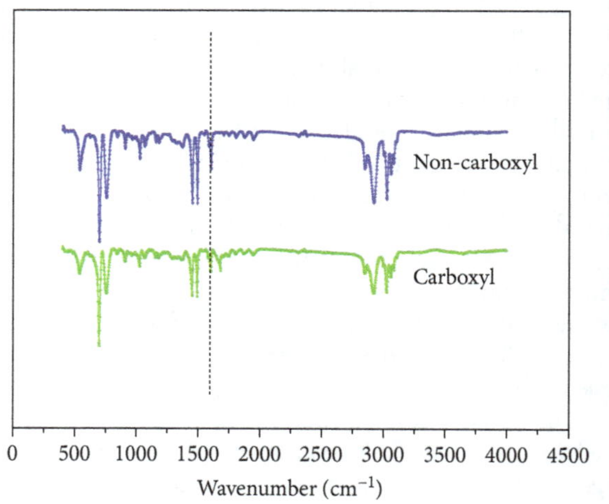


FIGURE 2: FTIR spectra of carboxyl and noncarboxyl PS microspheres prepared.

*3.3. Effect of Solvent on Carboxyl Content of the Surface.* Table 3 presents a table with all reaction conditions in this part for the synthesis experiments conducted.

As can be seen from Figure 5, with the addition of an equal amount of acid, the carboxyl content on the surface of the microspheres is increased as the solvent carbon chain length increases. Xing [20] suggested that this was related to the polarity of the solvent. With increasing carbon chain

TABLE 1: Synthesis experiment conditions in investigation of the effect of first-step reaction time on the carboxyl content of the surface.

| | Solvent | $m_{PVP}:m_{St}$ | $m_{AMBN}:m_S$ | St wt% | Reaction time of the first step | AA wt% |
|---|---|---|---|---|---|---|
| 1 | Ethanol | 0.20 | 0.02 | 20% | 1 h | 7% |
| 2 | Ethanol | 0.20 | 0.02 | 20% | 2 h | 7% |
| 3 | Ethanol | 0.20 | 0.02 | 20% | 3 h | 7% |
| 4 | Ethanol | 0.20 | 0.02 | 20% | 4 h | 7% |
| 5 | Ethanol | 0.20 | 0.02 | 20% | 5 h | 7% |

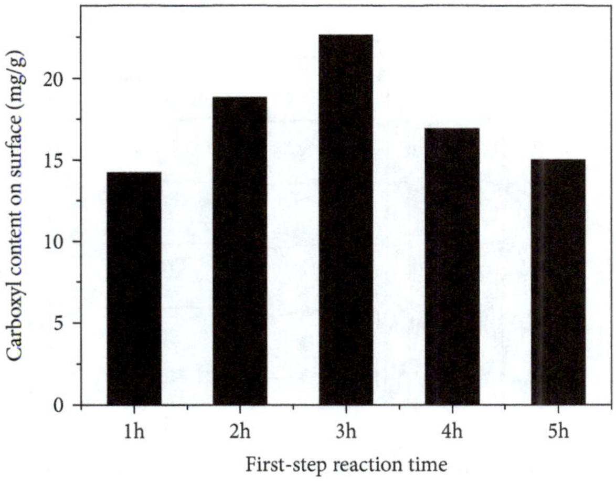


FIGURE 3: The effect of first-step reaction time on the carboxyl content on the surface (reaction conditions: $m_{PVP}:m_{St} = 0.20$; $m_{AMBN}:m_{St} = 0.02$; St wt% = 20%; ethanol as solvent; AA wt% = 7).

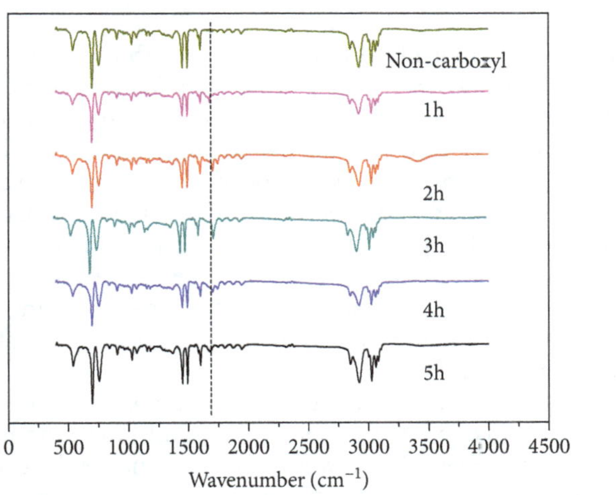


FIGURE 4: FTIR spectra of PS microspheres prepared with different times of the first step.

length in the solvent, the polarity of the solvent decreases, the solubility of the long chain of the polymer increases, and the critical length of the long chain of the polymer becomes longer, so that the size of the primary nucleus increases and the newly generated long-chain carboxyl-containing polymer is more exposed on the surface of microspheres, thus increasing the surface carboxyl content.

The results of the determination of the molecular weight and the molecular weight distribution of the PS microspheres are shown in Table 4. It demonstrates that, as the carbon

TABLE 2: Wavenumbers of functional groups in IR spectra of microspheres.

| Number | Wavenumber ($cm^{-1}$) | Absorption peak |
|---|---|---|
| 1 | ~3100–3000 | The stretching vibration peak of -C–H- on benzene ring |
| 2 | 2922 | The stretching vibration peak of -C–H- on saturated carbon |
| 3 | 1703 | The stretching vibration peak of the C=O bond on the carboxyl group |
| 4 | 1452<br>1601<br>1493 | The stretching vibration peak of -C=C- on benzene ring |

chain of the solvent is increased, the solubility of the solvent to the long chain of the polymer increases and the molecular weight of the polymer increases. It can be further speculated that larger polymer molecular weights correspond to larger formed primary nuclei, larger specific surface areas of the nuclei, and increased exposure of carboxyl groups; therefore, the carboxyl content on the microsphere surface increases with PS molecular weight.

Figure 6 is the FTIR spectra of carboxylated PS prepared in different solvents. From Figure 6, it can be seen that as the length of the carbon chain in the solvent increases, the intensity of the characteristic peak of the carboxyl group at $1703\ cm^{-1}$ is increased, indicating that the carboxyl content on the surface of the microsphere is increased. This rule is consistent with the results of the conductivity method shown in Figure 5.

*3.4. Effect of AA Concentration on Carboxyl Content of Microspheres.* Table 5 presents a table with all reaction conditions in this part for the synthesis experiments conducted.

Carboxylated PS microspheres were prepared in ethanol as the solvent, and the effect of the AA concentration on the carboxyl content on the surface of microspheres was investigated. When the concentration of AA is 7 wt% of the total amount of the monomers, microspheres with regular appearance are obtained as shown in Figure 7(a), and the surface carboxyl group content of the microspheres is 14.2 mg/g as measured by the conductivity method. When the AA concentration is increased to 10 wt%, the carboxyl content on the surface of the microspheres increases to 17.1 mg/g. When the AA concentration is increased to 14 wt% (the concentration limit for the preparation of PS microspheres), microspheres are not obtained; the product is a transparent mass, with

TABLE 3: Synthesis experiments conditons in investigation of the effect of solvent on carboxyl content of the surface.

| | Solvent | $m_{PVP}$:$m_{St}$ | $m_{AMBN}$:$m_S$ | St wt% | Reaction time of the first step | AA wt% |
|---|---|---|---|---|---|---|
| 1 | Methanol | 0.20 | 0.02 | 20% | 1 h | 7% |
| 2 | Ethanol | 0.20 | 0.02 | 20% | 1 h | 7% |
| 3 | Propanol | 0.20 | 0.02 | 20% | 1 h | 7% |
| 4 | Butanol | 0.20 | 0.02 | 20% | 1 h | 7% |
| 5 | Pentanol | 0.20 | 0.02 | 20% | 1 h | 7% |

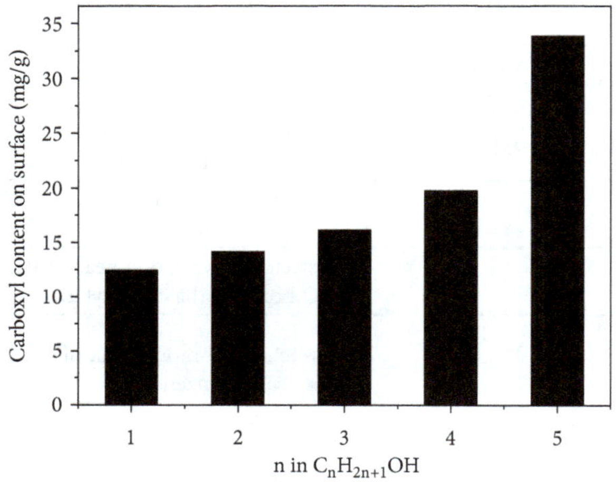


FIGURE 5: Effect of solvent on carboxyl content of the surface (reaction conditions: $m_{PVP}$:$m_{St}$ = 0.20; $m_{AMBN}$:$m_{St}$ = 0.02; St wt% = 20%; the first-step reaction time = 1 h; AA wt% = 7%).

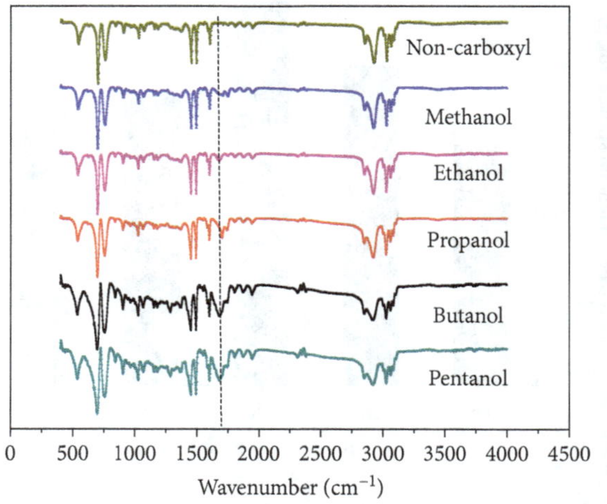


FIGURE 6: FTIR spectra of PS microspheres prepared in different solvents.

TABLE 4: Molecular weight of PS microspheres in different solvents.

| Solvent | $M_n$ | $M_w$ | $M_w/M_n$ |
|---|---|---|---|
| Methanol | 99407 | 160244 | 1.612 |
| Ethanol | 116312 | 202382 | 1.740 |
| Propanol | 129899 | 246937 | 1.901 |
| Butanol | 225064 | 514496 | 2.286 |
| Pentanol | 460272 | 1168630 | 2.539 |

microscopic morphology as shown in Figure 7(b). When the concentration of AA as a comonomer increases, the probability of the self-polymerization of the acrylic monomer increases; the influence of the polyacrylic acid (PAA) chain segment in the copolymer cannot be ignored. PAA has a greater affinity for ethanol than PS does; thus, increasing the content of PAA in the PS-PAA copolymer impedes product precipitation from the solvent. The polymerization system has a reaction limit in the concentration of AA. When the reaction limit is exceeded, the system becomes unstable, resulting in failure to yield spherical particles. Therefore, the effect of other solvents on the maximum concentration of AA in the solvent and the effect of AA on the surface carboxyl content in each solvent is examined.

Figure 8 shows the relationship between the surface carboxyl content of PS microspheres and the AA concentration in different solvents. From Figure 8, it can be seen that as the concentration of AA is increased, the carboxyl content on the surface of the microspheres increases. With the same AA concentration, the surface carboxyl content of the microspheres is increased with the increasing of the carbon chain length in the solvent. This can be attributed to the polarity of the solvent. With the increasing of the carbon number in the solvent, the polarity decreases, and the affinity of the PAA segment in the PS-PAA copolymer decreases. This is favorable for the precipitation of the polymer from the solvent, resulting in the increase of the surface carboxyl content of the microspheres.

Figure 9 is an infrared spectrum of carboxylated polystyrene prepared in different solvents with the limit concentration of AA. From Figure 9, it can be seen that as the length of the carbon chain in the solvent increases, the intensity of the characteristic peak of the carboxyl group at 1703 cm-1 increases first and then decreased, indicating that the carboxyl content on the surface of the microspheres increases first and then decreased.

Figure 10 shows the limit concentration of AA for the different solvents. If the limit concentration is exceeded, PS microspheres cannot be successfully prepared. As can be seen from Figure 10, the limit concentration of the AA for preparing the PS microspheres is maximized in butanol. For alcohol solvents with carbon numbers between one and four, the limit concentration of the AA concentration increases with increasing of the carbon number in the solvent. It is consistent with the results of the FTIR spectra shown in Figure 9.

TABLE 5: Synthesis experiment conditions in investigation of effect of acrylic acid concentration on carboxyl content of microspheres.

| | Solvent | $m_{PVP}$:$m_{St}$ | $m_{AMBN}$:$m_S$ | St wt% | Reaction time of the first step | AA wt% |
|---|---|---|---|---|---|---|
| 1 | Methanol | 0.20 | 0.02 | 20% | 1 h | 7% |
| 2 | Ethanol | 0.20 | 0.02 | 20% | 1 h | 7%<br>10.5 |
| 3 | Propanol | 0.20 | 0.02 | 20% | 1 h | 7%<br>14%<br>21% |
| 4 | Butanol | 0.20 | 0.02 | 20% | 1 h | 7%<br>14%<br>21%<br>28% |
| 5 | Pentanol | 0.20 | 0.02 | 20% | 1 h | 7%<br>14% |

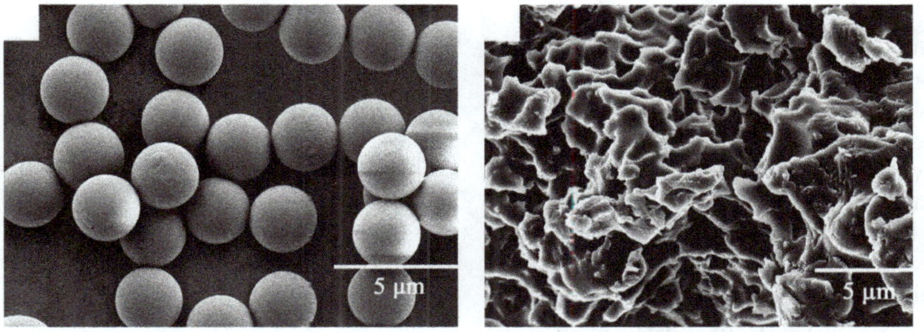


(a) (b)

FIGURE 7: SEM micrographs of PS microspheres with different amount of AA (a). 7 wt% AA; (b). 14 wt% AA) (reaction conditions: $m_{PVP}$:$m_{St}$ = 0.20; $m_{AMBN}$:$m_{St}$ = 0.02; St wt% = 20%; the first-step reaction time = 1 h; ethanol as solvent).

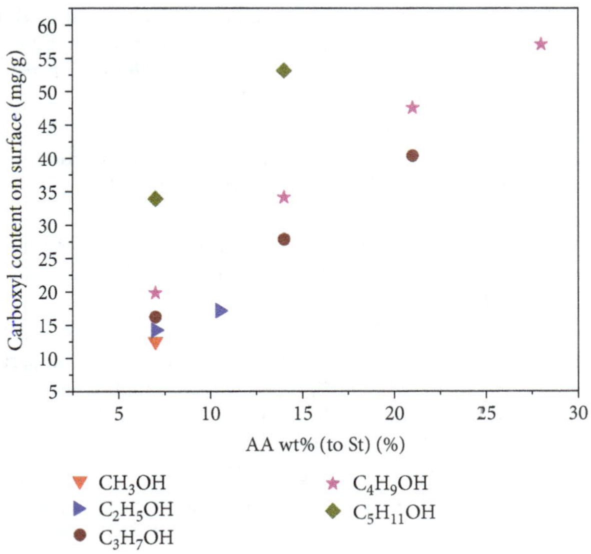


FIGURE 8: The effects of solvents on the surface carboxyl content (reaction conditions: $m_{PVP}$:$m_{St}$ = 0.20; $m_{AMBN}$:$m_{St}$ = 0.02; St wt% = 20%; the first-step reaction time = 1 h).

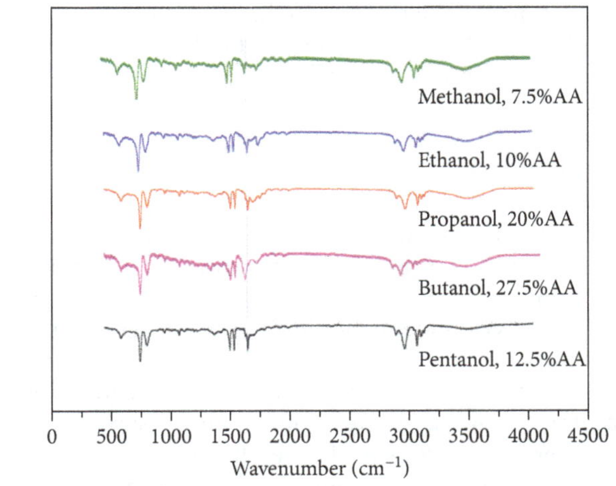


FIGURE 9: FTIR spectra of PS microspheres prepared in different solvents with limit concentration of acrylic acid.

This is also related to the polarity of the solvent; lower solvent polarity corresponds to the lower affinity for the PAA segment in the PS-PAA copolymer. It can be considered that

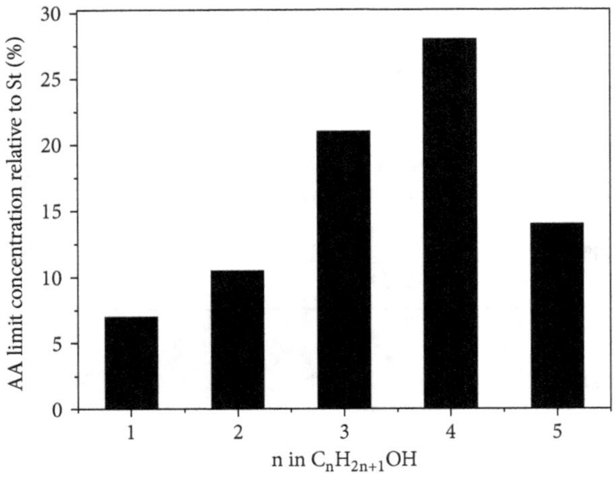


FIGURE 10: The limit concentration of AA for different solvents (reaction conditions: $m_{PVP}$:$m_{St}$ = 0.20; $m_{AMBN}$:$m_{St}$ = 0.02; b. 20 wt%; the first-step reaction time = 1 h).

lower solvent polarity is related to the weakened influence of the PAA segment on the precipitation of the PS-PAA copolymer; thus, the PAA content can be further increased. Table 6 is the molecular weight of PS microspheres synthesized in different solvents with limit concentration of acrylic acid, and this result can also confirm the above conclusions.

Wang et al. [21] believed that, because of the ionization of carboxyl groups on the surface of the microspheres, the repulsion among microspheres was increased and the stability of the particles was enhanced. As the concentration of AA increased, the reaction rate of the entire system increased. When the concentration of AA reaches a certain value, a large amount of AA is polymerized onto the long chain, and excessive hydrophilic groups are introduced onto the polymer chain, thus prolonging the nucleation period. At the same time, the particle surface becomes sticky because of the tendency of the medium to dissolve. It is easy to coalesce; when the AA concentration is too high, maintaining sphericity in the product is difficult.

## 4. Conclusion

Carboxylated PS microspheres are prepared by a two-step dispersion method using C1-C5 hydrocarbon alcohols as the solvents. By changing the process conditions, the surface carboxyl content of the PS microspheres can be well controlled.

The effect of the first-step reaction time, solvent, and AA concentration on the carboxyl content on the surface of PS microspheres is investigated. The results show that the effect of solvent on the carboxyl content of the PS microspheres is significant. When n-butanol is used as the solvent, the carboxyl content of PS microspheres can reach 57.05 mg/g. The first-step reaction time of the two-step method also has some influence on the carboxyl content on the surface of the microspheres, mainly reflected in the influence of the first reaction time on the nucleation process.

TABLE 6: Molecular weight of PS microspheres in different solvents with limit concentration of AA.

| Solvent | AA (%) | $M_n$ | $M_w$ | $M_w/M_n$ |
|---|---|---|---|---|
| Methanol | 7.5 | 99902 | 166244 | 1.664 |
| Ethanol | 10 | 135810 | 265780 | 1.957 |
| Propanol | 20 | 174992 | 385507 | 2.203 |
| Butanol | 27.5 | 371524 | 1563001 | 4.207 |
| Pentanol | 12.5 | 502101 | 1226633 | 2.443 |

Carboxylated PS microspheres have important applications in biomedical detection, and achieving controllable surface carboxyl contents for microspheres establishes a good foundation for advancing their application in biomedical dry slides.

## Conflicts of Interest

The authors declare that they have no conflicts of interest.

## Acknowledgments

This research is supported by the National Natural Science Foundation of China (91634101, 21601016), the Beijing Municipal Natural Science Foundation (2174073), and the Project of Construction of Innovative Teams and Teacher Career Development for Universities and Colleges under Beijing Municipality (IDHT20180508).

## References

[1] P. Lapekas, "New technologies for laboratory productivity," *Journal of Healthcare Materiel Management*, vol. 12, no. 3, pp. 30–32, 1994.

[2] R. H. Ng, K. M. Sparks, and B. E. Statland, "Colorimetric determination of potassium in plasma and serum by reflectance photometry with a dry-chemistry reagent," *Clinical Chemistry*, vol. 38, no. 7, pp. 1371-1372, 1992.

[3] Y. H. Yu, H. Y. Yang, X. Zhang, Y. H. Li, and Y. Y. Niu, "Monolayer dry film for renal function examination and preparation method of monolayer dry film," China Patent CN104020283A, 2014.

[4] A. Yoshihiko, M. Kishi, and I. Hiroko, "Method for measuring high density lipoprotein cholesterol," Japan Patent CN101078729A, 2007.

[5] J. Jacobs, A. Byrne, N. Gathergood, T. E. Keyes, J. P. A. Heuts, and A. Heise, "Facile synthesis of fluorescent latex nanoparticles with selective binding properties using amphiphilic glycosylated polypeptide surfactants," *Macromolecules*, vol. 47, no. 21, pp. 7303–7310, 2014.

[6] P. M. Dietrich, A. Hennig, M. Holzweber et al., "Surface analytical study of poly(acrylic acid)-grafted microparticles (beads): characterization, chemical derivatization, and quantification of surface carboxyl groups," *The Journal of Physical Chemistry C*, vol. 118, no. 35, pp. 20393–20404, 2014.

[7] S. Bousalem, A. Yassar, T. Basinska et al., "Synthesis, characterization and biomedical applications of functionalized

polypyrrole-coated polystyrene latex particles," *Polymers for Advanced Technologies*, vol. 14, no. 11-12, pp. 820–825, 2003.

[8] R. H. Utama, M. Drechsler, S. Forster, P. B. Zetterlund, and M. H. Stenzel, "Synthesis of pH-responsive nanocapsules via inverse miniemulsion periphery RAFT polymerization and post-polymerization reaction," *ACS Macro Letters*, vol. 3, no. 9, pp. 935–939, 2014.

[9] X. Liu, Y. Xu, S. Ma et al., "Encapsulation of quantum dot clusters in stimuli-responsive spherical polyelectrolyte brushes," *Industrial & Engineering Chemistry Research*, vol. 53, no. 28, pp. 11326–11332, 2014.

[10] J. Sun, C. Yi, W. Wei, D. Zhao, Q. Hu, and X. Liu, "Nanohybrids from direct chemical self-assembly of poly(styrene-alt-maleic anhydride) as pH-responsive particulate emulsifiers," *Langmuir*, vol. 30, no. 49, pp. 14757–14764, 2014.

[11] R. Perrier-Cornet, V. Heroguez, A. Thienpont, O. Babot, and T. Toupance, "Functional crosslinked polymer particles synthesized by precipitation polymerization for liquid chromatography," *Journal of Chromatography A*, vol. 1179, no. 1, pp. 2–8, 2008.

[12] T. J. Romack, E. E. Maury, and J. M. DeSimone, "Precipitation polymerization of acrylic acid in supercritical carbon dioxide," *Macromolecules*, vol. 28, no. 4, pp. 912–915, 1995.

[13] D. Wang and Y. Liu, "Mono-dispersed cross-linked polystyrene micro-spheres prepared by seed swelling polymerization method," *Frontiers of Chemistry in China*, vol. 3, no. 4, pp. 440–444, 2008.

[14] X. Zhao, S. Zhou, M. Chen, and L. Wu, "Effective encapsulation of Sudan Black B with polystyrene using miniemulsion polymerization," *Colloid and Polymer Science*, vol. 287, no. 8, pp. 969–977, 2009.

[15] W. Yang, D. Yang, J. Hu, C. Wang, and S. Fu, "Dispersion copolymerization of styrene and other vinyl monomers in polar solvents," *Journal of Polymer Science Part A: Polymer Chemistry*, vol. 39, no. 4, pp. 555–561, 2001.

[16] J. S. Song, L. Chagal, and M. A. Winnik, "Monodisperse micrometer-size carboxyl-functionalized polystyrene particles obtained by two-stage dispersion polymerization," *Macromolecules*, vol. 39, no. 17, pp. 5729–5737, 2006.

[17] S. G. Lee and J.-W. Ha, "Synthesis of highly carboxylated monodisperse polystyrene microspheres by dispersion polymerization in fluorinated alcohol," *Macromolecular Research*, vol. 24, no. 8, pp. 675–683, 2016.

[18] F. Bai, X. Yang, R. Li, B. Huang, and W. Huang, "Monodisperse hydrophilic polymer microspheres having carboxylic acid groups prepared by distillation precipitation polymerization," *Polymer*, vol. 47, no. 16, pp. 5775–5784, 2006.

[19] X. Hu, *Preparation of carboxyl-functional polystyrene microspheres and its application in emulsion polymerization, [M.S. thesis]*, University of Science and Technology of China, 2009.

[20] Q. L. Xing, *Preparation of monodisperse micron-sized pplystyrene microspheres and its self-assembly*, Harbin Institute of Technology, 2012.

[21] Y. J. Wang, J. F. Liu, Z. L. Yao, C. D. Yuan, and T. Y. Cao, "Preparation and study of magnetic poly (methyl-methacrylate) microspheres," *Journal of Tianjin University(Science and Technology)*, vol. 34, no. 1, pp. 64–68, 2001.

# 21

# Curing Reaction and Dielectric Properties of Rigid and Elastic Liquid Crystal Epoxy Networks Modified with Nanofillers

**Magdalena Włodarska**

*Institute of Physics, Lodz University of Technology, Wólczańska 219, 90-924 Łódź, Poland*

Correspondence should be addressed to Magdalena Włodarska; magdaw@p.lodz.pl

Academic Editor: Ulrich Maschke

The aim of the paper is to compare the progress of the curing reaction and the dielectric properties of liquid crystalline epoxy networks of different elasticity and to investigate how modification with nanofillers influences their properties. The study focuses on a liquid crystalline epoxy monomer with four aromatic rings in the mesogen, cross-linked with 4,4′-diaminodiphenylmethane (DDM) and pimelic acid (PA) to produce rigid and elastic polymer networks. The obtained results are compared to networks modified with nanofillers (diphenyl aluminum phosphate nanorods). The curing process is monitored in situ with broadband dielectric spectroscopy, which is a good tool to view the progress of the reaction. Two stages with different reaction dynamics are observed in the studied cases. Dielectric properties of the products cured with two chosen agents show significant differences in the obtained parameters (activation energy, glass transition, and fragility index) as well as in the dynamics of the $\alpha$ and $\beta$ relaxations. Although the curing agent is the main factor determining the properties of the product, the nanofiller additive also modifies the values of specific parameters that characterize both the process of the reaction and the properties of the final composites. Particularly, adding the nanofiller raises the glass transition temperature in both the cases. The obtained results are in good agreement with the earlier calorimetric study.

## 1. Introduction

Although epoxy resins have already found a wide range of industrial applications, there is a continuing interest in these materials, aiming at the development of advanced composites with specific thermal, electrical, and mechanical properties [1–14]. For example, epoxy materials have high thermal, mechanical, and electrical stability, which is highly demanded by the industry [1–4]. Another big advantage of epoxy resins is the prospect of modifying their selected properties by using organic and inorganic materials or nanoparticles as fillers and building advanced composites [4, 7–11]. A separate group of such materials are epoxy resins based on liquid crystalline mesogens, which are used to produce liquid crystalline epoxy networks (LCEN) [1–3, 5, 6]. Due to the structure of LCEN, their properties and curing conditions are different and more complex than those of traditional resins [6, 12–14]. Both the curing process and the properties of the cured products depend on various factors such as the curing agent, curing conditions, or addition of nanofillers, as well as the properties of the pure epoxy resin such as the type of mesophase [5, 6, 12].

Many techniques can be used to describe the properties of the final resin; one of them is the dielectric relaxation spectroscopy (DRS). This method is very sensitive to changes in the molecular properties of the material. In consequence, it is frequently used for observation of phase transitions or monitoring of the reaction progress. The dielectric response is usually represented in terms of the complex electric permittivity $\varepsilon^*$ as a function of frequency and temperature (or frequency and time, in the in situ experiments). The principle of dielectric spectroscopy is based on measuring changes in the dielectric properties of the sample in response to the applied small, alternating electric field. As a result, several parameters can be observed—such as the electric conductivity, the dielectric constant connected with location of diverse dipole moments and their cooperation in the material, and some relaxation processes (with characteristic times) which result from delayed reaction on sample polarization. These parameters, describing the electrical behavior of the sample, can also be linked with some mechanical properties. For example, in polymers, at least two relaxation processes (denoted with Greek letters $\alpha$ and $\beta$) are most frequently observed [15–21]. The structural $\alpha$-process usually corresponds to the glass transition. It is thereby related to hindering of the motion of polymer chain segments, caused by approaching the vitrification temperature $T_g$. The $\alpha$-process is typically the slowest one in polymers (appearing at the lowest frequencies/the highest temperatures). The $\beta$-process occurs at higher frequencies/lower temperatures and is believed to involve rotational motions of polar group fragments containing single bonds within the molecule [17–19]. These motions can be treated as almost free rotations of dipolar groups. The $\alpha$- and $\beta$-processes exhibit different temperature dependencies. The relaxation time of the molecular $\beta$-process can usually be described with the exponential Arrhenius formula, but deviations from the exponential dependency may be observed near the vitrification temperature, where the structural ($\alpha$) process appears. For the description of dynamic properties of glass, such as the structural relaxation time, an empirical formula was proposed by Vogel, Fulcher, and Tammann (VFT) [17–21]. Although the complex permittivity $\varepsilon^*$ is most commonly used to discuss the results, another representation was proposed for polymers with high ionic conductivity: the electric modulus [22]. It is defined as the inverse of the complex permittivity [22–24]:

$$M^* = \frac{1}{\varepsilon^*}. \quad (1)$$

This quantity was introduced by analogy with the modulus observed in mechanical studies. The modulus representation is particularly convenient when no relaxation processes occur at a given temperature and ionic conductivity is high. Then, the imaginary component of the electric modulus ($M''$) displays a characteristic peak which is directly related to ionic conductivity [22]:

$$\tau_\sigma = \frac{\varepsilon_0 \varepsilon_s}{\sigma_0} = \frac{\varepsilon_0}{M_S \cdot \sigma_0}. \quad (2)$$

Here, $\tau_\sigma$ can be interpreted as the conductivity relaxation time (the inverse of the frequency of the maximum of $M''$), $\varepsilon_0$ is the electric permittivity of free space ($8.854^* 10^{-14}$ F/cm), and $\varepsilon_s$ is the frequency-independent relative permittivity which is linked with the real part of the electric modulus:

$$\lim_{\omega\tau_\sigma \gg 1} M' = M_s = \frac{1}{\varepsilon_s}. \quad (3)$$

On the other side, the real part of the modulus approaches zero at low frequencies:

$$\lim_{\omega\tau_\sigma \ll 1} M' = 0. \quad (4)$$

In this work, the modulus representation is used to monitor the reaction. This approach has already been successfully applied to in situ observation of curing reactions. Early papers, in which dielectric spectroscopy was employed to study curing of commercially available materials, demonstrated its usefulness [25, 26]. Nowadays, the technique continues to be developed with the aim of extracting more information, from the dielectric response in the course of the curing reaction to its connection with characteristic parameters of the reaction, such as the reaction rate.

Dielectric response of liquid crystalline epoxy compounds is quite complex due to the presence of polar, rigid mesogenic units. It is known that dielectric spectra of a liquid crystal consist of different relaxation processes which depend on the direction of the electric field with respect to the orientation of the mesogens. In unordered samples of low-molecular mass liquid crystals, one (broadened) or two relaxation modes are usually observed by virtue of overlapping of all the existing relaxation types [20, 27–30]. Hence, in cured liquid crystalline epoxies, various relaxation processes connected with different relaxation modes may appear; this has a potential for providing more information about the reaction and the molecular behavior in the final material. Dielectric studies on materials such as liquid crystalline epoxy networks (LCEN) not only facilitate a better understanding of the processes occurring in the material but also enable correlating distinct features of the dielectric response with specific properties of the sample. Additionally, building composites based on LCEN allows for modification of specific properties of the obtained network while keeping other properties unchanged.

In this work, one liquid crystalline epoxy monomer with a long mesogenic group (containing four aromatic rings) was chosen for the investigation. The synthetic route and the best curing conditions for this material were established in a previous work [31]. 4,4′-Diaminodiphenylmethane (DDM) and pimelic acid (PA) were selected as curing agents to obtain rigid and elastic networks. The size of the mesogen causes the phase transitions to occur at relatively high temperatures, which is usually a disadvantage; but it is beneficial for the curing process—which also takes place at elevated temperatures—because it facilitates carrying out of the process in the liquid crystalline state. The results were supplemented by a comparison of the studied materials with the same polymer matrices containing a nanofiller. Dielectric spectroscopy was applied to study both the curing process and the properties of the cured products. The aim of the author was to show differences in the dielectric responses of rigid and more elastic LCEN, by using two different curing agents. An additional modification of the LCEN with nanofillers resulted in increased rigidity of both the products. However, these changes had a different character than those related to the selection of the curing agent.

## 2. Materials

The molecular structures of the monomer and the curing agents are shown in Figure 1. All the reagents and curing agents (4,4′-diaminodiphenylmethane (DDM) and pimelic acid (PA)) were commercial (Sigma-Aldrich, Fluka, POCH, Gliwice) and were used without further purification.

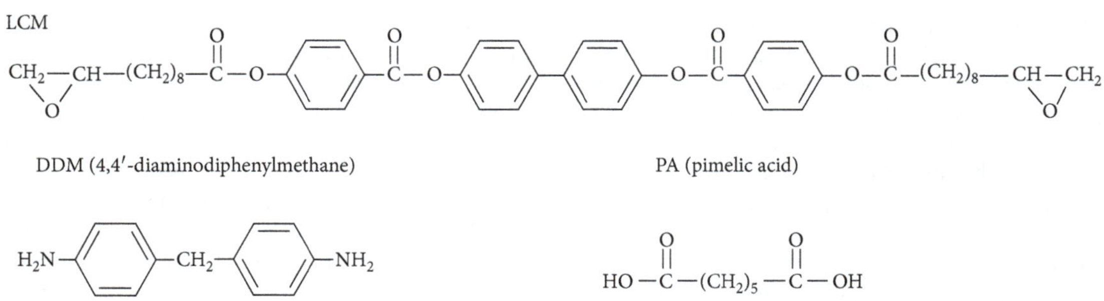


Figure 1: Molecular structure of the epoxy monomer and the curing agents.

Table 1: Temperatures of phase transitions in the pure liquid crystal epoxy monomer.

| | Smectic | Nematic | Isotropic/ degradation |
|---|---|---|---|
| $T$ (°C) | 166 | 212 | 290 |
| Heating | 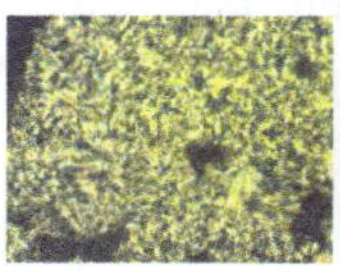 | 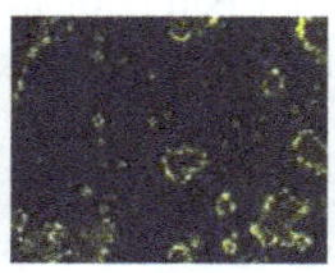 | 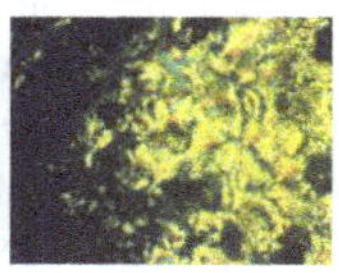 |

Temperatures of phase transitions found in the previous study were confirmed by optical data and are presented in Table 1. The curing conditions chosen for the dielectric study were the same for all the materials—isothermal curing at 170°C for 2.5 hours followed by postcuring at 210°C for 1.5 hours—and comparable with those for the previous study [31]. The earlier DSC study of the investigated mixtures also enabled determination of the optimal curing conditions. DSC analysis of the cured products did not reveal any phase transitions (except vitrification) [31]. Both compositions were subsequently cured with the addition of a nanofiller, in the same conditions. Diphenyl aluminum phosphate nanorods (NR) having 100–200 nm in diameter and 1–5 $\mu$m in length were used as the filler. The details of the synthesis of the nanorods and their properties were described elsewhere [32, 33].

## 3. Experimental

The dielectric data during isothermal curing was recorded in the frequency range from $10^{-2}$ to $10^5$ Hz using Solartron 1260 Impedance Analyzer with Chelsea Dielectric Interface. The mixtures were placed between two parallel metal electrodes with diameters of 5 and 10 mm, separated by 50 $\mu$m silica spacers. The temperature of the samples was stabilized using Unipan 620 temperature controller with the accuracy of 0.05 K. The duration of a single-frequency sweep at the curing temperature of 170°C was about 3 minutes. The next series of measurements was started after 4 minutes (in most of the cases), yielding a 3D plot with a 4-minute separation between subsequent data points at a given frequency.

Measurements in a wide range of temperatures for all the cured products were performed with an alpha high-resolution dielectric analyzer coupled with a Quatro Cryo-system, providing isothermal conditions and temperature stabilization of 0.1 K (both devices are manufactured by Novocontrol). Data was collected in the cooling route in a wide range of frequency ($10^6$–$10^{-1}$ Hz).

The optical observations were made using a polarizing microscope with crossed polarizers and a BIC-2S 1/2″ CMOS digital camera. The material was placed between plain glass plates in a Linkam HFS600E-PB4 hot stage.

## 4. Results and Discussion

*4.1. Process of Curing.* The electric modulus representation can be used for in situ monitoring of the curing process of epoxy resins. The data collected in the course of curing the investigated monomer with DDM is shown in Figure 2(a).

For comparison, Figure 2(b) shows the process of curing with PA. The shift in the frequency of the electric modulus peak is connected, as mentioned earlier, with the changes in the electric conductivity. In principle, the curing reaction in epoxy materials is biphasic. The first, highly exothermic stage, characterized with rapid changes of the conductivity, lasts between a few minutes and a few tens of minutes. However, the mixture does not attain its final state and properties in that phase yet. In the second stage, slow evolution of certain properties can still be observed, until they finally stabilize after several hours. Sometimes, postcuring treatment at an elevated temperature is required for complete stabilization of the product. Dielectric studies allow for monitoring of the full progress of curing, since both phases of the reaction are accompanied by changes in the dielectric response. A comparison of the two plots shows that the initial, rapid phase ends much sooner in the LCM/DDM mixture (Figure 2(a)) than that in the case when pimelic acid was used as the curing agent (Figure 2(b)). Different reaction kinetics is related to different functionalities of the reagents (DDM is tetrafunctional and PA is difunctional), which results in a higher cross-link density and a more rigid network cured with DDM. An additional explanation of the different reaction kinetics is the presence of steric constraints in the cured product, which may significantly slow down the process: when PA is used, no extra aromatic rings are added to the composition and the reaction can progress rapidly for a longer time. This observation, based on comparing the reaction

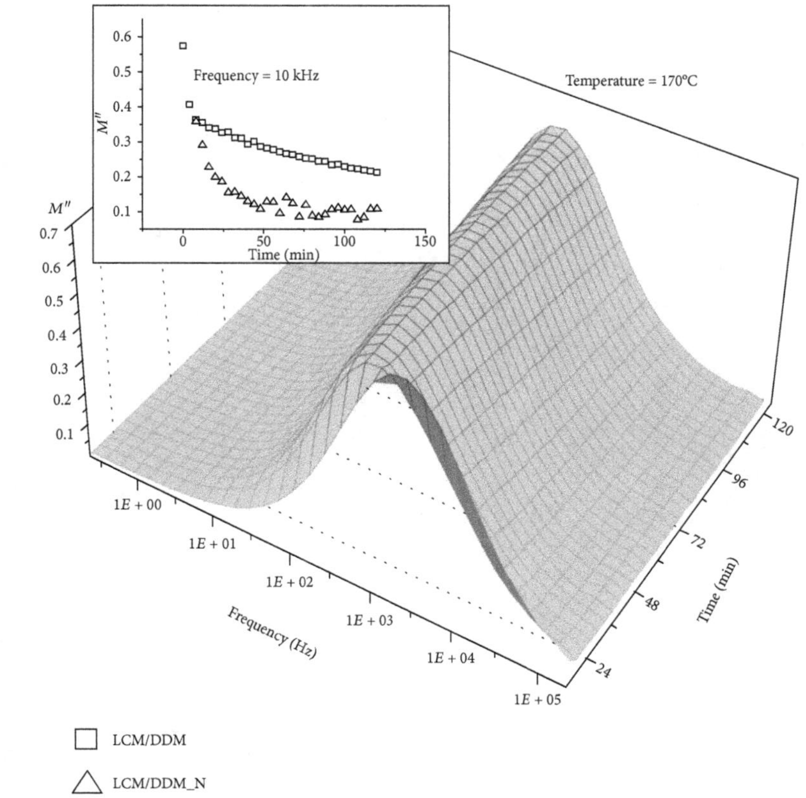


(a)

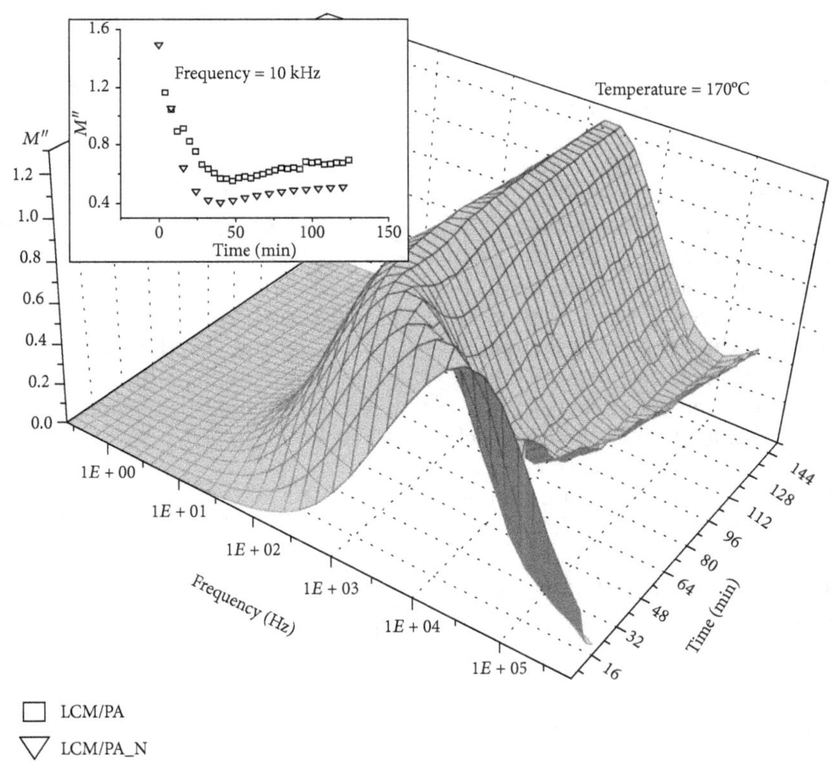


(b)

FIGURE 2: Isothermal curing of the epoxy resin at $T = 170^{\circ}C$, in the electric modulus representation. Imaginary part of the electric modulus plotted versus time in a wide frequency range: (a) curing with the aromatic amine (DDM); (b) curing with pimelic acid (PA). Insets: modulus evolution with time at one specific frequency, for mixtures cured with and without nanofillers.

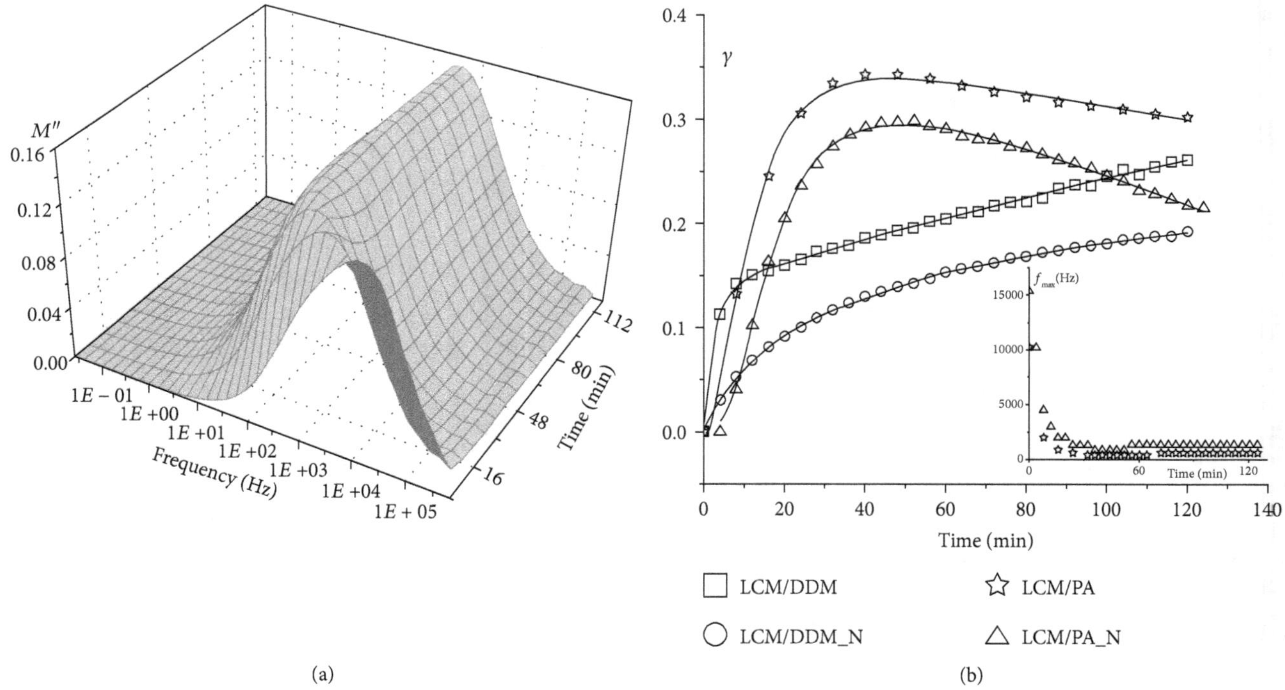


FIGURE 3: Curing process of the epoxy resin and its composites: (a) 3D representation (electric modulus versus frequency and time) of curing of the epoxy resin with pimelic acid and nanorods; (b) normalized real part of permittivity ($\gamma$) versus time for all the investigated mixtures, with solid lines representing the best fit of the chosen kinetic model to the data (inset: estimated frequency of the $M''$ peak versus time for composites cured with pimelic acid).

profiles, was confirmed by the study of the reaction products: indeed, more elastic materials were obtained from the mixtures cured with PA.

Introducing the nanofiller does not fundamentally change the kinetics of the curing reaction in either case, which depends mainly on the properties of the curing agent. The reaction profiles are almost identical for the composites cured with DDM, with and without the nanofiller. Small differences are visible between the composites cured with PA—as shown for comparison in Figure 3(a). Adding nanorods to the mixture slightly shortens the first stage of the reaction (cf. Figure 2(b)) which is likely caused by the appearance of additional steric constraints in the mixture, but the fundamental character of the reaction does not change.

In the second stage of the reaction (usually, final products are obtained from epoxy materials only after several hours of curing or after postcuring at an elevated temperature), changes are no longer visible so well in the modulus representation, in any of the studied cases. A parameter found to be more sensitive in that phase is the real part of the electric permittivity at low frequencies. It does not display a relaxation profile that is so convenient for observation, but its value changes steadily during the entire observation time—reflecting the progress of the reaction—even though the rate of those changes varies with the curing progress. For further analysis, the normalized value ($\gamma$) of the real part of the permittivity ($\varepsilon$) was chosen:

$$\gamma = \frac{\varepsilon - \varepsilon_0}{\varepsilon}, \tag{5}$$

where $\varepsilon$ is the current value and $\varepsilon_0$ is the initial value. This representation enables seeing the two-phase character of the reaction, both for curing with DDM and for curing with PA (Figure 3(b)). Curve fitting provides a tool to quantify and follow the rate changes of both the processes—-not only in the initial phase but during the whole observation interval. It also enables comparing the curing parameters for mixtures with and without nanorods. However, the fundamental mechanisms of the curing reactions are different for PA and DDM and they require different fitting curves. Generally, a single exponential function appears in the description of first-order reactions. But higher-order reactions (including reactions in epoxy materials) are described by a somewhat more complex differential equation [34]:

$$\frac{d\alpha}{dt} = k(T) \cdot f(\alpha) = k(T) \cdot \alpha^m \cdot (1 - \alpha)^n, \tag{6}$$

where the conversion factor ($\alpha$) depends on the reaction rate $k(T)$ and the exponents $n + m$ which determine the kinetic order. For curing with DDM, a relatively simple exponential model enables identification of two stages with clearly different reaction rates. The reaction process can be modelled by a double-decay function with two exponential terms describing the observed fast and slow phases of curing:

$$\gamma = \gamma_0 + A_1 e^{-t/t_1} + A_2 e^{-t/t_2}, \tag{7}$$

TABLE 2: Kinetic parameters of the reaction progress in various systems. Values determined from the best fit of the double-decay function (7) for mixtures with DDM and the Hill equation (8) for systems with PA.

| Parameters | $y_0$ | $A_1$ | $t_1$ | $A_2$ | $t_2$ |
|---|---|---|---|---|---|
| LCM/DDM | 0.545 | −0.138 | 2.6 | −0.107 | 334 |
| LCM/DDM_N | 0.217 | −0.057 | 9.0 | −0.160 | 66 |
| Parameters | $P_m$ | $K_a$ | $K_i$ | $H_a$ | $H_i$ |
| LCM/PA | 0.33 | 16 | 160 | 2.6 | 2.2 |
| LCM/PA_N | 0.48 | 13 | 250 | 1.8 | 0.7 |

where $y_0$ is offset, $A_1$ and $A_2$ are amplitudes, and $t_1$ and $t_2$ are time constants. The obtained fitting parameters are listed in Table 2. It is easy to notice that the parameter $A_1$ describing the first phase of the reaction is almost an order of magnitude smaller after addition of the nanofiller. Also, the characteristic times for the first stage of the reaction ($t_1$) are much smaller than those for the second stage ($t_2$), in both the mixtures.

Similar conclusions can be drawn from the observation of curing with pimelic acid as the curing agent, but the reaction model is quite different. The apparent two-stage reaction progress can be approximated in this case by biphasic Hill equation, used in biochemistry to describe the enzyme kinetics [35–37]:

$$y = \frac{P_m}{[1 + (K_a/x)H_a] \cdot [1 + (x/K_i)H_i]}, \tag{8}$$

where $P_m$ is the maximum activity, $K_a$ is the half-maximal activity, $K_i$ is the half-maximal inhibitory coefficient, $H_a$ is the activation Hill coefficient, and $H_i$ is the inhibitory Hill coefficient. To apply this formula to the experimental data from Figure 3(b), the generic variables $x$ and $y$ need to be replaced with the reaction time $t$ and the normalized permittivity $\gamma$, respectively. After adding nanorods, there is a visible growth of the $K_i$ coefficient (connected with the factors inhibiting the reaction) which has an influence on deceleration of changes in the second phase of the reaction. The other parameters do not differ substantially. Thanks to the proposed representation one can easily follow not only the curing reaction kinetics but also the influence of the introduced modifiers on the reaction progress.

*4.2. Properties of the Cured Resin and Its Composites.* In the next part of the work, the dielectric properties of the products curing the investigated compound with DDM and PA are compared, including products modified by the addition of nanorods. Electrical and mechanical changes in the material at different temperatures can be described by analyzing the dielectric data represented in terms of the complex permittivity. Here, we chose to compare the properties of the studied composites by looking at the imaginary part of the permittivity, shown in a wide range of frequency and temperature in Figure 4.

The imaginary part of the permittivity is generally associated with dielectric losses in the material. The large growth of the imaginary part of the permittivity in both materials at low frequencies and high temperatures is due to the electric conductivity of the material (Figure 4). With decreasing temperature, the conductivity drops substantially and relaxation processes become visible. Along with further temperature drop, the $\alpha$-process vanishes and a broad $\beta$-process appears. In all the studied materials, the vanishing of the $\alpha$-process (which is related to the glass transition) is accompanied by a rapid change in the conductivity, which decreases by several orders of magnitude in the glassy phase.

The response of the cured material varies substantially for different curing agents, mirroring the differences already observed in the course of the curing process. The dielectric response of the pure epoxy matrix is generally similar to the response of the modified product; they only differ in detail, which will be analyzed below. In the mixtures of the epoxy resin with DDM, modified by adding the nanofiller, the $\alpha$-process is barely seen in the imaginary part of the dielectric permittivity at high temperatures because it is partially masked by high conductivity (Figure 4(a)). In the mixtures with nanorods, a growth of the imaginary part of the permittivity occurs at very low temperatures and high frequencies, next to the $\beta$-process—this hints towards a possible occurrence of another relaxation. However, the collected range of data is not sufficient for definitive confirmation of such an additional process in the materials with nanorods, or its closer analysis (Figure 4).

In the mixtures of the investigated resin with PA, both processes ($\alpha$ and $\beta$) are well visible; the $\alpha$-process is observed at temperatures lower than those of the products cured with DDM (Figure 4(b)). At low temperatures, a similar $\beta$-process appears in the mixtures both with and without the addition of nanorods.

As it was mentioned in Introduction, the $\alpha$-process is generally connected with the glass transition and enables estimation of the fragility index; the $\beta$-process reflects the mobility of molecular groups with large dipole moments and appears nearly always together with the $\alpha$-process. These relaxation processes can be analyzed more quantitatively by using the Havriliak-Negami (HN) formula with a conductivity term [38]:

$$\varepsilon^*(\omega) - \varepsilon_\infty = -i\frac{\sigma_0}{\varepsilon_0\omega^s} + \sum_k \frac{\Delta\varepsilon_k}{\left(1 + (i\omega\tau_k)^{\alpha_k}\right)^{\beta_k}}. \tag{9}$$

Here, $\varepsilon_0$ is the electric permittivity of vacuum, $\omega = 2\pi f$ is the radial frequency, and $\varepsilon_\infty$ is $\varepsilon'$ at $f \to \infty$; the dielectric strength $\Delta\varepsilon$ is the difference between the high-frequency and low-frequency electric permittivities corresponding to the relaxation process under consideration, $\tau$ is the characteristic relaxation time, and $\alpha$ and $\beta$ are shape parameters of the relaxation time spectra; $\sigma_0$ denotes the DC conductivity and the parameter $s$ ($s \le 1$) accounts for non-ohmic effects. The relaxation times for all the processes were

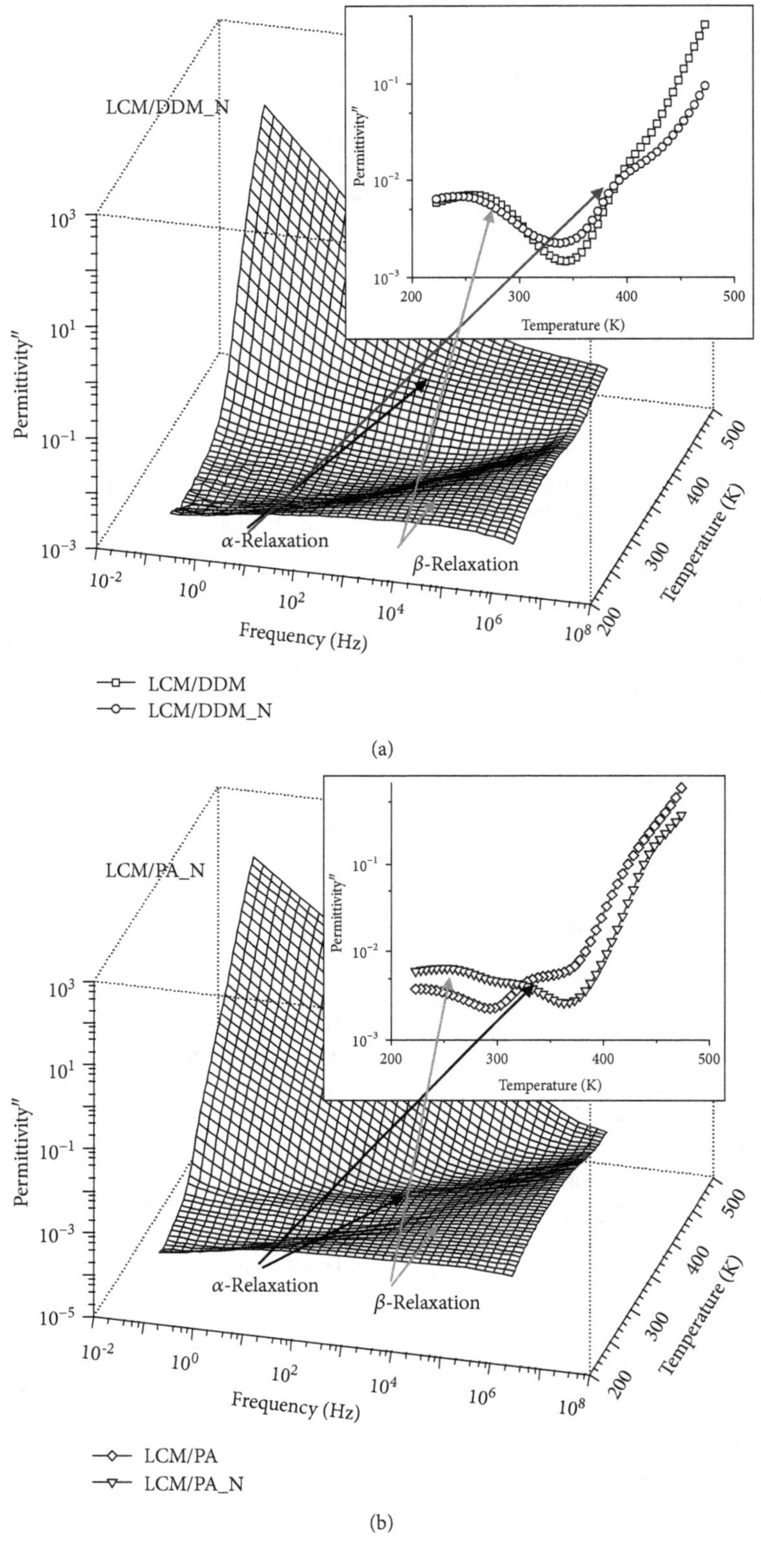


FIGURE 4: Imaginary part of the complex permittivity, at various temperatures: (a) LCM/DDM with nanorods; (b) LCM/PA with nanorods. Insets: comparative plots of $\varepsilon''(T)$ at $f = 4$ kHz for mixtures cured with and without nanofillers.

determined by fitting the HN formula to experimental data, for each temperature at which these processes were visible. The use of the HN formula enables extraction of individual processes with their characteristic parameters. The relaxation times for all the mixtures, both cured with the acid and the amine, are shown in Figure 5.

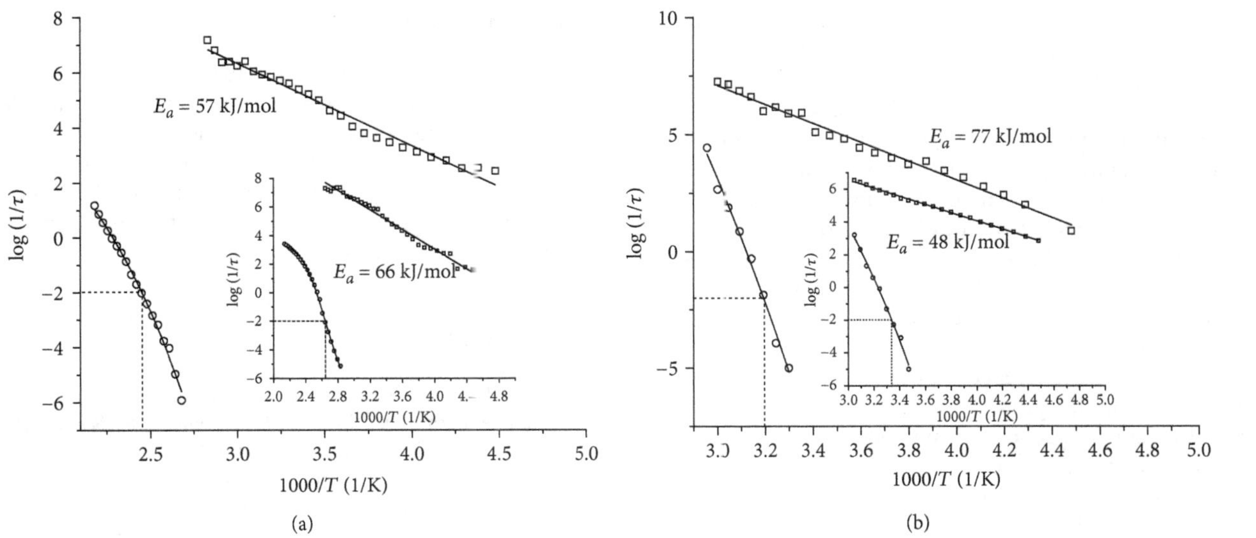


FIGURE 5: Temperature dependency of the relaxation times for all the observed processes, with the VFT formula fitted to the $\alpha$-processes and the Arrhenius formula fitted to the $\beta$-processes: (a) monomer cured with DDM and nanofiller additive (inset: pure LCM/DDM mixture); (b) monomer cured with PA and nanofiller additive (inset: pure LCM/PA mixture).

TABLE 3: Glass transition temperature $T_g(\tau_M{}'' = 10^2 s)$ and fragility index $m$ for each of the investigated mixtures. Fragility index values obtained from VFT equation (11) and formula (13).

| Parameters | $T_g$ (K) | $D$ | $T_0$ (K) | $m$ |
|---|---|---|---|---|
| LCM/DDM | 378 | 2 | 337 | 76 |
| LCM/DDM_N | 408 | 12 | 275 | 35 |
| LCM/PA | 298 | 33 | 188 | 68 |
| LCM/PA_N | 313 | 48 | 194 | 90 |

It is noteworthy that the $\beta$-process realizes the typical Arrhenius dependency [17–21]:

$$\frac{1}{\tau} = \frac{1}{\tau_0} \cdot e^{-E_a/kT}. \tag{10}$$

The $\alpha$-process, which deviates from that dependency, can be modelled by the VFT relation:

$$\frac{1}{\tau} = \frac{1}{\tau_0} \cdot e^{-DT_0/(T-T_0)}, \tag{11}$$

where $D$ is a dimensionless quantity called the strength parameter and $T_0$ is the Vogel temperature [39].

The glass transition temperature $T_g$ can be estimated from the vanishing of the $\alpha$-process. By convention, it is extracted from the VFT plot at the frequency of $10^{-2}$ Hz; that approach was confirmed in many cases [17–21]. The parameters obtained from Figure 5 were collected in Table 3. The estimated $T_g$ for the mixture of the monomer with DDM is 105°C, whereas addition of the nanorods results in the growth of $T_g$ to the value of 135°C (Figure 5(a)). Much lower temperatures were obtained for the mixtures cured with pimelic acid, which produced elastomers with $T_g$ of 25°C for the plain mixture and 40°C for the mixture with the nanofiller (Figure 5(b)). The above conclusions are in good agreement with those obtained earlier from calorimetric data, though the values of $T_g$ are not identical; this is linked with different cooling rates in the calorimetric studies and isothermal nature of dielectric measurements. Nevertheless, the addition of the nanofiller increased the vitrification temperature by 15°C in the materials cured with PA and by 30°C in the materials cured with DDM; this behavior was also noticed in the calorimetric study [31].

Another parameter describing the properties of the material is the fragility index, whose values typically lie in the range from about 16 for particularly hard substances to about 200 for brittle materials [23]. The fragility index is defined as [23, 39, 40]:

$$m = \left[\frac{d(\log_{10}\tau)}{d(T_g/T)}\right]_{T=T_g}. \tag{12}$$

Knowing the parameters of the VFT equation, the fragility index can be calculated directly from the relation:

$$m = \frac{DT_0}{2.3 \cdot T_g\left(1 - T_0/T_g\right)^2}. \tag{13}$$

The parameters calculated by fitting the VFT formula to the experimental data for all the mixtures are shown in Table 3. The apparent effect of adding nanorods is the increase of $T_g$, regardless of the hardener used. On the contrary, the fragility index $m$ behaves differently for the two curing agents. It decreases after introduction of the nanofiller in the mixture cured with DDM; together with the increase of $T_g$, this can be interpreted as the material is becoming harder. The value of $m$ increases after adding the nanorods in the more elastic mixtures, cured with PA; this suggests that the material becomes more brittle and less elastic.

The dielectric response of all the mixtures displays also a second relaxation, the $\beta$-process (Figure 4). The activation energy of that process changes after introduction of the nanofiller (Figure 5). The $\beta$-process is commonly associated with the motions of local dipolar groups. In the studied case, it is most likely linked with the dipolar carboxy groups in the monomer, because it appears in all the compositions, regardless of the curing agent. Although the activation energy of the observed process varies slightly among the studied mixtures, such variations may be due to differences in the energy barrier, reflecting differences in the internal structure of the network. A similar process was also observed in materials with different lengths of the mesogen [24]. An interesting case was found in the LCM/PA mixtures cured with the addition of nanorods, where an additional growth of the imaginary part of the permittivity was observed at high frequencies and low temperatures (Figure 4(b), inset). It may suggest that an additional process with very low activation energy, overlapping with the $\beta$-process, emerges in the samples containing the nanofiller. The possibility of such a process in the dielectric response can be explained by increased rigidity of the material after adding the nanofiller, shifting a fast relaxation process from high frequencies into the frequency range covered by the experiment. However, unequivocal interpretation of this growth of permittivity would require more detailed studies at low temperatures or at high frequencies.

## 5. Conclusions

The dynamics of the curing process observed in the dielectric spectroscopy was different in the materials cured with pimelic acid (PA) and 4,4′-diaminodiphenylmethane (DDM). The addition of the nanofiller did not change that dynamics but slightly slowed down the initial phase of the reaction.

Hard products were obtained from mixtures cured with DDM, whereas curing with PA yielded elastic materials. Dielectric observation of the $\alpha$-relaxation in the investigated epoxy matrix confirmed that the main factor determining the elasticity of the cured product is the choice of the hardener. The observation of the $\alpha$-relaxation in products with the addition of nanorods confirmed the increase of the vitrification temperature $T_g$, regardless of the hardener used. The product obtained by adding nanorods to the mixture of the monomer with DDM was characterized by the highest rigidity.

The $\beta$-relaxation process was observed in all the mixtures. The suggested origin of the process is the motion of the dipolar carboxy groups in the mesogen.

## Conflicts of Interest

The author declares that there is no conflict of interest regarding the publication of this paper.

## Acknowledgments

The author is indebted to Beata Mossety-Leszczak, Krzysztof Lokaj, and Zbigniew Florjańczyk for providing the samples of liquid crystal monomer and the nanorods, and to Konrad Słaby for technical support. This research was performed as part of the employment of the author at Lodz University of Technology, supported by the Ministry of Science and Higher Education, Republic of Poland.

## References

[1] E. P. Douglas, "Structure-property relationships in liquid crystalline thermosets," *Journal of Macromolecular Science, Part C: Polymer Reviews*, vol. 46, no. 2, pp. 127–141, 2006.

[2] A. Shiota and C. K. Ober, "Rigid rod and liquid crystalline thermosets," *Progress in Polymer Science*, vol. 22, no. 5, pp. 975–1000, 1997.

[3] T. S. Chung, *Thermotropic Liquid Crystal Polymers: Thin-film Polymerization, Characterization, Blends, and Applications*, Technomic Pub Co, Lancaster, PA, USA, 2001.

[4] A. Shalwan, M. Alajmi, and A. Alajmi, "Insulation characteristics of sisal fibre/epoxy composites," *International Journal of Polymer Science*, vol. 2017, Article ID 7312609, 6 pages, 2017.

[5] B. Mossety-Leszczak and M. Wlodarska, "Liquid-crystalline epoxy thermosets as matrices for ordered nanocomposites—a summary of experimental studies," *Polymer Composites*, vol. 38, no. 2, pp. 277–286, 2017.

[6] J. Liu, C. Wang, G. A. Campbell, J. D. Earls, and R. D. Priester, "Effects of liquid crystalline structure formation on the curing kinetics of an epoxy resin," *Journal of Polymer Science Part A: Polymer Chemistry*, vol. 35, no. 6, pp. 1105–1124, 1997.

[7] H. Körner, A. Shiota, T. J. Bunning, and C. K. Ober, "Orientation-on-demand thin films: curing of liquid crystalline networks in ac electric fields," *Science*, vol. 272, no. 5259, pp. 252–255, 1996.

[8] J. Y. Lee and J. Jang, "Anisotropically ordered liquid crystalline epoxy network on carbon fiber surface," *Polymer Bulletin*, vol. 59, no. 2, pp. 261–268, 2007.

[9] N. Suzuki, S. Kiba, and Y. Yamauchi, "Fabrication of epoxy composites with large-pore sized mesoporous silica and investigation of their thermal expansion," *Journal of Nanoscience and Nanotechnology*, vol. 12, no. 2, pp. 983–987, 2012.

[10] N. Suzuki, S. Kiba, and Y. Yamauchi, "Fabrication of mesoporous silica KIT-6/polymer composite and its low thermal expansion property," *Materials Letters*, vol. 65, no. 3, pp. 544–547, 2011.

[11] W. Yu, M. Qian, and H. Li, "Elastic and plastic properties of epoxy resin syntactic foams filled with hollow glass microspheres and glass fibers," *Journal of Applied Polymer Science*, vol. 133, no. 46, 2016.

[12] C. Carfagna, E. Amendola, and M. Giamberini, "Liquid crystalline epoxy based thermosetting polymers," *Progress in Polymer Science*, vol. 22, no. 8, pp. 1607–1647, 1997.

[13] J. Y. Lee and J. Jang, "The effect of mesogenic length on the curing behavior and properties of liquid crystalline epoxy resins," *Polymer*, vol. 47, no. 9, pp. 3036–3042, 2006.

[14] P. Castell, A. Serra, M. Galià, M. Giamberini, and C. Carfagna, "Anisotropic thermosets from liquid-crystalline azomethynic epoxy resins and primary aromatic diamines," *Journal of*

*Polymer Science Part A: Polymer Chemistry*, vol. 41, no. 1, pp. 1–12, 2003.

[15] D. Prevosto, S. Capaccioli, S. Sharifi, K. Kessairi, M. Lucchesi, and P. A. Rolla, "Secondary dynamics in glass formers: relation with the structural dynamics and the glass transition," *Journal of Non-Crystalline Solids*, vol. 353, no. 47–51, pp. 4278–4282, 2007.

[16] K. Mohomed, T. G. Gerasimov, F. Moussy, and J. P. Harmon, "A broad spectrum analysis of the dielectric properties of poly(2-hydroxyethyl methacrylate)," *Polymer*, vol. 46, no. 11, pp. 3847–3855, 2005.

[17] J. Rault, "Relaxation of glasses: the Kohlrausch exponent," *Journal of Non-Crystalline Solids*, vol. 357, no. 2, pp. 339–345, 2011.

[18] J. Rault, "Origin of the Vogel–Fulcher–Tammann law in glass-forming materials: the $\alpha$–$\beta$ bifurcation," *Journal of Non-Crystalline Solids*, vol. 271, no. 3, pp. 177–217, 2000.

[19] J. Rault, "Remarks on the Kohlrausch exponent and the Vogel–Fulcher–Tamann law in glass-forming materials," *Journal of Non-Crystalline Solids*, vol. 260, no. 1-2, pp. 164–166, 1999.

[20] F. Kremer and A. Schönhals, *Broadband Dielectric Spectroscopy*, Springer, Berlin, Germany, 2003.

[21] G. P. Johari and M. Goldstein, "Viscous liquids and the glass transition. II. Secondary relaxations in glasses of rigid molecules," *The Journal of Chemical Physics*, vol. 53, no. 6, pp. 2372–2388, 1970.

[22] P. B. Macedo, C. T. Moynihan, and R. Bose, "The role of ionic diffusion in polarization in vitreous ionic conductors," *Physics and Chemistry of Glasses*, vol. 13, pp. 171–179, 1972.

[23] B. Gonzalo, J. L. Vilas, M. San Sebastián et al., "Electric modulus and polarization studies on piezoelectric polyimides," *Journal of Applied Polymer Science*, vol. 125, no. 1, pp. 67–76, 2012.

[24] M. Wlodarska, A. Maj, B. Mossety-Leszczak et al., "Liquid crystal epoxy resins based on biphenyl group cured with aromatic amines - studied by dielectric spectroscopy," *Journal of Polymer Research*, vol. 20, no. 9, 2013.

[25] S. D. Senturia and N. F. Sheppard, "Dielectric analysis of thermoset cure," in *Epoxy Resins and Composites IV*, vol. 80 of Advances in Polymer Science, pp. 1–47, 1986.

[26] D. Kranbuehl, S. Delos, E. Yi et al., "Dynamic dielectric analysis: nondestructive material evaluation and cure cycle monitoring," *Polymer Engineering and Science*, vol. 26, no. 5, pp. 338–345, 1986.

[27] W. H. Jeu, *Physical Properties of Liquid Crystalline Materials*, Gordon and Breach, London, UK, 1980.

[28] A. Kozak, J. K. Mościcki, and G. Williams, "On dielectric relaxation in liquid crystals," *Molecular Crystals and Liquid Crystals*, vol. 201, no. 1, pp. 1–12, 1991.

[29] H. Kresse, "Dielectric behavior of liquid crystals," *Advanced Liquid Crystal*, vol. 6, p. 109, 1983.

[30] J. Jadżyn, G. Czechowski, R. Douali, and C. Legrand, "Molecular model of the dielectric relaxation in nematic liquid crystals," in *Proceedings Volume 4147, Liquid Crystals: Chemistry, Physics, and Applications*, pp. 176–183, Krynica Zdroj, Poland, 2000.

[31] B. Mossety-Leszczak, H. Galina, M. Włodarska, M. Kowalik, K. Łokaj, and Z. Florjańczyk, "Żywice i sieci epoksydowe o właściwościach anizotropowych," *Polimery*, vol. 54, no. 639, pp. 719–726, 2009.

[32] Z. Florjańczyk, A. Wolak, M. Dębowski et al., "Organically modified aluminophosphates: transformation of boehmite into nanoparticles and fibers containing aluminodiethylphosphate tectons," *Chemistry of Materials*, vol. 19, no. 23, pp. 5584–5592, 2007.

[33] Z. Florjańczyk, A. Lasota, A. Wolak, and J. Zachara, "Organically modified aluminum phosphates: synthesis and characterization of model compounds containing diphenyl phosphate ligands," *Chemistry of Materials*, vol. 18, no. 7, pp. 1995–2003, 2006.

[34] E. Amendola, C. Carfagna, and M. Giamberini, "Role of curing agent on the nature of the mesophase and the properties of mesogenic epoxy resins," in *Liquid Crystalline Polymer Systems: Technological Advances*, A. I. Isayev, T. Kyu, and S. Z. D. Cheng, Eds., American Chemical Society, Washington, DC, USA, 1996.

[35] J. M. Hutzler and T. S. Tracy, "Atypical kinetic profiles in drug metabolism reactions," *Drug Metabolism and Disposition*, vol. 30, no. 4, pp. 355–362, 2002.

[36] B. Wu, R. Ako, and M. Hu, "A useful Microsoft Excel add-in program for modeling steady-state enzyme kinetics," *Pharmaceutica Analytica Acta*, vol. 1, no. S1, 2011.

[37] K. E. Kenworthy, S. E. Clarke, J. Andrews, and J. B. Houston, "Multisite kinetic models for CYP3A4: simultaneous activation and inhibition of diazepam and testosterone metabolism," *Drug Metabolism and Disposition*, vol. 29, no. 12, pp. 1644–1651, 2001.

[38] S. Havriliak and S. Negami, "A complex plane representation of dielectric and mechanical relaxation processes in some polymers," *Polymer*, vol. 8, pp. 161–210, 1967.

[39] R. Böhmer, K. L. Ngai, C. A. Angell, and D. J. Plazek, "Nonexponential relaxations in strong and fragile glass formers," *The Journal of Chemical Physics*, vol. 99, no. 5, pp. 4201–4209, 1993.

[40] R. Böhmer and C. A. Angell, "Correlations of the nonexponentiality and state dependence of mechanical relaxations with bond connectivity in Ge-As-Se supercooled liquids," *Physical Review B*, vol. 45, no. 17, pp. 10091–10094, 1992.

# Preparation of Polymer Membranes by In Situ Interfacial Polymerization

**Michał Adamczak, Gabriela Kamińska, and Jolanta Bohdziewicz**

*Institute of Water and Wastewater Engineering, Silesian University of Technology, Konarskiego St. 18, Gliwice 44-100, Poland*

Correspondence should be addressed to Michał Adamczak; michal.adamczak@polsl.pl

Academic Editor: Miriam H. Rafailovich

Membranes currently have a wide application in sewage treatment and water purification processes, in seawater desalination, and in various technological processes where high product purity is required. Deposition of an ultrathin skin layer of TFC (thin-film composite) and TFN (thin-film nanocomposite) onto the surface of membranes is discussed in this article. Their presence improves membrane properties such as retention of impurities and permeability. The aim of this paper is to present the current state of knowledge about the methods of preparing composite and nanocomposite membranes. The properties of the prepared TFC membranes can be modified by changing the type and concentration of the reacting monomers, and the physical conditions in which the membrane preparation process itself is carried out are also significant. Additionally, the properties of TFN membranes can be further modified with nanocomposites. The membranes are characterized by different properties not only because they have nanoparticles in their structure but also because their concentration and the way they are blended into the membrane structure were changed. This paper provides information on modifications of TFN membranes with nanoparticles, as well as modification by changes in polymerization reaction conditions and monomer concentration. Examples of the use of TFN and TFC membranes are also presented.

## 1. Introduction

Membranes are produced on a commercial scale for a wide range of applications including sewage treatment and water purification, seawater desalination, and various technological processes where high product purity is required. Most membranes are produced by the simple method of phase inversion. In this process, solvent and polymer solution are poured onto a flat surface to form a film of appropriate thickness, and then, a controlled process of exchanging solvent to a nonsolvent takes place and a solid, uniform membrane with a skin layer is formed. Membranes for microfiltration (MF) and ultrafiltration (UF) are most commonly produced with this phase inversion method. Both MF and UF processes are considered low-pressure processes since they operate at relatively low pressures that range from 0.1 to 1.0 MPa. Macromolecular impurities, such as emulsions, colloids, bacteria, proteins, and viruses, are removed through MF and UF. Due to various modifications, UF membranes can also be used to remove organic micropollutants, which are usually removed by the use of high-pressure processes like nanofiltration (NF) and reverse osmosis (RO) [1–3]. These processes and also low-pressure membrane processes are shown in Figure 1.

Apart from phase inversion, another equally important method of membrane formation is in situ interfacial polymerization (IPIS). This method is used to form membranes on an industrial scale, mainly for NF and RO processes. NF and RO are high-pressure processes involving applied pressure ranges from 0.5 to 5.0 MPa. Substances including dyes, polyvalent ions, monovalent ions, and simple sugars are removed in NF and RO [1]. The first known application of the interfacial polymerization process was published in 1959 by Wittbecker and Morgan [4]. The process was called "interfacial polycondensation," and it was conducted between diamine and acidic monomer-containing chlorine atoms by the Schotten-Bauman reaction and was based on the work from 1946 [4]. For preparation of the first

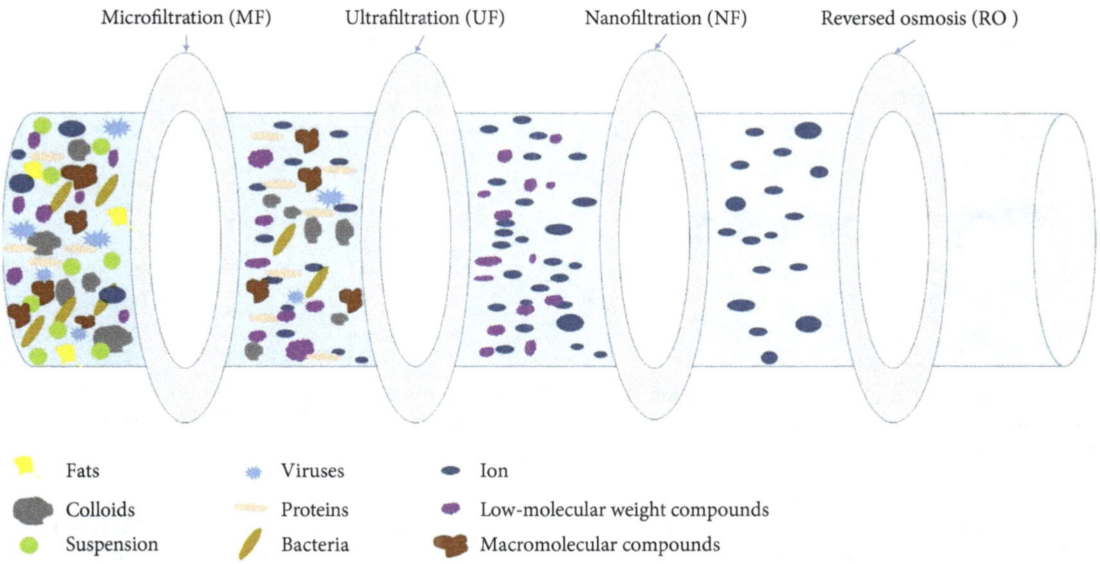


FIGURE 1: Separation in pressure membrane techniques depending on the size of the components [1–3].

thin-film composite membranes, this reaction was used by John Cadotte in 1985 for RO membrane preparation. This membrane showed 99% rejection of NaCl [5]. The next step for TFC membrane development was the use of nanocomposites for modification of these membranes. The first use of this technique was noted in the years 2005-2007 by Jeong et al. [6] and it used zeolite nanoparticles dispersed within a polyamide film.

Nowadays, a relatively new type of membrane process with the possibility of implementing TFC and TFN membranes is forward osmosis (FO) which occurs under the influence of the osmotic pressure of water. Over the last decade, this method has earned more attention. The first article applying this method was published in 2012. In FO, the osmotic pressure is the driving force that is created by the difference in concentrations between the solutions on both sides of the membrane. The process makes use of the so-called extraction solution, which consists of special dissolved substances. Water passes very slowly through the semipermeable membrane from the feed of lower pressure to the draw solution of higher pressure, which is gradually diluted [7–10].

Membranes formed by the IPIS method are referred to as TFC (thin-film composite) membranes, or in the case of membranes modified with nanoparticles, they are referred to as TFN or TFNC (thin-film nanocomposite) membranes. Additionally, most of the currently applied desalination plants in the world use nanofiltration or osmotic TFC membranes as the most significant part of the entire desalination process. They are also used for separation processes in food, pharmaceutical, and chemical industries [11, 12].

## 2. Thin-Film Composite Membranes

TFC membranes are essentially composed of two separate layers: a support layer prepared exactly like a classic membrane by means of phase inversion and an ultrathin skin layer. Each layer has a different structure and is composed of different polymers. The porous support provides the membrane with mechanical stability against cracking and tearing when operating at high pressures, whereas the ultrathin surface layer is directly involved in the rejection process. Compared to membranes produced by simple-phase inversion, TFC membranes allow high retention of substances in the feed while maintaining the highest possible flow rates [11]. TFC membranes are important in commercial applications of nanofiltration and reverse osmosis. TFC membranes also offer stabile at high pH (i.e., pH 5-13) and temperatures up to 70°C. Their widespread use is due to a higher retention of the feed components with a simultaneously increased permeate flux. Interestingly, there is a possibility of optimizing both the support and the skin layer independently of each other, depending on the desired effects. These properties make TFC membranes suitable for use in the removal of heavy metals, hardness, and micropollutants such as pesticides, endocrine disruptors, and pharmaceuticals from sewage water [9, 13].

*2.1. Thin-Film Composite Membrane Manufacturing.* An ultrathin layer of polymer is generally obtained by the method of forming a polyamide (PA) by polycondensation of two monomers, diamine and polyacyl chloride, on the surface of a porous support. These monomers should be dissolved in two immiscible liquids. One of them, with dissolved polyamine, should pretreat the support, and the reaction itself takes place at the interface [11, 14]. This process is schematically illustrated in Figure 2.

The support for composite membranes can be made of a variety of polymers, i.e., polyamides, polyethersulfone, polyurethanes, and even polyesters [15]. Poly(vinylidene fluoride) (PVDF), poly(vinyl alcohol) (PVA), polyacrylonitrile, or polytetrafluoroethylene (PTFE) can also be used [16]. In addition, the solvent that is used also affects membrane properties. The most commonly applied solvents are N,N-dimethylformamide (DMF), dimethyl sulfoxide (DMSO), and

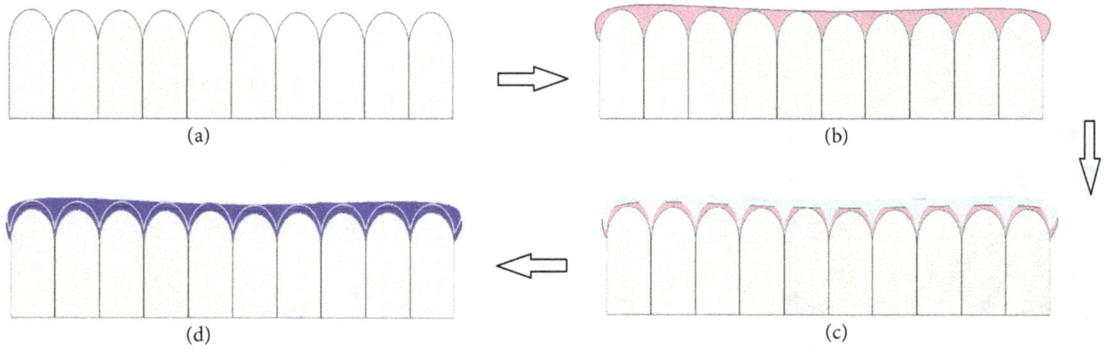


FIGURE 2: The process of in situ interfacial polymerization [11, 14]: (a) creation of a support layer by phase inversion; (b) distribution of diamine aqueous solution on the membrane surface; (c) distribution and mixing of polyacyl chloride dissolved in an organic solvent on the membrane surface; (d) polymerization reaction, formation of the polyamide skin layer.

N-methylpyrrolidone (NMP) [17]. However, the composition of the skin layer, i.e., the monomers that can be used during the in situ interfacial polymerization, is constantly being modified. The previously mentioned diamines are most commonly used—they are aliphatic or aromatic compounds and are reactive enough to form a layer of polyamide. Piperazine (PIP), m-phenylenediamine (MPD), and p-phenylenediamine (PPD) are used, among others, and they may react with acidic monomers containing chlorine atoms: polyacyl chloride (TMC), isophthaloyl chloride (IPC), and 5-isocyanatoisophthaloyl chloride (ICIC). Structural formulas of these substances are shown in Figure 3. The most commonly used reaction for commercial purposes is MPD/TMC, also shown in Figure 3, as it yields good results in seawater desalination [18].

### 2.2. Factors and Preparation Conditions Influencing TFC Membrane Properties

*2.2.1. Effect of Support.* As it was mentioned in the previous chapter, the first step of TFC membrane fabrication is the preparation of a support by mixing a polymer and organic solvent and casting the prepared solution. The main polymer type with the widest application is polysulfone (PSf) due to its high thermal and chemical stability, glass transition temperature, and its good ability for forming membranes. The only limit of this material is its hydrophilicity [20]. Higher glass transition temperature, important for further curing of membrane, has also uncommon polymers like poly(phtalazione ether sulfone ketone) and poly(phthalazinone ether amide) [21]. Most works include the use of polysulfone. PSf support was used in the study of Ding et al. [22] with concentrations ranging from 11 to 19 wt.%. Values of water fluxes for support layers dropped significantly from 1750 to 0 ($L/m^2*h$) with a decrease in PSf concentration. A less significant tendency was found for complete TFC membranes, i.e., from 40 to 10 ($L/m^2*h$). This can be explained by the low porosity of the support structure with higher concentrations of polymer. Contrariwise, salt rejection increased when polymer concentration increased from 11 to 16 wt.% and a further increase in polymer concentration did not affect salt rejection. Zhang et al. [23] proved that the size and pore density became more visible and distinct when polymer concentration in the support decreased from 20 wt.% to 12 wt.%. Moreover, supports with low porosity had less visible surface structures and lower thickness and had weak bonding to the active layer. Contrarily, membranes with higher porosity had more desirable properties.

Properties of thin-film membranes can be changed not only by use of different polymers but also by different organic solvents. Lu et al. [24] used dimethylformamide (DMF) and N-methyl-2-pyrrolidnone (NMP) to prepare casting solutions. The DMF-prepared membrane had a larger pore size and higher surface area. This membrane also had a greater amount of area available for water permeation at the polysulfone-polyamide interface [24]. Importantly, the fouling tendency was higher for the DMF-prepared membranes due to the greater hydrophilicity of NMP-prepared membranes.

A dependence between the pore size of the support and the retention of salts by TFC membranes was very well presented by Huang and McCutcheon [25] for commercially used supports and by Sharabati [26] for laboratory-prepared membranes. Both reported that an increase in support pore size was highly connected with a decrease in salt rejection and an increase in permeation. The porosity of the membrane support can be also changed by addition of a porogenic substance like polyvinylpyrrolidone (PVP). This is a water-soluble polymer and has excellent wetting properties in solutions This additive caused an increase in the hydrophilicity of membranes, rejection of NaCl, and an overall increase in porosity by a having a larger number of pores with a smaller size. This substance is widely recommended for modification of support. PVP is nontoxic and it can be used in processes for drinking water [22, 27].

*2.2.2. Effect of the Monomer Type.* For TFC membranes, an important factor is the selection of proper monomers. In the research paper written by Li et al. [28], PES was used as a support. The membrane obtained was submerged in an aqueous solution of one of the amines—diethylenetriamine (DETA), triethylenetetramine (TETA), tetraethylenepentamine (TEPA), or PIP—at a concentration of 0.2% by weight and left for 10 minutes. Excess solution was removed with

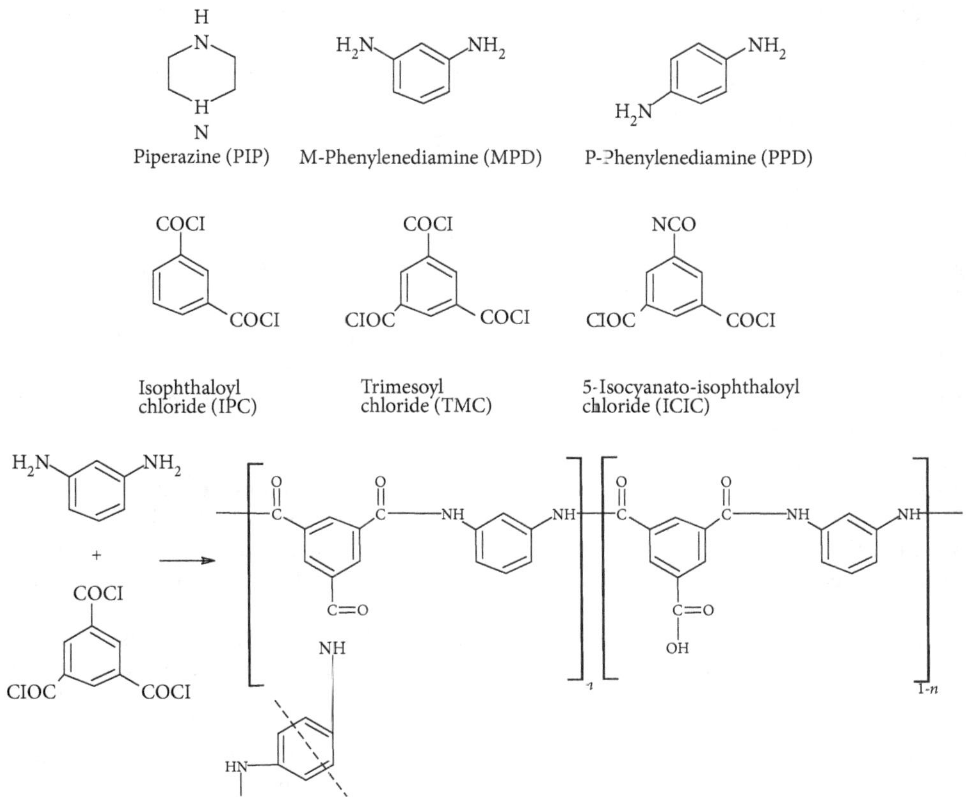


FIGURE 3: The examples of substances commonly used in the in situ interfacial polymerization process. PIP, MPD, and PPD are dissolved usually in deionised water and IPC, TMC, and ICIC in organic solvent like hexane. Beneth presented typical reaction between MPD and TMC [18, 19].

tissue paper. Then, a 0.2%-by-weight TMC solution in hexane was slowly poured onto the membrane for two minutes to fully cover its surface and it was left in place for 30 minutes to allow the hexane to evaporate and the skin layer to reach the desired stability. The membrane was then stored in deionized water until it was used. The membrane thus obtained had osmotic properties. Depending on the polymer used, different membrane properties were obtained. First of all, the contact angle values were significantly different. While the DETA/TMC membrane had almost the same contact angle as the unmodified pristine PES membrane (66°), the use of TEPA or PIP produced membranes that were hydrophilic. This positively affected permeate flux, which was directly dependent on the hydrophilicity of the membrane. The second parameter that was subjected to change was the value of the zeta potential. The name zeta potential represents the electrostatic charge of a membrane, which is associated with the presence of functional groups on its surface and is dependent on the pH value of the feed. Its value refers to the membrane's ability to repulse or attract positively or negatively charged substances present in the feed solution. In a study already mentioned [28], a strong negative charge was caused by the deprotonation of carboxyl groups from the hydrolysis of acyl chloride from TMC; however, it may be compensated by the protonation of amine groups from the same reaction. Although both hydrophilicity and the zeta potential of the membrane played a significant role in the separation of contaminants, each membrane has achieved the level of selected salt removal at over 80% (one ionic and two ionic). Permeate flux recovery ranged from 87% to 100%, depending on the selected monomer. The TFC membrane with PIP had the best properties both in terms of permeability and rejection [28, 29]. The highest flux of this membrane was related to its contact angle. Generally, membranes with high contact angle have hydrophobic properties that limit water permeation, and thus, a lower water flux is observed. But along with reduction of the contact angle, the hydrophilicity of membranes increases and their permeability is raised. Details regarding the membranes discussed above are given in Table 1.

Similar tests were also reported by Saha and Joshi [30] but with other types of monomers. In addition to piperazine and MPD, n-2-diaminethylpiperazine (AEPIP) and 3,5-diaminobenzoic acid (BA) were used. These monomers react with TMC and isophthaloyl chloride IPC. This undertaking did not yield very good results. However, interesting correlations were observed based on differences in the concentration of monomers. Namely, it was observed that an increase in PIP concentration from 0.5 to 5% by weight in relation to TMC caused an increase in NaCl retention from less than 10% to about 70% and $MgSO_4$ from about 25% to almost 100% retention with a slight decrease in the value of the permeate flux. A similar relationship was found for MPD-TMC membranes, where both NaCl retention and

Table 1: Parameters of the membranes formed, depending on the monomer used [17].

| Membrane name | Contact angle (°) | Zeta potential (mV) | Permeability (dm $^3$/m$^2$h*MPa) |
|---|---|---|---|
| DETA/TMC | 64.5 ± 0.5 | −13.7 ± 0.4 | 33.5 ± 2 |
| TETA/TMC | 47 | −18.9 ± 0.6 | 43.5 ± 3 |
| TEPA/TMC | 35.4 ± 2.2 | −17.7 ± 0.5 | 51 ± 4.5 |
| PIP/TMC | 39.6 ± 1.6 | −24.4 ± 0.8 | 75 ± 6 |

Table 2: Parameters of the membranes formed, depending on the organic solvent [5].

| Membrane symbol | Heptane temperature (°C) | Contact angle (°) | Flux (LMH) | NaCl retention (%) |
|---|---|---|---|---|
| TFC-1 | -20 | 53.3 ± 1.2 | 92.1 | 94.8 |
| TFC-2 | 1 | 56.9 ± 1.1 | 51 | 97 |
| TFC-3 | 25 | 81.2 ± 1.6 | 10.7 | 98.8 |
| TFC-4 | 50 | 66.2 ± 1 | 27.9 | 99.1 |

the flux increased with an increase in the MPD monomer concentration from 0.5% to 2% by weight. Sodium chloride retention was reported to increase from 75% to more than 90%. On the basis of the presented experiments, it can be concluded that TMC is characterized by more favourable properties than IPC, as it increased the hydrophilic properties of the skin layer. On the other hand, the use of AEPIP resulted in a significant decrease in the permeate flux [30].

*2.2.3. Effect of Reaction Parameters.* It has already been mentioned that the properties of composite membranes are also influenced by the conditions in which the processing is conducted. In a research undertaken by Khorshidi et al. [11], TFC membranes were prepared using PES as a support and MPD and TMC reactions were used to make the skin layer. Four membranes were made at four different temperatures of organic solvent: −20°C, 1°C, 25°C, and 50°C. At the temperature of 25°C, a membrane surface topography with classic "hills and valleys" was obtained. At elevated organic solvent temperatures, the skin layer of the membrane reached a greater thickness, as opposed to the membranes formed at −20 and 1°C, which had the smoothest surface and the thinnest layer on the PES membrane support. A higher temperature of skin formation caused the MPD/TMC polymerization reaction to be faster, whereas at lower temperatures, an opposite phenomenon was observed. Changes in the value of the contact angle were mainly related to different morphologies of these membranes. The membranes prepared at room temperature were characterized by the highest hydrophobicity, whereas the membranes formed at lower and higher temperatures were much more hydrophilic. The permeate flux was also temperature dependent. For TFC-1 and TFC-2, a lower temperature of acid chloride solution meant lower thickness of the membrane and thus higher permeability. At a temperature higher than room temperature, the permeate flux increased due to a larger and more extended surface which made the surface of contact between water molecules and the membrane surface larger [11]. Detailed parameters of these membranes have been demonstrated in Table 2.

*2.2.4. Permeability and Selectivity.* However, the most important preparation factors for permeability and selectivity of TFC membranes, according to Mah et al. [31], are the concentrations of both monomers, reaction time, presence, and time of curing membrane. Interestingly, the pH of the aqueous solution during the preparation process had no influence on increasing the permeability of the membrane. However, a slight increase in membrane permeation was observed when the first monomer concentration was increased and the reaction time in the interphase was reduced. From other works [32–34], an increase in the concentration of MPD and TMC in solutions caused an increase in selectivity with a decrease in permeability. High MPD concentrations can produce a dense and thick PA layer by promoting the IP reaction. The effect of this is reduction of water permeation and promotion of high salt rejection. The influence of higher TMC concentration is also connected with a higher amount of TMC that participates in the polymerization reaction [32]. Opposite results for TMC, achieved by Khorshidi et al. [34], were explained by a complicated influence of the polyamide thin-film thickness and density on the properties of membranes.

The dependence between TMC concentration and permeability and selectivity, appropriate for low concentrations of monomers, is shown in the flow chart (Figure 4). Generally, lower flux and higher selectivity of TFC membranes are directly connected with reduction of pore size and overall membrane thickness. This is a result of more cross-linking occurring between monomers during formation of the thin film. The hydrophilicity of a membrane depends on the functional groups attached to its surface. The presence of amino groups changes the hydrophilicity of membranes depending on pH [31]. The most significant influence on membrane permeability was curing temperature and time. Tests conducted by Mah et al. [35] showed that longer exposure to curing caused the top layer to become denser which decreased the flux of pure water and promoted the rejection of salt. The minimum time of hexane evaporation from the TFC membrane (during curing) was 5 minutes. Fifteen minutes of curing was needed to enhance the selectivity of a membrane [31]. Khorshidi et al. [34] found that increasing the curing temperature improved flux and rejection but temperatures higher than 75°C lowered the flux. A decrease in curing temperature resulted in thinner membranes and higher permeation. This factor contributed 40.7% to the impact on the water permeation. MPD and TMC concentration exerted a lower effect, around 28% of contribution. Changes in reaction time between MPD and TMC had the least influence on water permeation—less than 2% [34]. However, all mentioned factors influenced the porosity of membranes, including the properties of support used. A decrease in concentration of the first monomer caused an increase in pore size. A similar effect was caused by a reduction in the time of soaking and immersion in both

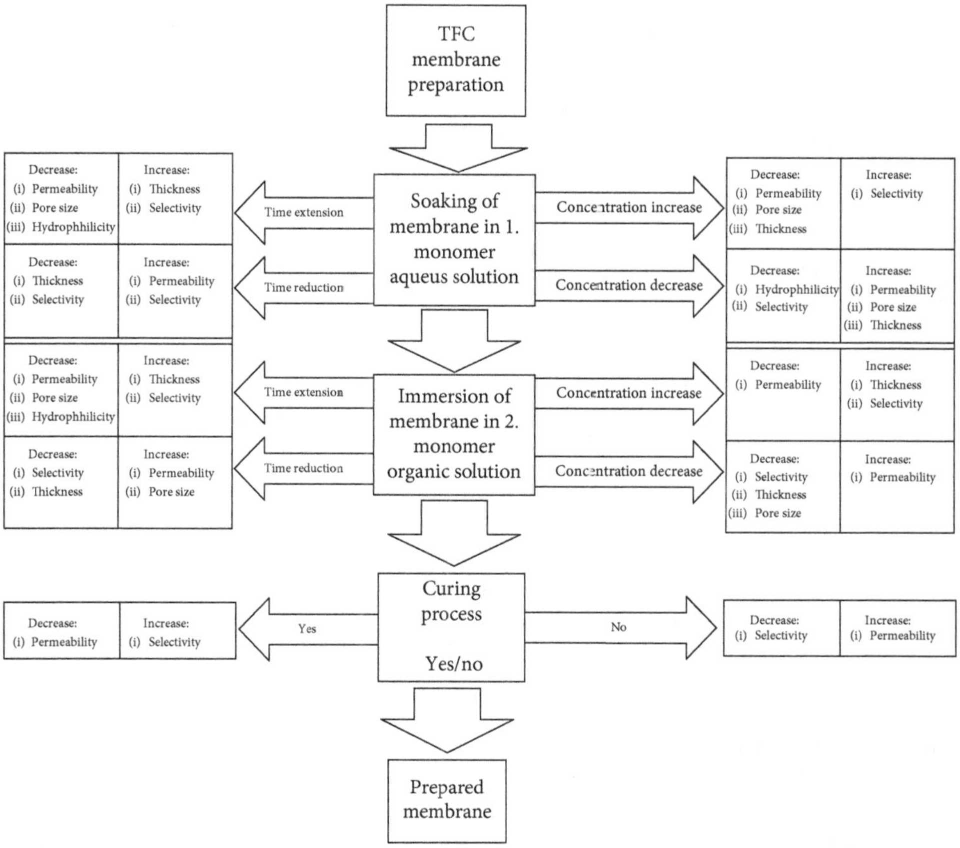


FIGURE 4: Flow chart of preparation of TFC membranes with general influence of the most important factors: time of reaction/process, concentration of monomers, and curing temperature. First and second monomer concentrations were in ranges 1-4 wt.% and 0.1-0.2 wt.%, respectively [31–34].

monomers. A decrease in TMC concentration may cause the pore size to decrease. The porosity of TFC membranes mostly depends on the porosity and pore size of the supports. Thin-film membrane properties are also affected by the addition of nanocomposites.

## 3. Thin-Film Nanocomposite Membranes

For TFN membranes, the "nano" prefix is the key aspect. It indicates the presence of a filler in the form of nanoparticles, nanotubes, or nanosheets. This addition is used to enhance the properties of the membranes. The specific features of the TFN membrane area significantly increase permeate flux combined with better separation properties, and an energy cost reduction is associated with the operation of these membranes. These are new types of membranes formed by means of the process of in situ interfacial polymerization. The first membrane of this kind appeared in 2007, when the first publications on the subject were published [10]. In 2010, the first desalination plant to use such membranes appeared. As in the case of TFC membranes, most applications are related to nanofiltration or reverse osmosis but they are increasingly being used in low-pressure processes such as UF and MF [10].

The preparation of TFN membranes is almost identical to that of TFC membranes. The only change in the membrane preparation process is the addition of a nanomodifier. The TFN membrane is made up of a thin skin layer of up to 500 nm thick which is placed on a much thicker support made with a method other than the skin layer, usually by means of phase inversion. Nanoparticles can be placed within the membrane structure in two ways: by adding them to the support layer to change the membrane characteristics or by adding them to the selective skin layer to improve roughness, selectivity and, above all, permeability [13]. Additionally, nanocomposites can also be added to the thin selective layer also in two ways. Namely, the nanoparticles can be placed on the TFN membrane either together with an aqueous solution or with an organic solvent solution. It is much more common to place the nanocomposite together with the TMC solution in hexane. This is mainly due to better results achieved from these types of membranes [36]. Figure 5 demonstrates the basic types of nanocomposite membranes.

*3.1. Modification of TFN Membranes.* Nanomaterials in TFN membranes belong to one of four basic groups. They can be classified as dimensionless hydrophilic metals or metal oxides, one-dimensional carbon materials such as nanotubes, two-dimensional materials such as graphene or graphene oxides, or also three-dimensional porous materials such as nanozeolites [37]. The types of the most common

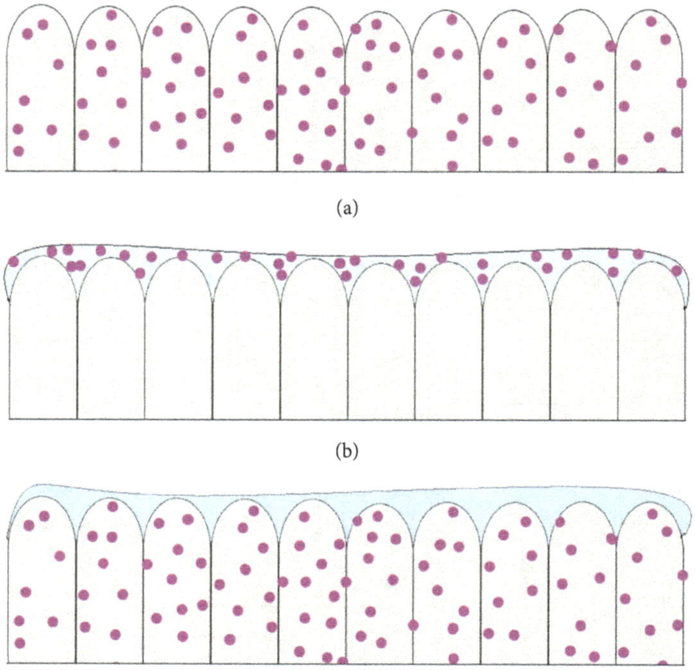

(a)

(b)

(c)

FIGURE 5: Types of membranes with nanoparticles added [13]: (a) typical nanocomposite membrane; (b) TFN membrane with nanoparticles in the skin layer; (c) TFN membrane with nanoparticles in the support.

nanoadditives are listed in Table 3. As mentioned before, these ingredients can be added together with an aqueous amine solution or in an organic solvent, together with TMC. Besides, they can also be placed in the support of the membrane [13].

The variety of nanoparticles provides numerous sizes and shapes. They occur in the form of crystals (cellulose), tubes (carbon nanotubes, halloysite nanotubes), ribbons, platelets, dots, and sheets (graphene). Their size is in the range of a few nanometers to dozens of micrometers Another important factor is the spontaneous and preferential aggregation of many of them which causes their useful size to be multiplied. As was already mentioned, filler-polymer interface greatly affects the overall TFN properties. Therefore, both the size and shape the nanoadditives play an important role in their fabrication. For example, Alberto et al. [40] found that the size of GO flakes affects the porosity of the membrane structure. Micrometer-sized flakes of GO generated large and nonselective voids between polymer and graphene, while smaller flakes were aligned with polymer blocks [40]. Another aspect is the shape of nanofillers. Halloysite nanotubes tended to egress from polymer matrices, leaving a hollow space inside the composite [41]. Chong et al. compared the topography and thickness of two TFN membranes modified with (1) titanium oxide and (2) titanate nanotube. Titanate nanotubes, due to their long length, preferred to lay horizontally on the surface, simultaneously creating a slightly lower membrane thickness than TFN with titanium oxide [42]. On the contrary, Karkooti et al. [43] demonstrated no effect of different shapes (sheet, platelet, or ribbon) of graphene oxide on TFN cross-section morphology. As they observed, all membranes had an asymmetric structure with a dense top layer and a porous finger-like support [43].

The most common types of carbon nanomaterials used in the preparation of nanocomposite membranes include carbon nanotubes (CNTs), graphene, graphene oxide (GO), MXenes (two-dimensional inorganic compounds made up of carbides but also nitrides or metal cyanides), and carbon materials obtained from carbides and fullerenes. Due to their unique physical properties, such as high thermal, chemical, and mechanical durability, high conductivity, good optical properties and low density, these compounds are widely used in material engineering and chemistry. However, the most popular nanoadditives are carbon nanotubes (CNTs), which we have mentioned before. Their wide application is due to their exceptional mechanical, electrical, and thermal properties. Their main advantage is wide chemical modifiability, as well as their large area in relation to the dimensions of the nanotube itself [17, 44]. A membrane with the addition of nanotubes can be as durable as a ceramic membrane, while retaining the flexibility of the polymer. Their special feature is the ability to modify the physicochemical properties of the membrane itself. Modification of membranes with hydrophilic particles increases the affinity of their surface to water, increasing permeability. The inner nanotube pores act as selective nanopores, which makes the modified membrane much more permeable than the unmodified membrane. However, the main disadvantage of nanotube use is their tendency to agglomerate. Besides, some carbon nanomaterials, e.g., CNTs, GO, and graphene, also have specific properties, and when incorporated into the membrane, they inhibit the growth of bacteria that is in direct contact with

TABLE 3: Division of nanoparticles used in the preparation of nanocomposite membranes [18, 20, 38, 39].

| Particle shapes | Particle types | Description and examples |
|---|---|---|
| Dimensionless | Metals | Metals and metal oxides: silver, titanium, zirconium, iron, gold, and others |
| One dimensional | Carbon nanotubes (CNTs) | Single walled (SWCNTs), multiwalled (MWCNTs), and modified with carboxyl (-COOH), amine, amide, and hydroxyl (-OH) residues |
| | Carbon nanofibers (CNFs) | Available in bundles and modifiable as CNTs |
| | Cellulose nanofibers (CNF) | Contains a lot of hydroxyl groups on the surface; possibility of modification similar to CNTs |
| | Cellulose nanocrystals (CNC) | Similar properties as CNF |
| Two dimensional | Graphene | In the form of hexagonal nanosheets |
| | Graphene oxide (GO) | Like graphene, with oxygen and hydrogen atoms attached to carbon atoms |
| | MXenes | A substance consisting of several layers of metal carbides or carbon nitrides; they have different conductivity and hydrophilicity depending on their composition |
| Two dimensional and three dimensional | Nanozeolites | A group of hydrated clay minerals whose composition includes silicon, aluminium, magnesium hydroxides, and metal ions; this group also includes synthetic materials; natural: montmorillonite and halloysite; synthetic: MCM-41 |
| Three dimensional | Fullerenes | Spherical allotropic form of carbon consisting of graphene platelets; fullerenes were components of the first nanocomposite membranes |
| | Metal organic frameworks (MOFs) | Nanoporous materials composed of inorganic subunits of metal ions and polytropic organic ligands |

the surface of the membrane. This may have a positive effect on the reduction of fouling. Moreover, graphene-modified membranes have the same durability as those with CNT addition but they are much more flexible, whereas the addition of carbon nanotubes makes them bactericidal—according to research, 61% of bacteria were killed in one hour [45, 46].

The presence of nanoparticles in the membrane structure results in the membrane having better properties, most notably making it much more permeable. For example, in tests conducted by Mahmood et al. [47], brackish water from a river mouth was filtered with TFC and TFN membranes with the addition of multiwalled carbon nanotubes at a concentration of 0.4% by weight. These were placed in the support of the membrane, and the MCM-41 substance (made of synthetic aluminosilicates of a special structure) was placed in the skin layer at the concentration of 0.05% by weight. The result of these studies [47] was to obtain an almost identical level of retention of salts for both membranes (about 93.7%) with a much higher permeate flux for the TFN membrane by about 54.85% in comparison to the TFC membrane. This increase in permeability was obtained thanks to the formation of macrochannels in the support layer and the strong antifouling properties of the nanocomposite membranes [47].

Membrane porosity depends mostly on the support structure and the kind of modifier on its surface. Adamczak et al. [48] prepared UF membranes with PES support and modified them with single-walled carbon nanotubes. The presence of nanotubes caused the decrease in porosity. In these tests, porosity of the pristine support was around 70-75%, while membranes containing nanotubes had porosity at around 30-50%, dependent on nanotube location in TFN membranes: in the support or the in thin film. In the Emadzadeh et al. [49] study, the overall porosity of membranes increased as the $TiO_2$ loading in the support increased; however, the average pore size decreased. Importantly, the water flux increased with increasing $TiO_2$ concentration. The membrane prepared from the PSf substrate bound with 0.5 wt.% $TiO_2$ has shown the best performance for the water desalination process. However, according to Dalvi et al. [50], in tests with polyhedral oligomeric silsesquioxane $TiO_2$-$SiO_2$, nanoparticles, regardless of size and shape, can cause formation of defects in the polyamide layer. The defects appeared more often with larger size and higher concentrations of nanoparticles. In this test, they lowered rejection of salts.

A membrane was formed with the use of graphene for the purpose of the Yin et al. study [51]. It was made in the same way as reported by Mahmood et al. [47] where graphene was added to the TMC solution. The presence of graphene allowed the membrane to maintain a similar retention value of $Na_2SO_4$ and NaCl salts at the level of 93-98% with a significant increase of the permeate flux from about 39 LMH to over 59 LMH at the concentration of 0.015% of GO. Similarly, with an increase in the concentration of nanosheets, there were slight fluctuations in the increased value of the membrane's contact angle, between 62° and 68°, indicating an increase in hydrophilicity and a decrease in the average roughness of the membrane from approximately 60 nm to approximately 40 nm. The most important aspect of this undertaking was to maintain retention with increased permeability through the introduction of nanoparticles [51].

Such tests, related to the method of introducing nanocomposites, were conducted in the work of Huang et al. [36]. Three membranes based on a PS membrane were formed with a skin layer based on MPD and TMC reactions with zeolite nanoparticles: one with NaA added in the aqueous phase, a second with hexane solution added during the phase, and a third with no nanoparticles added. In the test, a NaCl salt solution was filtered through membranes at a pressure of 1.6 MPa. It was demonstrated that the permeate flux increased with an increase of nanoparticle concentration (for concentrations from 0% to 0.2% by weight), while the degree of retention slightly decreased. This was explained by the presence of defects in the membrane structure, which allowed cations and anions to permeate and pass through the membrane. Salt retention on the NaA membrane in organic solvent solution was higher than that on the NaA membrane in water and was therefore considered to be more beneficial for particle dispersion on TFN membranes [36].

Less commonly, aluminosilicates, which are also known as halloysites, are used as nanocomposites. They are used in the shape of nanotubes (HNTs) and are extracted and formed from natural deposits of this substance. This substance has unique properties. The outer layer is negatively charged due to the presence of siloxane groups (Si-O-Si), while the inside of the nanotubes is filled with positively charged hydroxyl groups (Al-OH). Moreover, unlike carbon nanotubes, they do not have a tendency to agglomerate. Nowadays, they are increasingly used in research works. Also, their production cost is lower than CNTs [52]. The most recent aspect related to the addition of nanoparticles is the use of a new type of compound, the so-called metal organic frameworks (MOFs). These are new-class nanoporous materials that are composed of inorganic subunits of metal ions and polytropic organic ligands. The presence of the organic fragments makes the substance more compatible with the polymer that the membrane is composed of. Furthermore, these nanomaterials can be freely modified on account of the presence of free bonds that can be supplemented with functional groups. The presence of the alkyl chains reduces nanoparticle aggregation. A study conducted by Guo et al. [53], where nanoparticles were added to the TMC solution in hexane, confirmed the desired effect of this type of nanomaterial. The material maintained almost 100% tetracycline retention, and nearly 30% higher membrane permeability was achieved compared to the unmodified membrane [53].

A quite new aspect of membrane modification by nanocomposite is the application of nanofibers or nanocrystals of cellulose (CNFs or CNC). These fillers are reminiscent of carbon nanotubes and nanofibers, but cellulose materials are hydrophilic, naturally occurring in the environment, and they are a renewable resource. Similar to CNTs, they can be functionalized by adding functional groups, such as carboxyl groups (carboxycellulose nanofibers), on their surface. Nanofibers or nanocrystals may be extracted from cellulose, but this requires large energy consumption, and their processing uses a large amount of acid [38, 54]. Extraction of these nanofillers may be conducted from jute fibers, bamboo cellulose, or grass [8]. The use of these products from cellulose in membrane preparation is gaining more and more attention each year. For example, Asempour et al. [55] used CNC in a TFN membrane for RO application. Embedded nanocrystals improved antifouling properties, lowered the contact angle to 20°, and doubled water flux with only a slight decrease in salt rejection. Bai et al. [56] used CNC on the PES substrate and improved the removal of cationic and anionic dyes from around 96-97% to 99.83-99.9%. In the same work, the addition of CNC on the membrane surface caused a slight decrease in NaCl and $MgSO_4$ retention and a significant increase in water permeability. They also found that the contact angles of the pristine substrate and unmodified TFC were 66.5° and 59.2°, respectively, and CNC addition resulted in a decrease of the contact angle to 37.5°. For TFN membranes with cellulose nanofibers, Carpenter et al. [38] provide data about an improvement of the mechanical strength of membranes filled with CNF. Similar to CNC, membranes with CNF are characterized by an increase in water permeability and hydrophilicity with constant contaminant removal.

## 4. Advantages and Disadvantages of TFC/TFN Membranes

The main advantages of thin-film membranes produced in the form of TFC are related to their good separation performance and wide pH tolerance range. However, composite membranes with polyamide films are not free from limits. Typical disadvantages of unmodified polyamide films are the relatively low water flux, poor antifouling property, and low resistance to exposure to oxidizing agents such as chlorine [57, 58]. There are several mechanisms that have an effect on polyamide films that are described by Verbeke et al. [59], but the effects are inconsistent. This may also be a result of a tightening or swelling effect. In some cases, the PA layer may dissolve or separate from the support. In some cases, reaction of chlorine with the PA top layer enhances the hydrophobicity of the membrane simultaneously with water flux reduction [59]. Another limiting aspect of TFC is the low stability of PA films in acidic pH. In this environment ($pH < 5$), the carbonyl groups of amide linkages in the PA chains are attacked [60]. The degradation of PA chains with partial hydrolyzation of amide groups by the $H^+$ or $OH^-$ causes a reduction in the degree of cross-linking. This increases membrane hydrophilicity and at the same results in larger permeability and lower selectivity [61]. In Rezania et al. [62], filtration of solution of $Na_2SO_4$ conducted at pH 3 caused a decrease in rejection from 91.1% in pH 7 to 81.3%, and at pH 5, rejection decreased to 88.2%. To restrain this effect and also degradation by chlorine, nanocomposites are placed on the surface of the membrane [59].

Thin-film nanocomposite membranes have many advantages in comparison to TFC membranes such as higher water flux and better bacterial resistance without a significant reduction in salt rejection. This is due to the small particles changing the membrane performance. The presence of nanocomposites may improve surface properties of membranes such as hydrophilicity/hydrophobicity, porosity, zeta potential, and antibacterial properties [63, 64]. For example, Bano et al. [65] compared the influence of TFC and TFN properties on water flux and salt retention. They proved that addition of graphene oxide improves hydrophilicity, zeta potential, and roughness while simultaneously maintaining the same level of salt retention.

The major obstacles for the commercialization of the TFN membranes are high production cost, leaching out of toxic nanoparticles into the permeate and retentate, and low interaction between nanoparticles and polyamide thin films [55]. Leached nanoparticles may be toxic for the user and the environment. Elimination of this threat may be done in two ways: (1) by preventing the leaching out of the nanoparticles or (2) by using biocompatible nanoparticles [55]. For example, silver nanoparticles, taken orally, are distributed to all the organs and attack primarily the liver and intestinal tract [66]. Leaching of silver nanoparticles (AgNPs) was firstly proved and then limited by Yin et al. [67]. In a batch test, they measured that within 14 days, 12% of the total amount of AgNPs was leached from the membrane. In a flow test, leaching of silver was reduced to 3.5% by using cysteamine as a bridging agent. In a work presented by Wan et al. [68], ferrum/platinum nanoparticles were integrated into a membrane by functionalization with poly(acrylic acid). Another way to bypass the danger of nanoparticle release is by the use of nontoxic materials. An example of such a substance is cellulose nanocrystal that is obtained from acid hydrolysis of native cellulose [55]. The only limitation is the application of TFN membranes in acidic conditions. Some nanoparticles like aluminium-rich zeolites may be less structurally stable in low-pH conditions (pH of less than 2) particularly in the presence of phosphate (pH of less then 5) [69]. However, by overcoming these obstacles, both nanocomposite-free and modified membranes are being used more and more frequently in manufacturing and technological processes.

## 5. Application of TFC/TFN Membranes

TFC/TFN membranes have a great potential for water treatment. Due to their high-permeability and selectivity, good durability, and antifouling properties, they dominate over conventional polymer membranes. These membranes are mainly used in high-pressure membrane processes (i.e., nanofiltration and reverse osmosis for desalination of seawater). However, some cases of their use for micropollutants and organic matter removal have been also reported. Karkooti et al. [43] proved that a TFN membrane modified with 0.1 wt.% graphene nanoribbons had excellent properties in terms of both water flux and organic matter removal [43]. In the Bojnourd [70] study, the concentration of cephalexin was reduced by 98.8% by using a TFN membrane with 0.4 wt.% of modified montmorillonite [70].

While TFC membranes are widely employed in desalination plants and in the food industry, the use of TFN membranes on an industrial scale is negligible. The high production cost and the potential threats connected with nanoparticle release are obstacles to wider use. On the other hand, many studies have shown that TFN membranes are more effective than TFC membranes, especially in terms of the removal of toxic pesticides, arsenic salts, proteins, hormones, and micropollutants from wastewater (e.g., in the pharmaceutical industry) [49, 69, 71]. Moreover, in comparison to other sewage treatment technologies, such as biological methods, TFN results in a better quality of treated sewage at lower operating costs. Additionally, TFN membranes are used in laboratories to obtain high purity products [71].

Currently, TFC membranes made of polyamide and cellulose acetate are most commonly used for water desalination. They take part in both RO and FO processes. TFN membranes are also commercially available. One example is the RO TFN membrane modified with zeolite nanoparticles (100 nm in diameter) that is intended for seawater desalination. This has been available since August 2010. Its efficiency is approximately 10-20% higher than that of the classic RO membrane with a similar reduction in energy consumption. This type of membrane has been successfully installed on the island of Curacao for the purpose of seawater desalination [72, 73]. A similar cost reduction related to energy expenditure was observed in the processes of reverse osmosis with a TFN membrane [74].

## 6. Conclusions

Despite having many unquestionable advantages for both TFN and TFC membranes, there are still many challenges that modern science must face. The main problem is a poor dispersion of nanocomposites in inorganic solvents and, consequently, their agglomeration in the membrane structure. Surface modification of nanoparticles partially solves this problem, but it cannot eliminate this phenomenon completely [10]. Moreover, a problem of secondary contamination from nanoparticles leaching from membranes should be solved.

Compared to TFC membranes, TFN membranes have a higher potential for further development both in terms of eliminating fouling and increasing permeability while maintaining a high level of retention. Yet, another difference between these two membrane types is the higher production price of TFN membranes. However, the use of TFC/TFN membranes increases productivity and thus results in lower operating costs [74].

However, it should be remembered that no universal membrane exists that would be able to work with any kind of feed material. Moreover, the membrane manufacturing process needs to be optimized, depending on the needs, both on a laboratory scale and an industrial scale.

## Conflicts of Interest

The authors declare that there is no conflict of interest regarding the publication of this paper.

## Acknowledgments

The paper was created as a result of a research project financed from the funds of the National Science Centre titled "Modelling the Transport and Retention of Organic Micropollutants by New Generation of Ultrafiltration Membranes," decision number DEC-2016/21/B/ST8/03128.

## References


[1] K. Nath, *Membrane Separation Processes (Second Edition)*, PHI learning Pvt. Ltd., Delhi, 2017.

[2] M. Dudziak and E. Burdzik-Niemiec, "Ultrafiltration in the treatment of 17 beta-estradiol and bisphenol A-containing wastewaters through modified membranes," *Przemysł Chemiczny*, vol. 96, pp. 448–452, 2017.

[3] G. Kamińska, J. Bohdziewicz, J. I. Calvo, P. Pradanos, and A. Hernandez, "Fabrication and characterization of polyethersulfone nanocomposite membranes for the removal of endocrine disrupting micropollutants from wastewater. Mechanisms and performance," *Journal of Membrane Science*, vol. 493, pp. 66–79, 2015.

[4] E. L. Wittbecker and P. W. Morgan, "Interfacial Polycondensation I," *Journal of Polymer Science*, vol. 40, no. 137, pp. 289–297, 1959.

[5] A. Sagle and B. Freeman, "Fundamentals of membranes for water treatment," in *The Future of Desalination in Texas*, pp. 137–154, Texas Water Developement Board, Austin, TX, USA, 2004.

[6] B. H. Jeong, E. M. Hoek, Y. Yan et al., "Interfacial polymerization of thin film nanocomposites: a new concept for reverse osmosis membranes," *Journal of Membrane Science*, vol. 294, no. 1-2, pp. 1–7, 2007.

[7] M. Xie, J. Lee, L. Nghiem, and M. Elimelech, "Role of pressure in organic fouling in forward osmosis and reverse osmosis," *Journal of Membrane Science*, vol. 493, pp. 748–754, 2015.

[8] D. L. Shaffer, J. R. Werber, H. Jaramillo, S. Lin, and M. Elimelech, "Forward osmosis: where are we now?," *Desalination*, vol. 356, pp. 271–284, 2015.

[9] X. Jin, J. Shan, C. Wang, J. Wei, and C. Y. Tang, "Rejection of pharmaceuticals by forward osmosis membranes," *Journal of Hazardous Materials*, vol. 227-228, pp. 55–61, 2012.

[10] W. J. Lau, S. Gray, T. Matsuura, D. Emadzadeh, J. P. Chen, and A. F. Ismail, "A review on polyamide thin film nanocomposite (TFN) membranes: history, applications, challenges and approaches," *Water Research*, vol. 80, pp. 306–324, 2015.

[11] B. Khorshidi, T. Thundat, B. A. Fleck, and M. Sadrzadeh, "A novel approach toward fabrication of high performance thin film composite polyamide membranes," *Scientific reports*, vol. 6, no. 1, article 22069, 2016.

[12] A. S. Al-Hobaib, J. El Ghoul, and L. El Mir, "Synthesis and characterization of polyamide thin-film nanocomposite membrane reached by aluminum doped ZnO nanoparticles," *Materials Science in Semiconductor Processing*, vol. 42, no. 1, pp. 111–114, 2016.

[13] J. Yin and B. Deng, "Polymer-matrix nanocomposite membranes for water treatment," *Journal of Membrane Science*, vol. 479, pp. 256–275, 2015.

[14] Y. Song, P. Sun, L. L. Henry, and B. Sun, "Mechanisms of structure and performance controlled thin film composite membrane formation via interfacial polymerization process," *Journal of membrane science*, vol. 251, no. 1-2, pp. 67–79, 2005.

[15] B. Tang, Z. Huo, and P. Wu, "Study on a novel polyester composite nanofiltration membrane by interfacial polymerization of triethanolamine (TEOA) and trimesoyl chloride (TMC): I. Preparation, characterization and nanofiltration properties test of membrane," *Journal of Membrane Science*, vol. 320, no. 1-2, pp. 198–205, 2008.

[16] G.-d. Kang and Y.-m. Cao, "Application and modification of poly(vinylidene fluoride) (PVDF) membranes – a review," *Journal of Membrane Science*, vol. 463, pp. 145–165, 2014.

[17] G. Arthanareeswaran and V. M. Starov, "Effect of solvents on performance of polyethersulfone ultrafiltration membranes: investigation of metal ion separations," *Desalination*, vol. 267, no. 1, pp. 57–63, 2011.

[18] W. J. Lau, A. F. Ismail, N. Misdan, and M. A. Kassim, "A recent progress in thin film composite membrane: a review," *Desalination*, vol. 287, pp. 190–199, 2012.

[19] G.-d. Kang and Y.-m. Cao, "Development of antifouling reverse osmosis membranes for water treatment: a review," *Water Research*, vol. 46, no. 3, pp. 584–600, 2012.

[20] M. Ghasemi, W. R. Wan Daud, J. Alam et al., "Treatment of two different water resources in desalination and microbial fuel cell processes by poly sulfone/sulfonated poly ether ether ketone hybrid membrane," *Energy*, vol. 96, pp. 303–313, 2016.

[21] D. Li and H. Wang, "Recent developments in reverse osmosis desalination membranes," *Journal of Materials Chemistry*, vol. 20, no. 22, pp. 4551–4566, 2010.

[22] C. Ding, J. Yin, and B. Deng, "Effects of polysulfone (PSf) support layer on the performance of thin-film composite (TFC) membranes," *Journal of Chemical and Process Engineering*, vol. 1, no. 102, pp. 1–8, 2014.

[23] Q. Zhang, Z. Zhang, L. Dai, H. Wang, S. Li, and S. Zhang, "Novel insights into the interplay between support and active layer in the thin film composite polyamide membranes," *Journal of Membrane Science*, vol. 537, pp. 372–383, 2017.

[24] X. Lu, L. Arias Chavez, S. R.-V. Castrillón, J. Ma, and M. Elimelech, "Influence of active layer and support layer surface structures on organic fouling propensity of thin-film composite forward osmosis membranes," *Environmental Science & Technology*, vol. 49, no. 3, pp. 1436–1444, 2015.

[25] L. Huang and J. R. McCutcheon, "Impact of support layer pore size on performance of thin film composite membranes for forward osmosis," *Journal of Membrane Science*, vol. 483, pp. 25–33, 2015.

[26] J.-A.-D. Sharabati, S. Guclu, S. Erkoc-Ilter et al., "Interfacially polymerized thin-film composite membranes: impact of support layer pore size on active layer polymerization and seawater desalination performance," *Separation and Purification Technology*, vol. 212, pp. 438–448, 2019.

[27] M. Askarizadeh, B. Honarvar, and A. Mansousrizadeh, "The new method for making polyamide membranes for water treatment," *Journal of Middle East Applied Science and Technology*, vol. 13, pp. 514–519, 2014.

[28] Y. Li, Y. Su, Y. Dong et al., "Separation performance of thin-film composite nanofiltration membrane through interfacial polymerization using different amine monomers," *Desalination*, vol. 333, no. 1, pp. 59–65, 2014.

[29] A. Kowalik-Klimczak, A. Bednarska, M. Grądkowski, and P. Gierycz, "Analysis of polymeric nanofiltration membranes by modern techniques," *Polimery*, vol. 61, no. 5, pp. 339–346, 2016.

[30] N. K. Saha and S. V. Joshi, "Performance evaluation of thin film composite polyamide nanofiltration membrane with variation in monomer type," *Journal of Membrane Science*, vol. 342, no. 1-2, pp. 60–69, 2009.

[31] K. H. Mah, H. W. Yussof, M. N. Abu Seman, and A. W. Mohammad, "Polyester thin film composite nanofiltration

membranes via interfacial polymerization: influence of five synthesis parameters on water permeability," *Journal of Mechanical Engineering and Sciences*, vol. 12, no. 1, pp. 3387–3398, 2018.

[32] S.-J. Park, S. J. Kwon, H.-E. Kwon et al., "Aromatic solvent-assisted interfacial polymerization to prepare high performance thin film composite reverse osmosis membranes based on hydrophilic supports," *Polymer*, vol. 144, pp. 159–167, 2018.

[33] M. Namvar-Mahboub and M. Pakizeh, "Optimization of preparation conditions of polyamide thin film composite membrane for organic solvent nanofiltration," *Korean Journal of Chemical Engineering*, vol. 31, no. 2, pp. 327–337, 2014.

[34] B. Khorshidi, T. Thundat, B. A. Fleck, and M. Sadrzadeh, "Thin film composite polyamide membranes: parametric study on the influence of synthesis conditions," *RSC Advances*, vol. 5, no. 68, pp. 54985–54997, 2015.

[35] K. H. Mah, H. W. Yussof, M. N. Abu Seman, and A. W. Mohammad, "Thin-film composite nanofiltration membrane: a study on the curing time and its performance evaluation," *International Journal of Biomass and Renewables*, vol. 5, no. 3, pp. 1–7, 2016.

[36] H. Huang, X. Qu, H. Dong, L. Zhang, and H. Chen, "Role of NaA zeolites in the interfacial polymerization process towards a polyamide nanocomposite reverse osmosis membrane," *RSC Advances*, vol. 3, no. 22, article 8203, 8207 pages, 2013.

[37] H. Dong, L. Wu, L. Zhang, H. Chen, and C. Gao, "Clay nanosheets as charged filler materials for high-performance and fouling-resistant thin film nanocomposite membranes," *Journal of Membrane Science*, vol. 494, pp. 92–103, 2015.

[38] A. W. Carpenter, C. F. de Lannoy, and M. R. Wiesner, "Cellulose nanomaterials in water treatment technologies," *Environmental Science and Technology*, vol. 49, no. 9, pp. 5277–5287, 2015.

[39] S. Janakiram, M. Ahmadi, Z. Dai, L. Ansaloni, and L. Deng, "Performance of nanocomposite membranes containing 0D to 2D nanofillers for $CO_2$ separation: a review," *Membranes*, vol. 8, no. 2, p. 24, 2018.

[40] M. Alberto, R. Bhavsar, J. Luque-Alled et al., "Study on the formation of thin film nanocomposite (TFN) membranes of polymers of intrinsic microporosity and graphene-like fillers: effect of lateral flake size and chemical functionalization," *Journal of Membane Science*, vol. 565, pp. 390–401, 2018.

[41] J. Zhang, L. Ye, and S. Deng, "Halloysite–epoxy nanocomposites with improved particle dispersion through ball mill homogenisation and chemical treatments," *Composites Science and Technology*, vol. 69, no. 14, pp. 2497–2505, 2009.

[42] C.-Y. Chong, W.-J. Lau, N. Yusof, G.-S. Lai, and A. F. Ismail, "Roles of nanomaterial structure and surface coating on thin film nanocomposite membranes for enhanced desalination," *Composites Part B: Engineering*, vol. 160, pp. 471–479, 2019.

[43] A. Karkooti, A. Z. Yazdi, P. Chen, M. McGregor, N. Nazemifard, and M. Sadrzadeh, "Development of advanced nanocomposite membranes using graphene nanoribbons and nanosheets for water treatment," *Journal of Membrane Science*, vol. 560, pp. 97–107, 2018.

[44] X. Qu, P. J. Alvarez, and Q. Li, "Applications of nanotechnology in water and wastewater treatment," *Water Research*, vol. 47, no. 12, pp. 3931–3946, 2013.

[45] A. S. Ahmad, M. I. A. Qureshi, S. Anum, and G. Yaqub, "Applications of carbon nanotubes (CNTs) for the treatment of drinking and waste water-a brief review," *International Journal of Advanced Research and Development*, vol. 1, no. 12, pp. 11–16, 2016.

[46] Y. Manawi, V. Kochkodan, M. A. Hussein, M. A. Khaleel, M. Khraisheh, and N. Hilal, "Can carbon-based nanomaterials revolutionize membrane fabrication for water treatment and desalination?," *Desalination*, vol. 391, pp. 69–88, 2016.

[47] A. K Mahmood, R. H. Alanbari, and F. A. Rasin, "Applications of synthesized nanocomposite membranes for water purification in Iraqi brackish water," *Civil and Environmental Research*, vol. 7, no. 7, pp. 6–12, 2015.

[48] M. Adamczak, G. Kamińska, and J. Bohdziewicz, "Modification of PES ultrafiltration membranes for removal of caffeine," *Desalination and Water Treatment*, vol. 128, pp. 351–357, 2018.

[49] D. Emadzadeh, W. J. Lau, T. Matsuura, M. Rahbari-Sisakht, and A. F. Ismail, "A novel thin film composite forward osmosis membrane prepared from PSf–$TiO_2$ nanocomposite substrate for water desalination," *Chemical Engiennering Journal*, vol. 237, pp. 70–80, 2014.

[50] V. Dalvi, Y. P. Tang, C. Staudt, and T. S. Chung, "Influential effects of nanoparticles, solvent and surfactant treatments on thin film nanocomposite (TFN) membranes for seawater desalination," *Desalination*, vol. 420, pp. 216–225, 2017.

[51] J. Yin, Z. Guocheng, and B. Deng, "Graphene oxide (GO) enhanced polyamide (PA) thin-film nanocomposite (TFN) membrane for water purification," *Desalination*, vol. 379, pp. 93–101, 2016.

[52] M. Ghanbari, D. Emadzadeh, W. J. Lau, S. O. Lai, T. Matsuura, and A. F. Ismail, "Synthesis and characterization of novel thin film nanocomposite (TFN) membranes embedded with halloysite nanotubes (HNTs) for water desalination," *Desalination*, vol. 358, pp. 33–41, 2015.

[53] X. Guo, D. Liu, T. Han, H. Huang, Q. Yang, and C. Zhong, "Preparation of thin film nanocomposite membranes with surface modified MOF for high flux organic solvent nanofiltration," *AIChE Journal*, vol. 63, no. 4, pp. 1303–1312, 2017.

[54] P. Sharma, R. Joshi, and B. S. Hsiao, "A simple approach to prepare carboxycellulose nanofibers from untreated biomass," *Biomacromolecules*, vol. 18, no. 8, pp. 2333–2342, 2017.

[55] F. Asempour, D. Emadzadeh, T. Matsuura, and B. Kruczek, "Synthesis and characterization of novel cellulose nanocrystals-based thin film nanocomposite membranes for reverse osmosis applications," *Desalination*, vol. 439, pp. 179–187, 2018.

[56] L. Bai, Y. Liu, A. Ding, N. Ren, G. Li, and H. Liang, "Fabrication and characterization of thin-film composite (TFC) nanofiltration membranes incorporated with cellulose nanocrystals (CNCs) for enhanced desalination performance and dye removal," *Chemical Engineering Journal*, vol. 358, pp. 1519–1528, 2019.

[57] S. Xia, L. Yao, Y. Zhao, N. Li, and Y. Zheng, "Preparation of graphene oxide modified polyamide thin film composite membranes with improved hydrophilicity for natural organic matter removal," *Chemical Engineering Journal*, vol. 280, pp. 720–727, 2015.

[58] Y. K. Kim, S. Y. Lee, D. H. K. B. S. Lee, S. Y. Nam, and J. W. Rhim, "Preparation and characterization of thermally cross-linked chlorine resistant thin film composite polyamide membranes for reverse osmosis," *Desalination*, vol. 250, no. 2, pp. 865–867, 2010.

[59] R. Verbeke, V. Gómez, and I. F. Vankelecom, "Chlorine-resistance of reverse osmosis (RO) polyamide membranes," *Progress in Polymer Science*, vol. 72, pp. 1–15, 2017.

[60] L. Shen, F. Wang, X. Zhang, C. Ding, and Y. Wang, "High--performance thin-film composite membranes with surface functionalization by organic phosphonic acids," *Journal of Membrane Science*, vol. 563, pp. 284–297, 2018.

[61] L. Shen, X. Zhang, J. Zuo, and Y. Wang, "Performance enhancement of TFC FO membranes with polyethyleneimine modification and post-treatment," *Journal of Membrane Science*, vol. 534, pp. 46–58, 2017.

[62] J. Rezania, V. Vatanpour, A. Shockravi, and M. Ehsani, "Preparation of novel carboxylated thin-film composite polyamide-polyester nanofiltration membranes with enhanced antifouling property and water flux," *Reactive and Functional Polymers*, vol. 131, pp. 123–133, 2018.

[63] P. Cay-Durgun, C. McCloskey, J. Konecny, A. Khosravi, and M. L. Lind, "Evaluation of thin film nanocomposite reverse osmosis membranes for long-term brackish water desalination performance," *Desalination*, vol. 404, pp. 304–312, 2017.

[64] S. S. Shenvi, A. M. Isloor, and A. F. Ismail, "A review on RO membrane technology: developments and challenges," *Desalination*, vol. 368, pp. 10–26, 2015.

[65] S. Bano, A. Mahmood, K. Seong-Joong, and L. Kew-Ho, "Graphene oxide modified polyamide nanofiltration membrane with improved flux and antifouling properties," *Journal of Materials Chemistry A*, vol. 3, no. 5, pp. 2065–2071, 2015.

[66] S. Gaillet and J.-M. Rouanet, "Silver nanoparticles: their potential toxic effects after oral exposure and underlying mechanisms – a review," *Food and Chemical Toxicology*, vol. 77, pp. 58–63, 2015.

[67] J. Yin, Y. Yang, Z. Hu, and B. Deng, "Attachment of silver nanoparticles (AgNPs) onto thin-film composite (TFC) membranes through covalent bonding to reduce membrane biofouling," *Journal of Membrane Science*, vol. 441, pp. 73–82, 2013.

[68] H. Wan, N. J. Briot, A. Saad, L. Ormsbee, and D. Bhattacharyya, "Pore functionalized PVDF membranes with in-situ synthesized metal nanoparticles: material characterization, and toxic organic degradation," *Journal of Membrane Science*, vol. 530, pp. 147–157, 2017.

[69] P. Cay-Durgun and M. L. Lind, "Nanoporous materials in polymeric membranes for desalination," *Current Opinion in Chemical Engineering*, vol. 20, pp. 19–27, 2018.

[70] M. P. F. M. Bojnourd, "Preparation and characterization of a nanoclay/PVA/PSf nanocomposite membrane for removal of pharmaceuticals from water," *Applied Clay Science*, vol. 162, pp. 326–338, 2018.

[71] M. Khajouei, M. Peyravi, and M. Jahanshahi, "The potential of nanoparticles for upgrading thin film nanocomposite membranes – a review," *Journal of Membrane Science and Research*, vol. 3, no. 1, pp. 2–12, 2017.

[72] N. Voutchkov, "Energy use for membrane seawater desalination – current status and trends," *Desalination*, vol. 431, pp. 2–14, 2017.

[73] D. Kim-Hak, M. B. Dixon, M. A. Galan, F. Boisseau, J. Gallastegui, and R. Martina, "Santa Barbara, Curacao desalination plant expansion using NanoH2O thin film nanocomposite (TFN) SWRO membrane," *Desalination and Water Treatment*, vol. 55, no. 9, pp. 2446–2452, 2015.

[74] A. Subramani and J. G. Jacangelo, "Emerging desalination technologies for water treatment: a critical review," *Water Research*, vol. 75, pp. 164–187, 2015.

# Permissions

All chapters in this book were first published in IJPS, by Hindawi Publishing Corporation; hereby published with permission under the Creative Commons Attribution License or equivalent. Every chapter published in this book has been scrutinized by our experts. Their significance has been extensively debated. The topics covered herein carry significant findings which will fuel the growth of the discipline. They may even be implemented as practical applications or may be referred to as a beginning point for another development.

The contributors of this book come from diverse backgrounds, making this book a truly international effort. This book will bring forth new frontiers with its revolutionizing research information and detailed analysis of the nascent developments around the world.

We would like to thank all the contributing authors for lending their expertise to make the book truly unique. They have played a crucial role in the development of this book. Without their invaluable contributions this book wouldn't have been possible. They have made vital efforts to compile up to date information on the varied aspects of this subject to make this book a valuable addition to the collection of many professionals and students.

This book was conceptualized with the vision of imparting up-to-date information and advanced data in this field. To ensure the same, a matchless editorial board was set up. Every individual on the board went through rigorous rounds of assessment to prove their worth. After which they invested a large part of their time researching and compiling the most relevant data for our readers.

The editorial board has been involved in producing this book since its inception. They have spent rigorous hours researching and exploring the diverse topics which have resulted in the successful publishing of this book. They have passed on their knowledge of decades through this book. To expedite this challenging task, the publisher supported the team at every step. A small team of assistant editors was also appointed to further simplify the editing procedure and attain best results for the readers.

Apart from the editorial board, the designing team has also invested a significant amount of their time in understanding the subject and creating the most relevant covers. They scrutinized every image to scout for the most suitable representation of the subject and create an appropriate cover for the book.

The publishing team has been an ardent support to the editorial, designing and production team. Their endless efforts to recruit the best for this project, has resulted in the accomplishment of this book. They are a veteran in the field of academics and their pool of knowledge is as vast as their experience in printing. Their expertise and guidance has proved useful at every step. Their uncompromising quality standards have made this book an exceptional effort. Their encouragement from time to time has been an inspiration for everyone.

The publisher and the editorial board hope that this book will prove to be a valuable piece of knowledge for researchers, students, practitioners and scholars across the globe.

# List of Contributors

**Tingting Zhu and Bin Huang**
College of Petroleum Engineering, Northeast Petroleum University, China

**Cheng Fu**
College of Petroleum Engineering, Northeast Petroleum University, China
Post-Doctoral Scientific Research Station, Daqing Oilfield Company, China

**Di Wu**
Hangzhou Urban & Rural Construction Design Institute Limited Company, China

**Ying Wang**
Aramco Asia, China

**Bob A. Howell, Kendahl L. Oberdorfer and Eric A. Ostrander**
Center for Applications in Polymer Science, Department of Chemistry and Biochemistry, Central Michigan University, Mt. Pleasant, MI 48859-0001, USA

**Maria Osorio-Fortich**
Faculty of Pharmaceutical Sciences, University of Cartagena, Research in Technology Pharmaceutical, Cosmetic and of Food Group (GITFCA), Cartagena de Indias D.T. y C, Cartagena, Colombia

**Glicerio Leon-Méndez**
Faculty of Pharmaceutical Sciences, University of Cartagena, Research in Technology Pharmaceutical, Cosmetic and of Food Group (GITFCA), Cartagena de Indias D.T. y C, Cartagena, Colombia
Center of Commerce and Services, Regional Bolivar, SENA, Research in Innovation and Biotechnology Group (GIBEI), Cartagena de Indias D.T. y C, Cartagena, Colombia
Doctorate Student in Engineering, University of Cartagena, Cartagena, Bolívar 130015, Colombia

**Miladys Torrenegra-Alarcón**
Center of Commerce and Services, Regional Bolivar, SENA, Research in Innovation and Biotechnology Group (GIBEI), Cartagena de Indias D.T. y C, Cartagena, Colombia
Doctorate Student in Engineering, University of Cartagena, Cartagena, Bolívar 130015, Colombia

**Rodrigo Ortega-Toro**
Faculty of Engineering, University of Cartagena, Grupo de Investigación de Ingeniería de Fluidos Complejos y Reología de Alimentos (IFCRA), Cartagena de Indias D.T. y C, Cartagena, Colombia

**Nerlys Pajaro-Castro**
Department of Medicine, School of Health Sciences, Research in Medical and Pharmaceutical Sciences Group, University of Sucre, Sincelejo, Sucre 700003, Colombia

**Adriana Herrera-Barros**
Faculty of Engineering, University of Cartagena, Research in Multi-functional Nanomaterials Group, Cartagena de Indias D.T. y C., Cartagena 130015, Colombia

**Riccardo Checchetto**
Dipartimento di Fisica, Università di Trento, via Sommarive 14, Povo, 38123 Trento, Italy

**Jae-Yeong Jung, Da-Seul Kim and Kyoung-Hwan Kim**
Department of Energy Systems Research, Ajou University, Suwon 16499, Republic of Korea

**Tae Soup Shim and Ju-Hyung Kim**
Department of Energy Systems Research, Ajou University, Suwon 16499, Republic of Korea
Department of Chemical Engineering, Ajou University, Suwon 16499, Republic of Korea

**Seokwon Joo and Soonmin Seo**
College of BioNano Technology, Gachon University, Gyeonggi 13120, Republic of Korea

**N. I. M. Nawi, N. A. H. M. Nordin, Z. A. Putra and M. D. H. Wirzal**
Chemical Engineering Department, Universiti Teknologi PETRONAS, Bandar Seri Iskandar, 32610 Perak, Malaysia

**M. R. Bilad**
Chemical Engineering Department, Universiti Teknologi PETRONAS, Bandar Seri Iskandar, 32610 Perak, Malaysia
Jurusan Pendidikan Kimia, Institut Keguruan Ilmu Pendidikan, Jalan Pemuda No 59A, Mataram, Indonesia

**M. O. Mavukkandy**
Institute Center for Water and Environment (iWater), Department of Chemical and Environmental Engineering, Masdar Institute, Khalifa University of Science and Technology, Abu Dhabi, UAE

**J. Jaafar**
Advanced Membrane Technology Research Centre (AMTEC), Faculty of Chemical and Natural Resources Engineering, Universiti Teknologi Malaysia, 81310 Skudai, Johor, Malaysia

**Asim L. Khan**
Department of Chemical Engineering, COMSATS Institute of Information Technology, Lahore, Pakistan

**M. Berer, M. Halb and G. Pacher**
Polymer Competence Center Leoben GmbH, Roseggerstrasse 12, 8700 Leoben, Austria

**M. Feuchter and G. Pinter**
Material Science and Testing of Polymers, Montanuniversitaet Leoben, Otto Gloeckel-Strasse 2, 8700 Leoben, Austria

**Mostafa Jabbari, Mikael Skrifvars, Dan Åkesson, and Mohammad J. Taherzadeh**
Swedish Centre for Resource Recovery, University of Borås, Borås, Sweden

**Yu Tang and Zeyang Sun**
Key Laboratory of Concrete and Prestressed Concrete Structures of the Ministry of Education, Southeast University, Nanjing 210096, China

**Gang Wu**
Key Laboratory of Concrete and Prestressed Concrete Structures of the Ministry of Education, Southeast University, Nanjing 210096, China
Laboratory of Industrialized Structural and Bridge Engineering of Jiangsu Province, Nanjing 210096, China

**Genggongwo Shi and Jeseob Kim**
CCTech Inc., 72, Sicheng-Ro, 721Beon-Gil, Paltan-Myeon, Haseong-Si, Gyeonggi-Do 18524, Republic of Korea

**Sung Hoon Park, Minji Kim and Lee Soon Park**
School of Material Science and Engineering, Ulsan National Institute of Science and Technology (UNIST), Ulsan 44919, Republic of Korea

**Pablo G. Argudo**
Departamento de Química Física I, Universidad Complutense de Madrid, Ciudad Universitaria s/n, 28040 Madrid, Spain

**Eduardo Guzmán, Ramón G. Rubio and Francisco Ortega**
Departamento de Química Física I, Universidad Complutense de Madrid, Ciudad Universitaria s/n, 28040 Madrid, Spain
Instituto Pluridisciplinar, Universidad Complutense de Madrid, Paseo Juan XXIII 1, 28040 Madrid, Spain

**Alejandro Lucia**
Instituto de Ecología y Desarrollo Sustentable (INEDES, CONICET-UNLu), Ruta 5 y Avenida Constitución, 6700 Luján, Buenos Aires, Argentina

**Davi Rabelo de Oliveira, Selma Elaine Mazzetto and Diego Lomonaco**
Department of Organic and Inorganic Chemistry, Federal University of Ceara, 60440-900 Fortaleza CE, Brazil

**B. Mensah**
932, Yun-Tech, ITECO 762, Deokpung-dong, Hanam-si, Gyeonggi-do, Republic of Korea
Department of Materials Science and Engineering, CBAS, University of Ghana, Legon, Ghana

**B. Agyei-Tuffour, E. Nyankson, Y. D. Bensah, D. Dodoo-Arhin, B. Onwona-Agyeman and A. Yaya**
Department of Materials Science and Engineering, CBAS, University of Ghana, Legon, Ghana

**J. K. Bediako**
Division of Semiconductor and Chemical Engineering, Chonbuk National University, Jeonju, Jeonbuk 561-756, Republic of Korea

**Cristina Fornaguera**
Grup d'Enginyeria de Materials, Institut Químic de Sarrià (IQS), Universitat Ramon Llull and Sagetis-Biotech, Via Augusta 390, 08017 Barcelona, Spain

**Conxita Solans**
Institute of Advanced Chemistry of Catalonia (IQAC-CSIC) and Networking Centre on Bioengineering, Biomaterials and Nanomedicine (CIBER-BBN), C/ Jordi Girona 18-26, 08034 Barcelona, Spain

**Li Hong Huang**
College of Chemical Engineering, Fuzhou University, Fuzhou, Fujian 350108, China
Fuzhou University Zhicheng College, Fuzhou, Fujian 350002, China

**Xiaoxiang Yang and Jianhong Gao**
College of Chemical Engineering, Fuzhou University, Fuzhou, Fujian 350108, China
Quanzhou Normal University, Quanzhou, Fujian 362000, China

**Ricardo S. Souza and Ana L. N. da Silva**
Instituto de Macromoléculas Professora Eloisa Mano (IMA), Programa em Ciência e Tecnologia de Polímeros, Universidade Federal do Rio de Janeiro (UFRJ), Avenida Horácio Macedo 2.030, Centro de Tecnologia-Bloco J, Ilha do Fundão, 21941-598 Rio de Janeiro, Brazil

**Leila L. Y. Visconte**
Instituto de Macromoléculas Professora Eloisa Mano (IMA), Programa em Ciência e Tecnologia de Polímeros, Universidade Federal do Rio de Janeiro (UFRJ), Avenida Horácio Macedo 2.030, Centro de Tecnologia-Bloco J, Ilha do Fundão, 21941-598 Rio de Janeiro, Brazil
Escola Politécnica/Programa de Engenharia Ambiental, Universidade Federal do Rio de Janeiro (UFRJ), Av. Athos da Silveira Ramos 149, Centro de Tecnologia-Bloco A, Ilha do Fundão, 21941-909 Rio de Janeiro, RJ, Brazil

**Valéria G. Costa**
Departamento de Processamento e Caracterização de Materiais (DPCM), Laboratório de Materiais Poliméricos (LAMAP), Instituto Nacional de Tecnologia (INT), Av. Venezuela 82, Praça Mauá, 20081-312 Rio de Janeiro, RJ, Brazil

**Wei He**
School of Water Conservancy and Environment, Zhengzhou University, Zhengzhou, Henan Province, China
Zhengzhou Metro Co. Ltd., Zhengzhou, Henan Province, China

**Bing Fu**
School of Civil and Transportation Engineering, Guangdong University of Technology, Guangzhou, Guangdong Province, China
Department of Civil and Environmental Engineering, The Hong Kong Polytechnic University, Hung Hom, Hong Kong

**Feng-Chen An**
College of Petroleum Engineering, China University of Petroleum, Beijing, China

**Franco L. Redondo, Andrés E. Ciolino and Marcelo A. Villar**
Planta Piloto de Ingeniería Química, PLAPIQUI (UNS-CONICET), CP: 8000, Camino "La Carrindanga" Km 7, Bahía Blanca, Buenos Aires, Argentina
Departamento de Ingeniería Química, Universidad Nacional del Sur (UNS), CP: 8000, Av. Alem 1253, Bahía Blanca, Buenos Aires, Argentina

**Mario D. Ninago**
Departamento de Ingeniería Química, Facultad de Ciencias Aplicadas a la Industria (FCAI), Universidad Nacional de Cuyo (UNCuyo), CP: 5600, Bernardo de Irigoyen 375, San Rafael, Mendoza, Argentina
Consejo Nacional de Investigaciones Científicas y Técnicas (CONICET), Godoy Cruz, 2290 Ciudad Autónoma de Buenos Aires, Argentina

**Augusto G. O. de Freitas**
Universidade Federal do Pampa, Unipampa (97650-000), Itaqui, RS, Brazil

**Cristiano Giacomelli**
Departamento de Química, Universidade Federal de Santa María, 97105-900 Santa María, RS, Brazil

**Jiazhen Lu and Xin Liu**
College of Chemical Engineering, Beijing Institute of Petrochemical Technology, Beijing 102617, China

**Haibo Jin, Guangxiang He, Bin Zhang and Xiaoyan Guo**
College of Chemical Engineering, Beijing Institute of Petrochemical Technology, Beijing 102617, China
Beijing Key Laboratory of Fuels Cleaning and Advanced Catalytic Emission Reduction Technology, Beijing 102617, China

**Magdalena Włodarska**
Institute of Physics, Lodz University of Technology, Wólczańska 219, 90-924 Łódź, Poland

**Michał Adamczak, Gabriela Kamińska and Jolanta Bohdziewicz**
Institute of Water and Wastewater Engineering, Silesian University of Technology, Konarskiego St. 18, Gliwice 44-100, Poland

# Index

www.ingramcontent.com/pod-product-compliance
Lightning Source LLC
LaVergne TN
LVHW080919210926
848485LV00004B/685

* 9 7 8 1 6 3 9 8 9 1 8 9 4 *